IFGC®

2015

CODE AND COMMENTARY

The complete IFGC with commentary after each section

American Gas Association

INTERNATIONAL CODE COUNCIL

2015 International Fuel Gas Code® Commentary

First Printing: May 2015

ISBN: 978-1-60983-292-6

COPYRIGHT© 2015
by
INTERNATIONAL CODE COUNCIL, INC.

Material designated IFGS
by
AMERICAN GAS ASSOCIATION
400 N. Capitol Street, N.W. • Washington, DC 20001
(202) 824-7000
Copyright © American Gas Association, 2009. All rights reserved.

T021823
PRINTED IN THE U.S.A.

PREFACE

The principal purpose of the Commentary is to provide a basic volume of knowledge and facts relating to building construction as it pertains to the regulations set forth in the 2015 *International Fuel Gas Code*® (IFGC®). The person who is serious about effectively designing, constructing and regulating buildings and structures will find the Commentary to be a reliable data source and reference to almost all fuel gas components of the built environment.

As a follow-up to the *International Fuel Gas Code*, we offer a companion document, the *International Fuel Gas Code Commentary*. The basic appeal of the Commentary is that it provides in a small package and at reasonable cost thorough coverage of many issues likely to be dealt with when using the *International Fuel Gas Code*. It then supplements that coverage with historical and technical background. Reference lists, information sources and bibliographies are also included.

Throughout all of this, strenuous effort has been made to keep the vast quantity of material accessible and its method of presentation useful. With a comprehensive yet concise summary of each section, the Commentary provides a convenient reference for regulations applicable to the construction of buildings and structures. In the chapters that follow, discussions focus on the full meaning and implications of the code text. Guidelines suggest the most effective method of application, and the consequences of not adhering to the code text. Illustrations are provided to aid understanding; they do not necessarily illustrate the only methods of achieving code compliance.

The format of the Commentary includes the full text of each section, table and figure in the code, followed immediately by the commentary applicable to that text. At the time of printing, the Commentary reflects the most up-to-date text of the 2015 *International Fuel Gas Code*. Each chapter's narrative includes a section on "General Comments" and "Purpose," and each section's narrative usually includes a discussion about why the requirement commands the conditions set forth. Code text is reproduced as it appears in the *International Fuel Gas Code*, and commentary is indented below the code text, beginning with the symbol ❖. Code figures and tables are reproduced as they appear in the *International Fuel Gas Code*. Commentary figures and tables are identified in the text by the word "Commentary" (as in "see Commentary Figure 704.3"), and each has a full border.

The *International Fuel Gas Code* is segregated by section numbers into two categories: code and standard. Code sections are identified as IFGC; standard sections are identified as IFGS.

Commentary is to be used in conjunction with the *International Fuel Gas Code* and not as a substitute for the code. **The Commentary is advisory only;** the code official alone possesses the authority and responsibility for interpreting the code.

Comments and recommendations are encouraged, for through your input, we can improve future editions. Please direct your comments to the Codes and Standards Development Department at the Chicago District Office.

TABLE OF CONTENTS

Chapter 1:
Scope and Administration

General Comments

The law of building regulation is grounded on the police power of the state. It is used so that the state may legislate for the general welfare of its citizens. This power enables passage of such laws as a fuel gas code. It is from the police power delegated by the state legislature that local governments are able to enact building regulations. If the state legislature has limited this power in any way, the municipality may not exceed these limitations. Although the municipality may not further delegate its police power (e.g., by delegating the burden of determining code compliance with the building owner, contractor or architect), it may turn over the administration of building regulations to a municipal official, such as a code official, if he or she is given sufficient criteria to clearly establish the basis for decisions concerning whether or not a proposed building, including its fuel gas systems, conforms to the code.

Chapter 1 is largely concerned with maintaining "due process of law" in enforcing the performance criteria contained in the code. Only through careful observation of the administrative provisions can the code official reasonably hope to demonstrate that "equal protection under the law" has been established. Although it is generally assumed that the administrative and enforcement sections of a code are geared toward the code official, this is not entirely true. The provisions also establish the rights and privileges of the registered design professional, the contractor and the building owner. The position of the code official is merely to review the proposed and completed work and determine whether a fuel gas installation conforms to the code requirements. The registered design professional is responsible for the design of a safe, sanitary fuel gas system. The contractor is responsible for installing the system in strict accordance with the plans.

During the course of the construction of a fuel gas system, the code official reviews the activity to verify that the spirit and intent of the law are being met and that the fuel gas system provides adequate protection of public health. As a public servant, the code official enforces the code without bias. Every individual is guaranteed equal enforcement of the code. Furthermore, design professionals, contractors and building owners have the right of due process for any requirement in the code.

Purpose

A fuel gas code, like any other code, is intended to be adopted as a legally enforceable document to safeguard health, safety, property and public welfare. A fuel gas code cannot be effective without adequate provisions for its administration and enforcement. The official charged with the administration and enforcement of fuel gas regulations has a great responsibility, and with this responsibility goes authority. No matter how detailed the fuel gas code may be, the code official must, to some extent, exercise judgment in determining compliance. The code official has the responsibility of establishing that the homes in which the citizens of the community reside and the buildings in which they work are designed and constructed to be reasonably free from hazards associated with the presence and use of fuel gas appliances, appurtenances, fixtures and systems. The code intends to establish a minimum acceptable level of safety.

PART 1—SCOPE AND APPLICATION

SECTION 101 (IFGC)
GENERAL

[A] **101.1 Title.** These regulations shall be known as the *Fuel Gas Code* of **[NAME OF JURISDICTION]**, hereinafter referred to as "this code."

❖ This section identifies the adopted regulations by inserting the name of the adopting jurisdiction into the code.

[A] **101.2 Scope.** This code shall apply to the installation of fuel-gas *piping* systems, fuel gas appliances, gaseous hydro- gen systems and related accessories in accordance with Sections 101.2.1 through 101.2.5.

Exception: Detached one- and two-family dwellings and multiple single-family dwellings (townhouses) not more than three stories high with separate means of egress and their accessory structures shall comply with the *International Residential Code.*

❖ This section describes the types of fuel gas systems to which the code is intended to apply and specifically lists those systems to which the code does not apply. The applicability of the code spans from the initial design of fuel gas systems, through the installation and construction phases, and into the maintenance of operating systems. Chapter 24 of the *International Residential Code®* (IRC®) covers fuel gas systems and is a duplication of the applicable *International Fuel Gas Code®* (IFGC®) text.

[A] 101.2.1 Gaseous hydrogen systems. Gaseous hydrogen systems shall be regulated by Chapter 7.

❖ See the general comments for Chapter 7.

[A] 101.2.2 Piping systems. These regulations cover *piping* systems for natural gas with an operating pressure of 125 pounds per square inch gauge (psig) (862 kPa gauge) or less, and for LP-gas with an operating pressure of 20 psig (140 kPa gauge) or less, except as provided in Section 402.6. Coverage shall extend from the *point of delivery* to the outlet of the *appliance* shutoff valves. *Piping* system requirements shall include design, materials, components, fabrication, assembly, installation, testing, inspection, operation and maintenance.

❖ The code does not limit the operating pressure of systems, but rather limits the code's coverage of piping systems to those having pressures less than or equal to the stated pressures. Consistent with the definition, piping systems begin at the point of delivery and end at the outlet of the appliance shutoff valves (see Section 101.2.3 and definition of "Piping system").

[A] 101.2.3 Gas appliances. Requirements for gas appliances and related accessories shall include installation, combustion and ventilation air and venting and connections to *piping* systems.

❖ The piping and connectors between the appliance shutoff valves and the appliance served are covered by the code, although the piping and connectors are outside the scope of the definition of "Piping system."

[A] 101.2.4 Systems, appliances and equipment outside the scope. This code shall not apply to the following:

1. Portable LP-gas appliances and *equipment* of all types that is not connected to a fixed fuel *piping* system.

2. Installation of farm appliances and *equipment* such as brooders, dehydrators, dryers and irrigation *equipment*.

3. Raw material (feedstock) applications except for *piping* to special atmosphere generators.

4. Oxygen-fuel gas cutting and welding systems.

5. Industrial gas applications using gases such as acetylene and acetylenic compounds, hydrogen, ammonia, carbon monoxide, oxygen and nitrogen.

6. Petroleum refineries, pipeline compressor or pumping stations, loading terminals, compounding plants, refinery tank farms and natural gas processing plants.

7. Integrated chemical plants or portions of such plants where flammable or combustible liquids or gases are produced by, or used in, chemical reactions.

8. LP-gas installations at utility gas plants.

9. Liquefied natural gas (LNG) installations.

10. Fuel gas *piping* in power and atomic energy plants.

11. Proprietary items of *equipment*, apparatus or instruments such as gas-generating sets, compressors and calorimeters.

12. LP-gas *equipment* for vaporization, gas mixing and gas manufacturing.

13. Temporary LP-gas *piping* for buildings under construction or renovation that is not to become part of the permanent *piping* system.

14. Installation of LP-gas systems for railroad switch heating.

15. Installation of hydrogen gas, LP-gas and compressed natural gas (CNG) systems on vehicles.

16. Except as provided in Section 401.1.1, gas *piping*, meters, gas pressure regulators and other appurtenances used by the serving gas supplier in the distribution of gas, other than undiluted LP-gas.

17. Building design and construction, except as specified herein.

18. *Piping* systems for mixtures of gas and air within the flammable range with an operating pressure greater than 10 psig (69 kPa gauge).

19. Portable fuel cell appliances that are neither connected to a fixed *piping* system nor interconnected to a power grid.

❖ This section lists the specific installations and equipment that the code does not intend to regulate. Item 19 relates to Chapter 7 and addresses portable fuel cell appliances as defined in Chapter 2.

[A] 101.2.5 Other fuels. The requirements for the design, installation, maintenance, *alteration* and inspection of mechanical systems operating with fuels other than fuel gas shall be regulated by the *International Mechanical Code*.

❖ This section simply defers the coverage of all equipment other than gas-fired equipment to the *International Mechanical Code®* (IMC®). The IRC also regulates the installation of residential equipment that is not gas fired.

[A] 101.3 Appendices. Provisions in the appendices shall not apply unless specifically adopted.

❖ This section certifies that the appendices are not part of the code unless specifically included in the adopting ordinance of the jurisdiction. Otherwise, the appendices are not intended to be enforceable.

[A] 101.4 Intent. The purpose of this code is to establish minimum standards to provide a reasonable level of safety, health, property protection and public welfare by regulating and controlling the design, construction, installation, quality of materials, location, operation and maintenance or use of fuel gas systems.

❖ The intent of the code is to set forth requirements that establish the minimum acceptable level to safeguard life or limb, health, property and public welfare. The intent becomes important in the application of such sections as Sections 102, 104.2, 105.2 and 108, as well as any enforcement-oriented interpretive action or judgement. Like any code, the written text is subject to interpretation. Interpretations should not be affected by economics or the potential impact on any party. The only consideration should be protection of the public health, safety and welfare.

[A] 101.5 Severability. If a section, subsection, sentence, clause or phrase of this code is, for any reason, held to be unconstitutional, such decision shall not affect the validity of the remaining portions of this code.

❖ Once the code is adopted, only a court can set aside any provisions of the code. This is essential to safeguard the application of the code text if a provision of the code is declared illegal or unconstitutional. This section would preserve the legislative action that established the legal provisions.

SECTION 102 (IFGC)
APPLICABILITY

[A] 102.1 General. Where there is a conflict between a general requirement and a specific requirement, the specific requirement shall govern. Where, in a specific case, different sections of this code specify different materials, methods of construction or other requirements, the most restrictive shall govern.

❖ The scope of the code as described in Section 101 is referenced in this section. The most restrictive code requirement is to apply where different requirements may be specified in the code for a specific installation. The code is designed to regulate new construction and new work and is not intended to be applied retroactively to existing buildings except where existing fuel-gas-related systems are specifically addressed in this section and Section 108.

[A] 102.2 Existing installations. Except as otherwise provided for in this chapter, a provision in this code shall not require the removal, *alteration* or abandonment of, nor prevent the continued utilization and maintenance of, existing installations lawfully in existence at the time of the adoption of this code.

❖ Existing installations are generally considered to be "grandfathered" with code adoption if the system meets a minimum level of safety. Frequently the criteria for this level are the regulations (or code) under which the existing building was originally constructed. If there are no previous code criteria to apply, the code official is to apply those provisions of the code that are reasonably applicable to existing buildings. A specific level of safety is dictated by provisions dealing with hazard abatement in existing buildings and maintenance provisions as contained in this code, the *International Property Maintenance Code*® (IPMC®) and the *International Fire Code*® (IFC®).

[A] 102.2.1 Existing buildings. Additions, alterations, renovations or repairs related to building or structural issues shall be regulated by the *International Building Code*.

❖ This section states that the *International Building Code*® (IBC®) regulates construction related to building or structural issues.

[A] 102.3 Maintenance. Installations, both existing and new, and parts thereof shall be maintained in proper operating condition in accordance with the original design and in a safe condition. Devices or safeguards that are required by this code shall be maintained in compliance with the edition of the code under which they were installed. The owner or the owner's authorized agent shall be responsible for maintenance of installations. To determine compliance with this provision, the code official shall have the authority to require an installation to be reinspected.

❖ Fuel gas systems and equipment are subject to deterioration resulting from aging, wear, accumulation of dirt and debris, corrosion and other factors. Maintenance is necessary to keep these systems and equipment in proper operating condition. Required safety devices and controls must be maintained to continue providing the protection that they afford. Existing equipment and systems could be equipped with safety devices or other measures that were necessary because of the nature of the equipment, and such safeguards may have been required by a code that predates the current code. All such safeguards required by previous or present codes must be maintained for the life of the equipment or system served by those safeguards.

[A] 102.4 Additions, alterations or repairs. Additions, alterations, renovations or repairs to installations shall conform to that required for new installations without requiring the existing installation to comply with all of the requirements of this code. Additions, alterations or repairs shall not cause an existing installation to become unsafe, hazardous or overloaded.

Minor additions, alterations, renovations and repairs to existing installations shall meet the provisions for new construction, unless such work is done in the same manner and arrangement as was in the existing system, is not hazardous and is *approved*.

❖ Simply stated, new work must comply with the current requirements for new work. Any alteration or addition to an existing system involves some extent of new work, and such new work is subject to the requirements of the code. Additions and alterations must not cause an existing system to be any less in compliance with the code than it was before the changes.

[A] 102.5 Change in occupancy. It shall be unlawful to make a change in the *occupancy* of a structure that will subject the structure to the special provisions of this code applicable to the new *occupancy* without approval. The code official shall certify that such structure meets the intent of the provisions of law governing building construction for the proposed new *occupancy* and that such change of *occupancy* does not result in any hazard to the public health, safety or welfare.

❖ When a building undergoes a change of occupancy, the fuel gas systems must be evaluated to determine what effect the change of occupancy has on them. If an existing system serves an occupancy that is different from the occupancy it served when the code went into effect, the fuel gas system must comply with the applicable code requirements for a system serving the newer occupancy. Depending on the nature of the previous occupancy, changing a building's occupancy classification could result in a change to the fuel gas system. For example, if a mercantile building is converted to a restaurant, additional fuel gas piping system capacity and modifications may be required.

[A] 102.6 Historic buildings. The provisions of this code relating to the construction, *alteration*, repair, enlargement, restoration, relocation or moving of buildings or structures shall not be mandatory for existing buildings or structures identified and classified by the state or local jurisdiction as historic buildings where such buildings or structures are judged by the code official to be safe and in the public interest of health, safety and welfare regarding any proposed construction, *alteration*, repair, enlargement, restoration, relocation or moving of buildings.

❖ This section gives the code official the widest possible flexibility in enforcing the code when the building in question has historic value. This flexibility, however, is not without conditions. The most important criterion for application of this section is that the building must be specifically classified as being of historic significance by a qualified party or agency. Usually, a state or local authority does this after considerable scrutiny of the historical value of the building. The agencies with this authority typically exist at the state or local government level.

[A] 102.7 Moved buildings. Except as determined by Section 102.2, installations that are a part of buildings or structures moved into or within the jurisdiction shall comply with the provisions of this code for new installations.

❖ Buildings that have been relocated are subject to the requirements of the code as if they were new construction. Placing a building where one did not previously exist is the same as constructing a new building. This section requires that the existing fuel gas systems be altered to the extent necessary to bring them into compliance with the provisions of the code applicable to new construction or that the existing fuel gas system comply with Section 102.2.

[A] 102.8 Referenced codes and standards. The codes and standards referenced in this code shall be those that are *listed* in Chapter 8 and such codes and standards shall be considered as part of the requirements of this code to the prescribed extent of each such reference and as further regulated in Sections 102.8.1 and 102.8.2.

> **Exception:** Where enforcement of a code provision would violate the conditions of the listing of the *equipment* or *appliance*, the conditions of the listing and the manufacturer's installation instructions shall apply.

❖ A referenced standard or portion of one is an enforceable extension of the code as if the content of the standard were included in the body of the code.

The use and application of referenced standards are limited to those portions of the standards that are specifically identified. It is the intention of the code to be in harmony with the referenced standards. In the unlikely event that conflicts occur, the code text governs. The exception recognizes the extremely unlikely but possible occurrence of the code requiring or allowing something less restrictive or stringent than the product listing or manufacturer's instructions. If the code conflicts with or deviates from the conditions of the listing, this may or may not mean that the code violated the listing. For example, the listing for an appliance might allow a particular application of an appliance that is expressly prohibited by the code. In this case, the code has not violated the listing, but instead has simply limited the application allowed by the listing. The intent is for the highest level of safety to prevail.

[A] 102.8.1 Conflicts. Where conflicts occur between the provisions of this code and the referenced standards, the provisions of this code shall apply.

❖ There could be instances where the code is intentionally more stringent than a provision in a referenced standard; therefore, the code intends to prevail (see Section 102.8).

[A] 102.8.2 Provisions in referenced codes and standards. Where the extent of the reference to a referenced code or standard includes subject matter that is within the scope of this code, the provisions of this code, as applicable, shall take precedence over the provisions in the referenced code or standard.

❖ This code might reference a standard that has coverage for a subject that is also covered in this code. In such case, the coverage in this code is applicable, not the coverage in the referenced standard. Otherwise, there would be dual coverage of a subject and such coverage would likely conflict or prescribe differing and incompatible requirements. For example, this code references NFPA 211 and that standard contains many installation provisions that are already covered by this code, many of which will differ. Standards are often referenced for a particular or narrow purpose and this section prevents the unintended consequence of such a reference enacting other requirements in the standard that were not meant to be included. See Section 102.8 and the commentary to Section 401.2.

[A] 102.9 Requirements not covered by code. Requirements necessary for the strength, stability or proper operation of an existing or proposed installation, or for the public safety, health and general welfare, not specifically covered by this code, shall be determined by the code official.

❖ New technology and unusual installations will sometimes result in a situation or circumstance not specifically covered by the code. This section of the code gives the code official the authority to decide whether and how the code can be used to cover the new situation. Clearly such a section is needed, and the code official's reasonable application of the section is necessary. The purpose of the section, however, is not to impose requirements that may be preferred when the code provides alternative methods or is not silent on the circumstances. Additionally, the section can be used to implement the general performance-oriented language of the code to specific enforcement situations.

[A] 102.10 Other laws. The provisions of this code shall not be deemed to nullify any provisions of local, state or federal law.

❖ In some cases, other laws enacted by the jurisdiction or the state or federal government may be applicable to a condition that is also governed by a requirement in the code. In such circumstances, the requirements of the code are in addition to the other law that is still in effect, although the code official may not be responsible for its enforcement.

[A] 102.11 Application of references. Reference to chapter section numbers, or to provisions not specifically identified by number, shall be construed to refer to such chapter, section or provision of this code.

❖ In a situation where the code may make reference to a chapter or section number or to another code provision without specifically identifying its location in the code, assume that the referenced section, chapter or provision is in the code and not in a referenced code or standard.

PART 2—ADMINISTRATION AND ENFORCEMENT

SECTION 103 (IFGC)
DEPARTMENT OF INSPECTION

[A] 103.1 General. The Department of Inspection is hereby created and the executive official in charge thereof shall be known as the code official.

❖ The executive official in charge of inspections is named the "code official" by this section. In actuality, the person who is in charge of the department may hold a different title, such as building commissioner, plumbing inspector, mechanical inspector or construction official. For the purpose of the code, that person is referred to as the "code official."

[A] 103.2 Appointment. The code official shall be appointed by the chief appointing authority of the jurisdiction.

❖ This section establishes the code official as an appointed position and gives the circumstances under which the official can be removed from office.

[A] 103.3 Deputies. In accordance with the prescribed procedures of this jurisdiction and with the concurrence of the appointing authority, the code official shall have the authority to appoint a deputy code official, other related technical officers, inspectors and other employees. Such employees shall have powers as delegated by the code official.

❖ This section gives the code official the authority to appoint other individuals to assist with the administration and enforcement of the code. These individuals have the authority and responsibility as designated by the code official.

[A] 103.4 Liability. The code official, member of the board of appeals or employee charged with the enforcement of this code, while acting for the jurisdiction in good faith and without malice in the discharge of the duties required by this code or other pertinent law or ordinance, shall not thereby be rendered civilly or criminally liable personally, and is hereby relieved from all personal liability for any damage accruing to persons or property as a result of an act or by reason of an act or omission in the discharge of official duties.

❖ Any suit instituted against any officer or employee because of an act performed by that officer or employee in the lawful discharge of duties and under the provisions of this code must be defended by the legal representative of the jurisdiction until the final termination of the proceedings. The code official or any subordinate is not liable for costs in an action, suit or proceeding that is instituted in pursuance of the provisions of this code; and any officer of the department of mechanical inspection who is acting in good faith and without malice is free from liability for acts performed under any of its provisions or by reason of any act or omission in the performance of official duties.

[A] 103.4.1 Legal defense. Any suit or criminal complaint instituted against any officer or employee because of an act performed by that officer or employee in the lawful discharge of duties and under the provisions of this code shall be defended by the legal representatives of the jurisdiction until the final termination of the proceedings. The code official or any subordinate shall not be liable for costs in an action, suit or proceeding that is instituted in pursuance of the provisions of this code.

❖ See the commentary to Section 103.4.

SECTION 104 (IFGC)
DUTIES AND POWERS OF THE CODE OFFICIAL

[A] 104.1 General. The code official is hereby authorized and directed to enforce the provisions of this code. The code official shall have the authority to render interpretations of this code and to adopt policies and procedures in order to clarify the application of its provisions. Such interpretations, policies and procedures shall be in compliance with the intent and purpose of this code. Such policies and procedures shall not have the effect of waiving requirements specifically provided in this code.

❖ The duty of the code official is to enforce the code. Because the code official must also act on all questions related to this responsibility except as specifically exempted by law or elsewhere in the code, the code official is the "authority having jurisdiction" for all matters relating to the code and its enforcement. It is the duty of the code official to interpret the code and to determine compliance with the code. Code compliance will not always be easy to determine and will require the judgement and expertise of the code official. When the code is silent with respect to a particular mechanical installation or lack thereof, the code official is obligated to secure the intent of the code by using the best possible judgement in acting on the matter.

[A] 104.2 Applications and permits. The code official shall receive applications, review *construction documents* and issue permits for installations and alterations of fuel gas systems, inspect the premises for which such permits have been issued and enforce compliance with the provisions of this code.

❖ The code enforcement process normally begins with an application for a permit. The code official is responsible for processing the applications and issuing permits for the installation, replacement, addition to or modification of fuel gas systems in accordance with the code.

[A] 104.3 Inspections. The code official shall make all of the required inspections, or shall accept reports of inspection by *approved* agencies or individuals. Reports of such inspections shall be in writing and shall be certified by a responsible officer of such *approved* agency or by the responsible individual. The code official is authorized to engage such expert opinion as deemed necessary to report upon unusual technical issues that arise, subject to the approval of the appointing authority.

❖ The code official must make inspections as necessary to determine compliance with the code or to accept written reports of inspections by an approved agency. The inspection of the work in progress or accomplished is another significant element in determining code compliance. Although a department might not have the resources to inspect every aspect of all work, the required inspections are those that are dictated by administrative rules and procedures based on many factors, including available inspection resources. To expand the resources available for inspections, the code official may approve an inspection agency that, in the code official's opinion, possesses the proper qualifications to perform the inspections. When unusual, extraordinary or complex technical issues arise relative to either fuel gas instal-

lations or the safety of an existing fuel gas system, the code official has the authority to seek the opinion and advice of experts. A technical report from an expert requested by the code official can be used to assist the code official in the approval process.

[A] 104.4 Right of entry. Where it is necessary to make an inspection to enforce the provisions of this code, or where the code official has reasonable cause to believe that there exists in a building or upon any premises any conditions or violations of this code that make the building or premises unsafe, dangerous or hazardous, the code official shall have the authority to enter the building or premises at all reasonable times to inspect or to perform the duties imposed upon the code official by this code. If such building or premises is occupied, the code official shall present credentials to the occupant and request entry. If such building or premises is unoccupied, the code official shall first make a reasonable effort to locate the owner, the owner's authorized agent or other person having charge or control of the building or premises and request entry. If entry is refused, the code official has recourse to every remedy provided by law to secure entry.

Where the code official has first obtained a proper inspection warrant or other remedy provided by law to secure entry, an owner, the owner's authorized agent, occupant or person having charge, care or control of the building or premises shall not fail or neglect, after proper request is made as herein provided, to promptly permit entry therein by the code official for the purpose of inspection and examination pursuant to this code.

❖ The first part of this section establishes the right of the code official to enter premises to conduct the permit inspections required by Section 107. Permit application forms typically include a statement in the certification signed by the applicant (who is the owner or the owner's agent) granting the code official the authority to enter areas covered by the permit to enforce code provisions related to the permit. The right to enter other structures or premises is more limited. First, to protect the right of privacy, the owner or occupant must grant the code official permission before an interior inspection of the property can be conducted. Permission is not required for inspections that can be accomplished from within the public right-of-way. Second, the owner or occupant may deny access. Unless the inspector has reasonable cause to believe that a violation of the code exists, access may be unattainable. Third, code officials must present proper identification (see commentary, Section 104.5) and request admittance during reasonable hours—usually, the normal business hours of the establishment. Fourth, inspections must be aimed at securing or determining compliance with the provisions and intent of the regulations that are specifically within the established scope of the code official's authority.

Searches to gather information for the enforcement of other codes, ordinances or regulations are considered unreasonable and are prohibited by the Fourth

Amendment to the U.S. Constitution. "Reasonable cause" in the context of this section must be distinguished from "probable cause," which is required to gain access to property in criminal cases. The burden of proof establishing reasonable cause may vary among jurisdictions. Usually, an inspector must show that the property is subject to inspection under the provisions of the code; that the interests of the public health, safety and welfare outweigh the individual's right to maintain privacy; and that such an inspection is required solely to determine compliance with the provisions of the code.

Many jurisdictions do not recognize the concept of an administrative warrant and may require the code official to prove probable cause in order to gain access. This burden of proof often requires the code official to state in advance why access is needed (usually access is restricted to gathering evidence for seeking an indictment or making an arrest); what specific items or information is sought; the information's relevance to the case against the subject; how knowledge of the relevance of the information or items sought was obtained; and how the evidence sought will be used. In all such cases, the right to privacy must always be weighed against the right of the code official to conduct an inspection to verify that the public health, safety and welfare are not in jeopardy. Such important and complex constitutional issues should be discussed with the jurisdiction's legal counsel. Jurisdictions should establish procedures for securing the necessary court orders when an inspection is deemed necessary following a refusal.

The last paragraph in this section requires the owner or occupant to permit entry for inspection if a proper inspection warrant or other documentation required by law has been obtained.

[A] 104.5 Identification. The code official shall carry proper identification when inspecting structures or premises in the performance of duties under this code.

❖ This section requires the code official (including all authorized inspection personnel) to carry identification when conducting the duties of the position. The identification removes any question of the purpose and authority of the inspector.

[A] 104.6 Notices and orders. The code official shall issue all necessary notices or orders to ensure compliance with this code.

❖ An important element of code enforcement is the necessary advisement of deficiencies and corrections that is accomplished through notices and orders. The code official is required to issue orders to abate illegal or unsafe conditions. Sections 108.7, 108.7.1, 108.7.2 and 108.7.3 contain additional information for these notices.

[A] 104.7 Department records. The code official shall keep official records of applications received, permits and certificates issued, fees collected, reports of inspections and notices and orders issued. Such records shall be retained in the offi-

cial records for the period required for the retention of public records.

❖ Good business practice requires the code official to keep official records pertaining to permit applications, permits, fees collected, inspections, notices and orders issued. This documentation is a valuable source of information if questions arise "throughout the life of the building and its occupancy." The code does not require that the construction documents be kept after the project is complete, but other regulations or the administrative rules may state a minimum period for retaining the construction documents.

SECTION 105 (IFGC)
APPROVAL

[A] 105.1 Modifications. Where there are practical difficulties involved in carrying out the provisions of this code, the code official shall have the authority to grant modifications for individual cases, upon application of the owner or owner's authorized agent, provided that the code official shall first find that special individual reason makes the strict letter of this code impractical and that such modification is in compliance with the intent and purpose of this code and does not lessen health, life and fire safety requirements. The details of action granting modifications shall be recorded and entered in the files of the Department of Inspection.

❖ The code official may amend or make exceptions to the code as needed where strict compliance with the code is impractical. Only the code official has the authority to grant modifications. Consideration of a particular difficulty is to be based on the application of the owner and a demonstration that the intent of the code is satisfied. This section is not intended to permit setting a code provision aside or ignoring a provision; rather, it is intended to provide for the acceptance of equivalent protection. Such modifications do not, however, extend to actions that are necessary to correct violations of the code. In other words, a code violation or the expense of correcting a code violation cannot constitute a practical difficulty.

Details of the modification must be filed for later retrieval to substantiate the reasons for the modification. Comprehensive written records are an essential part of an effective administrative system. Unless clearly written records of the considerations and documentation used in the modification process are created and maintained, subsequent enforcement action cannot be supported.

[A] 105.2 Alternative materials, methods, appliances and equipment. The provisions of this code are not intended to prevent the installation of any material or to prohibit any method of construction not specifically prescribed by this code, provided that any such alternative has been *approved*. An alternative material or method of construction shall be *approved* where the code official finds that the proposed design is satisfactory and complies with the intent of the provisions of this code, and that the material, method or work offered is, for the purpose intended, not less than the equivalent of that prescribed in this code in quality, strength, effectiveness, fire resistance, durability and safety. Where the alternative material, design or method of construction is not *approved*, the *code official* shall respond in writing, stating the reasons why the alternative was not *approved*.

❖ The code is not intended to inhibit innovative ideas or technological advances. A comprehensive regulatory document, such as a fuel gas code, cannot envision and then address all future innovations in the industry. As a result, a performance code must be applicable to and provide a basis for the approval of an increasing number of newly developed, innovative materials, systems and methods for which no code text or referenced standards yet exist. The fact that a material, product or method of construction is not addressed in the code is not an indication that prohibition of the material, product or method is intended. The code official is expected to apply sound technical judgement in accepting materials, systems or methods that, although not anticipated by the drafters of the current code text, can be demonstrated to offer equivalent performance. By virtue of its text, the code regulates new and innovative construction practices while addressing the relative safety of building occupants. The code official is responsible for determining whether a requested alternative provides a level of protection of the public health, safety and welfare as required by the code.

[A] 105.2.1 Research reports. Supporting data, where necessary to assist in the approval of materials or assemblies not specifically provided for in this code, shall consist of valid research reports from *approved* sources.

❖ When an alternative material or method is proposed for construction, it is incumbent upon the code official to determine whether this alternative is, in fact, an equivalent to the methods prescribed by the code. Reports providing evidence of this equivalency are required to be supplied by an approved source, meaning a source that the code official finds to be reliable and accurate. The ICC Evaluation Service (ICC-ES) is an example of an agency that provides research reports for alternative materials and methods.

[A] 105.3 Required testing. Where there is insufficient evidence of compliance with the provisions of this code or evidence that a material or method does not conform to the requirements of this code, or in order to substantiate claims for alternative materials or methods, the code official shall have the authority to require tests as evidence of compliance to be made at no expense to the jurisdiction.

❖ Sufficient technical data, test reports and documentation must be submitted for evaluation by the code official to provide the basis on which the code official can make a decision regarding an alternative material or type of equipment. If evidence satisfactory to the code official proves that the alternative equipment, material or construction method is equivalent to that

required by the code, the code official is obligated to approve it. Any such approval cannot have the effect of waiving any requirements of the code. The burden of proof of equivalence lies with the applicant who proposes the use of alternative equipment, materials or methods.

[A] 105.3.1 Test methods. Test methods shall be as specified in this code or by other recognized test standards. In the absence of recognized and accepted test methods, the code official shall approve the testing procedures.

❖ The code official must require the submission of any appropriate information and data to assist in the determination of equivalency. This information must be submitted before a permit can be issued. The type of information required includes test data in accordance with the referenced standards, evidence of compliance with the referenced standard specifications and design calculations. A research report issued by an authoritative agency is particularly useful in providing the code official with the technical basis for evaluation and approval of new and innovative materials and components. The use of authoritative research reports can greatly assist the code official by reducing the time-consuming engineering analysis necessary to review materials and products. Failure to substantiate a request for the use of an alternative is a valid reason for the code official to deny a request.

[A] 105.3.2 Testing agency. Tests shall be performed by an *approved* agency.

❖ The code official must approve the testing agency, and the agency should have technical expertise, test equipment and quality assurance to properly conduct and report the necessary testing.

[A] 105.3.3 Test reports. Reports of tests shall be retained by the code official for the period required for retention of public records.

❖ Test reports are to be retained as required by public record laws. The attorney of the jurisdiction could be questioned concerning the specific time period of applicable laws of the localities.

[A] 105.4 Used material, appliances and equipment. The use of used materials which meet the requirements of this code for new materials is permitted. Used appliances, *equipment* and devices shall not be reused unless such elements have been reconditioned, tested and placed in good and proper working condition, and *approved* by the code official.

❖ The code criteria for materials, equipment and devices have changed over the years. Evaluation of testing and materials technology has permitted the development of new criteria that the old materials may not satisfy. As a result, used materials must be evaluated in the same manner as new materials. Used (previously installed) equipment must be equivalent to that required by the code if it is to be used again in a new installation.

[A] 105.5 Approved materials and equipment. Materials, *equipment* and devices *approved* by the code official shall be constructed and installed in accordance with such approval.

❖ The code is a compilation of criteria with which materials, equipment, devices and systems must comply to be suitable for a particular application. The code official has a duty to evaluate such materials, equipment, devices and systems for code compliance and, when compliance is determined, approve the same for use. The materials, equipment, devices and systems must be constructed and installed in compliance with, and all conditions and limitations considered as a basis for, that approval. For example, the manufacturer's instructions and recommendations are to be followed if the approval of the material was based even in part on those instructions and recommendations. The approval authority given to the code official is a significant responsibility and is a key to code compliance. The approval process is first technical and then administrative and must be approached as such. For example, if data to determine code compliance is required, such data should be in the form of test reports or engineering analysis and not simply taken from a sales brochure.

SECTION 106 (IFGC)
PERMITS

[A] 106.1 Where required. An owner, owner's authorized agent or contractor who desires to erect, install, enlarge, alter, repair, remove, convert or replace an installation regulated by this code, or to cause such work to be performed, shall first make application to the code official and obtain the required permit for the work.

> **Exception:** Where *appliance* and *equipment* replacements and repairs are required to be performed in an emergency situation, the permit application shall be submitted within the next working business day of the Department of Inspection.

❖ In general, a permit is required for all activities that are regulated by the code, and these activities cannot begin until the permit is issued.

A fuel gas permit is required for the installation, replacement, alteration or modification of fuel gas systems and components that are in the scope of applicability of the code. Replacement of an existing piece of equipment or related piping is treated no differently than a new installation in new building construction. A permit causes the work to be inspected to determine compliance with the intent of the code.

The exception provides for prompt permit applications for situations where equipment and appliance replacements and repairs are done to address an emergency. This action enables the department having jurisdiction to promptly inspect the work.

[A] 106.1.1 Annual permit. Instead of an individual construction permit for each alteration to an already *approved* system or *equipment* installation, the code official is authorized to issue an annual permit upon application therefor to any person, firm or corporation regularly employing one or more qualified tradespersons in the building, structure or on the premises owned or operated by the applicant for the permit.

❖ Sometimes the authority having jurisdiction (AHJ) will issue an annual permit for buildings, such as hospitals, school campuses, industrial and manufacturing plants, and the like, where work that requires a permit is ongoing throughout the year. Such occupancies typically have their own full-time engineering and trades employees to design, oversee and perform the work. Some buildings are altered continuously because of changes in technologies and changing operations within the occupancy; therefore, it would be burdensome to obtain numerous permits throughout the year on a continuous basis. It is the responsibility of the permit holder to make arrangements with the code official regarding periodic inspections for the work. This is especially important for work that will be concealed.

[A] 106.1.2 Annual permit records. The person to whom an annual permit is issued shall keep a detailed record of alterations made under such annual permit. The code official shall have access to such records at all times or such records shall be filed with the code official as designated.

❖ Annual permits need to be monitored to prevent abuse of the privilege and to allow for changes in the scope of the work or changes in who is performing the work. Accurate record keeping of the work performed will advise the code official of situations where a project-specific permit needs to be issued instead of being under the umbrella of an annual permit.

[A] 106.2 Permits not required. Permits shall not be required for the following:

1. Portable heating *appliances*.

2. Replacement of any minor component of an *appliance* or *equipment* that does not alter approval of such *appliance* or *equipment* or make such *appliance* or *equipment* unsafe.

Exemption from the permit requirements of this code shall not be deemed to grant authorization for work to be done in violation of the provisions of this code or of other laws or ordinances of this jurisdiction.

❖ The installations exempt from the requirement for a permit are very limited as evidenced by Items 1 and 2. Item 1 pertains to appliances and equipment that are used temporarily and that are not designed for permanent installation. Examples of portable heating appliances and equipment include space heaters and construction site heaters.

Item 2 applies to the replacement of minor equipment and appliance components. A permit would be required if the component replacement could potentially affect either the safety of the equipment or the conditions of approval of the equipment. For example, replacement of a defective control with a control of the same type and specifications as the original factory-supplied control would not require a permit. Replacement of a burner assembly with a burner assembly having a different input capacity or one that is designed to burn a different fuel would require a permit.

[A] 106.3 Application for permit. Each application for a permit, with the required fee, shall be filed with the code official on a form furnished for that purpose and shall contain a general description of the proposed work and its location. The application shall be signed by the owner or an owner's authorized agent. The permit application shall indicate the proposed *occupancy* of all parts of the building and of that portion of the site or lot, if any, not covered by the building or structure and shall contain such other information required by the code official.

❖ This section states that the building owner or an authorized agent of the owner is the only person who can apply for a building permit. An owner's authorized agent could be anyone who is given written permission to act in the owner's interest to obtain a permit, such as an architect, engineer, contractor or tenant. Permit forms will generally have enough space for a very brief description of the work to be accomplished, which is sufficient for small jobs. For larger projects, the description will be contained in construction documents.

[A] 106.3.1 Construction documents. *Construction documents*, engineering calculations, diagrams and other data shall be submitted in two or more sets with each application for a permit. The code official shall require *construction documents*, computations and specifications to be prepared and designed by a registered design professional where required by state law. *Construction documents* shall be drawn to scale and shall be of sufficient clarity to indicate the location, nature and extent of the work proposed and show in detail that the work conforms to the provisions of this code. *Construction documents* for buildings more than two stories in height shall indicate where penetrations will be made for installations and shall indicate the materials and methods for maintaining required structural safety, fire-resistance rating and fireblocking.

> **Exception:** The code official shall have the authority to waive the submission of *construction documents*, calculations or other data if the nature of the work applied for is such that reviewing of *construction documents* is not necessary to determine compliance with this code.

❖ When the work is of a "minor nature," either in scope or needed description, the code official may use judgement in determining the need for a detailed description of the work. An example of minor work that may not involve a detailed description is the replacement of an existing piece of equipment in a mechanical system or the replacement or repair of a defective portion of a piping system.

These provisions are intended to reflect the minimum scope of information needed to determine code compliance. A statement on the construction documents, such as "All fuel gas systems must comply

with the 2009 edition of the *International Fuel Gas Code* (IFGC)," is not an acceptable substitute for showing the required information.

This section also requires the code official to determine that state professional registration laws are complied with as they apply to the preparation of construction documents.

[A] 106.3.2 Time limitation of application. An application for a permit for any proposed work shall be deemed to have been abandoned 180 days after the date of filing, unless such application has been pursued in good faith or a permit has been issued; except that the code official shall have the authority to grant one or more extensions of time for additional periods not exceeding 180 days each. The extension shall be requested in writing and justifiable cause shall be demonstrated.

❖ Typically, an application for a permit is submitted and goes through a review process that ends with the issuance of a permit. If a permit has not been issued 180 days after the date of filing, the application is considered to be abandoned, unless the applicant was diligent in efforts to obtain the permit. The code official has the authority to extend this time limitation, provided that there is reasonable cause. This would cover delays beyond the applicant's control, such as prerequisite permits or approvals from other authorities within the jurisdiction or state. The intent of this section is to limit the time between the review process and the issuance of a permit.

[A] 106.4 Preliminary inspection. Before a permit is issued, the code official is authorized to inspect and evaluate the systems, *equipment*, buildings, devices, premises and spaces or areas to be used.

❖ The code official is granted authority to inspect the site before permit issuance. This may be necessary to verify existing conditions that impact the plan review and permit approval. This section provides the code official with the right-of-entry authority that otherwise does not occur until after the permit is issued.

[A] 106.5 Permit issuance. The application, *construction documents* and other data filed by an applicant for a permit shall be reviewed by the code official. If the code official finds that the proposed work conforms to the requirements of this code and all laws and ordinances applicable thereto, and that the fees specified in Section 106.6 have been paid, a permit shall be issued to the applicant.

❖ This section requires the code official to review submittals for a permit for compliance with the code and verify that the project will be carried out in accordance with other applicable laws. This may involve interagency communication and cooperation so that all laws are being obeyed. Once the code official finds this to be so, a permit may be issued upon payment of the required fees.

[A] 106.5.1 Approved construction documents. When the code official issues the permit where *construction documents* are required, the *construction documents* shall be endorsed in writing and stamped "APPROVED." Such *approved con-*

struction documents shall not be changed, modified or altered without authorization from the code official. Work shall be done in accordance with the *approved construction documents*.

The code official shall have the authority to issue a permit for the construction of part of an installation before the *construction documents* for the entire installation have been submitted or *approved*, provided adequate information and detailed statements have been filed complying with all pertinent requirements of this code. The holder of such permit shall proceed at his or her own risk without assurance that the permit for the entire installation will be granted.

❖ Construction documents that reflect compliance with code requirements form an integral part of the permit process. Successful completion of the work depends on these documents. This section requires the code official to stamp the complying construction documents as being "APPROVED" and fix the status of the document in time. Once approved, the documents may not be revised without the express authorization of the code official. This maintains the code-compliance level of the documents.

[A] 106.5.2 Validity. The issuance of a permit or approval of *construction documents* shall not be construed to be a permit for, or an approval of, any violation of any of the provisions of this code or of other ordinances of the jurisdiction. A permit presuming to give authority to violate or cancel the provisions of this code shall be invalid.

The issuance of a permit based upon *construction documents* and other data shall not prevent the code official from thereafter requiring the correction of errors in said *construction documents* and other data or from preventing building operations from being carried on thereunder when in violation of this code or of other ordinances of this jurisdiction.

❖ This powerful code section states the fundamental premise that the permit is only "a license to proceed with the work." It is not a license to "violate, cancel or set aside any provisions of the code." This statement is important because, despite any errors in the approval process, the permit applicant is responsible for code compliance.

[A] 106.5.3 Expiration. Every permit issued by the code official under the provisions of this code shall expire by limitation and become null and void if the work authorized by such permit is not commenced within 180 days from the date of such permit, or is suspended or abandoned at any time after the work is commenced for a period of 180 days. Before such work recommences, a new permit shall be first obtained and the fee therefor shall be one-half the amount required for a new permit for such work, provided that changes have not been and will not be made in the original *construction documents* for such work, and further that such suspension or abandonment has not exceeded one year.

❖ The permit becomes invalid in two distinct situations, but both are based on a 6-month period. The first situation is when no work has started 6 months from issuance of the permit. The second situation is when

there is no continuation of authorized work for 6 months. The person who was issued the permit should be notified in writing that the permit is invalid and what steps must be taken to allow work to begin again.

This section also gives the administrative authority a means of offsetting the costs associated with the administration of expired, reissued permits by charging a nominal fee for permit reissuance. If, however, the nature or scope of the work to be resumed is different from that indicated on the original permit, the permit process essentially starts all over again, and full fees are charged. The same procedure would also apply if the work has not commenced within 1 year of the date of permit issuance or if work has been suspended for a year or more.

[A] 106.5.4 Extensions. A permittee holding an unexpired permit shall have the right to apply for an extension of the time within which he or she will commence work under that permit when work is unable to be commenced within the time required by this section for good and satisfactory reasons. The code official shall extend the time for action by the permittee for a period not exceeding 180 days if there is reasonable cause. A permit shall not be extended more than once. The fee for an extension shall be one-half the amount required for a new permit for such work.

❖ Although it is customary for a project to begin immediately following issuance of a permit, there may be occasions when an unforeseen delay may occur. This section affords the permit holder an opportunity to apply for and receive a single, 180-day extension within which to begin a project under a still-valid permit (less than 180 days old). The applicant must, however, submit to the code official an adequate explanation for the delay in starting a project that could include such things as the need to obtain approvals or permits for the project from other agencies having jurisdiction. This section requires the code official to determine what constitutes "good and satisfactory" reasons for any delay and further allows the jurisdiction to offset its administrative costs for extending the permit by charging one-half the permit fee for the extension.

[A] 106.5.5 Suspension or revocation of permit. The code official shall have the authority to suspend or revoke a permit issued under the provisions of this code wherever the permit is issued in error or on the basis of incorrect, inaccurate or incomplete information, or in violation of any ordinance or regulation or any of the provisions of this code.

❖ A permit is, in reality, a license to proceed with the work. The code official, however, must revoke all permits shown to be based, all or in part, on any false statement or misrepresentation of fact. An applicant may subsequently reapply for a permit with the appropriate corrections or modifications made to the application and construction documents.

[A] 106.5.6 Retention of construction documents. One set of *approved construction documents* shall be retained by the code official for a period of not less than 180 days from date of completion of the permitted work, or as required by state or local laws. One set of *approved construction documents* shall be returned to the applicant, and said set shall be kept on the site of the building or work at all times during which the work authorized thereby is in progress.

❖ Once the code official has stamped or endorsed as approved the construction documents on which the permit is based (see commentary, Section 106.5.1), one set of approved construction documents must be kept on the construction site to serve as the basis for all subsequent inspections. To avoid confusion, the construction documents on the site must be precisely the documents that were approved and stamped because inspections are to be performed based on the approved documents. Additionally, the contractor cannot determine compliance with the approved construction documents unless those documents are readily available. If the approved construction documents are not available, the inspection should be postponed and work on the project halted.

[A] 106.5.7 Previous approvals. This code shall not require changes in the *construction documents*, construction or designated *occupancy* of a structure for which a lawful permit has been heretofore issued or otherwise lawfully authorized, and the construction of which has been pursued in good faith within 180 days after the effective date of this code and has not been abandoned.

❖ Under circumstances where a new edition of the code is adopted by the legislative body of the jurisdiction at the same time that construction is proceeding at an acceptable pace on a project where a permit has been issued, requiring that the building be constructed to conform to the new edition of the code is unreasonable. This section provides for the continuity of permits issued under previous codes, as long as such permits are being "actively prosecuted" subsequent to the effective date of the ordinance adopting this edition of the code.

[A] 106.5.8 Posting of permit. The permit or a copy shall be kept on the site of the work until the completion of the project.

❖ The permit, or copy thereof, is to be kept on the job site and made available to the code official or representative to conveniently make required entries thereon until the work is complete.

[A] 106.6 Fees. A permit shall not be issued until the fees prescribed in Section 106.6.2 have been paid, nor shall an amendment to a permit be released until the additional fee, if any, due to an increase of the installation, has been paid.

❖ All fees are to be paid prior to permit issuance. This requirement facilitates payment and also establishes that the permit applicant intends to proceed with the work.

[A] 106.6.1 Work commencing before permit issuance. Any person who commences work on an installation before

obtaining the necessary permits shall be subject to 100 percent of the usual permit fee in addition to the required permit fees.

❖ This section is intended to serve as a deterrent to proceeding with work on a fuel gas system without a permit (except as provided in Section 106.2). As a punitive measure, it doubles the cost of the permit fee to be charged. This section does not, however, intend to penalize a contractor called upon to do emergency work after hours if he or she promptly notifies the code official the next business day, obtains the requisite permit for the work done and has the required inspections performed.

[A] 106.6.2 Fee schedule. The fees for work shall be as indicated in the following schedule.

[JURISDICTION TO INSERT APPROPRIATE SCHEDULE]

❖ A published fee schedule must be established for plans examination, permits and inspections. Ideally, the department should generate revenues that cover operating costs and expenses. The permit fee schedule is an integral part of this process.

[A] 106.6.3 Fee refunds. The code official shall authorize the refunding of fees as follows.

1. The full amount of any fee paid hereunder that was erroneously paid or collected.

2. Not more than **[SPECIFY PERCENTAGE]** percent of the permit fee paid where work has not been done under a permit issued in accordance with this code.

3. Not more than **[SPECIFY PERCENTAGE]** percent of the plan review fee paid where an application for a permit for which a plan review fee has been paid is withdrawn or canceled before any plan review effort has been expended.

The code official shall not authorize the refunding of any fee paid, except upon written application filed by the original permittee not later than 180 days after the date of fee payment.

❖ The code official must not authorize the refunding of any fee paid, except upon written application filed by the original permittee not later than 180 days after the date of fee payment.

SECTION 107 (IFGC)
INSPECTIONS AND TESTING

[A] 107.1 General. The code official is authorized to conduct such inspections as are deemed necessary to determine compliance with the provisions of this code. Construction or work for which a permit is required shall be subject to inspection by the code official, and such construction or work shall remain accessible and exposed for inspection purposes until *approved*. Approval as a result of an inspection shall not be construed to be an approval of a violation of the provisions of this code or of other ordinances of the jurisdiction. Inspections presuming to give authority to violate or cancel the provisions of this code or of other ordinances of the jurisdiction shall not be valid.

❖ The inspection function is one of the more important aspects of building department operations. This section authorizes the code official to inspect the work for which a permit has been issued and requires that the work to be inspected remain accessible to the code official until inspected and approved. Any expense incurred in removing or replacing material that conceals an item to be inspected is not the responsibility of the code official or the jurisdiction. As with the issuance of permits (see Section 106.5), approval as a result of an inspection is not a license to violate the code and an approval in violation of the code does not relieve the applicant from complying with the code and is not valid.

[A] 107.2 Required inspections and testing. The code official, upon notification from the permit holder or the permit holder's agent, shall make the following inspections and other such inspections as necessary, and shall either release that portion of the construction or notify the permit holder or the permit holder's agent of violations that are required to be corrected. The holder of the permit shall be responsible for scheduling such inspections.

1. Underground inspection shall be made after trenches or ditches are excavated and bedded, *piping* is installed and before backfill is put in place. Where excavated soil contains rocks, broken concrete, frozen chunks and other rubble that would damage or break the *piping* or cause corrosive action, clean backfill shall be on the job site.

2. Rough-in inspection shall be made after the roof, framing, fireblocking and bracing are in place and components to be concealed are complete, and prior to the installation of wall or ceiling membranes.

3. Final inspection shall be made upon completion of the installation.

The requirements of this section shall not be considered to prohibit the operation of any heating *appliance* installed to replace an existing heating *appliance* serving an occupied portion of a structure in the event a request for inspection of such heating *appliance* has been filed with the department not more than 48 hours after replacement work is completed, and before any portion of such *appliance* is concealed by any permanent portion of the structure.

❖ Inspections are necessary to determine that an installation conforms to all code requirements. Because the majority of a fuel gas piping system could be hidden within the building enclosure, periodic inspections are necessary before portions of the system are concealed. The code official is required to determine that fuel gas systems and equipment are installed in accordance with the approved construction documents and the applicable code requirements. Inspections that are necessary to provide verification must be conducted. Generally, the administrative rules of a department may list the interim inspections judged to

be required. Construction that occurs in steps or phases may necessitate multiple inspections; therefore, an exact number of required inspections cannot be specified. Where violations are noted and corrections are required, reinspections may be necessary. As time permits, frequent inspections of some job sites, especially where the work is complex, can be beneficial if the inspector detects code-compliance problems or potential problems before they develop or become more difficult to correct. The contractor, builder, owner or other authorized party is responsible for arranging for the required inspections and coordinating inspections to prevent work from being concealed before it is inspected.

1. Inspection of underground piping is especially important because once it is covered, it is the most challenging part of a fuel gas system in which to detect a leak. If repairs are necessary, underground repairs are proportionally more expensive because of the need for heavy equipment and the more labor-intensive nature of working below ground. To reduce possible damage to pipe from rubble, rocks and other rough materials, excavations must be bedded and backfilled with clean fill materials spread and tamped to provide adequate support and protection for piping.

2. A rough-in inspection is an inspection of all parts of the fuel gas system that will eventually be concealed in the building structure. The inspection must be made before any of the system is closed up or hidden from view. To gain approval, the fuel gas systems must pass the required rough-in tests.

3. A rough-in inspection may be completed all at one time or as a series of inspections. This is administratively determined by the local inspections department and is typically dependent on the size of the job.

4. A final inspection may be done as a series of inspections or all at one time, similar to a rough-in inspection. A final inspection is required prior to the approval of mechanical work and fuel gas installations. For the construction of a new building, final approval is required prior to the issuance of the certificate of occupancy as specified in the building code. To verify that all previously issued correction orders have been complied with and to determine whether subsequent violations exist, a final inspection must be made. Violations observed during the final inspection must be noted, and the permit holder must be advised of them.

The final inspection is made after the completion of the work or installation. Typically, the final inspection is an inspection of all that was installed after the rough-in inspection and not concealed in the building construction. Subsequent reinspections are necessary if the final inspection has generated a notice of violation.

The last paragraph of this section is emergency related and provides for prompt operation of replacement heating equipment, allowing the occupied areas of a facility to be heated as soon as the new heating equipment is installed. Such installations are subject to all requirements of the code.

[A] 107.2.1 Other inspections. In addition to the inspections specified in Section 107.2, the code official is authorized to make or require other inspections of any construction work to ascertain compliance with the provisions of this code and other laws that are enforced.

❖ Any item regulated by the code is subject to inspection by the code official to determine compliance with the applicable code provision, and no list can include all items in a given project. This section, therefore, gives the code official the authority to inspect any regulated items.

[A] 107.2.2 Inspection requests. It shall be the duty of the holder of the permit or his or her duly authorized agent to notify the code official when work is ready for inspection. It shall be the duty of the permit holder to provide *access* to and means for inspection of such work that is required by this code.

❖ It is the responsibility of the permit holder or other authorized person, such as the contractor performing the work, to arrange for the required inspections when completed work is ready and to allow for sufficient time for the code official to schedule a visit to the site to prevent work from being concealed prior to being inspected. Access to the work to be inspected must be provided, including any special means such as a ladder.

[A] 107.2.3 Approval required. Work shall not be done beyond the point indicated in each successive inspection without first obtaining the approval of the code official. The code official, upon notification, shall make the requested inspections and shall either indicate the portion of the construction that is satisfactory as completed, or notify the permit holder or his or her agent wherein the same fails to comply with this code. Any portions that do not comply shall be corrected and such portion shall not be covered or concealed until authorized by the code official.

❖ This section establishes that work cannot progress beyond the point of a required inspection without the code official's approval. Upon making the inspection, the code official must either approve the completed work or notify the permit holder or other responsible party of that which does not comply with the code. Approvals and notices of noncompliance must be in writing, as required by Section 104.3, to avoid any misunderstanding as to what is required. Any item not approved cannot be concealed until it has been corrected and approved by the code official.

[A] 107.2.4 Approved inspection agencies. The code official is authorized to accept reports of *approved* agencies, pro-

vided that such agencies satisfy the requirements as to qualifications and reliability.

❖ As an alternative to conducting the inspection, the code official can accept inspections and reports by approved inspection agencies. Appropriate criteria on which to base approval of an inspection agency include competence, objectivity, certifications and experience.

[A] 107.2.5 Evaluation and follow-up inspection services. Prior to the approval of a prefabricated construction assembly having concealed work and the issuance of a permit, the code official shall require the submittal of an evaluation report on each prefabricated construction assembly, indicating the complete details of the installation, including a description of the system and its components, the basis upon which the system is being evaluated, test results and similar information and other data as necessary for the code official to determine conformance to this code.

❖ As an alternative to the physical inspection by the code official in the plant or location where prefabricated components are fabricated (such as modular homes and prefabricated structures), the code official has the option of accepting an evaluation report from an approved agency detailing such inspections. These evaluation reports can serve as the basis upon which the code official will determine code compliance.

[A] 107.2.5.1 Evaluation service. The code official shall designate the evaluation service of an *approved* agency as the evaluation agency, and review such agency's evaluation report for adequacy and conformance to this code.

❖ The code official must review all submitted reports for conformity to the applicable code requirements. If in the judgment of the code official the submitted reports are acceptable, the code official should document the basis for the approval.

[A] 107.2.5.2 Follow-up inspection. Except where ready *access* is provided to installations, appliances, service *equipment* and accessories for complete inspection at the site without disassembly or dismantling, the code official shall conduct the in-plant inspections as frequently as necessary to ensure conformance to the *approved* evaluation report or shall designate an independent, *approved* inspection agency to conduct such inspections. The inspection agency shall furnish the code official with the follow-up inspection manual and a report of inspections upon request, and the installation shall have an identifying label permanently affixed to the system indicating that factory inspections have been performed.

❖ The owner must provide special inspections of fabricated assemblies at the fabrication plant. The code official or an approved inspection agency must conduct periodic in-plant inspections to ensure conformance to the approved evaluation report described in Section 107.2.5. Such inspections are required if the fuel gas systems cannot be inspected completely at the job site.

[A] 107.2.5.3 Test and inspection records. Required test and inspection records shall be available to the code official at all times during the fabrication of the installation and the erection of the building; or such records as the code official designates shall be filed.

❖ All testing and inspection records related to a fabricated assembly must be filed with the code official to maintain a complete and legal record of the assembly and erection of the building.

[A] 107.3 Testing. Installations shall be tested as required in this code and in accordance with Sections 107.3.1 through 107.3.3. Tests shall be made by the permit holder and observed by the code official.

❖ Testing of fuel gas systems is required where testing is specified in the technical chapters of the code. See Section 406 for specific requirements and methods of testing gas piping systems.

[A] 107.3.1 New, altered, extended or repaired installations. New installations and parts of existing installations, which have been altered, extended, renovated or repaired, shall be tested as prescribed herein to disclose leaks and defects.

❖ Testing is necessary to make sure that the system is free from leaks and other defects. To the extent specified in the technical chapters of the code, testing is also required for portions of existing systems that have been altered, extended, renovated or repaired.

[A] 107.3.2 Apparatus, instruments, material and labor for tests. Apparatus, instruments, material and labor required for testing an installation or part thereof shall be furnished by the permit holder.

❖ The permit holder is responsible for performing tests, as well as for supplying all of the labor and apparatus necessary to conduct the tests. The code official observes, but never performs, the test.

[A] 107.3.3 Reinspection and testing. Where any work or installation does not pass an initial test or inspection, the necessary corrections shall be made so as to achieve compliance with this code. The work or installation shall then be resubmitted to the code official for inspection and testing.

❖ If a system or portion of a system does not pass the initial test or inspection, violations must be corrected and the system must be reinspected.

[A] 107.4 Approval. After the prescribed tests and inspections indicate that the work complies in all respects with this code, a notice of approval shall be issued by the code official.

❖ After the code official has performed the required inspections and observed any required equipment and system tests (or has received written reports of the results of such tests), he or she must determine whether the installation or work is in compliance with all applicable sections of the code. The code official must issue a written notice of approval if the subject work or installation is in apparent compliance with the code. This notice is given to the permit holder, and a copy of the notice is retained on file by the code official.

[A] 107.4.1 Revocation. The code official is authorized to, in writing, suspend or revoke a notice of approval issued under the provisions of this code wherever the notice is issued in error, or on the basis of incorrect information supplied or where it is determined that the building or structure, premise, or portion thereof is in violation of any ordinance or regulation or any of the provisions of this code.

❖ The code official is authorized to, in writing, suspend or revoke a certificate of occupancy or completion issued under the provisions of the code wherever the certificate is issued in error, on the basis of incorrect information supplied, or where it is determined that the building or structure or portion thereof is in violation of any ordinance, regulation or any of the provisions of this code.

This section is needed to give the code official the authority to revoke a certificate of occupancy for the reasons indicated in the code text. The code official may also suspend the certificate of occupancy until all of the code violations are corrected.

[A] 107.5 Temporary connection. The code official shall have the authority to allow the temporary connection of an installation to the sources of energy for the purpose of testing the installation or for use under a temporary certificate of *occupancy*.

❖ Typical procedure for a local jurisdiction is to withhold the issuance of the certificate of occupancy until approvals have been received from each code official responsible for inspection of the structure. The code official is permitted to issue a temporary authorization to make connections to the public utility system prior to the completion of all work. The certification is intended to acknowledge that, because of seasonal limitations, time constraints or the need for testing or partial operation of equipment, some building systems may be connected even though the building is not suitable for final occupancy. The intent of this section is that a request for temporary occupancy or the connection and use of mechanical equipment or systems should be granted when the requesting permit holder has demonstrated to the code official's satisfaction that the public health, safety and welfare will not be endangered. The code official should view the issuance of a "temporary authorization or certificate of occupancy" as substantial an act as the issuance of the final certificate. Indeed, the issuance of a temporary certificate of occupancy offers a greater potential for conflict because once the building or structure is occupied, it is very difficult to remove the occupants through legal means.

[A] 107.6 Connection of service utilities. A person shall not make connections from a utility, source of energy, fuel or power to any building or system that is regulated by this code for which a permit is required until authorized by the code official.

❖ This section establishes the authority of the code official to approve utility connections to a building for items such as water, sewer, electricity, gas and steam, and to require their disconnection when hazardous conditions or emergencies exist.

The approval of the code official is required before a connection can be made from a utility to a building system that is regulated by the code. This includes utilities supplying water, sewer, electricity, gas and steam services. For the protection of building occupants, including workers, such systems must have had final inspection approvals, except as allowed by Section 110.3 for temporary connections.

SECTION 108 (IFGC)
VIOLATIONS

[A] 108.1 Unlawful acts. It shall be unlawful for a person, firm or corporation to erect, construct, alter, repair, remove, demolish or utilize an installation, or cause same to be done, in conflict with or in violation of any of the provisions of this code.

❖ Violations of the code are unlawful. This is the basis for all citations and correction notices related to violations of the code.

[A] 108.2 Notice of violation. The code official shall serve a notice of violation or order to the person responsible for the erection, installation, *alteration*, extension, repair, removal or demolition of work in violation of the provisions of this code, or in violation of a detail statement or the *approved construction documents* thereunder, or in violation of a permit or certificate issued under the provisions of this code. Such order shall direct the discontinuance of the illegal action or condition and the abatement of the violation.

❖ The code official must notify the person responsible for the erection or use of a building found to be in violation of the code. The section that is allegedly being violated must be cited so that the responsible party can respond to the notice.

[A] 108.3 Prosecution of violation. If the notice of violation is not complied with promptly, the code official shall request the legal counsel of the jurisdiction to institute the appropriate proceeding at law or in equity to restrain, correct or abate such violation, or to require the removal or termination of the unlawful *occupancy* of the structure in violation of the provisions of this code or of the order or direction made pursuant thereto.

❖ The code official must pursue legal means to correct the violation through the use of legal counsel of the jurisdiction. This is not optional.

[A] 108.4 Violation penalties. Persons who shall violate a provision of this code, fail to comply with any of the requirements thereof or erect, install, alter or repair work in violation of the *approved construction documents* or directive of the code official, or of a permit or certificate issued under the provisions of this code, shall be guilty of a [SPECIFY OFFENSE], punishable by a fine of not more than [AMOUNT] dollars or by imprisonment not exceeding [NUMBER OF DAYS], or both such fine and imprisonment. Each day that a violation continues

after due notice has been served shall be deemed a separate offense.

❖ This section prescribes a standard fine or other penalty as deemed appropriate by the jurisdiction. Additionally, this section identifies a principle that "each day that a violation continues . . . shall be deemed a separate offense" for the purpose of applying the prescribed penalty to facilitate prompt resolution.

[A] 108.5 Stop work orders. Upon notice from the code official that work is being performed contrary to the provisions of this code or in a dangerous or unsafe manner, such work shall immediately cease. Such notice shall be in writing and shall be given to the owner of the property, the owner's authorized agent, or the person doing the work. The notice shall state the conditions under which work is authorized to resume. Where an emergency exists, the code official shall not be required to give a written notice prior to stopping the work. Any person who shall continue any work on the system after having been served with a stop work order, except such work as that person is directed to perform to remove a violation or unsafe condition, shall be liable for a fine of not less than [AMOUNT] dollars or more than [AMOUNT] dollars.

❖ Upon receipt of a violation notice from the code official, the builders must immediately cease construction activities identified in the notice, except as expressly permitted to correct the violation. A stop work order can result in inconvenience and monetary loss to the affected parties; therefore, justification must be evident and judgment must be exercised before a stop work order is issued. A stop work order can prevent a violation from becoming worse and more difficult or expensive to correct.

A stop work order may be issued where work is proceeding without a permit to perform the work. Hazardous conditions could develop when the code official is unaware of the nature of the work and a permit for the work has not been issued.

The issuance of a stop work order on a fuel gas system may result from work done by the contractor that affects a building component that is not fuel-gas related. For example, if a contractor cuts a structural element to install piping, the structure may be weakened enough to cause a partial or complete structural failure. As determined by the adopting jurisdiction, a penalty may be assessed for failure to comply with this section, and the dollar amount is to be inserted in the blanks shown.

[A] 108.6 Abatement of violation. The imposition of the penalties herein prescribed shall not preclude the legal officer of the jurisdiction from instituting appropriate action to prevent unlawful construction, restrain, correct or abate a violation, prevent illegal occupancy of a building, structure or premises, or stop an illegal act, conduct, business or utilization of the installations on or about any premises.

❖ Despite the assessment of a penalty in the form of a fine or imprisonment against a violator, the violation itself must still be corrected. Failure to make the necessary corrections will result in the violator being subject to additional penalties as described in the preceding section.

[A] 108.7 Unsafe installations. An installation that is unsafe, constitutes a fire or health hazard, or is otherwise dangerous to human life, as regulated by this code, is hereby declared an unsafe installation. Use of an installation regulated by this code constituting a hazard to health, safety or welfare by reason of inadequate maintenance, dilapidation, fire hazard, disaster, damage or abandonment is hereby declared an unsafe use. Such unsafe installations are hereby declared to be a public nuisance and shall be abated by repair, rehabilitation, demolition or removal.

❖ In the course of performing duties, the code official may identify a hazardous condition that must be declared in violation of the code and, therefore, must be abated.

[A] 108.7.1 Authority to condemn installations. Whenever the code official determines that any installation, or portion thereof, regulated by this code has become hazardous to life, health or property, he or she shall order in writing that such installations either be removed or restored to a safe condition. A time limit for compliance with such order shall be specified in the written notice. A person shall not use or maintain a defective installation after receiving such notice.

Where such installation is to be disconnected, written notice as prescribed in Section 108.2 shall be given. In cases of immediate danger to life or property, such disconnection shall be made immediately without such notice.

❖ When a system is determined to be unsafe, the code official must notify the owner or the agent of the building as the first step in correcting the difficulty. The notice is to describe the repairs and improvements necessary to correct the deficiency or require removal or replacement of the unsafe equipment or system. Such notices must specify a time frame in which the corrective actions must occur. Additionally, the notice should require the immediate response of the owner or the agent. If the owner or the agent is not available, public notice of the declaration should suffice for complying with this section. The code official may also determine that disconnecting the system is necessary to correct an unsafe condition and must give written notice to that effect (see commentary, Section 108.2), unless immediate disconnection is essential for public health and safety reasons (see commentary, Section 108.7.2).

[A] 108.7.2 Authority to disconnect service utilities. The code official shall have the authority to require disconnection of utility service to the building, structure or system regulated by the technical codes in case of emergency where necessary to eliminate an immediate hazard to life or property. The code official shall notify the serving utility and, where possible, the owner or the owner's authorized agent and occupant of the building, structure or service system of the decision to disconnect prior to taking such action. If not notified prior to disconnection, the owner or occupant of the building, structure or service system shall be notified in writing, as soon as practicable thereafter.

❖ Disconnecting the utility service is the most radical method of hazard abatement available to the code official and should be reserved for cases in which all other lesser remedies have proven ineffective. Such an action must be preceded by written notice to the owner and any occupants of the building being ordered to disconnect. Disconnection must be accomplished within the time frame established by the code official in the written notification to disconnect. When the hazard to the public health and welfare is so imminent as to mandate immediate disconnection, the code official has the authority and even the obligation to cause disconnection without notice.

[A] 108.7.3 Connection after order to disconnect. A person shall not make energy source connections to installations regulated by this code which have been disconnected or ordered to be disconnected by the code official, or the use of which has been ordered to be discontinued by the code official until the code official authorizes the reconnection and use of such installations.

Where an installation is maintained in violation of this code, and in violation of a notice issued pursuant to the provisions of this section, the code official shall institute appropriate action to prevent, restrain, correct or abate the violation.

❖ When any fuel gas system is maintained in violation of the code and in violation of any notice issued pursuant to the provisions of this section, the code official is to institute appropriate action to prevent, restrain, correct or abate the violation.

Once the reason for discontinuation of use or disconnection of the fuel gas system no longer exists, only the code official may authorize resumption of use or reconnection of the system after it is demonstrated to the code official's satisfaction that all repairs or other work are in compliance with applicable sections of the code. This section also requires the code official to take action to abate code violations (see commentary, Section 108.2).

SECTION 109 (IFGC)
MEANS OF APPEAL

[A] 109.1 Application for appeal. A person shall have the right to appeal a decision of the code official to the board of appeals. An application for appeal shall be based on a claim that the true intent of this code or the rules legally adopted thereunder have been incorrectly interpreted, the provisions of this code do not fully apply or an equally good or better form of construction is proposed. The application shall be filed on a form obtained from the code official within 20 days after the notice was served.

❖ This section holds that any aggrieved party with a material interest in the decision of the code official may appeal that decision before a board of appeals.

This provides a forum other than the court of jurisdiction in which to review the code official's actions.

This section allows any person to appeal a decision of the code official. In practice, this section has been interpreted to permit appeals only by those aggrieved parties with a material or definitive interest in the decision of the code official. An aggrieved party may not appeal a code requirement per se. The intent of the appeal process is not to waive or set aside a code requirement; rather, it is intended to provide a means of reviewing a code official's decision on an interpretation or application of the code or to review the equivalency of protection to the code requirements.

[A] 109.2 Membership of board. The board of appeals shall consist of five members appointed by the chief appointing authority as follows: one for 5 years; one for 4 years; one for 3 years; one for 2 years and one for 1 year. Thereafter, each new member shall serve for 5 years or until a successor has been appointed.

❖ The board of appeals is to consist of five members appointed on a rotating basis by the "chief appointing authority"—typically, the mayor or city manager. This method of appointment allows for a smooth transition of board of appeals members, thus ensuring continuity of action over the years.

[A] 109.2.1 Qualifications. The board of appeals shall consist of five individuals, one from each of the following professions or disciplines.

1. Registered design professional who is a registered architect; or a builder or superintendent of building construction with not less than 10 years' experience, 5 of which shall have been in responsible charge of work.

2. Registered design professional with structural engineering or architectural experience.

3. Registered design professional with fuel gas and plumbing engineering experience; or a fuel gas contractor with not less than 10 years' experience, 5 of which shall have been in responsible charge of work.

4. Registered design professional with electrical engineering experience; or an electrical contractor with not less than 10 years' experience, 5 of which shall have been in responsible charge of work.

5. Registered design professional with fire protection engineering experience; or a fire protection contractor with not less than 10 years' experience, 5 of which shall have been in responsible charge of work.

❖ The board of appeals consists of five persons with the qualifications and experience indicated in this section. One must be a registered design professional (see Item 2) with structural or architectural experience. The others must be registered design professionals, construction superintendents or contractors with experience in the various areas of building construction. These requirements are important in that technical people rule on technical matters. The board of appeals is not the place for policy or political deliberations. The intent is that these matters be decided purely on their technical merits with the appropriate regard for state-of-the-art construction technology.

[A] 109.2.2 Alternate members. The chief appointing authority shall appoint two alternate members who shall be called by the board chairman to hear appeals during the absence or disqualification of a member. Alternate members shall possess the qualifications required for board membership and shall be appointed for 5 years, or until a successor has been appointed.

❖ This section authorizes the chief appointing authority to appoint two alternate members who are to be available if the principal members of the board are absent or disqualified. Alternate members must possess the same qualifications as the principal members and are appointed for a term of 5 years, or until such time that a successor is appointed.

[A] 109.2.3 Chairman. The board shall annually select one of its members to serve as chairman.

❖ It is customary to determine chairmanship annually so that a regular opportunity is available to evaluate and either reappoint the current chairman or appoint a new one.

[A] 109.2.4 Disqualification of member. A member shall not hear an appeal in which that member has a personal, professional or financial interest.

❖ Members must not participate in the hearing of any appeal in which they have a personal, professional or financial interest.

[A] 109.2.5 Secretary. The chief administrative officer shall designate a qualified clerk to serve as secretary to the board. The secretary shall file a detailed record of all proceedings in the office of the chief administrative officer.

❖ The chief administrative officer is to designate a qualified clerk to serve as secretary to the board. The secretary is required to file a detailed record of all proceedings in the office of the chief administrative officer.

[A] 109.2.6 Compensation of members. Compensation of members shall be determined by law.

❖ Members of the board of appeals need not be compensated unless required by the local municipality or jurisdiction.

[A] 109.3 Notice of meeting. The board shall meet upon notice from the chairman, within 10 days of the filing of an appeal, or at stated periodic meetings.

❖ The board must meet within 10 days of the filing of an appeal or at regularly scheduled meetings.

[A] 109.4 Open hearing. Hearings before the board shall be open to the public. The appellant, the appellant's representative, the code official and any person whose interests are affected shall be given an opportunity to be heard.

❖ Hearings before the board must be open to the public. The person who filed the appeal, his or her representative, the code official and any person whose interests are affected must be heard.

[A] 109.4.1 Procedure. The board shall adopt and make available to the public through the secretary procedures under which a hearing will be conducted. The procedures shall not require compliance with strict rules of evidence, but shall mandate that only relevant information be received.

❖ The board is required to establish and make available to the public written procedures detailing how hearings are to be conducted. Additionally, this section provides that although strict rules of evidence are not applicable, the information presented must be deemed relevant.

[A] 109.5 Postponed hearing. When five members are not present to hear an appeal, either the appellant or the appellant's representative shall have the right to request a postponement of the hearing.

❖ When all five members of the board are not present, either the person making the appeal or his or her representative may request a postponement of the hearing.

[A] 109.6 Board decision. The board shall modify or reverse the decision of the code official by a concurring vote of three members.

❖ A concurring vote of three members of the board is needed to modify or reverse the decision of the code official.

[A] 109.6.1 Resolution. The decision of the board shall be by resolution. Certified copies shall be furnished to the appellant and to the code official.

❖ A formal decision in the form of a resolution is required to provide an official record. Copies of this resolution are to be furnished to both the person making the appeal and the code official. The code official is bound by the action of the board of appeals, unless it is the opinion of the code official that the board of appeals has acted improperly. In such cases, relief through the court having jurisdiction may be sought by corporate council.

[A] 109.6.2 Administration. The code official shall take immediate action in accordance with the decision of the board.

❖ To avoid any unnecessary delay in the progress of construction, the code official is required to act quickly on the board's decision. This action may be to enforce the decision or to seek legislative relief if the board's action can be demonstrated to be inappropriate.

[A] 109.7 Court review. Any person, whether or not a previous party to the appeal, shall have the right to apply to the appropriate court for a writ of certiorari to correct errors of law. Application for review shall be made in the manner and time required by law following the filing of the decision in the office of the chief administrative officer.

❖ This section allows any person to request a review by the court of jurisdiction if that person believes errors of law have occurred. Review must be applied for after the decision of the board is filed with the chief administrative officer. This helps to establish the observance of due process for all concerned.

SECTION 110 (IFGC)
TEMPORARY EQUIPMENT, SYSTEMS AND USES

[A] 110.1 General. The code official is authorized to issue a permit for temporary *equipment*, systems and uses. Such permits shall be limited as to time of service, but shall not be permitted for more than 180 days. The code official is authorized to grant extensions for demonstrated cause.

❖ In the course of construction or other activities, equipment or systems are sometimes needed or desired for a short time. This section contains the administrative provisions that permit such temporary equipment, systems, or uses without full compliance with the code requirements for permanently installed systems.

This section allows the code official to issue permits for temporary equipment, systems or uses. The applicant must specify the time period desired for the temporary equipment, system, or use, but the approval period cannot exceed 180 days. Equipment, systems, or uses that are temporary but are anticipated to be in existence for more than 180 days are required to conform to code requirements for permanent systems and uses. The section also authorizes the code official to grant extensions to this time period if the applicant can provide a valid reason for the extension, which typically includes circumstances beyond the applicant's control. This provision is not intended to be used to circumvent the 180-day limitation.

[A] 110.2 Conformance. Temporary *equipment*, systems and uses shall conform to the structural strength, fire safety, means of egress, accessibility, light, ventilation and sanitary requirements of this code as necessary to ensure the public health, safety and general welfare.

❖ This section prescribes those categories of the code requirements that must be complied with, despite the fact that the system will be removed or the use discontinued some time in the near future. These criteria are essential for measuring the safety of any system or use, temporary or permanent; therefore, the application of these criteria to a temporary system cannot be waived.

"Structural strength" refers to the ability of the temporary structure to resist anticipated live, environmental and dead loads (see Chapter 16 of the IBC). It also applies to anticipated live and dead loads imposed by a temporary equipment, system or use on the structure.

"Fire safety" provisions are those required by Chapters 7, 8 and 9 of the IBC or by provisions in the IMC and this code invoked by virtue of the type of temporary system or use, and the potential for increased fire risk.

"Means of egress" refers to full compliance with Chapter 10 of the IMC.

"Accessibility" refers to full compliance with Chapter 11 of the IBC for making buildings accessible to physically disabled persons.

"Light, ventilation and sanitary" requirements are those imposed by Chapter 12 of the code or applicable sections of the *International Plumbing Code*® (IPC®) or IMC.

[A] 110.3 Temporary utilities. The code official is authorized to give permission to temporarily supply utilities before an installation has been fully completed and the final certificate of completion has been issued. The part covered by the temporary certificate shall comply with the requirements specified for temporary lighting, heat or power in the code.

❖ Commonly, heating appliances/equipment are installed and operated during construction of a building. This section allows the code official to grant permission for fuel and power supplies to be connected to the building on a temporary basis.

[A] 110.4 Termination of approval. The code official is authorized to terminate such permit for a temporary structure or use and to order the temporary structure or use to be discontinued.

❖ This section provides the code official with the necessary authority to terminate the permit for temporary equipment, systems or uses. The code official can order that temporary equipment or systems be removed or temporary uses be discontinued if conditions of the permit have been violated or the structure or use poses an imminent hazard to the public. This text is important because it allows the code official to act quickly when time is of the essence in order to protect public health, safety and welfare.

Chapter 2:
Definitions

General Comments

The words or terms defined in this chapter are deemed to be of prime importance in either specifying the subject matter of code provisions or in giving meaning to certain terms used throughout the code for administration or enforcement. The user of the code should be familiar with the terms found in this chapter, because the definitions are essential to the correct interpretation of the code and because the user might not be aware that a particular term found in the text is defined.

Purpose

Codes, by their very nature, are technical documents. Every word, term and punctuation mark can alter a sentence's meaning and, if misused, muddy its intent.

Further, the code, with its broad scope of applicability, includes terms used in a variety of construction disciplines. These terms can often have multiple meanings depending on the context or discipline being used at the time.

For these reasons, a consensus must be reached on the specific meaning of terms contained in the code. Chapter 2 performs this function by stating clearly what specific terms mean for the purpose of the code.

SECTION 201 (IFGC)
GENERAL

201.1 Scope. Unless otherwise expressly stated, the following words and terms shall, for the purposes of this code and standard, have the meanings indicated in this chapter.

❖ In the application of the code, the terms used have the meanings given in Section 202.

201.2 Interchangeability. Words used in the present tense include the future; words in the masculine gender include the feminine and neuter; the singular number includes the plural and the plural, the singular.

❖ Although the definitions contained in Section 202 are to be taken literally, gender and tense are considered to be interchangeable; thus, any grammatical inconsistencies within the code text will not hinder the understanding or enforcement of the requirements.

201.3 Terms defined in other codes. Where terms are not defined in this code and are defined in the *International Building Code, International Fire Code, International Mechanical Code* or *International Plumbing Code*, such terms shall have meanings ascribed to them as in those codes.

❖ When a word or term appears in the code that is not defined in this chapter, other references may be used to find its definition. These include the *International Building Code*® (IBC®), the *International Fire Code*® (IFC®), the *International Mechanical Code*® (IMC®) and the *International Plumbing Code*® (IPC®). These codes contain additional definitions (some parallel and duplicative) that may be used in the enforcement of the code or in the enforcement of the other codes by reference.

201.4 Terms not defined. Where terms are not defined through the methods authorized by this section, such terms shall have ordinarily accepted meanings such as the context implies.

❖ Another resource for defining words or terms not defined in this chapter or in other codes is their "ordinarily accepted meanings." The intent of this statement is to indicate that a dictionary definition may suffice if the definition is in the proper context.

Construction terms used throughout the code might not be defined in Chapter 2 or in a dictionary. In such cases, one would first turn to the definitions contained in the referenced standards (see Chapter 8) and then to textbooks on the subject in question.

SECTION 202 (IFGC)
GENERAL DEFINITIONS

❖ This section contains definitions of terms that are associated with the subject matter of the code. Definitions of terms are necessary for the understanding and application of the code requirements. The meaning of a term as used in the code might be different than an ordinarily understood meaning of that term. Some terms are unique to the code. A code provision could be misinterpreted if the definition of a term, as used in the context of the code, is not understood.

[M] ACCESS (TO). That which enables a device, *appliance* or *equipment* to be reached by ready *access* or by a means that first requires the removal or movement of a panel, door or similar obstruction (see also "Ready *access*").

❖ Access to equipment means that the equipment can be physically reached without someone having to remove a permanent portion of the structure. Access to equipment and appliances is necessary to facilitate inspection, observation, maintenance, adjustment, repair or replacement. It is acceptable, for example, to install equipment in an interstitial space that would require removal of lay-in suspended ceiling panels to gain access. Equipment would not be considered as being provided with access if it were necessary to remove or open any portion of a structure other than panels, doors, covers or similar obstructions intended to be removed or opened [also see the definition of "Ready access (to)"].

Access can be described as the capability of being reached or approached for the purpose of inspection, observation, maintenance, adjustment, repair or replacement. Achieving access may first require the removal or opening of a panel, door or similar obstruction and may require overcoming an obstacle such as elevation.

AIR CONDITIONER, GAS-FIRED. A gas-burning, automatically operated *appliance* for supplying cooled or dehumidified air or chilled liquid.

❖ Gas-fired air conditioners are not specifically referred to in the code, but would include systems fueled with natural or propane gas including absorption-type units and reciprocating units powered by internal combustion engines.

[M] AIR CONDITIONING. The treatment of air so as to control simultaneously the temperature, humidity, cleanness and distribution of the air to meet the requirements of a conditioned space.

❖ Air conditioning is commonly referred to only in the context of cooling and dehumidifying air; however, the definition also indicates a much broader scope. In essence, the process of supplying ventilation air to a space constitutes air conditioning because the introduction of any ventilation air is an attempt to control the indoor environment.

[M] AIR, EXHAUST. Air being removed from any space or piece of *equipment* or *appliance* and conveyed directly to the atmosphere by means of openings or ducts.

❖ Exhaust air may be from a space, an appliance or a piece of equipment. Exhaust air systems are terminated outside of the building, in some cases after the exhaust air has been treated to remove any harmful emissions. Exhaust air is not recirculated.

[M] AIR-HANDLING UNIT. A blower or fan used for the purpose of distributing supply air to a room, space or area.

❖ In addition to blowers, air-handling units may contain heat exchangers, filters and means to control air volume.

[M] AIR, MAKEUP. Any combination of outdoor and transfer air intended to replace exhaust air and exfiltration.

❖ Makeup air is not to be confused with combustion air. Makeup air replaces the air being exhausted through such systems as bathroom and toilet exhausts, kitchen exhaust hoods and clothes dryer exhaust systems. Refer to Section 614.5 for specific requirements for makeup air. Exhaust systems cannot function at design capacity without adequate volumes of makeup air to replace the air being exhausted.

[A] ALTERATION. A change in a system that involves an extension, addition or change to the arrangement, type or purpose of the original installation.

❖ An alteration is any modification or change made to an existing installation. For example, increasing the size of piping for a portion of the system to accommodate different appliances would be an alteration.

ANODELESS RISER. A transition assembly in which plastic *piping* is installed and terminated above ground outside of a building.

❖ As the name implies, these riser assemblies protect the steel riser from corrosion by means other than cathodic protection that involves a sacrificial anode. Some anodeless risers allow the termination of plastic piping aboveground by encasing the piping in a steel conduit equipped with a plastic-to-steel transition fitting [see Commentary Figure 202(1)].

[M] APPLIANCE. Any apparatus or device that utilizes a fuel or raw material to produce light, heat, power, refrigeration or air conditioning.

❖ An appliance is a manufactured component or assembly of components that converts one source of energy into a different form of energy to serve a specific purpose. The term "appliance" generally refers to residential- and commercial-type utilization equipment that is manufactured in standardized sizes or types. The term "appliance" is generally not associated with industrial-type equipment. For the application of the code provisions, the terms "appliance" and "equipment" are not interchangeable.

Examples of appliances regulated by the code include furnaces; boilers; water heaters; room heaters; decorative gas log sets; cooking appliances; clothes dryers; pool, spa and hot tub heaters; unit heaters; ovens and similar gas-fired appliances (see the definition of "Equipment"). Note that the definition intends to state that a fuel can be used as a raw material to produce energy rather than the apparent statement that an appliance utilizes raw materials that are not fuels. For example, a fuel cell system could use natural gas as a raw material for the production of hydrogen for use in the fuel cell. The definition was revised to recognize that in a few instances in the code, the term "appliance" is used to describe appliances that are fueled by other than gaseous fuels, whereas generally, the term describes gas-fired appliances.

**Figure 202(1)
ANODELESS RISERS
(Photo courtesy of Perfection Corporation)**

APPLIANCE, AUTOMATICALLY CONTROLLED.
Appliances equipped with an automatic burner ignition and safety shutoff device and other automatic devices which accomplish complete turn-on and shutoff of the gas to the main burner or burners, and graduate the gas supply to the burner or burners, but do not affect complete shutoff of the gas.

❖ With respect to the code, an automatically controlled appliance is an appliance that is cycled through its operation by controls, such as thermostats, pressure switches and timers, without manual intervention. A residential water heater is automatically controlled; a four-burner cooktop is not.

APPLIANCE, FAN-ASSISTED COMBUSTION. An *appliance* equipped with an integral mechanical means to either draw or force products of combustion through the combustion chamber or heat exchanger.

❖ Fan-assisted appliances are a specific type of Category I appliance, typically furnaces and boilers. They are not to be confused with Category II, III or IV appliances; power-vented appliances; or appliances served by exhausters. As the name implies, these appliances use a fan (blower) to "assist" the combustion process by helping the flue gases to overcome the internal flow resistance of the appliance heat exchanger. Some fan-assisted appliances are field-convertible to Category III appliances.

APPLIANCE TYPE.

Low-heat appliance (residential appliance). Any *appliance* in which the products of combustion at the point of entrance to the flue under normal operating conditions have a temperature of 1,000°F (538°C) or less.

❖ Residential appliances, including solid-fuel appliances, are in this category.

Medium-heat appliance. Any *appliance* in which the products of combustion at the point of entrance to the flue under normal operating conditions have a temperature of more than 1,000°F (538°C), but not greater than 2,000°F (1093°C).

❖ This classification includes industrial-type equipment such as furnaces, kilns, ovens, dryers and incinerators.

APPLIANCE, UNVENTED. An *appliance* designed or installed in such a manner that the products of combustion are not conveyed by a vent or chimney directly to the outside atmosphere.

❖ Direct-fired makeup air heaters, gas-fired infrared radiant unit heaters, gas-fired cooking equipment and small gas-fired room heaters are examples of appliances that are sometimes unvented. With the exception of direct-fired equipment, the appliances listed herein can also be of the vented type, depending on the design and type of appliance and the particular application. Unvented appliances must be designed, installed and operated to avoid both the depletion of the oxygen supply and the accumulation of toxic combustion byproducts (see Sections 501.8, 621 and 630).

[M] APPLIANCE, VENTED. An *appliance* designed and installed in such a manner that all of the products of combustion are conveyed directly from the *appliance* to the outdoor atmosphere through an *approved* chimney or vent system.

❖ Most fuel-fired appliances are designed to vent the products of combustion to the outdoors through one or more specific types of vent or chimney (see Section 501.2).

[A] APPROVED. Acceptable to the code official.

❖ As related to the process of acceptance of fuel gas-related installations, including materials, equipment and construction systems, this definition identifies where ultimate authority rests. Whenever this term is used, it means that only the enforcing authority can accept a specific installation or component as complying with the code. Research reports prepared and published by the International Code Council® (ICC®) may be used by code officials to aid in their review and approval of the material or method described in the report. Publishing a report does not indicate automatic "approval" for the material or method described in the report. When the code states that an item or method "shall be approved," it does not mean that the code official is obligated to allow or accept it. Rather, it means that the code official must determine whether the item or method is acceptable; that is, the code official must make the decision to allow or disallow; accept or reject.

[A] APPROVED AGENCY. An established and recognized agency that is regularly engaged in conducting tests or furnishing inspection services, where such agency has been approved by the *code official*.

❖ The word "approved" means as approved by the code official. The basis for approval of an agency for a particular activity may include the capacity and capability of the agency to perform the work in accordance with Sections 301.4 through 301.4.2.3.

ATMOSPHERIC PRESSURE. The pressure of the weight of air and water vapor on the surface of the earth, approximately 14.7 pounds per square inch (psi) (101 kPa absolute) at sea level.

❖ This is the "standard" atmospheric pressure at sea level. The actual atmospheric pressure varies with climate conditions. Barometric pressure is a measure of the actual atmospheric pressure, usually given as inches of mercury. One standard atmosphere (14.7 psi) is equal to 29.9 inches of mercury (101.4 kPa) at 32°F (0°C). Absolute pressure includes atmospheric pressure. Gauge pressure is the pressure above atmospheric pressure. If not otherwise defined, stated pressure is the pressure above atmospheric pressure (gauge pressure). A vacuum (partial or complete) is a pressure below atmospheric pressure.

AUTOMATIC IGNITION. Ignition of gas at the burner(s) when the gas controlling device is turned on, including reignition if the flames on the burner(s) have been extinguished by means other than by the closing of the gas controlling device.

❖ As opposed to manual ignition, automatic ignition is accomplished by pilot burners, spark ignitors, hot surface ignitors and similar means.

BAFFLE. An object placed in an *appliance* to change the direction of or retard the flow of air, air-gas mixtures or flue gases.

❖ Baffles are used in gas-fired equipment to regulate the flow through the combustion chamber or flueways for various purposes, such as improving the mixing of fuel and air and affecting the speed and direction of flow of flue gases.

BAROMETRIC DRAFT REGULATOR. A balanced damper device attached to a chimney, vent connector, breeching or flue gas manifold to protect combustion appliances by controlling chimney draft. A double-acting barometric draft regulator is one whose balancing damper is free to move in either direction to protect combustion appliances from both excessive draft and backdraft.

❖ These units automatically open or close depending on the difference between the internal vent pressure and the atmospheric pressure. See the commentary for the definition of "Atmospheric pressure." Excessive negative pressure in the vent will cause excessive draft; therefore, the draft regulator will open as a result of the higher atmospheric pressure on the exterior of the damper, allowing air to enter the vent and thereby increase the internal vent pressure. With excessive positive internal vent pressure, a double-acting regulator will open to relieve the pressure, lowering the vent internal pressure [see Commentary Figure 202(2)].

BOILER, LOW-PRESSURE. A self-contained *appliance* for supplying steam or hot water.

❖ Low-pressure boilers operate at pressures less than or equal to 15 pounds per square inch (psi) (103 kPa) for steam and 160 psi (1103 kPa) for water. High-pressure boilers exceed those operating pressures.

Boilers are usually manufactured of steel, cast iron or copper and are used to transfer heat from the combustion of a fuel or from an electric-resistance element to water to make steam or pressurized hot water for heating or other process or power purposes.

Boilers are usually installed in closed systems where the heat transfer medium is recirculated and retained within the system.

Boilers must be labeled and installed in accordance with the manufacturer's instructions and the applicable sections of the code. Boilers are rated in

Figure 202(2)
BAROMETRIC DRAFT REGULATOR
(Photo courtesy of Tjernlund Products, Inc.)

accordance with standards published by the American Society of Mechanical Engineers (ASME), the Hydronics Institute, the Steel Boiler Institute (SBI), the Canadian Standards Association (CSA) and the American Boiler Manufacturers Association (ABMA). Boilers can be classified by working temperature, working pressure, type of fuel used (or electric boilers), materials of construction and whether or not the heat transfer medium changes phase from a liquid to a vapor [see Commentary Figure 202(3)].

Hot water heating boiler. A boiler in which no steam is generated, from which hot water is circulated for heating purposes and then returned to the boiler, and that operates at water pressures not exceeding 160 pounds per square inch gauge (psig) (1100 kPa gauge) and at water temperatures not exceeding 250°F (121°C) at or near the boiler *outlet*.

❖ Hot water heating boilers are normally part of a closed system in which the heated water is circulated through fan coils or radiators of various types, including convectors, finned tube units, radiant floor tubing systems and baseboard units.

Hot water supply boiler. A boiler, completely filled with water, which furnishes hot water to be used externally to itself, and that operates at water pressures not exceeding 160 psig (1100 kPa gauge) and at water temperatures not exceeding 250°F (121°C) at or near the boiler *outlet*.

❖ Hot water supply boilers are normally part of open systems in which the heated water is supplied and used externally to the boiler. Large domestic (potable) water heating systems often use hot water supply boilers.

WEIL-McLAIN
MODEL LGB SERIES 2 BOILER

MODEL NUMBER	NUMBER OF SECTIONS	JACKET LENGTH INCHES	MIN. RELIEF VALVE CAP. LBS/HR OR MBH	INPUT BTU/HR	MINIMUM INPUT BTU/HR	CSA GROSS OUTPUT BTU/HR	NET I=B=R OUTPUT		
							Steam Sq. Ft.	Steam MBH	Water MBH
☐ LGB-4	4	21	325	400,000	---	324,000	1013	243	282
☐ LGB-5	5	26	422	520,000	---	421,200	1317	316	366
☐ LGB-6	6	31	527	650,000	325,000	526,500	1646	395	458
☐ LGB-7	7	36	632	780,000	390,000	631,800	1975	474	549
☐ LGB-8	8	41	738	910,000	455,000	737,100	2304	553	641
☐ LGB-9	9	46	843	1,040,000	520,000	842,400	2633	632	733
☐ LGB-10	10	51	948	1,170,000	585,000	947,700	2965	711	824
☐ LGB-11	11	56	1053	1,300,000	650,000	1,053,000	3292	790	916
☐ LGB-12	12	61	1159	1,430,000	715,000	1,158,300	3621	869	1007
☐ LGB-13	13	66	1264	1,560,000	780,000	1,263,600	3954	949	1099
☐ LGB-14	14	71	1369	1,690,000	845,000	1,368,900	4313	1035	1190
☐ LGB-15	15	76	1475	1,820,000	910,000	1,474,200	4679	1123	1282
☐ LGB-16	16	81	1580	1,950,000	975,000	1,579,500	5046	1211	1373
☐ LGB-17	17	86	1685	2,080,000	1,040,000	1,684,800	5408	1298	1465
☐ LGB-18	18	91	1791	2,210,000	1,105,000	1,790,100	5775	1386	1557
☐ LGB-19	19	96	1896	2,340,000	1,170,000	1,895,400	6125	1470	1648
☐ LGB-20	20	101	2001	2,470,000	1,235,000	2,000,700	6471	1553	1740
☐ LGB-21	21	106	2106	2,600,000	1,300,000	2,106,000	6813	1635	1831
☐ LGB-22	22	111	2212	2,730,000	1,365,000	2,211,300	7155	1717	1923
☐ LGB-23	23	116	2317	2,860,000	1,430,000	2,316,600	7496	1799	2014

TO DETERMINE BOILER SIZE, COUNT THE NUMBER OF SECTIONS OR MEASURE THE JACKET LENGTH. CHECK BOX NEXT TO BOILER SIZE INSTALLED.

CERTIFIED BY
WEIL-McLAIN
523 S New Street
Eden, North Carolina
27288-3623

MAWP, WATER 50 PSI
MAWP, STEAM 15 PSI
MAX. WATER TEMP. 250°F

Electrical Input Less Than 12 Amperes, 120 Volts, & 60 Hertz.
Canadian Registration No. H7268.51234679T
MEA Number: 333-85-E
Design Certified Under ANSI Z21.13a 2005 • CSA 4.9a 2005
Low Pressure Boiler

IMPORTANT
Installation not complete unless gas information label attached here.
Labels are attached to gas train in gas control carton.

- For installation on noncombustible flooring.
- Provide service clearances and minimum 24" between jacket and any combustible wall(s) and ceiling. Install in space large in comparison to size of boiler.
- Vent Category I.

550-223-880 (0606)

Figure 202(3)
LABEL FOR LOW-PRESSURE BOILER
(Figure courtesy of Weil-McLain)

Steam heating boiler. A boiler in which steam is generated and that operates at a steam pressure not exceeding 15 psig (100 kPa gauge).

❖ Steam heating boilers are normally part of a closed system in which low-pressure steam can be circulated through a variety of terminal units, including natural convection units, forced-convection units and radiant panel systems.

BONDING JUMPER. A conductor installed to electrically connect metallic gas *piping* to the grounding electrode system.

❖ This term is used in the subsections of Section 310.1.1 relative to the bonding of gas piping systems. It is a number 6 AWG or larger copper conductor or equivalent that is used to electrically bond gas piping to the grounding electrodes that are a part of the electrical service for the structure.

[M] BRAZING. A metal-joining process wherein coalescence is produced by the use of a nonferrous filler metal having a melting point above 1,000°F (538°C), but lower than that of the base metal being joined. The filler material is distributed between the closely fitted surfaces of the joint by capillary action.

❖ Brazing is the act of producing a brazed joint and is often referred to as silver soldering. Silver soldering is more accurately described as silver brazing and uses high-silver-bearing alloys primarily composed of silver, copper and zinc. Silver soldering (brazing) typically requires temperatures in excess of 1,000°F (538°C), and these solders are classified as "hard" solders.

Confusion has always existed concerning the distinction between "silver solder" and "silver-bearing solder." Silver solders are unique and can be further subdivided into soft and hard categories that are determined by the percentages of silver and the other component elements of the particular alloy. The distinction is that silver-bearing solders [melting point below 600°F (316°C)] are used in soft-soldered joints, and silver solders [melting point above 1,000°F (538°C)] are used in silver-brazed joints.

Brazed joints are considered to have superior strength and stress resistance and, because of the high melting point, are less likely to fail when exposed to fire.

Brazing with a filler metal conforming to AWS A5.8 produces a strong joint that will perform under extreme service conditions. The surfaces to be brazed must be cleaned to be free from oxides and impurities. Flux should be applied as soon as possible after the surfaces have been cleaned. Flux helps to remove residual traces of oxides, to promote wetting and to protect the surfaces from oxidation during heating. Care should be taken to prevent flux from entering the piping system during the brazing operation because flux that remains may corrode the pipe or contaminate the system.

Air and any other residual products should be removed from the pipe being brazed by purging the piping with a nonflammable gas such as carbon dioxide or nitrogen. Purging the system has several benefits, such as preventing oxidation from occurring on the inside of the pipe and preventing the creation of toxic gases that can result from the chemical breakdown of other products. Additionally, purging will eliminate the possibility of an explosion from any other flammable gas in the pipe that could ignite when mixed with air.

BROILER. A general term including salamanders, barbecues and other appliances cooking primarily by radiated heat, excepting toasters.

❖ The food being cooked can be above or below the gas flame and is usually placed on an open grill to allow draining away of grease from meat products.

BTU. Abbreviation for British thermal unit, which is the quantity of heat required to raise the temperature of 1 pound (454 g) of water 1°F (0.56°C) (1 Btu = 1055 J).

❖ Btu (J) is a unit of energy measurement. Fuel-fired appliances and equipment are rated based on their Btu/h (W) input or output.

BURNER. A device for the final conveyance of the gas, or a mixture of gas and air, to the combustion zone.

❖ There are several types of burners that vary in the method of providing combustion air and the pressures of fuel/air mixtures. Burner types include atmospheric ribbon, ported, porous ceramic, slotted and cone types and atmospheric or powered inshot and upshot gun burners.

Induced-draft. A burner that depends on draft induced by a fan that is an integral part of the *appliance* and is located downstream from the burner.

❖ See the commentary to the definition of "Draft/mechanical or induced draft."

Power. A burner in which gas, air or both are supplied at pressures exceeding, for gas, the line pressure, and for air, atmospheric pressure, with this added pressure being applied at the burner.

❖ Power burners are typically used on very large nonresidential-type appliances and use a blower to introduce combustion air under pressure and mix it with the fuel.

[M] CHIMNEY. A primarily vertical structure containing one or more flues, for the purpose of carrying gaseous products of combustion and air from an *appliance* to the outside atmosphere.

❖ Chimneys differ from vents in their materials of construction and the type of appliance they are designed to serve. Chimneys are capable of venting much higher temperature flue gases than vents [see Commentary Figure 202(4)].

Factory-built chimney. A *listed* and *labeled* chimney composed of factory-made components, assembled in the field in accordance with manufacturer's instructions and the conditions of the listing.

❖ A factory-built chimney is a manufactured listed and labeled chimney that has been tested by an approved agency to determine its performance characteristics. Factory-built chimneys are manufactured in two basic designs: a double-wall mineral fiber insulated design and a triple-wall air-cooled design. Some chimneys use both convection cooling and mineral-fiber insulation. Both designs use stainless steel inner liners to resist the corrosive effects of combustion products [see Commentary Figure 202(5)].

Masonry chimney. A field-constructed chimney composed of solid masonry units, bricks, stones or concrete.

❖ Masonry chimneys can have one or more flues within them and are field constructed of brick, stone, concrete or fire-clay materials. Masonry chimneys can stand alone or be part of a masonry fireplace.

Metal chimney. A field-constructed chimney of metal.

❖ Metal chimneys, being field constructed, are unlisted and often referred to as smokestacks. Metal chimneys are associated with industrial installations and equipment.

[M] CLEARANCE. The minimum distance through air measured between the heat-producing surface of the mechanical *appliance*, device or *equipment* and the surface of the *combustible material* or *assembly*.

❖ Clearances between sources of heat and combustibles are always airspace clearances.

CLOTHES DRYER. An *appliance* used to dry wet laundry by means of heated air. Dryer classifications are as follows:

Type 1. Factory-built package, multiple production. Primarily used in family living environment. Usually the smallest unit physically and in function output.

❖ Type 1 clothes dryers are used in dwelling units and in residential settings.

Type 2. Factory-built package, multiple production. Used in business with direct intercourse of the function with the public. Not designed for use in individual family living environment.

❖ Type 2 clothes dryers are intended for commercial, public or institutional use.

[A] CODE. These regulations, subsequent amendments thereto or any emergency rule or regulation that the administrative authority having jurisdiction has lawfully adopted.

❖ The adopted regulations are generally referred to as "the code" and include not only this code, but also any adopted modifications to the code and all other related rules and regulations promulgated and enacted by the jurisdiction.

[A] CODE OFFICIAL. The officer or other designated authority charged with the administration and enforcement of this code, or a duly authorized representative.

❖ The statutory power to enforce the code is normally vested in a building department (or the like) of a state, county or municipality whose designated enforcement officer is termed the "code official."

Figure 202(4)
FACTORY-BUILT CHIMNEYS
(Photo courtesy of Selkirk, L.L.C.)

Figure 202(5)
AIR-COOLED FACTORY-BUILT CHIMNEY SECTION
(Figure courtesy of Simpson Dura-Vent Company, Incorporated)

[M] COMBUSTIBLE ASSEMBLY. Wall, floor, ceiling or other assembly constructed of one or more component materials that are not defined as noncombustible.

❖ A combustible assembly is made up of one or more combustible components. For example, a wall assembly with wood studs is combustible regardless of the nature of the sheathing attached to the studs. Likewise, a metal stud wall with gypsum board sheathing is a combustible assembly because the gypsum board is not noncombustible. A noncombustible assembly has no combustible elements. See the definition of "Noncombustible materials."

[M] COMBUSTIBLE MATERIAL. Any material not defined as noncombustible.

❖ See the definition of "Noncombustible materials."

[M] COMBUSTION. In the context of this code, refers to the rapid oxidation of fuel accompanied by the production of heat or heat and light.

❖ The primary components of combustion are fuel, oxygen and heat. The code regulates many aspects of combustion technology, including the process of combustion and providing sufficient air; the use of energy produced from combustion; the safe venting of the products of combustion; combustion efficiency; the containment and control of combustion; and the fuel supplies for combustion equipment and appliances.

[M] COMBUSTION AIR. Air necessary for complete combustion of a fuel, including theoretical air and excess air.

❖ The process of combustion requires a specific amount of oxygen to initiate and sustain the combustion reaction. Combustion air includes primary air, secondary air, draft hood dilution air and excess air. Combustion air includes the amount of atmospheric air required for complete combustion of a fuel and is related to the molecular composition of the fuel being burned, the design of the fuel-burning equipment and the percentage of oxygen in the combustion air. Too little combustion air will result in incomplete combustion of a fuel and the possible formation of carbon deposits (soot), carbon monoxide, toxic alcohols, ketones, aldehydes, nitrous oxides and other by-products. The required amount of combustion air is usually stated in terms of cubic feet per minute (m^3/s) or pounds per hour (kg/h).

[M] COMBUSTION CHAMBER. The portion of an *appliance* within which combustion occurs.

❖ Combustion chambers are either open to the atmosphere or are isolated (sealed) from the atmosphere in which the appliance is installed. Combustion chambers are often referred to as "fireboxes."

[M] COMBUSTION PRODUCTS. Constituents resulting from the combustion of a fuel with the oxygen of the air, including inert gases, but excluding excess air.

❖ Such products include water, carbon dioxide, carbon monoxide, nitrous oxides and various trace compounds.

[M] CONCEALED LOCATION. A location that cannot be accessed without damaging permanent parts of the building structure or finish surface. Spaces above, below or behind readily removable panels or doors shall not be considered as concealed.

❖ The space above a "drop-in" tile suspended ceiling system, for example, would not be considered as a concealed location.

CONCEALED PIPING. *Piping* that is located in a *concealed location* (see "*Concealed location*").

❖ This refers to gas piping in locations meeting the definition of "Concealed location."

CONDENSATE. The liquid that condenses from a gas (including flue gas) caused by a reduction in temperature or increase in pressure.

❖ Condensate forms when the temperature of a vapor is lowered to its dew point temperature. Air conditioning systems produce condensate when an airstream contacts cooling coils. The moisture in the air condenses on the cold surface of the coils and the air is "dehumidified." High-efficiency (84 percent and up) fuel-burning appliances produce condensate from the combustion gases. Condensate also forms within improperly designed chimneys and vents when the products of combustion (that contain water vapor) contact the colder inner walls of the vent or chimney. If the temperature of the products of combustion is lowered to the dew point temperature of the water vapor, condensate will form on the inside walls of the vent or chimney. Condensed steam in hydronic systems is also referred to as "condensate."

CONNECTOR, APPLIANCE (Fuel). Rigid metallic pipe and fittings, semirigid metallic tubing and fittings or a *listed* and *labeled* device that connects an *appliance* to the gas *piping* system.

❖ An appliance fuel connector is the piping, tubing or factory-built device that is installed downstream of the equipment shutoff to connect the appliance to the gas piping system. See the definition of "Piping system." Section 411.1 lists the types of appliance connectors that include both factory-built assemblies and field-constructed assemblies of pipe, tube and fittings. Note that rigid metallic piping, such as Schedule 40 black steel, used to connect an appliance to the gas piping system is considered to be an appliance connector because it is placed between the appliance and the piping system, downstream of the appliance shutoff valve. As defined, the gas piping ends at the appliance/equipment shutoff valve, therefore, everything downstream of that point is defined, sized and installed as a connector except as required by Sections 409.5.2, 409.5.3 and the exception to Section 411.1.3.1.

CONNECTOR, CHIMNEY OR VENT. The pipe that connects an *appliance* to a chimney or vent.

❖ In most cases, appliances are not located directly in line with the vertically rising chimney or vent; therefore, a vent or chimney connector is necessary to connect the appliance flue outlet to the vent or chimney. Vent and chimney connectors can be single- or double-wall pipes and are usually made from steel, galvanized steel, stainless steel or aluminum sheet metal, depending on the application. Vent connectors are also made of corrugated aluminum. In many installations, the vent connectors must be constructed of the same material as the vent, as is typically done with Type B vent systems. There are many requirements in the code (see Sections 503 and 504) regulating the location, length, size and type of connector.

[A] CONSTRUCTION DOCUMENTS. All of the written, graphic and pictorial documents prepared or assembled for describing the design, location and physical characteristics of the elements of the project necessary for obtaining a mechanical permit.

❖ To determine that proposed construction is in compliance with code requirements, sufficient information must be submitted to the code official for review. This typically consists of the drawings (floor plans, elevations, sections, piping diagrams, etc.), specifications and product information describing the proposed work.

[M] CONTROL. A manual or automatic device designed to regulate the gas, air, water or electrical supply to, or operation of, a mechanical system.

❖ A control is a device designed to respond to changes in temperature, pressure, liquid or gas flow rates, current, voltage, resistance, humidity and/or liquid levels.

CONVERSION BURNER. A unit consisting of a burner and its controls for installation in an *appliance* originally utilizing another fuel.

❖ Typical conversion burners are designed to convert an appliance, such as a boiler, from solid-fuel-fired to oil-fired or from oil-fired to gas-fired.

COUNTER APPLIANCES. Appliances such as coffee brewers and coffee urns and any appurtenant water-heating *appliance*, food and dish warmers, hot plates, griddles, waffle bakers and other appliances designed for installation on or in a counter.

❖ Counter appliances are small low-energy-input appliances and need not be vented unless the aggregate input rating of all the appliances exceeds that allowed in Section 501.8.

CUBIC FOOT. The amount of gas that occupies 1 cubic foot (0.02832 m³) when at a temperature of 60°F (16°C), saturated with water vapor and under a pressure equivalent to that of 30 inches of mercury (101 kPa).

❖ This definition applies specifically to a measurement of fuel gas.

[M] DAMPER. A manually or automatically controlled device to regulate draft or the rate of flow of air or combustion gases.

❖ Dampers act as restrictors or valves in a gaseous flow stream (see Section 503.13).

DECORATIVE APPLIANCE, VENTED. A vented *appliance* wherein the primary function lies in the aesthetic effect of the flames.

❖ Such appliances include so-called "gas fireplaces" that are designed to simulate a wood-burning fireplace. The title of the standard for vented decorative gas-fired appliances (ANSI Z21.50) has been changed to *Vented Gas Fireplaces*.

DECORATIVE APPLIANCES FOR INSTALLATION IN VENTED FIREPLACES. A vented *appliance* designed for installation within the fire chamber of a vented *fireplace*, wherein the primary function lies in the aesthetic effect of the flames.

❖ These appliances include gas log sets for installation in solid-fuel-burning fireplaces.

DEMAND. The maximum amount of gas input required per unit of time, usually expressed in cubic feet per hour, or Btu/h (1 Btu/h = 0.2931 W).

❖ Demand refers to the load that appliances place on the fuel-gas distribution system.

[BS] DESIGN FLOOD ELEVATION. The elevation of the "design flood," including wave height, relative to the datum specified on the community's legally designated flood hazard map. In areas designated as Zone AO, the *design flood elevation* shall be the elevation of the highest existing grade of the *building's* perimeter plus the depth number (in feet) specified on the flood hazard map. In areas designated as Zone AO where a depth number is not specified on the map, the depth number shall be taken as being equal to 2 feet (610 mm).

❖ The design flood elevation is the height to which the floodwaters are predicted to rise during passage or occurrence of the design flood. The datum specified on the flood hazard map is important because it may differ from that used locally for other purposes. Communities adopt the Flood Insurance Rate Maps (FIRMS) prepared by the Federal Emergency Management Agency (FEMA) or another flood hazard map that shows at least the same flood hazard areas. FEMA uses commonly accepted computer models that estimate hydrologic and hydraulic conditions to determine the 1-percent annual chance (base) flood. Along rivers and streams, statistical methods and computer models may have been used to estimate the runoff and to develop flood elevations. The models take into consideration watershed characteristics and the shape and nature of the floodplain, including natural ground contours and the presence of buildings, bridges and culverts. Along coastal areas, base flood elevations may be developed using models that take into account offshore bathymetry, historical storms and typical wind patterns. In many coastal areas, the base flood elevation includes wave heights.

DILUTION AIR. Air that is introduced into a draft hood and is mixed with the flue gases.

❖ Dilution air is associated only with draft-hood equipped Category I appliances. Fan-assisted Category I appliances do not involve dilution air because all of the air brought in by the fan passes directly through the combustion chamber of the appliance. Dilution air lowers the dew point of vent gases, reducing the possibility that condensation will occur inside the vent or chimney.

DIRECT-VENT APPLIANCES. Appliances that are constructed and installed so that all air for combustion is derived directly from the outdoor atmosphere and all flue gases are discharged directly to the outdoor atmosphere.

❖ These appliances have independent exhaust and intake pipes or have concentric pipes that vent combustion gases through the inner pipe and convey combustion air in the annular space between the inner and outer pipe walls [see Commentary Figures 202(6) and 202(7)].

[M] DRAFT. The pressure difference existing between the *appliance* or any component part and the atmosphere, that causes a continuous flow of air and products of combustion through the gas passages of the *appliance* to the atmosphere.

❖ Draft is the negative static pressure measured relative to atmospheric pressure that is developed in chimneys and vents and in the flue-ways of fuel-burning appliances. Draft can be produced by hot flue-gas buoyancy ("stack effect"), mechanically by fans and exhausters or by a combination of both natural and mechanical means.

Mechanical or induced draft. The pressure difference created by the action of a fan, blower or ejector that is located between the *appliance* and the chimney or vent termination.

❖ Induced draft systems use a fan or blower to boost or "induce" draft in a venting system that produces insufficient natural draft. Draft induction implies that the combustion gases are "pulled" through a passageway or conduit. Draft inducers produce negative pressures on the inlet (upstream) side of the fan or blower. They are separate field-installed units located between an appliance and its venting system. Draft inducers are used with natural draft venting systems to overcome the resistance of vent or chimney connectors and to compensate for the inability of the chimney or vent to produce sufficient and reliable draft.

Draft induction is also a design principle used in many fan-assisted appliances. Draft-inducer fans or blowers that are integral with fuel-fired appliances are necessary to overcome the internal resistance of the heat exchanger flue passageways. To attain higher thermal transfer efficiencies, some heat exchanger designs use flue passageways that retain the combustion gases longer over a greater surface area, thus extracting more heat. Such heat exchanger designs cannot rely on natural (gravity) venting to overcome the higher resistance to flow and, therefore, must rely on mechanical means to pull the combustion gases through the heat exchanger.

Induced draft systems are not to be confused with systems using power exhausters or other self-venting equipment.

Mechanical draft systems produce positive pressures on the discharge side of the fan or blower and can be said to "push" combustion gases through a passageway or conduit.

Figure 202(6)
DIRECT-VENT APPLIANCE
(Figure courtesy of Simpson Dura-Vent Company, Incorporated)

Figure 202(7)
DIRECT-VENT APPLIANCE
(Figure courtesy of Simpson Dura-Vent Company, Incorporated)

Natural draft. The pressure difference created by a vent or chimney because of its height, and the temperature difference between the flue gases and the atmosphere.

❖ Natural draft systems do not use mechanical devices such as fans or blowers, but instead rely on the principle of buoyancy to carry the products of combustion to the atmosphere. Because of the difference in temperature and the resultant difference in density between the hot products of combustion and the ambient atmosphere, the gases within the chimney or flue will rise, creating a buoyant "draft." The phenomenon of natural draft is sometimes referred to as "stack effect" and is measured in inches of water column (kPa). The amount of draft is affected by the height of the chimney or vent and also by the ability of the chimney or vent to maintain the temperature differential between the combustion gases and the ambient air.

DRAFT HOOD. A nonadjustable device built into an *appliance*, or made as part of the vent connector from an *appliance*, that is designed to (1) provide for ready escape of the flue gases from the *appliance* in the event of no draft, backdraft or stoppage beyond the draft hood, (2) prevent a backdraft from entering the *appliance*, and (3) neutralize the effect of stack action of the chimney or gas vent upon operation of the *appliance*.

❖ Draft hoods are integral to or supplied with natural draft atmospheric-burner gas-fired appliances other than fan-assisted appliances. When classified, appliances equipped with draft hoods are classified by the manufacturer as Category I appliances. Because of minimum efficiency standards and the popularity of mechanical and other special proprietary venting systems, draft-hood-equipped appliances are becoming increasingly rare in the marketplace [see Commentary Figure 202(8)].

DRAFT REGULATOR. A device that functions to maintain a desired draft in the *appliance* by automatically reducing the draft to the desired value.

❖ Excessive draft reduces the combustion and thermal efficiency of an appliance. Draft regulators automatically adjust the draft by admitting air into the vent, thereby reducing the draft [see Commentary Figure 202(2)].

[M] DRIP. The container placed at a low point in a system of *piping* to collect condensate and from which the condensate is removable.

❖ These reservoirs, also referred to as "drip legs," are made up of pipe and fittings and are intended to collect liquid in piping systems where condensables are possible. A "drip leg" is distinct from a "sediment trap" even though they may be constructed identically.

DRY GAS. A gas having a moisture and hydrocarbon dew point below any normal temperature to which the gas *piping* is exposed.

❖ The dew point referred to herein is the temperature at which fuel gas hydrocarbons or water vapor begin to condense into liquid fuel or liquid water. Condensate will not form if the gas piping system temperature is maintained above the dew point of the fuel gas and the water vapor.

The gas supplier (utility company) should be contacted to determine the nature of the gas supplied.

DUCT FURNACE. A warm-air furnace normally installed in an air distribution duct to supply warm air for heating. This definition shall apply only to a warm-air heating *appliance* that depends for air circulation on a blower not furnished as part of the furnace.

❖ Duct furnaces are vented appliances and consist of burners, heat exchangers and related controls. Duct furnaces depend on an external air handler or blower.

[M] DUCT SYSTEM. A continuous passageway for the transmission of air that, in addition to ducts, includes duct fittings, dampers, plenums, fans and accessory air-handling *equipment*.

❖ Duct systems are part of an air distribution system and include supply, return and relief/exhaust air systems.

[BG] DWELLING UNIT. A single unit providing complete, independent living facilities for one or more persons, includ-

Figure 202(8)
DRAFT-HOOD-EQUIPPED APPLIANCE
(Photo courtesy of A.O. Smith Water Products)

ing permanent provisions for living, sleeping, eating, cooking and sanitation.

❖ Contrast this definition with the definition of "Sleeping unit." All of the listed provisions must be present for the occupancy to be considered as a dwelling unit.

EQUIPMENT. Apparatus and devices other than appliances.

❖ The term "equipment" does not refer to or describe anything that is defined as an "Appliance," despite the fact that some appliances are commonly referred to in the field as "pieces of equipment." The term "equipment" includes control devices, pressure regulators, valves, appliance appurtenances, gas connectors, power exhausters, fans, air handlers and a multitude of other devices that do not fit the description of "appliance."

EXCESS FLOW VALVE (EFV). A valve designed to activate when the fuel gas passing through it exceeds a prescribed flow rate.

❖ Some utility companies install EFVs in gas service lines to provide protection against high-pressure leaks resulting from damage to the piping caused by excavation, hole boring or hand digging. The device is a type of shutoff valve that abruptly closes when the gas flow through it reaches an abnormal flow rate corresponding to its design setting. The excess flow pushes a spring loaded element against a seat in the EFV body. The upstream gas pressure holds the valve in the closed position. After the gas line is repaired and closed off on the load end, the EFV will reset itself by allowing a small flow of gas to pass through an orifice in the valve into the downstream piping, thereby equalizing the pressure on both sides of the valve and allowing the valve to reopen by the force of the spring. Service line EFVs do not fall within the scope of the code because the service line is upstream of the point of delivery. Appliance-type EFVs are addressed by the code, are designed to protect individual appliances and are installed in the supply piping to the appliance. Some types are installed on the supply end of a gas appliance connector. The basic operating principle is the same for all types although some do not allow bypass gas flow and are, therefore, not self-resetting. The moving component of EFVs, depending on the type, is held open by gravity, a spring or a magnet. The gravity type must be installed in a vertical position.

EFVs can be compared to electrical circuit breakers in that they stop gas flow when a predetermined flow rate is exceeded just as the circuit breakers interrupt electrical current flow when the amperage setting is exceeded. Imagine the scenario where an appliance gas connector is severed or a gas line breaks loose at a fitting and gas would be flowing into the space at a high rate. An EFV device installed upstream of the point of leakage would recognize the abnormal flow and either shut off the gas flow completely or reduce the flow to a small bypass flow (see Section 410.4).

EXTERIOR MASONRY CHIMNEYS. Masonry chimneys exposed to the outdoors on one or more sides below the roof line.

❖ The part of a chimney that extends above the roof will obviously be exposed to the outdoors, but some of the sizing methods in Section 504 are based on the chimney being indoors below the roof. If any one wall or multiple walls of a chimney are exposed to the outdoors, the chimney will be hard to warm up, will lose heat more rapidly, will likely produce condensation internally and may fail to produce the required draft. Some tables in Section 504 are specific to exterior masonry chimneys.

[M] FIREPLACE. A fire chamber and hearth constructed of *noncombustible material* for use with solid fuels and provided with a chimney.

❖ Fireplaces burn solid fuels (wood, coal, etc.) and are not referred to as appliances in the code. The IBC and the *International Residential Code®* (IRC®) cover construction of masonry fireplaces.

Factory-built fireplace. A *fireplace* composed of *listed* factory-built components assembled in accordance with the terms of listing to form the completed *fireplace*.

❖ Factory-built fireplaces are solid-fuel-burning units having a fire chamber that is intended to be either open to the room or, if equipped with doors, operated with the doors either open or closed. Fireplaces are not referred to as appliances. The term "fireplace" describes a complete assembly that includes the hearth, the fire chamber and a chimney. A factory-built fireplace is composed of factory-built components representative of the prototypes tested and is installed in accordance with the manufacturer's instructions to form the completed fireplace.

Masonry fireplace. A hearth and fire chamber of solid masonry units such as bricks, stones, *listed* masonry units or reinforced concrete, provided with a suitable chimney.

❖ Masonry fireplaces must be constructed in accordance with the requirements found in the IBC or the IRC. These specific requirements are based on tradition and field experience and describe the conventional fireplace that has proven to be reliable where properly constructed, used and maintained.

FIRING VALVE. A valve of the plug and barrel type designed for use with gas, and equipped with a lever handle for manual operation and a dial to indicate the percentage of opening.

❖ This term is not used in the code.

FLAME SAFEGUARD. A device that will automatically shut off the fuel supply to a main burner or group of burners when the means of ignition of such burners becomes inoperative, and when flame failure occurs on the burner or group of burners.

❖ These devices are primary safety controls and are provided on all automatically operated fuel-fired appliances.

FLASHBACK ARRESTOR CHECK VALVE. A device that will prevent the backflow of one gas into the supply system of another gas and prevent the passage of flame into the gas supply system.

❖ See the commentary to Section 410.5.

[BS] FLOOD HAZARD AREA. The greater of the following two areas:

1. The area within a floodplain subject to a 1 percent or greater chance of flooding in any given year.

2. This area designated as a *flood hazard area* on a community's flood hazard map, or otherwise legally designated.

❖ FEMA prepares FIRMS that delineate the land area that is subject to inundation by the 1-percent annual chance flood. Some states and local jurisdictions develop and adopt maps of flood hazard areas that are more extensive than the areas shown on FEMA's map. For the purpose of the code, the flood hazard area within which the requirements are to be applied is the greater of the two delineated areas (see Section 301.11).

FLOOR FURNACE. A completely self-contained furnace suspended from the floor of the space being heated, taking air for combustion from outside such space and with means for observing flames and lighting the *appliance* from such space.

❖ These units supply heat through a floor grille placed directly over the unit's heat exchanger.

Fan type. A floor furnace equipped with a fan that provides the primary means for circulating air.

❖ Floor furnaces may be equipped with factory-installed fans to circulate the air.

Gravity type. A floor furnace depending primarily upon circulation of air by gravity. This classification shall also include floor furnaces equipped with booster-type fans that do not materially restrict free circulation of air by gravity flow when such fans are not in operation.

❖ Typically floor furnaces are classified as gravity-type furnaces. Air circulates from a gravity-type floor furnace by convection.

FLUE, APPLIANCE. The passage(s) within an *appliance* through which combustion products pass from the combustion chamber of the *appliance* to the draft hood inlet opening on an *appliance* equipped with a draft hood or to the *outlet* of the *appliance* on an *appliance* not equipped with a draft hood.

❖ Appliance flues are the passages through an appliance from the combustion chamber, through the heat exchanger and through the vent outlet.

FLUE COLLAR. That portion of an *appliance* designed for the attachment of a draft hood, vent connector or venting system.

❖ The flue collar size will be a determining factor in vent connector sizing.

FLUE GASES. Products of combustion plus excess air in *appliance* flues or heat exchangers.

❖ The exact composition of flue gases will depend on the fuel being burned. The primary components of flue gases are nitrogen, carbon dioxide, water vapor, particulates and a myriad of compounds and trace elements that vary with the nature of the fuel and purity of the combustion air. Carbon monoxide is also a component of flue gas.

[M] FLUE LINER (LINING). A system or material used to form the inside surface of a flue in a chimney or vent, for the purpose of protecting the surrounding structure from the effects of combustion products and for conveying combustion products without leakage to the atmosphere.

❖ Flue liners must be resistant to heat and the corrosive action of the products of combustion. They provide insulation value to retard the transfer of heat to the chimney structure and limit exposure of the chimney structure to the harmful effects of combustion products. Flue liners are generally made of fire-clay tile, refractory brick, poured-in-place refractory materials, stainless steel alloys and aluminum. Flue liners used with gas-fired appliances are typically made of stainless steel or aluminum [see Commentary Figure 202(9)].

FUEL GAS. A natural gas, manufactured gas, liquefied petroleum gas or mixtures of these gases.

❖ The nature of fuel gases makes proper design, installation and selection of materials and devices necessary to minimize the possibility of fire or explosion. Bringing fuel gases into a building is in itself a risk. The provisions of the code are intended to reduce that risk to a level comparable to that associated with other energy sources such as electricity.

The two most commonly used fuel gases are natural gas and liquefied petroleum gas (LP-gas or LPG). These fuel gases have the following characteristics or properties.

Natural gas: The principal constituent of natural gas is methane (CH_4). It can also contain small quantities of nitrogen, carbon dioxide, hydrogen sulfide, water vapor, other hydrocarbons (such as ethane and propane) and various trace elements. Natural gas is colorless, tasteless and odorless; however, an odorant is added to the gas so that it can be readily detected. Natural gas is lighter than air (specific gravity of 0.60 typical) and has the tendency to rise when escaping to the atmosphere. Natural gas has a rather narrow flammability range (approximately 3 to 15 percent volume in air) above and below which the gas-to-air mixture ratio will be too rich or too lean to support combustion. The heating value of natural gas is approximately 1,050 Btu per cubic foot (39 MJ/m³).

LP-gas: Liquefied petroleum gases include commercial propane and commercial butane. LP-gas vapors are heavier than air (specific gravity of 1.52 typical)

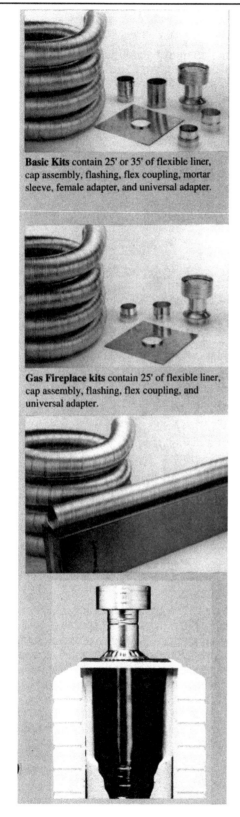

Basic Kits contain 25' or 35' of flexible liner, cap assembly, flashing, flex coupling, mortar sleeve, female adapter, and universal adapter.

Gas Fireplace kits contain 25' of flexible liner, cap assembly, flashing, flex coupling, and universal adapter.

For SI: 1 foot = 304.8 mm.

Figure 202(9)
CHIMNEY FLUE LINER SYSTEM FOR GAS APPLIANCE
(Photo courtesy of Simpson Dura-Vent Company, Incorporated)

and tend to accumulate in low areas and near the floor; however, a gas/air mixture that is within its flammability range will have a specific gravity very close to that of air. The ranges of flammability for LP-gases are narrower than those of natural gas (approximately 2 to 10 percent volume in air). Like natural gas, LP-gases are odorized to make them detectable. The heating value of propane is approximately 2,500 Btu per cubic foot (93 MJ/m^3) of gas. The heating value of butane is approximately 3,300 Btu per cubic foot (123 MJ/m^3) of gas.

[M] FURNACE. A completely self-contained heating unit that is designed to supply heated air to spaces remote from or adjacent to the *appliance* location.

❖ The single most distinguishing characteristic of furnaces is that they use air as the heat transfer medium. Furnaces can be fueled by gas, oil, solid fuel or electricity and use fans, blowers or gravity (convection) to circulate air to and from the unit. In the context of the code, the primary usage of the term "furnace" refers to heating appliances that combine a combustion chamber with related burner and control components, one or more heat exchangers, a flue gas conveying system and air-handling fans/blowers.

FURNACE, CENTRAL. A self-contained *appliance* for heating air by transfer of heat of combustion through metal to the air, and designed to supply heated air through ducts to spaces remote from or adjacent to the *appliance* location.

❖ The term "central furnace" has been adopted to identify a furnace that supplies conditioned air to remote locations from a central location. There are many types of furnaces. Most of these variations are a result of architectural design related to available space, desired location of the unit and size of the area to be conditioned.

Downflow furnace. A furnace designed with airflow discharge vertically downward at or near the bottom of the furnace.

❖ Downflow furnaces are typically installed with airflow directed into a plenum and ductwork installed in a crawl space.

Forced air furnace with cooling unit. A single-package unit, consisting of a gas-fired forced-air furnace of one of the types listed below combined with an electrically or fuel gas-powered summer air-conditioning system, contained in a common casing.

❖ A forced-air furnace with cooling unit is typically a package unit designed for outdoor installation on a rooftop or on a pad on grade.

Forced-air type. A central furnace equipped with a fan or blower that provides the primary means for circulation of air.

❖ All furnaces equipped with a fan/blower for the purpose of circulating air through the unit would be considered forced-air-type furnaces.

Gravity furnace with booster fan. A furnace equipped with a booster fan that does not materially restrict free circulation of air by gravity flow when the fan is not in operation.

❖ These units can be equipped with a booster fan that may or may not operate continually, depending on the application, to assist in the circulation of air.

Gravity type. A central furnace depending primarily on circulation of air by gravity.

❖ Gravity type units are used for relatively small spaces where the natural flow of air by convection will condition the space.

Horizontal forced-air type. A furnace with airflow through the *appliance* essentially in a horizontal path.

❖ Horizontal forced-air-type units are commonly installed in spaces above ceilings, suspended from ceiling and roof structures and installed in attics.

Multiple-position furnace. A furnace designed so that it can be installed with the airflow discharge in the upflow, horizontal or downflow direction.

❖ Many furnaces today are designed for multiple applications, allowing a single unit to be used in all of the common locations.

Upflow furnace. A furnace designed with airflow discharge vertically upward at or near the top of the furnace. This classification includes "highboy" furnaces with the blower mounted below the heating element and "lowboy" furnaces with the blower mounted beside the heating element.

❖ Upflow furnaces (typically made in high-boy and low-boy styles) are typically installed in basements and attics.

FURNACE, ENCLOSED. A specific heating, or heating and ventilating, furnace incorporating an integral total enclosure and using only outside air for combustion.

❖ Enclosed furnaces are believed to be extinct. Such furnaces had an integral housing that isolated the combustion chamber from the conditioned space, thus allowing all combustion air to be obtained directly from the housing that opened to the outdoors.

FURNACE PLENUM. An air compartment or chamber to which one or more ducts are connected and that forms part of an air distribution system.

❖ The term "plenum" is no longer defined or used in the code. A furnace plenum is a box made of sheet metal or other duct material that is attached directly to a furnace to facilitate connection of supply and return ducts to the furnace air inlet and outlet. This term does not refer to cavities within a building, such as stud and joist spaces and spaces above dropped ceilings.

GAS CONVENIENCE OUTLET. A permanently mounted, manually operated device that provides the means for connecting an *appliance* to, and disconnecting an *appliance* from, the supply *piping*. The device includes an integral, manually operated valve with a nondisplaceable valve member and is designed so that disconnection of an *appliance* only occurs when the manually operated valve is in the closed position.

❖ These devices are listed and labeled and designed to mate with specialized appliance connectors. A gas outlet is the point at which the gas distribution system connects to a gas appliance. The outlets are intended for use with portable appliances such as space heaters and outdoor cooking appliances [see Commentary Figures 202(10) and 202(11)].

GAS PIPING. An installation of pipe, valves or fittings installed on a premises or in a building and utilized to convey fuel gas.

❖ Gas piping includes all the components, fittings and piping needed to deliver the fuel gas from the point of delivery to the appliance or equipment connection. The point of delivery may be a regulator or meter that is typically installed by the gas utility. The point of delivery may be located at the user's property line, immediately outside the structure or in some instances in the structure.

Figure 202(10)
GAS CONVENIENCE OUTLET
(Photo courtesy of Gastite Division/Titeflex Corporation)

Figure 202(11)
GAS CONVENIENCE OUTLET
(Photo courtesy of Maxitrol Company)

[F] GASEOUS HYDROGEN SYSTEM. See Section 702.1.

[M] HAZARDOUS LOCATION. Any location considered to be a fire hazard for flammable vapors, dust, combustible fibers or other highly combustible substances. The location is not necessarily categorized in the building code as a high-hazard group classification.

❖ The environment in which mechanical equipment and appliances operate plays a significant role in the safe performance of the equipment installation. Locations that may contain ignitable or explosive atmospheres are classified as hazardous locations for the installation of mechanical equipment and appliances. For example, repair garages can be classified as hazardous locations because they can contain gasoline vapors from vehicles stored within them as well as other volatile chemicals. Public and private garages are not considered hazardous locations even though the presence of motor vehicles and the storage of fuel, paint, varnish, thinner, lawn- and home-maintenance products and other chemicals are a concern, as evidenced by Sections 305.3 and 305.5.

HOUSE PIPING. See "*Piping* system."

❖ House piping is the distribution piping downstream of the point of delivery. House piping is an antiquated term.

[F] HYDROGEN FUEL-GAS ROOM. See Section 702.1.

HYDROGEN-GENERATING APPLIANCE. See Section 702.1.

IGNITION PILOT. A pilot that operates during the lighting cycle and discontinues during main burner operation.

❖ Commonly referred to as "interrupted pilots," such pilots are ignited by a spark or hot surface device. Generally an electronic circuit controls the ignition system. When main burner ignition has been proven, the ignition pilot is shut off, and the main burner flame is supervised for the entire burn cycle.

[M] IGNITION SOURCE. A flame, spark or hot surface capable of igniting flammable vapors or fumes. Such sources include *appliance* burners, burner ignitors and electrical switching devices.

❖ This definition is important in the application of Section 305.3 regarding the elevation of ignition sources. By means of this definition, Section 305.3 applies to unintentional ignition sources, such as electrical switching devices, as well as intentional ignition sources, such as pilot lights, spark ignitors and hot surface ignitors.

In the context of Section 305.3, an "ignition source" is something capable of igniting flammable vapors that are present in the atmosphere in the locations listed in Section 305.3.

INCINERATOR. An *appliance* used to reduce combustible refuse material to ashes and that is manufactured, sold and installed as a complete unit.

❖ These appliances are believed to be extinct.

INDUSTRIAL AIR HEATERS, DIRECT-FIRED NON-RECIRCULATING. A heater in which all the products of combustion generated by the burners are released into the air stream being heated. The purpose of the heater is to offset building heat loss by heating only outdoor air.

❖ These heaters have traditionally been called makeup air heaters because they heat 100 percent of outdoor air without recirculation of any air across the burner. These heaters are unvented (direct-fired) appliances regulated by Section 611 (see Commentary Figure 611.1).

INDUSTRIAL AIR HEATERS, DIRECT-FIRED RECIRCULATING. A heater in which all the products of combustion generated by the burners are released into the air stream being heated. The purpose of the heater is to offset building heat loss by heating outdoor air, and, if applicable, indoor air.

❖ These heaters are used in large spaces, such as warehouses, and are often referred to as "air turnover units" because they are typically vertical air handlers designed to prevent air stratification. These heaters are unvented (direct-fired) appliances regulated by Section 612.

INFRARED RADIANT HEATER. A heater that directs a substantial amount of its energy output in the form of infrared radiant energy into the area to be heated. Such heaters are of either the vented or unvented type.

❖ These heaters include low-intensity tubular types and high-intensity ceramic burner element types. They heat objects (such as people) by direct radiation and do not heat the ambient air.

[M] JOINT, FLANGED. A joint made by bolting together a pair of flanged ends.

❖ Flanges are commonly used with large piping at locations where the piping must be disassembled periodically and at connections to valves, regulators, devices and equipment. Full-face gaskets must be used with bronze and cast-iron flange fittings. Typically, materials for flange gaskets include metal or metal-jacketed asbestos, asbestos and aluminum "O" rings or spiral-wound metal gaskets. Gaskets must be replaced whenever a joint is opened. The gasket material must be compatible with the piping contents to prevent chemical reaction.

[M] JOINT, FLARED. A metal-to-metal compression joint in which a conical spread is made on the end of a tube that is compressed by a flare nut against a mating flare.

❖ Because the pipe end is expanded in a flared joint, only annealed and bending-tempered soft-drawn copper tubing may be flared. Commonly used flaring tools use a screw yoke and block assembly or an expander tool that is driven into the tube with a hammer. The flared tubing end is compressed between a fitting seat and a threaded nut to form a metal-to-metal seal.

JOINT, MECHANICAL. A general form of gas-tight joints obtained by the joining of metal parts through a positive-holding mechanical construction, such as press joint, flanged joint, threaded joint, flared joint or compression joint.

❖ Mechanical joints can take many forms, but most share a common characteristic, which is applying a radial pressure to the pipes or fittings they join. Mechanical joining methods can be proprietary, which means that they are specific to one manufacturer of the fittings and devices and such manufacturer must provide adequate instructions for assembling the joint. Press joint fittings must be assembled with a special tool that is unique to the fitting. Press joints are irreversible because the fitting and pipe are permanently deformed by the pressing operation. Some press joint systems are manufactured for use with copper piping and some are manufactured for use with steel piping.

[M] JOINT, PLASTIC ADHESIVE. A joint made in thermoset plastic *piping* by the use of an adhesive substance that forms a continuous bond between the mating surfaces without dissolving either one of them.

❖ Unlike solvent-welded joints, adhesive (glue) joints are surface bonded. Often people refer to solvent-welded joints, such as used for PVC, CPVC and ABS piping, as "glued" joints, when actually, they are chemically welded, not glued.

[M] JOINT, PLASTIC HEAT FUSION. A joint made in thermoplastic *piping* by heating the parts sufficiently to permit fusion of the materials when the parts are pressed together.

❖ Heat-fusion joints for plastic pipe are analogous to welded joints for steel pipe. Only polyethylene and polybutylene can be joined by heat fusion and only in accordance with the pipe manufacturer's instructions. The process involves heating the pipe and fittings with a special iron. Some fittings have integral heating elements that are connected to a special power supply. When the parts to be joined reach their melting points, they are assembled and allowed to fuse together.

[M] JOINT, WELDED. A gas-tight joint obtained by the joining of metal parts in molten state.

❖ A welded joint is similar to a brazed joint. The primary differences are the temperature at which the joint is made, the type of filler metal used, and the fact that welding reaches the melting point of the base metal, whereas brazing temperatures are well below the melting point of the pipe and fittings.

[A] LABELED. Equipment, materials or products to which have been affixed a label, seal, symbol or other identifying mark of a nationally recognized testing laboratory, inspection agency or other organization concerned with product evaluation that maintains periodic inspection of the production of the above-labeled items and whose labeling indicates either that the *equipment*, material or product meets identified standards or has been tested and found suitable for a specified purpose.

❖ When a product is labeled, the label indicates that the material has been tested for conformance to an applicable standard and that the component is subject to third-party inspection to verify that the minimum level of quality required by the standard is maintained. Labeling provides a readily available source of information that is useful for field inspection of installed products. The label identifies the product or material and provides other information that can be further investigated if there is question concerning the suitability of the product or material for the specific installation. The labeling agency performing the third-party inspection must be approved by the code official. The basis for this approval may include, but is not necessarily limited to, the capacity and capability of the agency to perform the specific testing and inspection.

The referenced standard often states the minimum identifying information that must be on a label. The data contained on a label typically includes, but is not necessarily limited to, name of the manufacturer; product name or serial number; installation specifications; applicable tests and standards; and the approved testing and labeling agency.

This definition correlates with the definitions provided in other *International Codes®* (I-Codes®) [see Commentary Figure 202(12)].

LEAK CHECK. An operation performed on a gas *piping* system to verify that the system does not leak.

❖ A leak check is not the same as the pressure test required by Sections 406.1 through 406.5.2 (see Section 406.6).

LIMIT CONTROL. A device responsive to changes in pressure, temperature or level for turning on, shutting off or throttling the gas supply to an *appliance*.

❖ Limit controls are safety devices used to protect equipment, appliances, property and persons. The controls act at their setpoint to limit a condition, such as temperature or pressure, and include high and low limits. This definition applies to controls for all types of fuel, including gas, liquid and solid.

LIQUEFIED PETROLEUM GAS or LPG (LP-GAS). Liquefied petroleum gas composed predominately of propane, propylene, butanes or butylenes, or mixtures thereof that is gaseous under normal atmospheric conditions, but is capable of being liquefied under moderate pressure at normal temperatures.

❖ Liquefied petroleum gases are usually obtained as a byproduct of oil refinery operations or by stripping natural gas and are commercially available as butane, propane or a mixture of the two. Propane has a boiling point of -40°F (-40°C) at atmospheric pressure and a heating value of approximately 2,500 Btu per cubic foot (93 MJ/m^2). Butane has a boiling point of 32°F (0°C) at atmospheric pressure and a heating value of approximately 3,300 Btu per cubic foot (123 MJ/m^3). LP-gas is odorized so that gas leakage can be detected.

[A] LISTED. Equipment, materials, products or services included in a list published by an organization acceptable to the code official and concerned with evaluation of products or services that maintains periodic inspection of production of *listed equipment* or materials or periodic evaluation of services and whose listing states either that the *equipment*, material, product or service meets identified standards or has been tested and found suitable for a specified purpose.

❖ This definition correlates with the definition of "Listed" in other I-Codes. When a product is listed, it indicates that it has been tested for conformance to an applicable standard and was subject to a third-party inspection quality assurance (QA) program. The QA verifies that the minimum level of quality required by the appropriate standard is maintained. The agency performing the third-party inspection must be approved by the code official, and the basis for this approval may include, but is not limited to, the capacity and capability of the agency to perform the specified testing and inspection.

[M] LIVING SPACE. Space within a *dwelling unit* utilized for living, sleeping, eating, cooking, bathing, washing and sanitation purposes.

❖ The code uses this term to define locations that are not to be considered as part of a private garage for the purpose of requiring the elevation of ignition sources (see Section 305.3).

LOG LIGHTER. A manually operated solid fuel ignition *appliance* for installation in a vented solid fuel-burning *fireplace*.

❖ Log lighters are simple, manually operated burners used to start wood fires. Log lighters are not considered as decorative but rather as functional appliances. The heat produced by the log lighter flame raises the temperature of the wood fuel to its ignition temperature. A log lighter is designed to be turned off manually after the wood fire is capable of sustaining combustion.

LUBRICATED PLUG-TYPE VALVE. A valve of the plug and barrel type provided with means for maintaining a lubricant between the bearing surfaces.

❖ Valves of this type are rarely used today. They became rare with the advent of ball valves. This term is currently not used in the body of the code.

AMERICAN STANDARD INC.
THE TRANE COMPANY
TRENTON, N.J. 08619   MADE IN U.S.A.

FORCED AIR FURNACE   CATEGORY I

ANS Z21.47 - 1990 CENTRAL FURN

FOR INDOOR INSTALLATION IN A BUILDING
CONSTRUCTED ON SITE.   **NRTL**

MODEL NO. TUD080R936A1	SERIAL NO. G36519785	EQUIPPED FOR NAT. GAS
INPUT 80,000 BTU/HR.	LIMIT SETTING 190 °F	MFRD. 09/92
TEMP. RISE °F FROM 30 TO 60	MAX. EXT. STATIC PRESS .50 INCHES WATER	MAX. DESIGN AIR TEMP. 160 °F
VOLTS/PHASE/HERTZ 115/1/60	TOTAL AMPS 8.5	SERVICE CODE 1

FLAME ROLLOUT SWITCH - REPLACE IF BLOWN WITH CATALOG NO. WG09XC33 (333 F CUTOFF TEMP.) ONE TIME THERMAL FUSE.

**MANIFOLD PRESSURE
(IN INCHES OF WATER)**
NAT. 3.5 LP 10.5
**SUPPLY PRESSURE
(IN INCHES OF WATER)**
MAX. NAT. 10.5, L P 13.0
MIN. NAT. 4.5 L P 11.0 FOR PURPOSE OF INPUT ADJUSTMENT.

LOW INPUT 52,000 BTU/HR

MINIMUM CLEARANCE COMBUSTIBLE MATERIALS:

FOR CLOSET	INSTALLATION AS FOLLOWS:			
SIDES	0	IN. W/SINGLE WALL VENT		
FLUE	6	IN. W/SINGLE WALL VENT	1 IN. W/TYPE B-1 VENT
FRONT	6	IN. BACK 0 IN. TOP 1 IN.

UPFLOW UNITS, FOR INSTALLATION COMBUSTIBLE FLOORING.   21D340159 P01

**Figure 202(12)
TYPCIAL APPLIANCE LABEL
(Figure courtesy of the Trane Company, an Amercian Standard Company)**

MAIN BURNER. A device or group of devices essentially forming an integral unit for the final conveyance of gas or a mixture of gas and air to the combustion zone, and on which combustion takes place to accomplish the function for which the *appliance* is designed.

❖ With the exception of small amounts of fuel for the pilot or ignition burners, the main burner consumes the entire fuel input into an appliance. Common types of atmospheric burners are the drilled port, slotted port, ribbon and single port.

METER. The instrument installed to measure the volume of gas delivered through it.

❖ The meter is the actual point of commerce and is usually associated with the point of delivery. The code does not require or regulate meters. The gas-supplying utility generally installs meters for natural gas. Occasionally in locations where there are several tenants (large office buildings, mobile home parks and apartment complexes, for example), the gas utility installs one meter for the property owner and the owner installs individual tenant meters. Liquefied petroleum may be metered to individual tenants from a storage tank in the same manner [see Commentary Figure 202(13)].

Figure 202(13)
TYPICAL RESIDENTIAL GAS METER SETTING
(Photo courtesy of Washington Gas)

MODULATING. Modulating or throttling is the action of a control from its maximum to minimum position in either predetermined steps or increments of movement as caused by its actuating medium.

❖ In the context of the code, modulating is associated with burner control valves. One type of automatic control is a modulating control that can infinitely vary the controlled variable over the range being controlled. Another common type of control is a two-position action control, which simply turns the variable being controlled, such as fuel gas, on and off. Another type of control is the floating action control. It regulates the controlled device by adjusting the con-

trolled variable between the open or closed position using a sensing element, which must react faster than the actuating drive time or the system would be essentially the same as a two-position control.

[M] NONCOMBUSTIBLE MATERIALS. Materials that, when tested in accordance with ASTM E136, have at least three of four specimens tested meeting all of the following criteria:

1. The recorded temperature of the surface and interior thermocouples shall not at any time during the test rise more than 54°F (30°C) above the furnace temperature at the beginning of the test.

2. There shall not be flaming from the specimen after the first 30 seconds.

3. If the weight loss of the specimen during testing exceeds 50 percent, the recorded temperature of the surface and interior thermocouples shall not at any time during the test rise above the furnace air temperature at the beginning of the test, and there shall not be flaming of the specimen.

❖ If the material is not determined to be noncombustible, it is, by default, a combustible material. ASTM E136 is a test standard used to determine the combustion characteristics of building materials. The test method in the standard is intended to indicate whether a material will assist in combustion or add appreciable heat to an ambient fire. The listed criteria are taken from ASTM E136. In the standard, Items 1 and 2 are applicable if the weight loss of the specimen during the test is 50 percent or less. Item 3 applies if such weight loss is greater than 50 percent. The standard intends for three out of the four samples to comply with either Items 1 and 2 or Item 3.

[A] OCCUPANCY. The purpose for which a building, or portion thereof, is utilized or occupied.

❖ The occupancy classification of a building is an indication of the level of hazard to which the occupants are exposed to as a function of the actual building use. Occupancy in terms of a group classification is one of the primary considerations in the development and application of many code requirements that are designed to offset the hazards specific to each group designation.

[M] OFFSET (VENT). A combination of *approved* bends that makes two changes in direction bringing one section of the vent out of line but into a line parallel with the other section.

❖ An offset results in the lateral displacement of a vertical vent. Vent offsets are measured in degrees of angle from vertical and are limited in venting systems because they restrict the flow of flue gas. An offset angle is measured between a line that is an extension of the original vertical vent and the new line of direction of the vent [see Commentary Figure 202(14)]. Note that an offset is more than just a single change in direction and always involves two or more fittings and two changes of direction.

Figure 202(14)
VENT OFFSET
(Photo courtesy of Selkirk, L.L.C.)

ORIFICE. The opening in a cap, spud or other device whereby the flow of gas is limited and through which the gas is discharged to the burner.

❖ An orifice is a precisely sized hole through which gas flows. The orifice is located in the spud or cap that is generally attached to a manifold or supply line. The orifice should be installed in accordance with the manufacturer's instructions to ensure proper sizing to prevent underfiring or overfiring of the gas appliance.

OUTLET. The point at which a gas-fired *appliance* connects to the gas *piping* system.

❖ A gas outlet is analogous to an electrical receptacle outlet. See Section 404.16 for requirements covering the location of gas piping outlets.

OXYGEN DEPLETION SAFETY SHUTOFF SYSTEM (ODS). A system designed to act to shut off the gas supply to the main and pilot burners if the oxygen in the surrounding atmosphere is reduced below a predetermined level.

❖ An Oxygen Depletion Safety Shutoff System (ODS) is intended to protect the occupants from carbon monoxide build-up. The ODS measures the oxygen level in the air and shuts off the appliance if the level falls below a preset level, typically 18 percent. Monitoring oxygen levels will inversely monitor carbon monoxide levels because of the natural relationship between a decreasing oxygen level and an increasing carbon monoxide level [see Commentary Figure 202(15)].

PILOT. A small flame that is utilized to ignite the gas at the main burner or burners.

❖ Pilot burners (pilot lights) can be continuously burning or can be intermittent or interrupted depending upon the type of appliance. Continuously burning (standing) pilots are becoming more rare and are typically found only in certain hearth appliances, water heaters and commercial ranges. Intermittent pilots burn only while the main burner(s) are firing and interrupted pilots burn only until the main burner(s) are ignited and proven.

[M] PIPING. Where used in this code, "*piping*" refers to either pipe or tubing, or both.

Pipe. A rigid conduit of iron, steel, copper, brass or plastic.

Tubing. Semirigid conduit of copper, aluminum, plastic or steel.

❖ Piping includes tubing and pipe used to convey fuel gases. See the definition for "Piping system."

PIPING SYSTEM. All fuel *piping*, valves and fittings from the outlet of the *point of delivery* to the outlets of the *appliance* shutoff valves.

❖ A piping system includes tubing, pipe, fittings, valves and line pressure regulators used to convey fuel gas from the point of delivery to the appliance. This definition was revised to state that the piping system ends at the outlets of the appliance shutoff valves. This revision allows the last 6 feet (1829 mm) or less of gas conduit between the shutoff valve and the appliance inlet connection to be sized as a connector (see Section 409.5 and the definition for "Point of delivery").

[M] PLASTIC, THERMOPLASTIC. A plastic that is capable of being repeatedly softened by increase of temperature and hardened by decrease of temperature.

❖ Polyvinyl chloride (PVC) is a type of thermoplastic.

POINT OF DELIVERY. For natural gas systems, the *point of delivery* is the outlet of the service meter assembly or the outlet of the service regulator or service shutoff valve where a meter is not provided. Where a valve is provided at the outlet of the service meter assembly, such valve shall be considered

Figure 202(15)
OXYGEN DEPLETION SAFETY SHUTOFF
(Photo by D.F. Noyes Studios, courtesy of DESA International)

to be downstream of the *point of delivery*. For undiluted liquefied petroleum gas systems, the point of delivery shall be considered to be the outlet of the service pressure regulator, exclusive of line gas regulators, in the system.

❖ It is necessary to understand where the point of delivery begins with respect to fuel gas systems because that is the location where the enforcement of Chapter 4 begins. The point of delivery is a point of demarcation that determines what is or is not covered by the chapter. For natural gas systems, the utility-owned service piping, service shutoff valve, service pressure regulator and meter are all upstream of the point of delivery and thus are not covered by the code. Any valve located on the outlet of the service meter assembly is considered to be part of the piping system and under the scope of the code.

For liquefied petroleum (LP) applications, the point of delivery is the outlet of the service pressure regulator (see definition). NFPA 58 regulates the piping and components upstream of the outlet of the service pressure regulator, including the regulator itself. In typical two-stage LP systems, the service pressure regulator is the second-stage regulator. Two-stage systems use a first-stage regulator located at the tank to reduce tank pressure down to 10 psig, and a second-stage regulator to reduce the 10 psig pressure down to 11 inches water column (w.c.). Some two-stage systems use separate first- and second-stage regulators and some use integral two-stage regulators, located at the tank, in which both the first- and second-stage regulators are contained within a single unit. A 2-psig system is becoming more popular in LP applications. These systems use three stages of regulation, which include a first-stage regulator to reduce tank pressure to 10 psig, a 2-psi regulator (the service pressure regulator) to reduce the 10 psig pressure to 2 psig, and line gas pressure regulators (see definition) used to reduce the 2 psig pressure to 11 inches w.c. for delivery to the appliances. In these systems, the point of delivery is the outlet of the 2-psi service pressure regulator.

PORTABLE FUEL CELL APPLIANCE. A fuel cell generator of electricity, which is not fixed in place. A portable fuel cell *appliance* utilizes a cord and plug connection to a grid-isolated load and has an integral fuel supply.

❖ As stated in Section 101.2.4, Item 19, the code does not regulate these appliances because they do not connect to a fuel piping system and are completely self-contained.

PRESSURE DROP. The loss in pressure due to friction or obstruction in pipes, valves, fittings, regulators and burners.

❖ The pressure in a fuel gas system is reduced as the gas flows in the system. This is caused by a loss of energy resulting from friction and turbulence. Fuel gas systems are designed so that the pressure drop does not result in the system pressure falling below the minimum pressure required for proper equipment operation.

PRESSURE TEST. An operation performed to verify the gas-tight integrity of gas *piping* following its installation or modification.

❖ Installed piping systems are tested on the job site to verify that the system is free of leaks.

PURGE. To free a gas conduit of air or gas, or a mixture of gas and air.

❖ Piping systems are purged (flushed) to remove gaseous, liquid or solid contaminants that could be harmful to the piping contents, the piping system or the system components or that could create a fire or explosion hazard.

QUICK-DISCONNECT DEVICE. A hand-operated device that provides a means for connecting and disconnecting an *appliance* or an *appliance* connector to a gas supply and that is equipped with an automatic means to shut off the gas supply when the device is disconnected.

❖ These devices are commonly installed with gas appliance connectors and used with commercial cooking appliances and other gas appliances that are routinely moved for cleaning or used in another location [see Commentary Figure 202(16)].

Figure 202(16)
QUICK-DISCONNECT DEVICE
(Photo courtesy of Dormont Manufacturing Company)

[M] READY ACCESS (TO). That which enables a device, *appliance* or *equipment* to be directly reached, without requiring the removal or movement of any panel, door or similar obstruction (see "*Access*").

❖ Ready access can be described as the capability of being quickly reached or approached for the purpose of operation, inspection, observation or emergency action. Ready access does not require the removal or movement of any access door, panel or similar obstruction, nor does it require surmounting physical obstructions or obstacles, including differences in elevation. Note that this definition does not refer to the egress/ingress door through which a person enters a room or building.

[A] REGISTERED DESIGN PROFESSIONAL. An individual who is registered or licensed to practice their respective design profession as defined by the statutory requirements of the professional registration laws of the state or jurisdiction in which the project is to be constructed.

❖ Each state establishes legal qualifications for engineers and architects. Licensing and registration of engineers and architects is based on written or oral examinations offered by states or by reciprocity (licensing in other states).

REGULATOR. A device for controlling and maintaining a uniform supply pressure, either pounds-to-inches water column (MP regulator) or inches-to-inches water column (*appliance* regulator).

❖ In the context of the code, these devices are fuel-gas pressure regulating devices that reduce a higher pressure to a lower pressure.

REGULATOR, GAS APPLIANCE. A pressure regulator for controlling pressure to the manifold of the *appliance*. Types of *appliance* regulators are as follows:

Adjustable.

1. Spring type, limited adjustment. A regulator in which the regulating force acting upon the diaphragm is derived principally from a spring, the loading of which is adjustable over a range of not more than 15 percent of the outlet pressure at the midpoint of the adjustment range.

2. Spring type, standard adjustment. A regulator in which the regulating force acting upon the diaphragm is derived principally from a spring, the loading of which is adjustable. The adjustment means shall be concealed.

Multistage. A regulator for use with a single gas whose adjustment means is capable of being positioned manually or automatically to two or more predetermined outlet pressure settings. Each of these settings shall be adjustable or nonadjustable. The regulator may modulate outlet pressures automatically between its maximum and minimum predetermined outlet pressure settings.

Nonadjustable.

1. Spring type, nonadjustable. A regulator in which the regulating force acting upon the diaphragm is derived principally from a spring, the loading of which is not field adjustable.

2. Weight type. A regulator in which the regulating force acting upon the diaphragm is derived from a weight or combination of weights.

❖ These regulators are integral to the appliance and serve to stabilize the appliance input pressure and supply fuel gas to the appliance at the desired pressure. Most appliances come equipped with appliance regulators because the appliances are designed to operate at a pressure lower than the delivery pressure, even when low pressure gas [0.5 psig or lower (3.4 kPa)] is supplied. Commonly, appliance regulators reduce a pressure of 5 to 8 inches water column down to a manifold pressure of 3.5 inches w.c.

REGULATOR, LINE GAS PRESSURE. A device placed in a gas line between the service pressure regulator and the

appliance for controlling, maintaining or reducing the pressure in that portion of the *piping* system downstream of the device.

❖ Line gas pressure regulators are used to reduce the gas pressure supplied by the service (point of delivery) regulator. They are typically located at the connection to each gas appliance and are commonly used as "pounds to inches" regulators in systems having elevated [14 inches (3.5 kPa) wc to 5 psi typical (34.5 kPa)] supply pressures.

REGULATOR, MEDIUM-PRESSURE (MP Regulator). A line pressure regulator that reduces gas pressure from the range of greater than 0.5 psig (3.4 kPa) and less than or equal to 5 psig (34.5 kPa) to a lower pressure.

❖ With the increased use of high-pressure gas distribution systems, systems are designed with one or more regulators in addition to the service regulators. Medium-pressure (MP) regulators are most commonly associated with 2 psi (13.8 kPa) manifold distribution systems used with copper tubing or corrugated stainless steel tubing (CSST) systems. Appliances are also equipped with factory-installed regulators that control burner input (manifold) pressure. MP regulators are used to reduce the service pressure to a pressure suitable for delivery to the gas appliances where the appliance regulator may further reduce the pressure (see Section 410.2). MP regulators are used where the distribution line pressure is in the range of 5 psig (34.5 kPa) down to 0.5 psig (3.4 kPa) and such pressure needs to be reduced to be compatible with input pressure needs of the load.

REGULATOR, PRESSURE. A device placed in a gas line for reducing, controlling and maintaining the pressure in that portion of the *piping* system downstream of the device.

❖ See the definition and commentary to "Regulator."

REGULATOR, SERVICE PRESSURE. For natural gas systems, a device installed by the serving gas supplier to reduce and limit the service line pressure to delivery pressure. For undiluted liquefied petroleum gas systems, the regulator located upstream from all line gas pressure regulators, where installed, and downstream from any first stage or a high pressure regulator in the system.

❖ The utility company (gas supplier) installs this regulator to reduce the gas pressure from the utility distribution mains to that required for delivery to the meter and customer. For the typical house service, the pressure is reduced from 60 psig to 2 psig or less. The devices are usually located near the meter and outdoors [see Commentary Figure 202(17)]. For LP-gas systems, the service regulator is the regulator between the first-stage regulator and the line pressure regulators or between the first-stage regulator and the appliances served where there are no line pressure regulators. The service regulator is upstream of the point of delivery for natural and LP-gases (see "Point of delivery").

Figure 202(17)
SERVICE REGULATOR
(Photo courtesy of Washington Gas)

RELIEF OPENING. The opening provided in a draft hood to permit the ready escape to the atmosphere of the flue products from the draft hood in the event of no draft, back draft or stoppage beyond the draft hood, and to permit air into the draft hood in the event of a strong chimney updraft.

❖ Dilution air, a component of combustion air, enters the venting system through the draft hood opening. Combustion gases spilling from (exiting) a draft hood relief opening create an abnormal and potentially hazardous condition except for brief periods of spillage that might occur at the start of an appliance firing cycle. Typical periods of spillage at start-up last only a few seconds.

RELIEF VALVE (DEVICE). A safety valve designed to forestall the development of a dangerous condition by relieving either pressure, temperature or vacuum in the hot water supply system.

❖ A relief valve is a safety device designed to prevent the development of a potentially damaging or dangerous condition in a closed system. This is usually associated with a closed system that has a heat source as part of the system. Requirements for the installation of relief valves in water distribution systems and water heaters are given in Section 504 of the IPC.

RELIEF VALVE, PRESSURE. An automatic valve that opens and closes a relief vent, depending on whether the pressure is above or below a predetermined value.

❖ Pressure relief valves are intended to prevent harm to life and property. These valves are safety devices used to relieve abnormal pressures and to prevent the overpressurization of the vessel or system to which the valves are connected. Pressure relief valves are designed to operate (discharge) only when an abnormal pressure exists in a system. They are not to be used as regulating valves, nor are they intended to control or regulate flow or pressure. Pressure relief valves are set at the factory to begin opening at a predetermined pressure and are rated according to the maximum energy discharge per unit of time, usually in units of British thermal units per hour (Btu/h)(W).

Pressure relief valves must be properly sized and rated for the boiler, pressure vessel or system served; otherwise, a system malfunction could cause the pressure within the system to continue to rise, thereby creating a hazardous condition. Pressure relief valves do not open fully when they reach the factory-preset pressure; rather, the valves open an amount proportional to the forces produced by the pressure in the system. The valves open fully at a certain percentage above the preset pressure, then close when the pressure drops below the preset pressure. Pressure relief valves are sometimes combined in the same body with temperature relief valves, creating combination temperature and pressure relief valves.

RELIEF VALVE, TEMPERATURE.

Manual reset type. A valve that automatically opens a relief vent at a predetermined temperature and that must be manually returned to the closed position.

Reseating or self-closing type. An automatic valve that opens and closes a relief vent, depending on whether the temperature is above or below a predetermined value.

❖ Temperature relief valves are designed to open in response to excessive temperatures and to discharge heated water to limit the temperature of the water in the vessel, tank or system. Water heater control failure or improper adjustments could allow excessive temperature rises, which can heat water above the temperature at which it would vaporize at atmospheric pressure. Water heated to above 212°F (100°C) is referred to as superheated water and will flash to steam when the pressure is reduced below its vapor pressure. Very hot water and the presence of steam are hazardous; however, more importantly, higher temperatures produce higher pressures and superheated water increases the potential for, and the magnitude of, an explosion.

RELIEF VALVE, VACUUM. A valve that automatically opens and closes a vent for relieving a vacuum within the hot water supply system, depending on whether the vacuum is above or below a predetermined value.

❖ The vacuum relief valve is intended to prevent the possible reduction in pressure within the system (water heater tanks being of main concern) to below atmospheric pressure (that is, a partial vacuum). Many tanks are not designed to resist external pressures exceeding internal pressures; therefore, a vacuum relief valve is necessary to prevent atmospheric pressure from possibly collapsing or otherwise damaging the tank. A partial vacuum could also cause water to be siphoned from the tank, thereby creating an overheating hazard in a water heater.

The vacuum relief valve operates by automatically opening the system to the atmosphere when a partial vacuum is created, thereby permitting air to enter and maintaining the internal pressure at atmospheric pressure.

RISER, GAS. A vertical pipe supplying fuel gas.

❖ A vertical gas pipe that distributes gas from a building main to upper floor levels is commonly referred to as a "riser."

ROOM HEATER, UNVENTED. See "Unvented room heater."

❖ See the commentary to "Unvented room heater" and Commentary Figure 202(18).

Figure 202(18)
UNVENTED ROOM HEATER
(Photo courtesy of Empire Comfort Systems Inc.)

ROOM HEATER, VENTED. A free-standing heating unit used for direct heating of the space in and adjacent to that in which the unit is located (see also "Vented room heater").

❖ These heaters are typically small space-heating appliances designed to heat a single room.

SAFETY SHUTOFF DEVICE. See "Flame safeguard."

❖ See the commentary to "Flame safeguard."

[BF] SHAFT. An enclosed space extending through one or more stories of a building, connecting vertical openings in successive floors, or floors and the roof.

❖ A shaft is an enclosed, vertical passage that passes through stories and floor levels within a building. Shafts are sometimes referred to as "chases" and are usually constructed of fire-resistance-rated assemblies.

[BG] SLEEPING UNIT. A room or space in which people sleep, which can also include permanent provisions for living, eating and either sanitation or kitchen facilities, but not both. Such rooms and spaces that are also part of a *dwelling unit* are not sleeping units.

❖ Contrast this definition with the definition of "Dwelling unit."

SPECIFIC GRAVITY. As applied to gas, specific gravity is the ratio of the weight of a given volume to that of the same volume of air, both measured under the same condition.

❖ Analogous to density, specific gravity values are used to describe how a gaseous substance will behave in, for example, air or piping systems. Specific gravity indicates the density of a gas with respect to air, with air being assigned a value of 1. For example, natural gas has a specific gravity of less than 1, therefore it is less dense and lighter than air. LP-gas has a specific gravity of 1.5, therefore it is one and one-half times the density of air.

STATIONARY FUEL CELL POWER PLANT. A self-contained package or factory-matched packages that constitute an automatically operated assembly of integrated systems for generating electrical energy and recoverable thermal energy that is permanently connected and fixed in place.

❖ See the commentary to Chapter 7.

THERMOSTAT.

Electric switch type. A device that senses changes in temperature and controls electrically, by means of separate components, the flow of gas to the burner(s) to maintain selected temperatures.

Integral gas valve type. An automatic device, actuated by temperature changes, designed to control the gas supply to the burner(s) in order to maintain temperatures between predetermined limits, and in which the thermal actuating element is an integral part of the device.

1. Graduating thermostat. A thermostat in which the motion of the valve is approximately in direct proportion to the effective motion of the thermal element induced by temperature change.

2. Snap-acting thermostat. A thermostat in which the thermostatic valve travels instantly from the closed to the open position, and vice versa.

❖ Thermostats most often consist of one or more thermally actuated on/off switches, but also include types that transmit varying voltage and signals. A thermostat is also referred to as a controller. The set point or desired unit of temperature is adjusted or set at the thermostat. The thermostat transmits a signal to the controlled device or appliance.

[P] THIRD-PARTY CERTIFICATION AGENCY. An approved agency operating a product or material certification system that incorporates initial product testing, assessment and surveillance of a manufacturer's quality control system.

❖ These agencies evaluate products, verify compliance with specific standards or criteria and conduct periodic inspections of the manufacturer's production facilities to verify ongoing compliance with such standards or criteria (see commentary, Section 401.10).

[P] THIRD-PARTY CERTIFIED. Certification obtained by the manufacturer indicating that the function and performance characteristics of a product or material have been determined by testing and ongoing surveillance by an approved third-party certification agency. Assertion of certification is in the form of identification in accordance with the requirements of the third-party certification agency.

❖ A third-party certification agency certifies that a product meets specific standards or criteria and identifies the product with a certification mark. This mark is an indication to the installer and inspector that the product has been evaluated and tested and the manufacturing process has been monitored so as to verify compliance with the relative product standards or criteria. Additionally, the agency typically provides a list of the products that it certifies, which provides information relative to the certification (see Section 401.10).

[P] THIRD-PARTY TESTED. Procedure by which an approved testing laboratory provides documentation that a product, material or system conforms to specified requirements.

❖ A third-party testing laboratory or agency tests a product to verify that it meets specific standards or criteria and provides documentation to that fact. Having a product tested is not the same as having it listed, as listing entails additional requirements such as ongoing monitoring of the manufacturing process. Third-party tested products can be marked to indicate the testing agency.

[M] TRANSITION FITTINGS, PLASTIC TO STEEL. An adapter for joining plastic pipe to steel pipe. The purpose of this fitting is to provide a permanent, pressure-tight connection between two materials that cannot be joined directly one to another.

❖ Piping joined of dissimilar materials, such as plastic and steel, requires specialized transition fittings. For example, such fittings are used with underground plastic fuel-gas piping and steel meter setting risers [see Commentary Figure 202(19)].

UNIT HEATER.

High-static pressure type. A self-contained, automatically controlled, vented *appliance* having integral means for circulation of air against 0.2 inch (15 mm H_2O) or greater static pressure. Such *appliance* is equipped with provisions for attaching an outlet air duct and, where the *appliance* is for indoor installation remote from the space to be heated, is also equipped with provisions for attaching an inlet air duct.

Low-static pressure type. A self-contained, automatically controlled, vented *appliance*, intended for installation in the space to be heated without the use of ducts, having integral means for circulation of air. Such units are allowed to be equipped with louvers or face extensions

made in accordance with the manufacturer's specifications.

❖ Unit heaters are similar to warm-air furnaces, except that ducts are not usually associated with unit heaters. These heaters are typically suspended from ceilings or roof structures.

UNLISTED BOILER. A boiler not *listed* by a nationally recognized testing agency.

❖ Boilers in new installations must be listed, and gas equipment installed in existing unlisted boilers must be installed according to the manufacturers' instructions and the IMC (see the commentary to "Boiler, low-pressure").

UNVENTED ROOM HEATER. An unvented heating *appliance* designed for stationary installation and utilized to provide comfort heating. Such appliances provide radiant heat or convection heat by gravity or fan circulation directly from the heater and do not utilize ducts.

❖ As opposed to vented heaters, unvented heaters discharge all products of combustion into the space being heated [see Commentary Figure 202(18)].

VALVE. A device used in *piping* to control the gas supply to any section of a system of *piping* or to an *appliance*.

❖ The code requires valves to perform various functions in a fuel gas system as described in the definitions that follow. These valves are designed for the specific use intended and many are factory installed on equipment and are part of the listed equipment. However, there are valves that are not listed or manufactured in accordance with the standards listed in the code and if allowed must be evaluated in accordance with Section 105.2.

Figure 202(19)
TRANSITION FITTINGS
(Photo courtesy of Perfection Corporation)

Appliance shutoff. A valve located in the *piping* system, used to isolate individual appliances for purposes such as service or replacement.

❖ The definition makes it clear that the primary purpose of an appliance shutoff valve is not related to the emergency shutoff of the appliance.

Automatic. An automatic or semiautomatic device consisting essentially of a valve and operator that control the gas supply to the burner(s) during operation of an *appliance*. The operator shall be actuated by application of gas pressure on a flexible diaphragm, by electrical means, by mechanical means, or by other *approved* means.

Automatic gas shutoff. A valve used in conjunction with an automatic gas shutoff device to shut off the gas supply to a water-heating system. It shall be constructed integrally with the gas shutoff device or shall be a separate assembly.

Individual main burner. A valve that controls the gas supply to an individual main burner.

Main burner control. A valve that controls the gas supply to the main burner manifold.

Manual main gas-control. A manually operated valve in the gas line for the purpose of completely turning on or shutting off the gas supply to the *appliance*, except to pilot or pilots that are provided with independent shutoff.

Manual reset. An automatic shutoff valve installed in the gas supply *piping* and set to shut off when unsafe conditions occur. The device remains closed until manually reopened.

Service shutoff. A valve, installed by the serving gas supplier between the service meter or source of supply and the customer *piping* system, to shut off the entire *piping* system.

VENT. A pipe or other conduit composed of factory-made components, containing a passageway for conveying combustion products and air to the atmosphere, *listed* and *labeled* for use with a specific type or class of *appliance*.

❖ In code terminology, vents are distinguished from chimneys and usually are constructed of factory-made listed and labeled components intended to function as a system. Type B and BW vents are constructed of galvanized steel and aluminum sheet metal and are double wall and air insulated. Such vents are designed to vent gas-fired appliances and equipment that are equipped with draft hoods or are specifically listed (labeled) for use with Type B or BW vents. Type L vents are typically constructed of sheet steel and stainless steel. They are double wall and air insulated and are designed to vent gas- and oil-fired appliances and equipment. Some appliances are designed for use with corrosion-resistant special vents, such as those made of plastic pipe and special alloys of stainless steel [see Commentary Figure 202(20)].

Special gas vent. A vent *listed* and *labeled* for use with *listed* Category II, III and IV appliances.

❖ These vents include high-temperature plastic pipe, stainless steel pipe and low-temperature (PVC and CPVC) plastic pipes.

Type B vent. A vent *listed* and *labeled* for use with appliances with draft hoods and other Category I appliances that are *listed* for use with Type B vents.

❖ See the commentary to "Vent."

Type BW vent. A vent *listed* and *labeled* for use with wall furnaces.

❖ See the commentary to "Vent."

Type L vent. A vent *listed* and *labeled* for use with appliances that are *listed* for use with Type L or Type B vents.

❖ See the commentary to "Vent."

VENT CONNECTOR. See "Connector."

❖ See the commentary to "Connector."

VENT GASES. Products of combustion from appliances plus excess air plus dilution air in the vent connector, gas vent or chimney above the draft hood or draft regulator.

❖ It is critical that the materials used in venting systems be compatible with the vent gases being generated by the connecting appliances, equipment and venting system. The most important area of concern is the generation of condensate, which will be acidic and will corrode most metal venting systems. As newer equipment is produced and designed for higher effi-

Figure 202(20)
TYPE B VENT
(Photo courtesy of Selkirk, L.L.C.)

ciencies, which will result in lower vent-gas temperatures, the potential for generating condensate in the vent gases increases. Combustion gases that have not exited the appliance are called "flue gases."

VENT PIPING.

Breather. *Piping* run from a pressure-regulating device to the outdoors, designed to provide a reference to atmospheric pressure. If the device incorporates an integral pressure relief mechanism, a breather vent can also serve as a relief vent.

❖ A breather vent is distinct from a relief vent and allows the cavity on the air side of the regulator diaphragm to "breathe" by letting air in and out as the diaphragm moves within the regulator housing (see commentary, Section 410.3). Although not addressed in the definition, breather vents are also part of diaphragm-operated gas controls such as high- and low-pressure gas cutoffs.

Relief. *Piping* run from a pressure-regulating or pressure-limiting device to the outdoors, designed to provide for the safe venting of gas in the event of excessive pressure in the gas *piping* system.

❖ Relief vent piping discharges fuel gas from a pressure relief valve or pressure regulator with an integral relief valve. Unlike breather vents that would discharge fuel gas only in the event of diaphragm failure, relief vents will always discharge fuel gas, if they discharge at all. Relief devices release fuel gas under emergency conditions to prevent over pressurizing the piping system being monitored and protected.

VENTED APPLIANCE CATEGORIES. Appliances that are categorized for the purpose of vent selection are classified into the following four categories:

❖ Gas appliances are categorized for the purpose of matching the appliance to the required type of venting. Not all gas appliances are assigned a category.

Category I. An *appliance* that operates with a nonpositive vent static pressure and with a vent gas temperature that avoids excessive condensate production in the vent.

❖ These appliances include draft-hood-equipped and fan-assisted types.

Category II. An *appliance* that operates with a nonpositive vent static pressure and with a vent gas temperature that is capable of causing excessive condensate production in the vent.

❖ The condition of nonpositive pressure (draft dependent) combined with low flue-gas temperatures makes it difficult to design such an appliance. At the time of this publication, no such appliances are known to exist in the marketplace.

Category III. An *appliance* that operates with a positive vent static pressure and with a vent gas temperature that avoids excessive condensate production in the vent.

❖ These appliances include mid-efficiency (approximately 78 to 83 percent) appliances that are typically sidewall-vented with stainless steel special vents.

Category IV. An *appliance* that operates with a positive vent static pressure and with a vent gas temperature that is capable of causing excessive condensate production in the vent.

❖ These appliances include high-efficiency (84 percent and higher) condensing-type furnaces, boilers and water heaters.

VENTED ROOM HEATER. A vented self-contained, free-standing, nonrecessed *appliance* for furnishing warm air to the space in which it is installed, directly from the heater without duct connections.

❖ These appliances are typically small space heaters that discharge all products of combustion to the outdoors.

VENTED WALL FURNACE. A self-contained vented *appliance* complete with grilles or equivalent, designed for incorporation in or permanent attachment to the structure of a building, mobile home or travel trailer, and furnishing heated air circulated by gravity or by a fan directly into the space to be heated through openings in the casing. This definition shall exclude floor furnaces, unit heaters and central furnaces as herein defined.

❖ Wall furnaces are designed to occupy very little area of the room and provide space heating for one or more rooms in small occupancies, such as dwelling units, cottages and the like.

VENTING SYSTEM. A continuous open passageway from the flue collar or draft hood of an *appliance* to the outdoor atmosphere for the purpose of removing flue or vent gases. A venting system is usually composed of a vent or a chimney and vent connector, if used, assembled to form the open passageway.

❖ Venting systems fall into one of two categories: natural draft or mechanical draft systems. Natural draft chimneys and vents do not rely on any mechanical means to convey combustion products to the outdoors. Draft is produced by the temperature difference between the vent gases (combustion gases) and the ambient atmosphere. Hot gases are less dense and more buoyant; therefore, they are displaced by cooler (more dense) ambient gases, causing the hotter gases to rise and produce a draft.

Mechanical draft systems use fans or other mechanical means to cause the removal of flue or vent gases and also fall into one of two categories, based on positive or nonpositive static vent pressure. The positive vent pressure systems are referred to as forced-draft or power venting systems, and the nonpositive vent pressure systems are referred to as induced-draft venting systems.

Forced-draft venting system. A portion of a venting system using a fan or other mechanical means to cause the removal of flue or vent gases under positive static vent pressure.

❖ Power exhausters, including power draft venting systems, and some power burner systems are examples of forced-draft systems. Vents and chimneys must be listed for positive-pressure applications where used with forced-draft systems.

Induced draft venting system. A portion of a venting system using a fan or other mechanical means to cause the removal of flue or vent gases under nonpositive static vent pressure.

❖ Induced-draft venting is commonly accomplished with field-installed inducer fans designed to supplement natural draft chimneys or vents.

Mechanical draft venting system. A venting system designed to remove flue or vent gases by mechanical means, that consists of an induced draft portion under nonpositive static pressure or a forced draft portion under positive static pressure.

❖ These systems do not depend on draft, but rather, use fans or blowers.

Natural draft venting system. A venting system designed to remove flue or vent gases under nonpositive static vent pressure entirely by natural draft.

❖ Natural draft chimneys and vents do not rely on any mechanical means to convey combustion products to the outdoors. Draft is produced by the temperature difference between the combustion gases (flue gases) and the ambient atmosphere. Hot gases are less dense and more buoyant; therefore, they rise as they are displaced by more dense ambient air.

WALL HEATER, UNVENTED-TYPE. A room heater of the type designed for insertion in or attachment to a wall or partition. Such heater does not incorporate concealed venting arrangements in its construction and discharges all products of combustion through the front into the room being heated.

❖ See the definition and commentary to "Unvented room heater."

[M] WATER HEATER. Any heating *appliance* or *equipment* that heats potable water and supplies such water to the potable hot water distribution system.

❖ A water heater is a closed pressure vessel or heat exchanger that has a heat source and that supplies potable (drinkable) water to the building's hot water distribution system. Large commercial water heaters are sometimes referred to as "hot water supply boilers." Water heaters can be of the storage type with an integral storage vessel, circulating type for use with an external storage vessel, tankless instantaneous type without storage capacity and point-of-use type with or without storage capacity. The code and the IPC regulate water heaters because these appliances have elements and installation requirements related to both mechanical and plumbing systems.

Bibliography

The following resource materials were used in the preparation of the commentary for this chapter of the code.

IBC-2015, *International Building Code*. Washington, DC: International Code Council, 2014.

IFC-2015, *International Fire Code*. Washington, DC: International Code Council, 2014.

IMC-2015, *International Mechanical Code*. Washington, DC: International Code Council, 2014.

IPC-2015, *International Plumbing Code*. Washington, DC: International Code Council, 2014.

Chapter 3:
General Regulations

General Comments

A fundamental principle of the code is its dependence on the listing and labeling method of approval for appliances and equipment. Section 301.3 prohibits the installation of unlisted appliances except where approved in accordance with Section 105.

Purpose

Chapter 3 contains requirements for the safe and proper installation of gas-fired equipment and appliances to help ensure protection of life and property.

SECTION 301 (IFGC)
GENERAL

301.1 Scope. This chapter shall govern the approval and installation of all *equipment* and appliances that comprise parts of the installations regulated by this code in accordance with Section 101.2.

❖ This section states that this chapter governs the approval and installation of all gas-fired equipment and appliances that are regulated by the code. Section 101.2 establishes the scope of application of the code (see commentary, Section 101.2).

301.1.1 Other fuels. The requirements for combustion and dilution air for gas-fired appliances shall be governed by Section 304. The requirements for combustion and dilution air for appliances operating with fuels other than fuel gas shall be regulated by the *International Mechanical Code*.

❖ This code and the *International Mechanical Code®* (IMC®) each have a combustion air chapter that is specific to the fuels addressed in the respective code.

301.2 Energy utilization. Heating, ventilating and air-conditioning systems of all structures shall be designed and installed for efficient utilization of energy in accordance with the *International Energy Conservation Code*.

❖ This section states that all appliances and equipment must be designed and installed to use depletable energy sources efficiently. The *International Energy Conservation Code®* (IECC®) is the applicable document for regulating the efficiency and performance of appliances and heating, ventilating and air-conditioning (HVAC) systems. Special applications such as process heating or cooling should be designed for the maximum energy efficiency attainable.

301.3 Listed and labeled. Appliances regulated by this code shall be *listed* and *labeled* for the application in which they are used unless otherwise *approved* in accordance with Section 105. The approval of unlisted appliances in accordance with Section 105 shall be based on *approved* engineering evaluation.

❖ Gas-fired appliances must be listed and labeled by an approved agency to show that they comply with the applicable national standards. The code requires listing and labeling for appliances, such as boilers, furnaces, space heaters, direct-fired heaters, cooking appliances, clothes dryers, rooftop HVAC units, etc. The code also requires listing for system components as specifically stated in the text addressing those components. The label is the primary, if not the only, assurance to the installer, the inspector and the end user that a representative sample of an appliance model has been tested and evaluated by an approved agency and has been determined to perform safely and efficiently when installed and operated in accordance with its listing.

Appliances must be listed and labeled for the application in which they are used, otherwise the installation would be a misapplication of the appliance. For example, if an appliance is listed for indoor use only and is installed outdoors, this installation is a misapplication of the appliance and serious malfunctions and/or conditions could result. An appliance might be marketed and installed for a particular purpose for which it was not tested and listed and this is what this section intends to prohibit. Verifying that an appliance has a testing agency label is only part of the code official's responsibility. He or she must also verify that the listing from the testing agency includes the application at hand. The bottom line is, the use of an appliance must match the use for which the appliance was tested.

The presence of a label is part of the information that the code official considers when approving appliances. The only exception to the labeling requirement occurs when the code official approves a specific appliance in accordance with the authority granted in Section 105.2.

Approval of unlabeled appliances must be based on documentation that demonstrates compliance with applicable standards or, where no product standards exist, that the appliance is appropriate for the intended use and will provide the same level of performance as would be provided by listed and labeled appliances. A fundamental principle of the code is the reliance on the listing and labeling process to ensure appliance performance; approvals granted in accordance with Section 105.2 must be well justified with supporting documentation. To the code official, the installer and the end-user, very little is known about the performance of an appliance that is not tested and built to an appliance standard.

301.4 Labeling. Labeling shall be in accordance with the procedures set forth in Sections 301.4.1 through 301.4.2.3.

❖ This section establishes the requirements for testing and labeling appliances by an approved agency. Included within this section are the requirements for testing the product and for approval of the testing agency, the testing equipment and the personnel who conduct the test. Also included is the information that must appear on a label.

301.4.1 Testing. An *approved* agency shall test a representative sample of the appliances being *labeled* to the relevant standard or standards. The *approved* agency shall maintain a record of all of the tests performed. The record shall provide sufficient detail to verify compliance with the test standard.

❖ When an approved agency labels an appliance, the agency is ensuring that a representative sample of the appliance has been tested in accordance with an appropriate standard and has been determined to perform acceptably when installed and operated in accordance with the appliance's listing.

The basis for a label is the requirement for testing a representative, perhaps identical, sample of the appliance to indicate conformance to a required standard. This is an important premise in the code because a code official will consider the presence of a label in the approval of an appliance. For this reason, the appliance must meet the requirements of the standard. Because the appliance tested is installed and operated in accordance with the manufacturer's instructions, these instructions must provide for proper installation and operation. This is important because the code requires that the labeled appliance be installed in accordance with the manufacturer's instructions, and operating instructions must be either attached to or shipped with each appliance.

There are numerous standards, not all of which are specifically referenced in the code, applicable to various appliances and equipment. For this reason, the approved agency determines the applicable standards to be used for testing and then, in turn, as the basis for labeling. Each standard contains safety requirements for a given appliance or piece of equipment and specifies tests that must be performed. The labeling agency is required to maintain sufficient documentation to demonstrate compliance with the test standard. The code official may require that copies of the test reports be submitted to determine the validity of the label.

Examples of standards that are used as a basis for testing and labeling include:

- ANSI Z21.47, *Gas-fired Central Furnaces*;
- ANSI Z83.8, *Gas Unit Heaters*; and

- UL 795, *Commercial—Industrial Gas Heating Equipment*.

301.4.2 Inspection and identification. The *approved* agency shall periodically perform an inspection, which shall be in-plant if necessary, of the appliances to be *labeled*. The inspection shall verify that the *labeled* appliances are representative of the appliances tested.

❖ The approved agency whose identification insignia appears on the label must perform periodic in-plant inspections. The primary objective of these inspections is to determine that the manufactured product is equivalent to the sample that was tested. Because the label is good only for the products that were tested, the in-plant inspections are intended to discover any design changes or production quality control problems. If any discrepancies are found, that the labeling agency would discontinue labeling of that particular product, and the manufacturer would be required to resolve the problem and, if necessary, have the redesigned product retested before the labeling process is resumed.

301.4.2.1 Independent. The agency to be *approved* shall be objective and competent. To confirm its objectivity, the agency shall disclose all possible conflicts of interest.

❖ As a part of the basis for a code official's approval of a particular labeling agency, the agency must demonstrate both its independence from the manufacturer of the product and its competence to perform the required tests. The judgment of objectivity is linked to the financial and fiduciary independence of the agency. The competence of the agency is judged by its experience, organization and the experience of its personnel. As a hypothetical example, the Acme Inspection Agency is performing testing for gas-fired furnaces for the Real Hot Furnace Company. After some investigation, both Acme and Real Hot are found to be the subsidiaries of the same parent company. The inspection agency and the manufacturer clearly have a relationship that presents the potential for conflict of interest, and the objectivity of the inspection agency is sufficiently questionable for the code official to justify not approving Acme as a testing and labeling agency for equipment produced by the Real Hot Furnace Company.

301.4.2.2 Equipment. An *approved* agency shall have adequate *equipment* to perform all required tests. The *equipment* shall be periodically calibrated.

❖ Referring to the example in the commentary for Section 301.4.2.1, if the Acme Inspection Agency had only the facilities to test and label fire doors, the agency would not be qualified to test and label a gas-fired furnace. Although this example is oversimplified, the point is that the inspection agency must have all of the necessary equipment to perform the testing required by the applicable standard.

In addition to having the proper equipment, the agency must maintain records of the maintenance and calibration of their equipment to demonstrate that the equipment can be relied on to produce accurate, consistent and reproducible results. Testing apparatus, instruments and equipment must often be capable of measurements using very small units of measure within a specified tolerance. To produce accurate, dependable readings and reliable test results, testing apparatus, equipment and instruments must be routinely calibrated to a fixed reference. Having the proper testing equipment can be just as important as the competence of the testing personnel.

301.4.2.3 Personnel. An *approved* agency shall employ experienced personnel educated in conducting, supervising and evaluating tests.

❖ The competence of an inspection agency is based on the agency having the proper equipment to perform the test, as stated in Section 301.4.2.2, and also on the experience and abilities of its personnel. The best calibrated equipment can produce accurate results only when operated by experienced personnel who are trained to conduct, supervise and evaluate tests. For example, consider a newly formed agency that has employed individuals who do not have experience related to the testing to be conducted and have not been adequately trained. The capabilities and experience of supervisory personnel overseeing their work is also important.

301.5 Label information. A permanent factory-applied nameplate(s) shall be affixed to appliances on which shall appear in legible lettering, the manufacturer's name or trademark, the model number, serial number and, for *listed* appliances, the seal or mark of the testing agency. A label shall include the hourly rating in British thermal units per hour (Btu/h) (W); the type of fuel *approved* for use with the *appliance*; and the minimum *clearance* requirements.

❖ This section requires that the label be a metal plate, tag or other permanent label. In general, label materials other than metal tags or plates usually consist of material that is similar in appearance to a decal, and the label, its adhesive and the printed information must be durable and water resistant. Because of the important information given by a label, the label is intended to be permanent, not susceptible to damage and legible for the life of the appliance to which it is attached. The standards that appliances are tested to usually specify the required label performance criteria, the method of attachment and the required label information. The code requires that the label be affixed permanently and prominently on the appliance or equipment and specifies the information that must appear on the label. The manufacturer may be required by the relevant standard or may voluntarily provide additional information on the label. Commentary Figures 301.5(1) and 301.5(2) show typical appliance labels.

301.6 Plumbing connections. Potable water supply and building drainage system connections to appliances regulated

by this code shall be in accordance with the *International Plumbing Code*.

❖ Plumbing connections to appliances and equipment regulated by the code must be in accordance with the *International Plumbing Code®* (IPC®).

Section 624.2 of the code requires that combination domestic water heating and hydronic supply water heating units be listed and installed according to their listing and manufacturer's instructions.

Hydronic systems normally require a means of supplying fill and makeup water to replace any water lost to evaporation, leakage or intentional draining. Where direct connections are made to the potable water supply, the connections must be isolated from the potable water source. This requirement is intended to protect the potable water system from contamination by backflow when a direct connection is made to a hydronic system.

Hydronic systems are normally pressurized, contain nonpotable water and fluids and can contain conditioning chemicals or antifreeze solutions. Low-temperature hydronic fluids and cooling towers have also been associated with disease-causing organisms such as the Legionnaires' disease bacterium. The potable water system must be protected from potential contamination resulting from connection to hydronic systems, water-wash filter systems, cooling towers, solar systems, water-cooled heat exchangers, cooking appliances, ice makers, humidifiers, evaporative coolers, etc.

In addition, water heaters are part of the potable water distribution system and, therefore, must comply with both the code and the IPC. A water heater installation is complex in that it has a fuel or power supply; a chimney or vent connection, if fuel-fired; a combustion air supply, if fuel-fired; connections to the plumbing potable water distribution system and controls and devices to prevent a multitude of potential hazards from conditions such as excessively high temperatures, pressures and ignition failure.

It is not uncommon for jurisdictions to issue both plumbing and mechanical permits for water heater installations or to require that the installer be licensed in both the plumbing and mechanical trades when performing such installations (see commentary, Section 624). Water heaters are clearly under the purview of both plumbing and fuel gas codes. This section also refers to the IPC for the drainage associated with mechanical appliances and equipment, such as those addressed in Section 307.

301.7 Fuel types. Appliances shall be designed for use with the type of fuel gas that will be supplied to them.

❖ Appliances are usually designed by the manufacturer to operate on one specifically designated type of fuel. An element of information used for the approval of appliances is the label, which ensures that the appliance has been tested in accordance with a valid stan-

American Standard Inc.
The Trane Company
Trenton, N.J. 08619   Made In U.S.A.

FORCED AIR FURNACE   CATEGORY I

ANS Z21.47 - 1990 CENTRAL FURN
FOR INDOOR INSTALLATION IN A BUILDING
CONSTRUCTED ON SITE.   **NRTL**

MODEL NO.	SERIAL NO.	EQUIPPED FOR
TUD080R936A1	G36519785	NAT. GAS
INPUT	**LIMIT SETTING**	
80,000 BTU/HR.	190 °F	MFRD. 09/92
TEMP. RISE °F	**MAX. EXT. STATIC PRESS**	**MAX. DESIGN AIR**
FROM 30 TO 60	.50 INCHES WATER	TEMP. 160 °F
VOLTS/PHASE/HERTZ	**TOTAL AMPS**	**SERVICE CODE**
115/1/60	8.5	1

MANIFOLD PRESSURE
(IN INCHES OF WATER)
NAT. 3.5 L P 10.5
SUPPLY PRESSURE
(IN INCHES OF WATER)
MAX. NAT. 10.5, L P 13.0

MIN. NAT. 4.5L P 11.0 FOR PURPOSE OF INPUT ADJUSTMENT.

FLAME ROLLOUT SWITCH - REPLACE
IF BLOWN WITH CATALOG NO.
WG09X033 (333 F CUTOFF TEMP.)
ONE TIME THERMAL FUSE.

LOW INPUT 52,000 BTU/HR

MINIMUM CLEARANCE COMBUSTIBLE MATERIALS:

FOR CLOSET   INSTALLATION AS FOLLOWS:
SIDES 0   IN. W/SINGLE WALL VENT
FLUE   6   IN. W/SINGLE WALL VENT   1 IN. W/TYPE B-1 VENT
FRONT 6   IN.   BACK 0 IN.   TOP 1 IN.

UPFLOW UNITS. FOR INSTALLATION COMBUSTIBLE FLOORING.   21D340159 P01

Figure 301.5(1)
TYPCIAL LABEL FOR A CATEGORY I GAS-FIRED FURNACE
(Figure courtesy of the Trane Company, an American Standard Company)

dard and determined to perform acceptably when installed and operated in accordance with the appliance listing (see commentary, Section 301.5). The fuel used in the appliance test must be the type of fuel specified by the manufacturer.

301.7.1 Appliance fuel conversion. Appliances shall not be converted to utilize a different fuel gas except where complete instructions for such conversion are provided in the installation instructions, by the serving gas supplier or by the *appliance* manufacturer.

❖ A mismatch between an appliance and a fuel will create an extremely hazardous condition that can be avoided by examining the appliance label. When an appliance is converted to a different type of fuel, the original label that appears on the appliance is no longer valid. Because the original approval of the appliance is based in part on the label, the appliance is no longer approved for use. Fuel conversion kits will include a replacement label.

Many gas appliances are listed and labeled for more than one fuel, most commonly natural gas and liquefied petroleum (LP) (propane). Such appliances can be converted from one fuel to the other using a manufacturer's conversion kit and following instructions provided by the manufacturer. These conversions must be done only with the approval of the code official to make sure that the instructions are complied with and the safety of the original installation is maintained. Fuel conversions that are not performed correctly can adversely affect the performance of burners, the venting of combustion gases and the proper clearance to combustibles.

Once a conversion has been completed, a supplemental label must be installed to update the information contained on the original label, thereby alerting any service personnel to the modifications that have been made.

Fuel-fired appliances are designed to operate with a maximum and minimum heat energy input capacity.

Figure 301.5(2)
ROOFTOP HVAC UNIT LABEL

In accordance with the appliance manufacturer's instructions, this capacity might need to be field adjusted to suit the elevation because of the change in air density at different elevations. Alteration of heat energy input beyond the allowable limits can result in hazardous overfiring or underfiring. Either condition can cause operation problems that include overheating, vent failure, corrosion, poor draft and poor combustion. Input adjustments for different elevations are dictated by the appliance manufacturer and are not addressed in the code, other than by Section 305.1.

301.8 Vibration isolation. Where means for isolation of vibration of an *appliance* is installed, an *approved* means for support and restraint of that *appliance* shall be provided.

❖ Where vibration isolation connections are used in ducts and piping and where equipment is mounted with vibration dampers, support is required for the ducts, piping and equipment to maintain positioning and alignment and to prevent stress and strain on the vibration connectors and dampers.

301.9 Repair. Defective material or parts shall be replaced or repaired in such a manner so as to preserve the original approval or listing.

❖ Repair work must not alter the nature of appliances and equipment in a way that would invalidate the listing or conditions of approval. For example, replacement of safety control devices with different devices could alter the design and operation of an appliance from that intended by the manufacturer and the listing agency.

301.10 Wind resistance. Appliances and supports that are exposed to wind shall be designed and installed to resist the wind pressures determined in accordance with the *International Building Code*.

❖ Installations of equipment and appliances that are subject to wind forces must be designed to resist those forces. The wind pressures must be based on the wind provisions in the code. The wind pressure requirements are based on the exposure of the building and wind speeds for that region.

[BS] 301.11 Flood hazard. For structures located in flood hazard areas, the appliance, equipment and system installations regulated by this code shall be located at or above the elevation required by Section 1612 of the *International Building Code* for utilities and attendant equipment.

> **Exception:** The appliance, equipment and system installations regulated by this code are permitted to be located below the elevation required by Section 1612 of the *International Building Code* for utilities and attendant equipment provided that they are designed and installed to prevent water from entering or accumulating within the components and to resist hydrostatic and hydrodynamic loads and stresses, including the effects of buoyancy, during the occurrence of flooding to such elevation.

❖ In flood hazard areas, mechanical systems, equipment and appliances must be elevated above the elevation specified in the *International Building Code®* (IBC®). Exposure to water can damage most mechanical system components as well as cause serious appliance and equipment malfunctions. For example, the majority of appliance manufacturers require the replacement of safety controls or entire appliances that have been submerged in flood waters. See FEMA 348 for additional guidance.

The exception to this section provides criteria for placing specific equipment below the required elevation. To do so, the equipment must be designed to prevent the entry or accumulation of water. Standard equipment that typically is installed at-grade is not designed to withstand the entry of water and would not meet the requirements of this exception.

301.12 Seismic resistance. When earthquake loads are applicable in accordance with the *International Building Code*, the supports shall be designed and installed for the seismic forces in accordance with that code.

❖ This code requires building systems to be designed for specified seismic forces. This section references the detailed equipment and component seismic support requirements contained in the code to bring these requirements to the attention of the design professional and the permit applicant.

Equipment piping must be braced for earthquake loads as stated in the code. The failure of the supports for these components has been shown to be a threat to health and safety in geographical areas where moderate- to high-magnitude earthquakes occur. The code specifies the geographical locations where earthquake design is required for certain piping and equipment and the size of the components that must be braced.

301.13 Ducts. Ducts required for the installation of systems regulated by this code shall be designed and installed in accordance with the *International Mechanical Code*.

❖ Chapter 6 of the IMC governs duct systems used for the movement of environmental air, governing the construction, installation, alteration, maintenance and repair of these systems (see the definition of "Duct system" in Chapter 2).

301.14 Rodentproofing. Buildings or structures and the walls enclosing habitable or occupiable rooms and spaces in which persons live, sleep or work, or in which feed, food or foodstuffs are stored, prepared, processed, served or sold, shall be constructed to protect against rodents in accordance with the *International Building Code*.

❖ This section states the requirements to prevent rodent infestation of a building. Efforts must be made to protect the annular spaces around openings in exterior walls. The annular spaces can be sealed by any effective method, but the primary methods used are to fill the annular space with a sealant material or to place a metal collar around the penetrating pipe, duct, etc. The collar material must be durable for the weather exposure and strong enough to prevent the rodents from chewing through and entering the building.

Effort must also be made to cover ventilation and combustion air openings with wire mesh material to prevent the entry of rodents through the opening.

301.15 Prohibited location. The appliances, *equipment* and systems regulated by this code shall not be located in an elevator shaft.

❖ The code views an elevator shaft as an unnecessarily risky location for fuel-gas system components. There is a potential for leaks in fuel-gas system equipment and piping, which would release fuel gas into the shaft, creating a hazardous condition. Because the elevator is often relied on for fire-fighting access as well as egress for those with disabilities, the shaft must be maintained free from contaminants and potential hazards that could render the elevator inoperative or unsafe for the occupants. The location of fuel-gas-fired equipment in an elevator shaft could also create access problems for service of the gas-fired equipment and the elevator equipment.

SECTION 302 (IFGC)
STRUCTURAL SAFETY

[BS] 302.1 Structural safety. The building shall not be weakened by the installation of any gas *piping*. In the process of installing or repairing any gas *piping*, the finished floors, walls, ceilings, tile work or any other part of the building or premises which is required to be changed or replaced shall be left in a safe structural condition in accordance with the requirements of the *International Building Code*.

❖ The installation of fuel gas systems must not adversely affect the structural integrity of the building components. The code dictates the structural safety requirements that must be applied to any structural portion of the building that is penetrated, altered or removed during the installation, replacement or repair of fuel gas systems.

[BS] 302.2 Penetrations of floor/ceiling assemblies and fire-resistance-rated assemblies. Penetrations of floor/ceiling assemblies and assemblies required to have a fire-resistance rating shall be protected in accordance with the *International Building Code*.

❖ Penetrations of fire-resistance-rated assemblies can diminish the integrity of the assembly, allowing smoke and fire to pass through and cause it to fail prematurely.

[BS] 302.3 Cutting, notching and boring in wood members. The cutting, notching and boring of wood members shall comply with Sections 302.3.1 through 302.3.4.

❖ The sections referenced provide prescriptive sizes and locations of acceptable cuts, notches and bored holes in wood framing members. Ideally, framing members should not be altered in any way, but because this is not always practical, the code permits alterations that do not significantly weaken the members.

[BS] 302.3.1 Engineered wood products. Cuts, notches and holes bored in trusses, structural composite lumber, structural glued-laminated members and I-joists are prohibited except where permitted by the manufacturer's recommendations or where the effects of such alterations are specifically considered in the design of the member by a registered design professional.

❖ This section applies to engineered lumber products that are manufactured structural members. These products often have installation instructions that specifically state if, where and how the member can be altered, drilled, notched, etc. What appears to be a harmless cut, notch or boring might actually be structurally damaging to the member (see Section 302.4).

[BS] 302.3.2 Joist notching and boring. Notching at the ends of joists shall not exceed one-fourth the joist depth. Holes bored in joists shall not be within 2 inches (51 mm) of the top and bottom of the joist and their diameters shall not exceed one-third the depth of the member. Notches in the top or bottom of the joist shall not exceed one-sixth the depth and shall not be located in the middle one-third of the span.

❖ The code recognizes that at times floor and ceiling joists must be cut, notched or bored. The provisions are based on location and size limitations to ensure that the structural member can support the load.

A notch or cut will decrease the allowable stress of the member. A notch or cut might place a lumber defect that was previously in a low-stress location in the interior of the board into a high-stress area at the edge of the notch or cut. Cutting and notching also cause shear stress concentrations in the member at the corners of the notch or cut. Because of these shear stress concentrations, the full extent of the damage to the member cannot be calculated. In consideration of the detrimental effects of cutting and notching, it is a good practice to avoid cutting, notching or boring adjacent to areas of the member that contain knots and, where possible, to use rounded or sloped edged notches to reduce the stress concentrations at the corners [see Commentary Figure 302.3.2(1)].

Structural elements of engineered wood systems, such as metal-plate-connected parallel-chord wood trusses, premanufactured I-joists and metal-plate-connected wood roof trusses cannot be cut or notched unless structural analysis is provided by a qualified design professional.

Notches are not to exceed those permitted in this section unless a structural analysis is completed by a qualified design professional.

Holes cut and bored in joists and studs have the same effect as notches. Holes reduce the strength of the member. This section permits holes to be drilled in wood structural members within certain limitations. The limitations given in this section are not to be exceeded unless a structural analysis is submitted by a qualified design professional [see Commentary Figure 302.3.2(2)].

[BS] 302.3.3 Stud cutting and notching. In exterior walls and bearing partitions, any wood stud is permitted to be cut or notched to a depth not exceeding 25 percent of its width. Cutting or notching of studs to a depth not greater than 40 percent of the width of the stud is permitted in nonload-bearing partitions supporting no loads other than the weight of the partition.

❖ The load-sharing characteristics of a wall with studs, sheathing and plates allow for the stated limitation for notches in load-bearing studs. The maximum notch in a stud is not to exceed 25 percent of its depth unless the stud is reinforced to resist the anticipated load. See the commentary to Section 302.3.2 for further discussion of the effects of notching and cutting (see Commentary Figure 302.3.3).

[BS] 302.3.4 Bored holes. The diameter of bored holes in wood studs shall not exceed 40 percent of the stud depth. The diameter of bored holes in wood studs shall not exceed 60 percent of the stud depth in nonbearing partitions. The diameter of bored holes in wood studs shall not exceed 60 percent of the stud depth in any wall where each stud is doubled, provided that not more than two such successive doubled studs are so bored. The edge of the bored hole shall be not closer than $^5/_8$ inch (15.9 mm) to the edge of the stud. Bored holes shall not be located at the same section of stud as a cut or notch.

❖ This section specifies the limits of bored holes in wood studs. The 40-percent limit is intended for studs in exterior walls and for load-bearing walls. The 60-percent limit for doubled studs is also intended to apply to studs in exterior walls and other load-bearing walls. Bored holes not greater than 60 percent are allowed in any nonload-bearing stud.

The bored-hole edge location limitations maintain the load-carrying capacity of the stud. These limits are intended to apply to load-bearing and nonload-bearing partitions and studs in exterior walls. A bored hole in the same cross section as a cut or notch could cause the stud to fail (see Commentary Figure 302.3.3).

For SI: 1 inch = 25.4 mm.

Figure 302.3.2(1)
CUTTING, NOTCHING AND BORING IN JOIST

For SI: 1 inch = 25.4 mm.

Figure 302.3.2(2)
BORING JOISTS

[BS] 302.4 Alterations to trusses. Truss members and components shall not be cut, drilled, notched, spliced or otherwise altered in any way without the written concurrence and approval of a registered design professional. Alterations resulting in the addition of loads to any member, such as HVAC *equipment* and water heaters, shall not be permitted without verification that the truss is capable of supporting such additional loading.

❖ A truss is an engineered system of components that function as a structural unit. Trusses are susceptible to weakening by almost any type of alteration, such as boring, notching and cutting. Unlike dimensional solid lumber members, such as studs, rafters and joists, trusses cannot be altered in a code-prescribed way. The only exception is obtaining written approval of a specific alteration from a registered design professional.

The placement of a water heater in an attic, for example, can result in abnormal loads on roof trusses that must be accounted for structurally. A simple 40-gallon water heater weighs well over 400 pounds and

is a point load. Subjecting trusses to loads that they were not designed to carry is considered to be an alteration and is prohibited without a structural analysis to verify that the truss can bear the additional weight.

[BS] 302.5 Cutting, notching and boring holes in structural steel framing. The cutting, notching and boring of holes in structural steel framing members shall be as prescribed by the registered design professional.

❖ The sizes and locations of acceptable cuts, notches and bored holes in steel framing members must be specified by the designer of the structure; that is, the architect or structural engineer.

[BS] 302.6 Cutting, notching and boring holes in cold-formed steel framing. Flanges and lips of load-bearing, cold-formed steel framing members shall not be cut or notched. Holes in webs of load-bearing, cold-formed steel framing members shall be permitted along the centerline of the web of the framing member and shall not exceed the dimensional limitations, penetration spacing or minimum

TOP PLATES

STUD

BORED HOLE MAX.
DIAMETER 40%
OF STUD DEPTH

⁵/₈ IN. MIN. TO EDGE

⁵/₈ IN. MIN. TO EDGE

NOTCH MUST NOT EXCEED 25%
OF STUD DEPTH

BORED HOLES MUST NOT BE
LOCATED IN THE SAME SECTION
OF STUD AS A CUT OR NOTCH

IF HOLE IS BETWEEN 40% AND 60%
OF STUD DEPTH, THEN STUD MUST BE
DOUBLE AND NO MORE THAN TWO
SUCCESSIVE STUDS ARE PERMITTED TO
BE DOUBLED AND SO BORED

For SI: 1 inch = 25.4 mm.

Figure 302.3.3
BORING AND NOTCHING OF STUDS

hole edge distance as prescribed by the registered design professional. Cutting, notching and boring holes of steel floor/roof decking shall be as prescribed by the registered design professional.

❖ This section does not allow any structural steel framing member to be cut, notched or bored without approval by a registered design professional. Unlike Section 302.3, which allows field cutting, notching and boring of wood studs and joists within the prescribed limits, this section requires that alterations to steel structural framing members be allowed only as dictated by a registered design professional prior to the alteration. The code official should not approve any alterations to structural members unless a signed drawing or revision is on the job site showing the allowed size and location of the cut, notch or bored hole.

[BS] 302.7 Cutting, notching and boring holes in non-structural cold-formed steel wall framing. Flanges and lips of nonstructural cold-formed steel wall studs shall be permitted along the centerline of the web of the framing member, shall not exceed $1^1/_2$ inches (38 mm) in width or 4 inches (102 mm) in length, and the holes shall not be spaced less than 24 inches (610 mm) center to center from another hole or less than 10 inches (254 mm) from the bearing end.

❖ This section does not allow any cutting or notching of the flanges and lips of any cold-formed steel framing member. However, holes may be made in the web of nonstructural cold-formed steel wall studs without the involvement of a registered design professional if the prescribed size, location and spacing requirements are met.

SECTION 303 (IFGC) APPLIANCE LOCATION

303.1 General. Appliances shall be located as required by this section, specific requirements elsewhere in this code and the conditions of the *equipment* and *appliance* listing.

❖ Section 303 is a consolidation of the code's generally applicable location requirements and limitations. The listing for an appliance or equipment will often contain location requirements parallel with or in addition to these sections.

303.2 Hazardous locations. Appliances shall not be located in a *hazardous location* unless *listed* and *approved* for the specific installation.

❖ Gas-fired appliances that are to be installed in a hazardous location must be approved and listed for use in that location. A hazardous location is defined as any location considered to be a potential fire hazard from flammable vapors, dust, combustible fibers or other highly combustible substances. This location is not necessarily classified in the IBC or NFPA 70 as a high-hazard occupancy classification. Appliances located in a hazardous location require special design, construction and installation because of the potential for an explosion or fire caused by the presence of dust, flammable vapors and highly combustible materials.

This section does not specify the type of appliance to be used in any particular hazardous location; therefore, the code official must evaluate information submitted by the designer on the intended occupancy of a space, the specification for the appliance and the location of the appliance in the space to determine the requirements. For example, these requirements may include spark-resistant materials for moving parts, such as fan blades; explosionproof construction for electrical switches, starters and motors; or sealed combustion chambers for fuel-burning appliances (see commentary, Sections 305.3 and 305.6).

Examples of appliances installed in a hazardous location are fuel-fired appliances in an industrial baking facility where fine flour dust is present and fuel-fired appliances in a paint shop where paint fumes and vapors are present.

303.3 Prohibited locations. Appliances shall not be located in sleeping rooms, bathrooms, toilet rooms, storage closets or surgical rooms, or in a space that opens only into such rooms

or spaces, except where the installation complies with one of the following:

1. The *appliance* is a direct-vent *appliance* installed in accordance with the conditions of the listing and the manufacturer's instructions.

2. Vented room heaters, wall furnaces, vented decorative appliances, vented gas fireplaces, vented gas fireplace heaters and decorative appliances for installation in vented solid fuel-burning fireplaces are installed in rooms that meet the required volume criteria of Section 304.5.

3. A single wall-mounted unvented room heater is installed in a bathroom and such unvented room heater is equipped as specified in Section 621.6 and has an input rating not greater than 6,000 Btu/h (1.76 kW). The bathroom shall meet the required volume criteria of Section 304.5.

4. A single wall-mounted unvented room heater is installed in a bedroom and such unvented room heater is equipped as specified in Section 621.6 and has an input rating not greater than 10,000 Btu/h (2.93 kW). The bedroom shall meet the required volume criteria of Section 304.5.

5. The *appliance* is installed in a room or space that opens only into a bedroom or bathroom, and such room or space is used for no other purpose and is provided with a solid weather-stripped door equipped with an *approved* self-closing device. All *combustion air* shall be taken directly from the outdoors in accordance with Section 304.6.

❖ The intent of this section is to prevent fuel-fired appliances from being installed in rooms and spaces where the combustion process could pose a threat to the occupants. Potential threats include depleted oxygen levels; elevated levels of carbon dioxide, nitrous oxides, carbon monoxide, and other combustion gases; ignition of combustibles and elevated levels of flammable gases.

In small rooms, such as bedrooms and bathrooms, the doors are typically closed when the room is occupied, which could allow combustion gases to build up to life-threatening levels. In bedrooms, sleeping occupants would not be alert to or aware of impending danger.

If an appliance obtains combustion air from a room or space, it communicates with the atmosphere in that room or space whether or not it is installed in that room or space. An appliance might be in a room, closet or alcove and obtain combustion air from an adjacent room, so Section 303.3 is worded to address the location of an appliance in the rooms listed and in spaces that open only into such rooms. In other words, an appliance in a closet accessed from a bedroom is no different from an appliance located within the bedroom. It is not the intent of this section to prevent combustion air from being taken from a bedroom, bathroom, etc., as evidenced in

Items 2, 3 and 4. For example, the volume of a bedroom could be added to the volume of other rooms for the purpose of providing indoor combustion air for an appliance not installed in a location prohibited by this section if openings are installed to conjoin the space volumes in accordance with Section 304.5.3. If an appliance obtains combustion air from a room, the appliance combustion chamber would be open to the room and the appliance must be considered to be in that room.

Item 1 recognizes that direct-vent appliances have sealed combustion chambers and obtain all combustion air directly from the outdoors. The appliance combustion chambers do not communicate with the room atmosphere.

Item 2 requires that the room be able to supply the necessary combustion air by infiltration as specified in Section 304.5.3. Note that Item 2 includes both vented decorative appliances and vented gas fireplaces which are both addressed by ANSI Z21.50 and are essentially the same appliance, known by different names. Vented gas fireplace heaters (ANSI Z21.88) were also added to Item 2.

Items 3 and 4 allow the installation of a single wall-mounted unvented room heater in bathrooms and bedrooms if the heaters are equipped with oxygen depletion safety shutoff systems; are limited in Btu input rating; and the space is capable of supplying indoor combustion air in accordance with Section 304.5. These exceptions specify "wall-mounted" heaters, which are fully enclosed and less susceptible to tampering and other conditions that might affect the combustion process. Items 3 and 4 would not apply to room heaters that stand on the floor or fasten to a fireplace hearth or ventless firebox hearth [see Commentary Figure 303.3(1)].

Item 5 would allow installation of fuel-fired appliances within a separate dedicated space that is accessed from the rooms and spaces listed in this section. A separated space containing the appliance must be open to the outdoors in accordance with Section 304.6, and the access door to the space must be solid and weather-stripped to prevent communication between atmospheres in the separated spaces. The door must also be self-closing and not rely on occupants to keep it closed. The enclosure must not be used for storage or any other purpose. The intent is to isolate the appliance(s) from the rooms listed in this section and to obtain all combustion air directly from the outdoors [see Commentary Figure 303.3(2)]. This item can be used to avoid relocating an appliance when an existing appliance installed in a prohibited location needs to be replaced [see Commentary Figure 303.3(2)].

303.4 Protection from vehicle impact damage. Appliances shall not be installed in a location subject to vehicle impact damage except where protected by an *approved* means.

❖ Appliances must be installed with consideration for potential damage. Damage can result in property loss and can cause hazardous malfunctions, fire or explo-

sions. Potential damage includes impact from vehicles such as fork-lifts, farm machinery and passenger cars and trucks. For example, an appliance installed in a garage might be subject to vehicle impact. Means of protection of appliances must be approved and effective and if barriers are used, they must be strong enough to resist the impact anticipated. Locating appliances out of the vehicle's normal travel path can be a simple and effective means of protection (see Commentary Figure 303.4).

303.5 Indoor locations. Furnaces and boilers installed in closets and alcoves shall be *listed* for such installation.

❖ Fuel-fired appliances installed in spaces such as alcoves and closets can pose a potential fire hazard because of the heat emanating from the appliance and the inadequate ventilation afforded by the boundaries of small enclosures.

For appliances installed in closets or alcoves, the clearances stated in the listing (label) are necessary for ventilation cooling of the appliance and must not be reduced as provided for in Section 308.

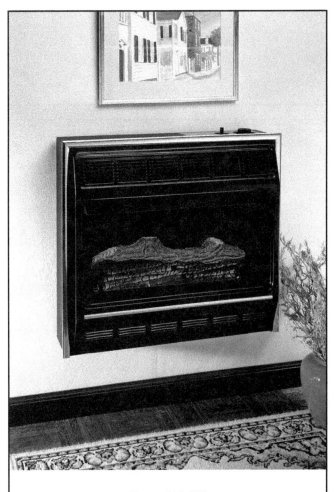

Figure 303.3(1)
UNVENTED WALL-MOUNTED ROOM HEATER
(Photo by D.F. Noyes Studios, courtesy of DESA International)

Figure 303.3(2)
APPLIANCE ENCLOSURE

Figure 303.4
APPLIANCE PROTECTED FROM VEHICLE IMPACT

303.6 Outdoor locations. Appliances installed in outdoor locations shall be either *listed* for outdoor installation or provided with protection from outdoor environmental factors that influence the operability, durability and safety of the appliances.

❖ Appliances installed outdoors must be specifically listed for outdoor installation or be protected from outdoor conditions that will affect the "operability, durability and safety of the equipment."

The concern is for weather and ambient temperatures. The manufacturer's instructions and listing must be consulted to determine whether a particular appliance is designed for or can be made suitable for outdoor installation. For example, furnaces cannot be installed outdoors in cold climates regardless of any weatherproof enclosure, unless the heat exchanger, burner assemblies and venting system are designed for exposure to temperatures below normal indoor temperatures. Cold ambient temperatures can cause harmful condensation to occur on heat exchanger surfaces of fuel-fired appliances.

Additionally, there may be local ordinances that govern the outdoor installation of appliances. Before installing an appliance outdoors, consult local zoning regulations, ordinances and subdivision covenants. Many of these regulations strictly limit the location of outdoor mechanical equipment and appliances. Also, for roof installations, the roof structure must be able to support all imposed static and dynamic structural loads. Substantiating data on the structural adequacy of the entire installation must be submitted and approved.

303.7 Pit locations. Appliances installed in pits or excavations shall not come in direct contact with the surrounding soil. The sides of the pit or excavation shall be held back a minimum of 12 inches (305 mm) from the *appliance*. Where the depth exceeds 12 inches (305 mm) below adjoining grade, the walls of the pit or excavation shall be lined with concrete or masonry, such concrete or masonry shall extend a minimum of 4 inches (102 mm) above adjoining grade and shall have sufficient lateral load-bearing capacity to resist collapse. The *appliance* shall be protected from flooding in an *approved* manner.

❖ Where installed in a depression, appliances could be damaged by corrosion or a malfunction could be caused by blockage of an opening by the entry of soil. Because a depression could hold water, protection from flooding is also necessary.

SECTION 304 (IFGS)
COMBUSTION, VENTILATION AND DILUTION AIR

304.1 General. Air for combustion, ventilation and dilution of flue gases for appliances installed in buildings shall be provided by application of one of the methods prescribed in Sections 304.5 through 304.9. Where the requirements of Section 304.5 are not met, outdoor air shall be introduced in accordance with one of the methods prescribed in Sections 304.6 through 304.9. *Direct-vent appliances*, gas appliances of other than natural draft design, vented gas appliances not designated as Category I and appliances equipped with power burners shall be provided with combustion, ventilation and dilution air in accordance with the *appliance* manufacturer's instructions.

Exception: Type 1 clothes dryers that are provided with makeup air in accordance with Section 614.6.

❖ The provisions of Section 304 describe requirements for the combustion air necessary for the complete combustion of fuel gas, dilution of flue gases, and ventilation of gas-fired appliances and the space in which they are installed. An inadequate combustion air supply to gas-fired appliances can compromise safety by causing incomplete combustion, resulting in appliance malfunction and production of excess carbon monoxide.

Complete combustion of fuel gas is essential for the proper operation of gas-fired appliances. If insufficient quantities of oxygen are supplied, the combustion process will be incomplete, creating hazardous byproducts [see Commentary Figures 304.1(1) and 304.1(2)].

Although not implied in the term, combustion air also serves other purposes in addition to supplying oxygen. Combustion air ventilates and cools appliances and the rooms or spaces that enclose them. Combustion air also plays an important role in producing and controlling draft in vents and chimneys.

Despite the fact that an adequate combustion air supply is extremely important, it is one aspect of gas-fired equipment installations that is often overlooked, ignored or compromised. Depending on the appliance type, location and building construction, supplying combustion air can either be easy or can involve complex designs and extraordinary methods. In any case, the importance of a proper combustion air supply cannot be overemphasized.

The methods of supplying combustion air range from simple (inherently more dependable) methods to more complex methods. This section offers five methods for providing combustion air [see Commentary Figure 304.1(3)]:

1. All indoor air.

2. All outdoor air.

3. Combination indoor and outdoor air.

4. Mechanical combustion air supply.

5. Engineered design.

The last sentence of this section clarifies that the section is intended to apply only to natural-draft atmospheric-burner-design appliances; Category I appliances; nondirect-vent appliances; and appliances not equipped with power burners (see Commentary, Section 304.8). For example, the application of Section 304.6 with large power-burner-equipped (nonnatural-draft, nonatmospheric-burner designs) boilers could result in excessively large openings in an outside wall, causing environmental problems within the boiler room [see Commentary Figure 304.1(6)].

Section 304.1 states that the methods of providing combustion air as prescribed in Sections 304.5 through 304.9 are applicable only to certain types of appliances, namely natural-draft appliances with draft hoods and fan-assisted appliances. The other types of appliances, including direct-vent appliances (direct-vent appliances can be natural-draft type but have no draft hoods), power-vented appliances and Category II, III and IV appliances, must be provided with combustion air as dictated in the appliance manufacturer's instructions. Another type of appliance has been added to the types that are subject only to the requirements in the manufacturer's instructions—appliances equipped with power burners. Power burners pull in combustion air under fan power and mix it with the fuel gas before conveying it to the burner. Natural-draft appliances take in combustion air by means of the weak force produced by the natural draft in the venting system only. Clearly there is a significant difference between the mechanical force of power burners with combustion air fans and natural-draft appliances that must rely on the venting sys-

tem draft to cause combustion air to enter the appliance. The methods in Sections 304.5 through 304.9 were based on the physics of natural-draft, fan-assisted and Category I appliances and, therefore, failed to account for the nature of powered systems like power burners. This fact has perplexed designers for many years as they tried to design combustion air openings for large input boilers and similar appliances that are commonly equipped with power burners. When they used the sizing rules in the code for combustion air openings in the exterior wall of a boiler room, for example, the opening size would be so large that it would be impossible to prevent the freezing of pipes and equipment in colder climates. The openings to the outdoors were prohibitively large for such large input power-burner appliances and out of proportion with the need when considering that power burners mechanically draw in combustion air. When natural draft is depended on, larger openings are necessary to reduce the pressure drop across the openings; however, power burners can tolerate much larger pressure drops, thus allowing smaller openings. The manufacturers' instructions for large input power-burner appliances specify the size of combus-

Figure 304.1(1)
THE CHEMISTRY OF IDEAL COMBUSTION
(Figure courtesy of Reznor/Homas & Betts Corporation)

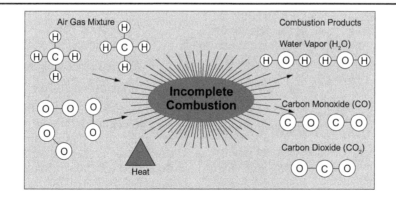

Figure 304.1(2)
THE CHEMISTRY OF INCOMPLETE COMBUSTION
(Figure courtesy of Reznor/Homas & Betts Corporation)

3-14   **2015 INTERNATIONAL FUEL GAS CODE® COMMENTARY** tion air openings by their own methods. Typically those methods result in much smaller openings than prescribed in the code; however, some manufacturers simply refer back to the code instead of offering their own method. A research study demonstrated that the code-prescribed opening sizing methods were overkill for power-burner appliances.

Direct-vent appliances supply their own combustion air through an outdoor air intake pipe or duct [see Commentary Figures 304.1(4) and 304.1(5)].

The exception recognizes that clothes dryers receive the required combustion air from the makeup air that compensates for the exhaust airflow from the appliance.

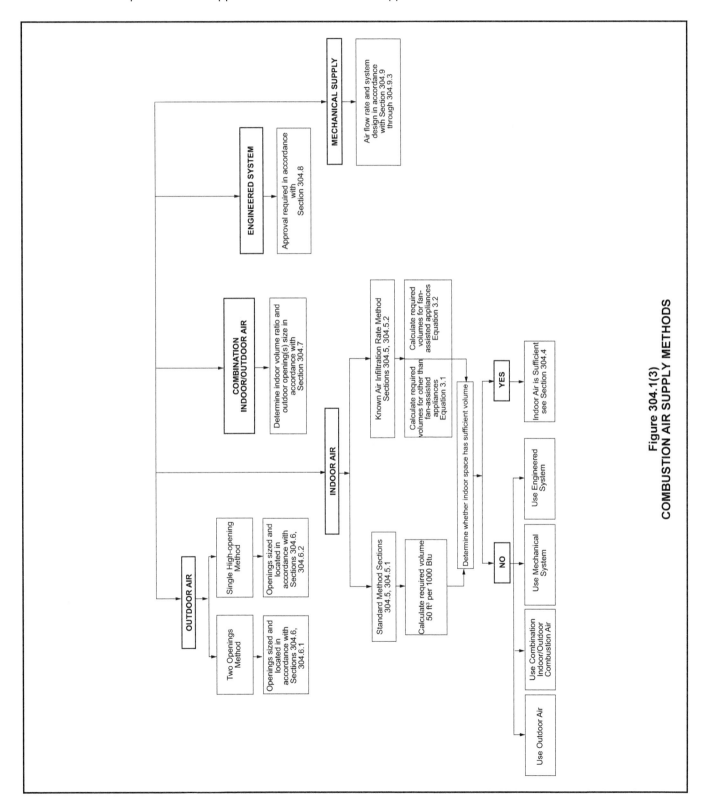

Figure 304.1(3)
COMBUSTION AIR SUPPLY METHODS

Figure 304.1(4)
DIRECT-VENT APPLIANCE

Figure 304.1(5)
DIRECT COMBUSTION AIR CONNECTION TO THE
OUTDOORS

Figure 304.1(6)
BOILER WITH POWER BURNER

304.2 Appliance location. Appliances shall be located so as not to interfere with proper circulation of combustion, ventilation and dilution air.

❖ The provisions of Section 304 rely on the natural (gravity) movement of air; therefore, installations must allow unimpeded flow of combustion air from the source to the appliances being supplied. Other mechanical systems, appliances or equipment in the same room or building can adversely affect the combustion air supply. For example, when placed too close to the combustion air openings, gas-fired appliances or equipment will restrict the free circulation of air.

304.3 Draft hood/regulator location. Where used, a draft hood or a barometric draft regulator shall be installed in the same room or enclosure as the *appliance* served to prevent

any difference in pressure between the hood or regulator and the *combustion air* supply.

❖ It is important that draft hoods and barometric draft regulators be located in the same pressure zone as the appliance combustion chamber.

Draft hoods and barometric regulators both serve to stabilize draft through an appliance and, in some cases, act to relieve backpressure/downdraft in the venting system. It is the draft in a Category I appliance that moves combustion air into the appliance and conveys the combustion gases to the venting system. Pressure differences between the atmospheres in which the draft hood/regulator and the combustion chamber are located could interfere with the function of the draft hood/regulator, resulting in poor combustion, excess draft, insufficient draft, combustion gas spillage and/or hazardous appliance operation. A draft hood/regulator is installed under the assumption that the combustion chamber it serves sees the same atmospheric pressure.

304.4 Makeup air provisions. Where exhaust fans, clothes dryers and kitchen ventilation systems interfere with the operation of appliances, makeup air shall be provided.

❖ The introduction of makeup air is critical to the proper operation of all fuel-burning appliances located in areas subject to the effects of exhaust systems. Too little makeup air will cause excessive negative pressures to develop, not only reducing the exhaust airflow but also interfering with appliance venting. Too little makeup air can cause loss of draft in appliance vents and chimneys or cause combustion byproducts to discharge into the building. It is the draft produced in an appliance that causes combustion air to enter the combustion chamber of natural-draft atmospheric-burner appliances. It is the draft in the venting system that causes the combustion gases and dilution air to move through the venting system toward the vent terminal. A lack of combustion air can result in incomplete fuel combustion, appliance malfunction and flue gas spillage.

A significant reduction in building pressure could be created by fireplaces, exhaust fans, ventilation systems, clothes dryers and similar appliances and equipment. Gas-fired appliances are often in competition with other equipment or systems for the available combustion air and makeup air that flows through (infiltrates) the building envelope or is otherwise introduced into a building, room or space. The competition between powered exhaust equipment and natural-draft gas-fired appliance venting systems is an unfair contest at best. Eventually, the powered equipment will starve the natural-draft appliances unless provisions are made to compensate for the effect of the powered exhaust equipment/appliances. Supplying makeup air will prevent negative pressures within a space, thus negating the effect of exhaust fans and clothes dryers. In many cases, makeup air will be required because this is the only way to prevent exhaust systems from negatively affecting the appliance venting systems that draw combustion air from the same space from which the exhaust systems remove air.

Natural-draft appliances also compete among themselves for combustion air. The appliance venting system that produces the strongest draft, such as a solid-fuel appliance or fireplace, can cause combustion air shortages for the appliance venting systems that produce a weaker draft.

Exhaust fans and similar equipment and appliances can produce significant negative building pressures that can interfere with the operation of vents and chimneys. This interference can cause reverse flow as outdoor air enters the building through the vents and chimneys because of the pressure difference. Any such interference with vents or chimneys could cause products of combustion to be discharged into the building and, therefore, must be avoided. The

provisions of Section 304 could be woefully inadequate or nullified where compensating features are not installed to mitigate the effect of appliances or equipment removing air from any room or space.

When any system depends on infiltration as the source of combustion air, it will probably be necessary to supply makeup air to offset the deficiency in many buildings. It is obvious that natural draft appliance installations cannot compete with exhaust fans and clothes dryers for the available infiltration air. Without makeup air, the gas-fired appliance venting systems and the mechanical exhaust systems will indeed compete for the same infiltration air, which could be insufficient to satisfy either need (see Commentary Figure 304.4). Note that exhaust fans, clothes dryers and similar equipment that remove air from a space will have no negative impact on unvented appliances because unvented appliances have no venting system. Direct-vent appliances are also immune in most cases.

304.5 Indoor combustion air. The required volume of indoor air shall be determined in accordance with Section 304.5.1 or 304.5.2, except that where the air infiltration rate is known to be less than 0.40 air changes per hour (ACH), Section 304.5.2 shall be used. The total required volume shall be the sum of the required volume calculated for all appliances located within the space. Rooms communicating directly with the space in which the appliances are installed through openings not furnished with doors, and through *combustion air* openings sized and located in accordance with Section 304.5.3, are considered to be part of the required volume.

❖ The terms "confined space," "unconfined space" and "unusually tight construction" are no longer used in the code. The required room volume is determined by two methods: one method based on the actual air infiltration rate of the building and the other based on the familiar fixed ratio of 50 cubic feet per 1,000 Btu/h (4.8 m³/kW). Air taken directly from inside the building is an acceptable source of combustion air if a sufficient volume of air is available for the appliances served and the building construction allows sufficient infiltration. The provisions of this section rely on building envelope infiltration as the only source of combustion air.

The method given in Section 304.5.1 is based on a conservative assumed air infiltration rate of at least 0.40 air change per hour (ACH); therefore, if the ACH is known to be less than 0.40, this method cannot be used. The air infiltration rate for most buildings is not known, and it is possible for the ACH rate of a building to be actually less than 0.40; however, some research suggests that buildings having less than 0.40 ACH are rare. It is believed that 80 to 95 percent of U.S. homes have infiltration rates of 0.35 ACH or greater; however, those numbers are changing because of more stringent energy codes and sustainable "green" building practices. A rate of 0.40 ACH or less is the quantitative expression of "tight construction." Note that the 0.40 ACH rate has an assumed pressure differential between the indoors and out-

For SI: 1 cubic foot per minute = 0.0004710 m³/s.

**Figure 304.4
DEPLETION OF COMBUSTION AIR**

doors of that equal to the draft produced by natural draft-vented appliances, not the typical 50 Pa used for determining the energy efficiency of buildings. Testing also indicates that the 50 cubic feet per 1,000 Btu/h (4.8 m³/kW) convention is somewhat liberal; thus, a built-in safety factor exists. If the designer suspects that the building in question is extraordinarily tight (less than 0.40 ACH), calculations and/or testing should be done to verify the ACH rate. A conservative approach could be to use the method in Section 304.5.2 with a conservative ACH rate such as 0.25, 0.30 or 0.35. Commentary Figure 304.5.2 shows calculated required volumes for appliances up to 300,000 Btu/h (87.9 kW) input rating based on the standard method and known ACH method for multiple ACH rates.

The standard method can be calculated for each appliance and then all the volumes are added or can be calculated by adding all of the appliance inputs first and then calculating the total volume. The known ACH-rate method must be performed in at least two distinct calculations where both fan-assisted and draft-hood-equipped appliances are present.

To increase the available volume, the code allows rooms and spaces to be coupled together by doorways without doors and by openings installed in accordance with Section 304.5.3.

304.5.1 Standard method. The minimum required volume shall be 50 cubic feet per 1,000 Btu/h (4.8 m³/kW) of the *appliance* input rating.

❖ This is the optional default method intended for use when the ACH rate is unknown. If the actual ACH rate is known to be greater than 0.40, the method of Section 304.5.2 will yield smaller required volumes.

If the ACH rate is known to be less than 0.40, this method must not be used because it is based on an assumed ACH rate of at least 0.40. The rate of 50 ft³/1,000 Btu/h (4.8 m³/kW) is a carryover from the previous editions of the code and was used for buildings that were not of unusually tight construction.

304.5.2 Known air-infiltration-rate method. Where the air infiltration rate of a structure is known, the minimum required volume shall be determined as follows:

For appliances other than fan-assisted, calculate volume using Equation 3-1.

$$Required\ Volume_{other} \geq \frac{21\ \text{ft}^3}{ACH} \left(\frac{I_{other}}{1,000\ \text{Btu/h}} \right)$$

(Equation 3-1)

For fan-assisted appliances, calculate volume using Equation 3-2.

$$Required\ Volume_{fan} \geq \frac{15\ \text{ft}^3}{ACH} \left(\frac{I_{fan}}{1,000\ \text{Btu/h}} \right)$$

(Equation 3-2)

where:

I_{other} = All appliances other than fan assisted (input in Btu/h).

I_{fan} = Fan-assisted *appliance* (input in Btu/h).

ACH = Air change per hour (percent of volume of space exchanged per hour, expressed as a decimal).

For purposes of this calculation, an infiltration rate greater than 0.60 *ACH* shall not be used in Equations 3-1 and 3-2.

❖ This method considers the actual or calculated ACH rate and requires space volumes to be commensurate with that rate. This method can also be used when the ACH rate is unknown by simply picking a conservative ACH rate (0.40 ACH or less) representing the lowest anticipated ACH rate for the given building. Equation 3-1 is for draft-hood-equipped appliances and reflects recent research and test results that show that such appliances need less airflow than previously assumed. The standard ratio of 50 ft³/1,000 Btu/h (4.8 m³/kW) that has been in codes for many years was based on an assumed combustion airflow need of 25 ft³/1,000 Btu/h (2.4 m³/kW) with an assumed ACH rate of 0.50. The 25 ft³/1,000 Btu/h airflow consisted of the air required for stoichiometric combustion (10 ft³/ft³ of natural gas) plus excess air to ensure complete combustion, draft hood dilution air and a safety factor allowance. The volume of 25 cubic feet (0.7 m³) divided by an ACH rate of 0.50 yields the familiar 50 ft³/1,000 Btu/h ratio. Equation 3-1 is based on a total airflow need of 21 cubic feet (0.6 m³) instead of 25 (0.7 m³) because research supports a less conservative revised volume.

Equation 3-2 is for fan-assisted appliances only and accounts for the fact that fan-assisted appliances have no draft hood and, therefore, do not need dilution air. Fan-assisted appliances need only air for stoichiometric combustion (10 ft³/ft³ of natural gas) and excess air (5 ft³/ft³ of natural gas). Dividing by the ACH rate adjusts the required volume to correspond to the available infiltration rate.

For example, if the ACH rate is 0.35, 42.8 cubic feet (1.2 m³) would be required for each 1,000 Btu/h (ft³ natural gas), and if the ACH rate is 0.50, 30 cubic feet (0.8 m³) would be required for each 1,000 Btu/h. The greater the ACH rate, the lower the room volume needed.

Research and testing results indicate that the absolute minimum volume for draft-hood-equipped appliances must be 35 ft³/1,000 Btu/h (3.4 m³/kW) and for fan-assisted appliances it must be 25 ft³/1,000 Btu/h (2.4 m³/kW). This is why this section limits the infiltration rate to a maximum of 0.60 ACH. Limiting the ACH rate to 0.60 results in a lower limit safety factor for the required volume, thereby preventing appli-ances from being installed in exceedingly small spaces. Commentary Figure 304.5.2 is based on the methodology of Sections 304.5.1 and 304.5.2 and is provided for convenience.

304.5.3 Indoor opening size and location. Openings used to connect indoor spaces shall be sized and located in accordance with Sections 304.5.3.1 and 304.5.3.2 (see Figure 304.5.3).

❖ Sections 304.5.3.1 and 304.5.3.2 prescribe the size and location of openings used to conjoin (couple) spaces for the purpose of increasing the available volume. The last sentence of Section 304.5 speaks of this. These openings must be permanently open, except where interlocked motorized dampers are used in accordance with Section 304.10.

FIGURE 304.5.3
ALL AIR FROM INSIDE THE BUILDING
(see Section 304.5.3)

304.5.3.1 Combining spaces on the same story. Each opening shall have a minimum free area of 1 square inch per 1,000 Btu/h (2,200 mm²/kW) of the total input rating of all appliances in the space, but not less than 100 square inches (0.06 m²). One opening shall commence within 12 inches (305 mm) of the top and one opening shall commence within 12 inches (305 mm) of the bottom of the enclosure. The minimum dimension of air openings shall be not less than 3 inches (76 mm).

❖ This section is applicable to adjacent spaces on the same floor level and provides for the familiar high and low openings. The opening configuration creates a thermosiphon airflow with the bottom opening acting

Standard Method Required Volume All Appliances		Known Air Infiltration Rate Method, Minimum Space Volume for Appliance Other Than Fan-Assisted, For Specified Infiltration Rates (ACH*)				Known Air Infiltration Rate Method, Minimum Space Volume for Fan-Assisted Appliance, For Specified Infiltration Rates (ACH*)			
Appliance Input Btu/h	Space Volume (ft³)	Appliance Input Btu/h	Space Volume (ft³) 0.25 ACH	Space Volume (ft³) 0.30 ACH	Space Volume (ft³) 0.35 ACH	Appliance Input Btu/h	Space Volume (ft³) 0.25 ACH	Space Volume (ft³) 0.30 ACH	Space Volume (ft³) 0.35 ACH
5,000	250	5,000	420	350	300	5,000	300	250	214
10,000	500	10,000	840	700	600	10,000	600	500	429
15,000	750	15,000	1,260	1,050	900	15,000	900	750	643
20,000	1,000	20,000	1,680	1,400	1,200	20,000	1,200	1,000	857
25,000	1,250	25,000	2,100	1,750	1,500	25,000	1,500	1,250	1,071
30,000	1,500	30,000	2,520	2,100	1,800	30,000	1,800	1,500	1,286
35,000	1,750	35,000	2,940	2,450	2,100	35,000	2,100	1,750	1,500
40,000	2,000	40,000	3,360	2,800	2,400	40,000	2,400	2,000	1,714
45,000	2,250	45,000	3,780	3,150	2,700	45,000	2,700	2,250	1,929
50,000	2,500	50,000	4,200	3,500	3,000	50,000	3,000	2,500	2,143
55,000	2,750	55,000	4,620	3,850	3,300	55,000	3,300	2,750	2,357
60,000	3,000	60,000	5,040	4,200	3,600	60,000	3,600	3,000	2,571
65,000	3,250	65,000	5,460	4,550	3,900	65,000	3,900	3,250	2,786
70,000	3,500	70,000	5,800	4,900	4,200	70,000	4,200	3,500	3,000
75,000	3,750	75,000	6,300	5,250	4,500	75,000	4,500	3,750	3,214
80,000	4,000	80,000	6,720	5,600	4,800	80,000	4,800	4,000	3,429
85,000	4,250	85,000	7,140	5,950	5,100	85,000	5,100	4,250	3,643
90,000	4,500	90,000	7,500	6,300	5,400	90,000	5,400	4,500	3,857
95,000	4,750	95,000	7,900	6,650	5,700	95,000	5,700	4,750	4,071
100,000	5,000	100,000	8,400	7,000	6,000	100,000	6,000	5,000	4,286
105,000	5,250	105,000	8,820	7,350	6,300	105,000	6,300	5,250	4,500
110,000	5,500	110,000	9,240	7,700	6,600	110,000	6,600	5,500	4,714
115,000	5,750	115,000	9,660	8,050	6,900	115,000	6,900	5,750	4,929
120,000	6,000	120,000	10,080	8,400	7,200	120,000	7,200	6,000	5,143
125,000	6,250	125,000	10,500	8,750	7,500	125,000	7,500	6,250	5,357
130,000	6,500	130,000	10,920	9,100	7,800	130,000	7,800	6,500	5,571
135,000	6,750	135,000	11,340	9,450	8,100	135,000	8,100	6,750	5,786
140,000	7,000	140,000	11,760	9,800	8,400	140,000	8,400	7,000	6,000
145,000	7,250	145,000	12,180	10,150	8,700	145,000	8,700	7,250	6,214
150,000	7,500	150,000	12,600	10,500	9,000	150,000	9,000	7,500	6,429
160,000	8,000	160,000	13,440	11,200	9,600	160,000	9,600	8,000	6,857
170,000	8,500	170,000	14,280	11,900	10,200	170,000	10,200	8,500	7,286
180,000	9,000	180,000	15,120	12,600	10,800	180,000	10,800	9,000	7,714
190,000	9,500	190,000	15,960	13,300	11,400	190,000	11,400	9,500	8,143
200,000	10,000	200,000	16,800	14,000	12,000	200,000	12,000	10,000	8,571
210,000	10,500	210,000	17,640	14,700	12,600	210,000	12,600	10,500	9,000
220,000	11,000	220,000	18,480	15,400	13,200	220,000	13,200	11,000	9,429
230,000	11,500	230,000	19,320	16,100	13,800	230,000	13,800	11,500	9,857
240,000	12,000	240,000	20,160	16,800	14,400	240,000	14,400	12,000	10,286
250,000	12,500	250,000	21,000	17,500	15,000	250,000	15,000	12,500	10,714
260,000	13,000	260,000	21,840	18,200	15,600	260,000	15,600	13,000	11,143
270,000	13,500	270,000	22,680	18,900	16,200	270,000	16,200	13,500	11,571
280,000	14,000	280,000	23,520	19,600	16,800	280,000	16,800	14,000	12,000
290,000	14,500	290,000	24,360	20,300	17,400	290,000	17,400	14,500	12,429
300,000	15,000	300,000	25,200	21,000	18,000	300,000	18,000	15,000	12,857

*ACH=Air Change per Hour   *ACH=Air Change per Hour

For SI: 1 British thermal unit per hour = 0.293 W, 1 cubic foot = 0.028 m³.

Figure 304.5.2
CALCULATED VOLUMES
(Figure Courtesy of American Gas Association)

as an inlet and the top opening acting as an outlet. This provision originated as a means to conjoin a small appliance enclosure (such as a furnace or boiler room) with other spaces, thereby effectively increasing the volume of the enclosures.

304.5.3.2 Combining spaces in different stories. The volumes of spaces in different stories shall be considered as communicating spaces where such spaces are connected by one or more openings in doors or floors having a total minimum free area of 2 square inches per 1,000 Btu/h (4402 mm^2/kW) of total input rating of all appliances.

❖ This section is a new concept that allows spaces on different floor levels to be conjoined, thereby effectively increasing the available volume for supplying combustion air. Installing a high- and low-opening configuration in accordance with the previous section would serve no purpose when vertically connecting adjacent stories; thus, a single opening is permitted. The opening or openings can be in a floor or in a door opening to an unenclosed stairway that connects the two stories. This method of conjoining spaces is particularly useful for dwelling units where the volume of a basement in which the appliances are located can be conjoined with the open spaces on upper stories by means of a louvered door at the basement stairs. The opening size requirement is simply the addition of the areas of the two openings required by Section 304.5.3.1. Bear in mind that openings in floors and stairway doors in other than dwelling units are strictly regulated by the IMC and IBC and would be prohibited or would require fire dampers in most cases. A combustion air opening in a floor would be considered by the IBC as a transfer opening in a horizontal assembly (see Commentary Figure 304.5.3.2).

Example:

Determine whether indoor combustion air is adequate for the building shown in Commentary Figure 304.5.3.2.

1. Starting with the fan-assisted boilers, use Equation 3-2 to calculate the required volume for each boiler.

$$\frac{15 \text{ ft}^3}{0.35} \times \frac{125,000}{1,000} = 42.8 \times 125 = 5,350 \text{ ft}^3$$

2. Use Equation 3-1 to calculate the required volume for each water heater.

$$\frac{21 \text{ ft}^3}{0.35} \times \frac{75,000}{1,000} = 60 \times 75 = 4,500 \text{ ft}^3$$

3. Total volumes required for all appliances.

Boiler	1	5,350
Boiler	2	5,350
Water Heater	1	4,500
Water Heater	2	4,500
		19,700 ft^3

4. Determine whether louvered door opening area is adequate.
   - In accordance with Section 304.10, determine actual louvered opening area.

30 in. × 72 in. = 2,160 in.^{2} × 0.25 = 540 in.^{2}

   - In accordance with Section 304.5.3.2, determine required area.

$$\frac{400,000}{1,000} \text{ Btu/h} \times 2 = 800 \text{ in.}^2$$

5. Determine available (actual) volume in building.

Basement volume = 48 ft × 25 ft × 9 ft = 10,800 ft^3

Grade floor volume = 48 ft × 25 ft × 9 ft = 10,800 ft^3

21,600 ft^3

The combined volume of both stories is greater than required; however, the louvered door opening area connecting the stories is inadequate. Possible solutions include adding outdoor openings in the basement in accordance with Section 304.7 or providing metal louvers in the stairway door instead of wood.

For SI: 1 inch = 25.4 mm, 1 foot = 305.8 mm, 1 cubic foot = 0.0283 m^3, 1 Btu per hour = 0.2931W

304.6 Outdoor combustion air. Outdoor *combustion air* shall be provided through opening(s) to the outdoors in accordance with Section 304.6.1 or 304.6.2. The minimum dimension of air openings shall be not less than 3 inches (76 mm).

❖ This section describes two methods for supplying combustion air from the outdoors: the traditional method of two direct openings or ducts to the outdoors and a newer method using one opening or duct to the outdoors.

Openings to spaces that are naturally ventilated with outdoor air, such as attics or crawl spaces, are considered as an acceptable alternative to a direct connection to the outdoors. Attics and crawl spaces can be acceptable sources of combustion air only where such spaces have adequate natural ventilation openings directly to the outdoors. Attics and crawl spaces ventilated by mechanical means are not an acceptable source of combustion air.

Combustion air ducts and openings that penetrate components of wall, floor, ceiling and roof assemblies must be installed as required by this code.

When designing combustion air installations, the effect that openings to the outdoors can have on appliances, plumbing systems and building occupants must be considered. For example, depending on the location, openings to the outdoors can:

- Cause drafts that can blow out pilot lights or otherwise interfere with appliance ignition and operation.
- Cause freezing of plumbing piping or other water-containing components.
- Cause objectionable cold drafts that encourage the occupants to block or cover the openings.

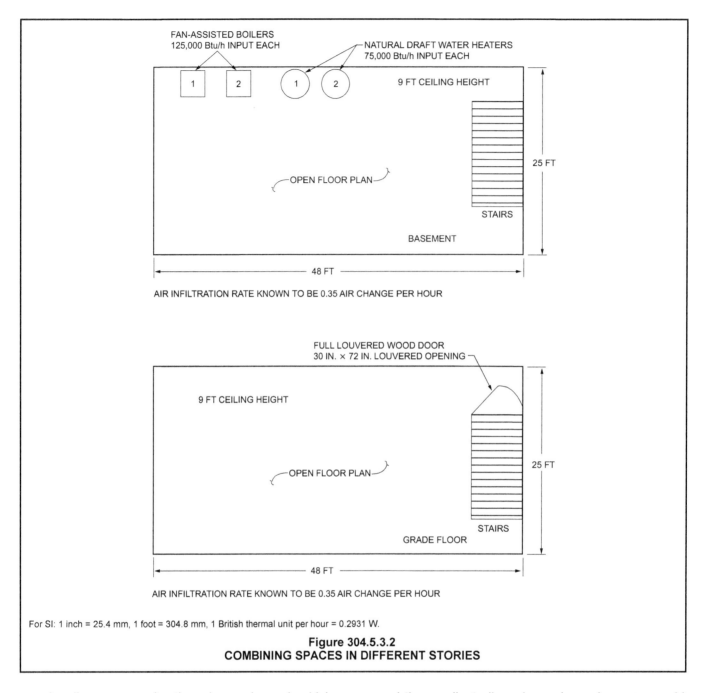

FAN-ASSISTED BOILERS
125,000 Btu/h INPUT EACH

NATURAL DRAFT WATER HEATERS
75,000 Btu/h INPUT EACH

9 FT CEILING HEIGHT

OPEN FLOOR PLAN

STAIRS

BASEMENT

25 FT

48 FT

AIR INFILTRATION RATE KNOWN TO BE 0.35 AIR CHANGE PER HOUR

FULL LOUVERED WOOD DOOR
30 IN. × 72 IN. LOUVERED OPENING

9 FT CEILING HEIGHT

OPEN FLOOR PLAN

STAIRS

GRADE FLOOR

25 FT

48 FT

AIR INFILTRATION RATE KNOWN TO BE 0.35 AIR CHANGE PER HOUR

For SI: 1 inch = 25.4 mm, 1 foot = 304.8 mm, 1 British thermal unit per hour = 0.2931 W.

Figure 304.5.3.2
COMBINING SPACES IN DIFFERENT STORIES

In all cases, combustion air openings should be located to reduce the likelihood that they will be accidentally or intentionally blocked or covered. A ceiling transfer opening that connects a furnace room with an attic is an example of a combustion air opening that is likely to be intentionally blocked by an occupant because of the drafts that can occur in cold climates.

Building occupants typically do not understand the need for or the importance of combustion air openings.

No side dimension of a square or rectangular opening and no diameter of a round opening can be less than 3 inches (76 mm). The smallest allowed square opening area would be 9 square inches (5806 mm²)

and the smallest allowed round opening area would be 7 square inches (4516 mm²) (see commentary, Section 304.10).

304.6.1 Two-permanent-openings method. Two permanent openings, one commencing within 12 inches (305 mm) of the top and one commencing within 12 inches (305 mm) of the bottom of the enclosure, shall be provided. The openings shall communicate directly or by ducts with the outdoors or spaces that freely communicate with the outdoors.

Where directly communicating with the outdoors, or where communicating with the outdoors through vertical ducts, each opening shall have a minimum free area of 1 square inch per 4,000 Btu/h (550 mm²/kW) of total input rating of all appliances in the enclosure [see Figures 304.6.1(1) and 304.6.1(2)].

Where communicating with the outdoors through horizontal ducts, each opening shall have a minimum free area of not less than 1 square inch per 2,000 Btu/h (1,100 mm²/kW) of total input rating of all appliances in the enclosure [see Figure 304.6.1(3)].

❖ Two openings located as prescribed in this section are intended to induce a convective air current in the room or space by admitting cooler, denser air in the lower opening and allowing the escape of warmer, less dense air through the upper opening. The farther apart the openings, the greater the temperature differential and the greater the convective force behind the current. A component of combustion air is cooling (ventilation) air for the appliance enclosure. The two-opening method was created to ventilate the appliance enclosure in addition to supplying combustion air. This ventilation cools the appliances and would help remove any combustion gases that spilled from the appliances.

304.6.2 One-permanent-opening method. One permanent opening, commencing within 12 inches (305 mm) of the top of the enclosure, shall be provided. The *appliance* shall have clearances of not less than 1 inch (25 mm) from the sides and back and 6 inches (152 mm) from the front of the *appliance*. The opening shall directly communicate with the outdoors, or through a vertical or horizontal duct, to the outdoors or spaces that freely communicate with the outdoors (see Figure 304.6.2) and shall have a minimum free area of 1 square inch per 3,000 Btu/h (734 mm²/kW) of the total input rating of all

appliances located in the enclosure and not less than the sum of the areas of all vent connectors in the space.

❖ Research has shown that for modern appliances, a single opening to the outdoors will perform as well as the traditional two-opening method. The one-opening method described in this section depends on a reduced pressure being created in the enclosure by the draft created by the venting system. This reduced pressure causes combustion air to enter the enclosure through the single opening. The opening must

For SI: 1 foot = 304.8 mm.

FIGURE 304.6.1(2)
ALL AIR FROM OUTDOORS THROUGH VENTILATED ATTIC
(see Section 304.6.1)

FIGURE 304.6.1(3)
ALL AIR FROM OUTDOORS
(see Section 304.6.1)

FIGURE 304.6.1(1)
ALL AIR FROM OUTDOORS—INLET AIR FROM VENTILATED
CRAWL SPACE AND OUTLET AIR TO VENTILATED ATTIC
(see Section 304.6.1)

be properly sized considering both sizing criteria: the square-inch-area-per-Btu/h ratio and the area minimum based on the sum of the areas of all vent connectors in the enclosure. This method allows for fewer openings, fewer ducts and fewer objections by the owners/occupants.

**FIGURE 304.6.2
SINGLE COMBUSTION AIR OPENING,
ALL AIR FROM THE OUTDOORS
(see Section 304.6.2)**

304.7 Combination indoor and outdoor combustion air. The use of a combination of indoor and outdoor *combustion air* shall be in accordance with Sections 304.7.1 through 304.7.3.

❖ This method of supplying combustion air is a combined application of Sections 304.5 and 304.6. This method allows credit for the amount of infiltration that exists and makes up for the shortage with supplemental outdoor air. In other words, in addition to obtaining combustion air directly from the outdoors, this method relies on building infiltration for a portion of the total combustion air. In spaces where the volume is insufficient to satisfy the method of Section 304.5 or where smaller outdoor air openings than required by Section 304.6 are desired, this method allows infiltration and outdoor openings to supplement each other. If the appliances are enclosed in a small room such as a closet, openings to adjacent spaces as prescribed by Section 304.5.3 must be provided to couple the appliance room volume with any other space volume counted on to provide combustion air. Frequently, sufficient volume cannot be obtained in the appliance enclosure or by opening the appliance enclosure to adjacent spaces. This method

of combining indoor and outdoor air is an alternative solution. Simply stated, this section uses ratios of what is required to what is actually supplied so that when combined, the indoor air component and the outdoor air component add up to the whole required.

304.7.1 Indoor openings. Where used, openings connecting the interior spaces shall comply with Section 304.5.3.

❖ See the commentary to Section 304.5.3.

304.7.2 Outdoor opening location. Outdoor opening(s) shall be located in accordance with Section 304.6.

❖ See the commentary to Section 304.6.

304.7.3 Outdoor opening(s) size. The outdoor opening(s) size shall be calculated in accordance with the following:

1. The ratio of interior spaces shall be the available volume of all communicating spaces divided by the required volume.

2. The outdoor size reduction factor shall be one minus the ratio of interior spaces.

3. The minimum size of outdoor opening(s) shall be the full size of outdoor opening(s) calculated in accordance with Section 304.6, multiplied by the reduction factor. The minimum dimension of air openings shall be not less than 3 inches (76 mm).

❖ Although the principle has not changed, this method has been simplified compared to the same provision in the 2000 edition of the code. The intent is still the same; that is, the fraction of the required indoor volume plus the fraction of outdoor openings must be equal to or greater than 1. The indoor volume method and the outdoor air method can both be stand-alone methods; therefore, half of one and half of the other will work, as will three fourths of one and a quarter of the other, etc.

This section is expressed in the following equation:

$$\left[1-\left(\frac{\text{available indoor volume}}{\substack{\text{volume required} \\ \text{by Section 304.5}}}\right)\right] \times \substack{\text{full size opening} \\ \text{required by} \\ \text{Section 304.6}} = \substack{\text{reduced size} \\ \text{outdoor air} \\ \text{openings}}$$

Example:

Given:

7,500 ft³ of indoor volume is required if all indoor air is to be used.

4,950 ft³ of indoor volume is available.

One opening to the outdoors is desired and must be 50 in.² if all outdoor air is to be used.

$$\left[1-\left(\frac{4,950 \text{ ft}^3}{7,500 \text{ ft}^3}\right)\right] \times 50 \text{ in.}^2 = \text{reduced outdoor opening size}$$

0.34 x 50 in.² = 17 in.² outdoor opening size

Either of the provisions of Section 304.6 when combined with the provision of this section can be used to satisfy the equation.

CHIMNEY OR GAS VENT, OPENING, ALTERNATE OPENING LOCATION, FURNACE, WATER HEATER.

Example 1:

Room A is within a dwelling unit and contains a fuel-fired furnace and a fuel-fired water heater with input ratings of 150,000 and 50,000 Btu/h, respectively.

The size of Room A is 20 feet by 30 feet, and adjacent Room B is also 20 feet by 30 feet. Both rooms have 8-foot-high ceilings. Rooms A and B communicate through two openings located in accordance with Section 304.5.3, each having dimensions of 10 inches by 20 inches. Two 3-inch-diameter (round) direct openings to the outdoors are installed in Room A and are located in accordance with Section 304.6.1.

Question: Do the openings to the outdoors, when combined with the volumes of Rooms A and B, meet the combustion air demand of the appliances?

For SI: 1 inch = 25.4 mm, 1 foot = 305 mm, 1 Btu/h = 0.2931 W, 1 cubic foot = 0.0283 m^3, 1 square inch = 645 mm^2.

Given:

The air infiltration rate of the building is unknown.

Volume of Room A = 4,800 ft^3

Volume of Room B = 4,800 ft^3

Area of 3-inch-diameter direct openings = 7.07 in.^{2}

Area of each opening between Rooms A and B = 200 in.^{2}

Total input rating of appliances = 200,000 Btu/h

Step 1: Determine whether Rooms A and B meet the volume requirements of Section 304.5.

Available volume of Rooms A and B = 4,800 ft^3 + 4,800 ft^3 = 9,600 ft^3

The required volume is:

$$\frac{200,000 \text{ Btu/h}}{1,000 \text{ Btu/h}} \times 50 \text{ ft}^3 = 10,000 \text{ ft}^3$$

Rooms A and B combined do not meet the volume requirements of Section 304.5.

Step 2: Determine whether the openings between Rooms A and B meet the requirements of Section 304.5.3.

The minimum required area is:

$$\frac{200,000 \text{ Btu/h}}{1,000 \text{ Btu/h}} \times 1 \text{ in}^2 = 200 \text{ in}^2$$

Actual area of each opening = 200 in.^{2}

The area of each opening meets the size requirement of Section 304.5.3.

Step 3: Determine the required area of each direct outdoor opening in accordance with Section 304.6.1.

Required area of each opening:

$$\frac{200,000 \text{ Btu/h}}{4,000 \text{ Btu/h}} \times 1 \text{ in}^2 = 50 \text{ in}^2$$

Step 4: Determine whether the 3-inch-diameter direct openings combined with the volumes of Rooms A and B comply with Section 304.7.3.

$$\left[1 - \left(\frac{9,600 \text{ ft}^3}{10,000 \text{ ft}^3}\right)\right] \times 50 = (1 - 0.96) \times 50 = 0.04 \times 50 = 2 \text{ in.}^2$$

The actual area of each direct opening (7.07 square inches) exceeds the required area of 2 square inches.

Therefore, the combination of the 3-inch-diameter direct openings and the volumes of Rooms A and B satisfies the combustion air demand of the appliances.

Example 2:

A combination of indoor and outdoor combustion air is to be supplied for two natural draft boilers in a room as shown in Commentary Figure 304.7.3 (the air infiltration rate is not known).

1. In accordance with Section 304.5.1, the required volume of the room containing the boilers would be:

$$\frac{90,000 \text{ Btu/h} + 90,000 \text{ Btu/h}}{1,000} \times 50 \text{ ft}^3$$

$$= \frac{180,000}{1,000} \times 50 \text{ ft}^3 = 180 \times 50 \text{ ft}^3 = 9,000 \text{ ft}^3$$

2. The actual room volume is 20 ft × 50 ft × 8 ft = 8,000 ft^3.

3. In accordance with Section 304.7.3 Item 1, determine the ratio of available (actual) volume to the required volume.

$$\frac{\text{Actual Volume}}{\text{Required Volume}} \quad \frac{8,000 \text{ ft}^3}{9,000 \text{ ft}^3} = 0.89$$

4. In accordance with Section 304.7.3 Item 2, determine outdoor opening size reduction factor.

    1 - 0.89 = 0.11

5. In accordance with Section 304.7.3 Item 3, determine the size of the outdoor opening required as if all combustion air is to be supplied through the outdoor opening. Because a single high opening is built into an exterior wall, Section 304.6.2 applies and would require an area of 1 in.^{2}/3000 Btu/h of total appliance input rating (Assuming 6-in. vent connectors, the total area of the boiler vent connectors is 56.6 square inches).

$$\frac{90,000 \text{ Btu/h} + 90,000 \text{ Btu/h}}{3,000} = 60 \text{ in}^2$$

6. The required size of the outdoor air opening is 60 square inches × 0.11 = 6.6 square inches. Because the minimum dimensions must be not less than 3 inches, the minimum area of a square opening must be 9 square inches. This combination satisfies the combustion air demand of the appliances; however, Section 304.6.2 requires an area of not less than the combined area of all vent connectors, thus 56.6 square inches is the minimum area of the opening.

304.8 Engineered installations. Engineered *combustion air* installations shall provide an adequate supply of combustion, ventilation and dilution air and shall be *approved*.

❖ Potential sources of combustion air are the indoor atmosphere in which the appliance is located; outdoor openings and ducts that communicate with the outdoor atmosphere; a combination of these; or any other approved arrangement for the dependable introduction of outdoor air into the building or directly to the fuel-burning appliance.

The provisions of Section 304 are intended for application only to appliances and equipment using natural draft atmospheric burners. Gas-fired appliances, especially those that are not of the atmospheric-burner type, are commonly supplied with combustion air by engineered systems. For example, special engineered systems are usually designed for appliances equipped with power burners. Power burners force combustion air into the burner under pressure and are, therefore, not dependent on gravity airflow. In some applications, engineered combustion air supplies have advantages and can be preferable to large gravity (natural) openings to the outdoors. Engineered designs must be evaluated for code compliance, approved and installed in accordance with the appliance manufacturer's instructions [see Commentary Figures 304.8(1) through 304.8(3)].

304.9 Mechanical combustion air supply. Where all *combustion air* is provided by a mechanical air supply system, the *combustion air* shall be supplied from the outdoors at a rate not less than 0.35 cubic feet per minute per 1,000 Btu/h (0.034 m³/min per kW) of total input rating of all appliances located within the space.

❖ This method supplies combustion air by means of a fan/blower that runs when any of the served appliances are in a firing cycle. The fan/blower must be sized to supply the required airflow based on the simultaneous operation of all appliances that are served by the fans/blowers. The total Btu/h input rating of all fuel-burning appliances located in the room or space must be used because the potential exists for all of the appliances to be operating at the same time. This method is used where gravity openings to the outdoors are impractical or undesirable. A small fan-powered intake opening can substitute for comparatively large gravity openings, especially where freezing temperatures are a problem or there is no room for the required size gravity openings. Commentary Figures 304.8(1), 304.8(2) and 304.8(3) illustrate factory-engineered mechanical combustion air supply units (see Commentary Figure 304.9).

The prescribed airflow rate of 0.35 cfm/1,000 Btu/h is equivalent to 1 cfm/2,857 Btu/h and is based on a combustion air need of 21 cubic feet of volume for each cubic foot of natural gas (1,000 Btu/h). The rate is derived as follows:

$$\frac{2857 \text{ Btu/h}}{1,000 \text{ Btu/ft}^3} = 2.857 \text{ ft}^3 \text{ of gas per hour}$$

2.857 ft³ of gas per hour × 21 ft³ of air per ft³ of gas = 60 ft³ of air per/hour = 60 cfh

$$\frac{60 \text{ cfh}}{60 \text{ min/hr}} = 1 \text{ cfm}$$

Thus, 1 cfm is required for each 2,857 Btu/h or 0.35 cfm/1,000 Btu/h.

For SI: 1 foot = 304.8 mm, 1 British thermal unit per hour = 0.2931 W.

Figure 304.7.3
ROOM CONFIGURATION USED IN EXAMPLE 2

Figure 304.8(1)
MECHANICAL COMBUSTION AIR/MAKEUP AIR SUPPLY SYSTEM
(Figure courtesy of Tjernlund Products, Inc.)

Figure 304.8(2)
MECHANICAL COMBUSTION AIR SUPPLY SYSTEM
(Photo courtesy of Tjernlund Products, Inc.)

Figure 304.8(3)
MECHANICAL COMBUSTION AIR/MAKEUP AIR SUPPLY SYSTEM
(Photo courtesy of Tjernlund Products, Inc.)

FIVE BOILERS (CATEGORY I)
STAGED OPERATION
EACH BOILER
150,000 Btu/h INPUT

BOILER BOILER BOILER BOILER BOILER

COMBUSTION
AIR FAN

BOILER ROOM

(COMBUSTION AIR FAN INTERLOCKED WITH ALL BOILERS)

Example:

1. Total appliance input rating: 750,000 Btu/h

2. $\dfrac{\text{Total input Btu/h}}{1,000\ \text{Btu/h}} \times 0.35 = 262.5$ cfm

For SI: 1 British thermal unit per hour = 0.2931 W, 1 cubic foot per minute = 0.000472 m³/s.

Figure 304.9
MECHANICALLY SUPPLIED COMBUSTION AIR

304.9.1 Makeup air. Where exhaust fans are installed, makeup air shall be provided to replace the exhausted air.

❖ This section expresses the exact same concern as Section 304.4 (see commentary, Section 304.4). If the air that is supplied by the combustion air fan becomes makeup air for an exhaust fan, the combustion air will not be available to the appliance(s) for which the air was intended.

304.9.2 Appliance interlock. Each of the appliances served shall be interlocked with the mechanical air supply system to prevent main burner operation when the mechanical air supply system is not in operation.

❖ This section requires an interlock circuit as opposed to simple parallel operation wiring. In other words, the combustion air fan/blower must start when initiated by the means that controls the appliance operation cycles (e.g., thermostat), and only after the combustion air fan/blower is proven to be operating will the appliance be allowed to start a firing cycle. If the combustion air fan/blower fails during a firing cycle of an appliance, the appliance firing cycle must be terminated (see Commentary Figure 304.9.2).

The following steps must occur in this order:

1. Call for heat.
2. Combustion air fan/blower starts.
3. Combustion air fan/blower is proven to be operating.
4. Proving controls allow appliance to fire.

304.9.3 Combined combustion air and ventilation air system. Where *combustion air* is provided by the building's mechanical ventilation system, the system shall provide the specified *combustion air* rate in addition to the required ventilation air.

❖ If combustion air is mechanically supplied through an HVAC system, the system must deliver the required combustion air rate above and beyond any outdoor air rate supplied for ventilation of the occupied spaces. The HVAC system must also be interlocked with the appliance(s) as required by Section 304.9.2. Neither continuous operation nor parallel operation of the HVAC system satisfies the requirement for a true interlock as required by Section 304.9.2.

304.10 Louvers and grilles. The required size of openings for combustion, ventilation and dilution air shall be based on the net free area of each opening. Where the free area through a design of louver, grille or screen is known, it shall be used in calculating the size opening required to provide the free area specified. Where the design and free area of louvers and grilles are not known, it shall be assumed that wood louvers will have 25-percent free area and metal louvers and grilles will have 75-percent free area. Screens shall have a mesh size not smaller than $^1/_4$ inch (6.4 mm). Nonmotorized louvers and grilles shall be fixed in the open position. Motorized louvers shall be interlocked with the *appliance* so that they are proven to be in the full open position prior to main burner ignition and during main burner operation. Means shall be provided to prevent the main burner from igniting if the louvers fail to open during burner start-up and to shut down the main burner if the louvers close during operation.

❖ This section recognizes that louvers and grilles are usually installed over air inlets and outlets to prevent rain, snow and animals from entering the building. When louvers or grilles are used, the solid portion of the louver or grille must be considered when determining the unobstructed (net clear) area of the opening.

Combustion air openings are sized based on there being a free, unobstructed area for the passage of air into the space where the fuel-burning appliances are located. Louvers or grilles placed over these openings reduce the area of the openings because of the area occupied by the solid portions of the grille or louver. The reduction in area must be considered

Figure 304.9.2
SIMPLIFIED EXAMPLE SCHEMATIC OF INTERLOCK
BETWEEN APPLIANCE AND POWER COMBUSTION AIR UNIT
(Shown for Illustrative Purpose Only)

because only the unobstructed area can be credited toward the required opening size.

The reduction in opening area caused by the presence of grilles or louvers will always require openings to be larger than determined from the sizing ratios of this chapter and larger than any duct of the minimum required size that might connect to these openings. Once the required size of a combustion air duct or opening is determined, any increase in opening size that must be made as a result of the effect of louvers or grilles must be determined. If a metal louver is used and based on the default opening area of 75 percent, the designer must calculate what size opening times 75 percent will yield the previously determined combustion air opening or duct area.

For example, assume that a combustion air duct needs to have a cross-sectional area of 200 square inches (0.129 m^2) and this duct connects to an exterior wall opening with a metal louver protecting it.

χ = The required opening area adjusted to account for the louver.

$$0.75\chi = 200 \text{ in}^2$$
$$\chi = \frac{200}{75}$$
$$\chi = 266.7 \text{ in}^2$$

This can be simplified to multiplying by the reciprocal of the percentage as follows:

For metal louvers and grilles, the required combustion air duct or opening size multiplied by 1.33 equals the adjusted opening size.

For wooden louvers and grilles, the required combustion air duct or opening size multiplied by 4 equals the adjusted opening size.

The actual free opening of a louver or grille should be obtained from the manufacturer rather than relying on default assumptions.

This section does not apply to grilles, louvers or screens that are an integral component of a labeled appliance that is installed in accordance with the manufacturer's instructions. Where the manufacturer states the free, unobstructed area of a grille or louver, it is acceptable to consider that area as actual without requiring further calculation.

The code also requires the free area of screens to be considered in the opening sizing calculations. If it is unknown, it will have to be calculated because there is no stated default area for screens as there is for louvers and grilles. The code prohibits screens having mesh sizes smaller than $^1/_4$ inch (6 mm). Screens with a mesh size of $^1/_4$ inch (6 mm) and larger are not insect screens and would be suitable only for keeping out large debris, rodents and other small animals. Smaller size mesh is likely to become obstructed by lint, plant fibers, cottonwood seeds, etc. At best, placing screens in combustion air openings is risky, and grilles and louvers are preferred to keep out the unwanted.

Metal louvers are constructed so that the solid portion of the louver occupies approximately 25 percent of the opening area, leaving 75 percent of this area unobstructed. Wood louvers occupy approximately 75 percent of the opening area, leaving 25 percent of the area unobstructed. When the required amount of unobstructed area is determined, the effect that a louver has on that area must be taken into account to make certain the minimum amount of unobstructed opening is achieved. The material type, thickness and spacing all have a significant effect on the actual net free area of any louver or grille.

Note that the 75 percent and 25 percent default numbers might be too liberal for some louvers or grilles and every effort should be made to consult the louver or grille manufacturer's specifications.

To prevent freezing of pipes, to save energy and maintain comfortable temperatures in appliance rooms, outdoor openings often have motorized dampers that close during the appliance(s) off cycle. The damper motors and/or damper blades must be supervised so that an interlock can be established between the dampers and the appliances, thereby preventing appliance operation when the dampers are closed.

Combustion air openings must not be closable unless they open and close automatically using a mechanism that is interlocked with the appliances (see Commentary Figure 304.10). Damper supervision is easily accomplished by damper motor end switches or blade position travel switches that monitor the position of the drive motor shaft, the damper/motor linkage hardware or the damper blades. These switches provide feedback to the appliance control circuitry to allow firing only when the dampers are proven to be fully open.

304.11 Combustion air ducts. *Combustion air* ducts shall comply with all of the following:

1. Ducts shall be constructed of galvanized steel complying with Chapter 6 of the *International Mechanical Code* or of a material having equivalent corrosion resistance, strength and rigidity.

   Exception: Within dwellings units, unobstructed stud and joist spaces shall not be prohibited from conveying *combustion air*, provided that not more than one required fireblock is removed.

2. Ducts shall terminate in an unobstructed space allowing free movement of *combustion air* to the appliances.

3. Ducts shall serve a single enclosure.

4. Ducts shall not serve both upper and lower *combustion air* openings where both such openings are used. The separation between ducts serving upper and lower *combustion air* openings shall be maintained to the source of *combustion air*.

5. Ducts shall not be screened where terminating in an attic space.

6. Horizontal upper *combustion air* ducts shall not slope downward toward the source of *combustion air*.

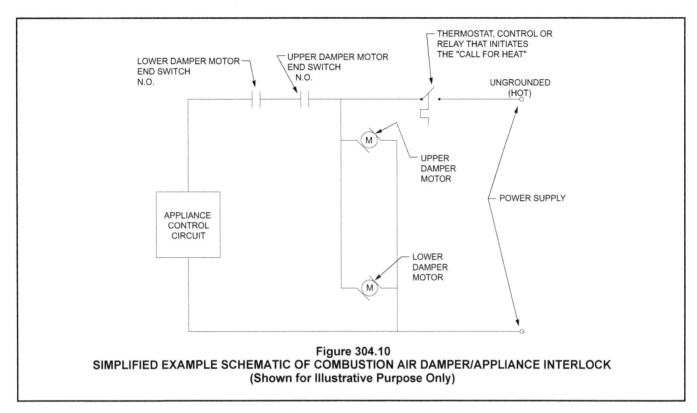

Figure 304.10
SIMPLIFIED EXAMPLE SCHEMATIC OF COMBUSTION AIR DAMPER/APPLIANCE INTERLOCK
(Shown for Illustrative Purpose Only)

7. The remaining space surrounding a chimney liner, gas vent, special gas vent or plastic *piping* installed within a masonry, metal or factory-built chimney shall not be used to supply *combustion air*.

   Exception: Direct-vent gas-fired appliances designed for installation in a solid fuel-burning *fireplace* where installed in accordance with the manufacturer's instructions.

8. *Combustion air* intake openings located on the exterior of a building shall have the lowest side of such openings located not less than 12 inches (305 mm) vertically from the adjoining finished ground level.

❖ This section addresses the construction and installation of combustion air ducts.

Item 1 requires that combustion air ducts be constructed of galvanized sheet steel or an approved equivalent material. The intent is to cause ducts to be constructed of a material that is resistant to corrosion and physical damage, which would allow the ducts to remain in place undamaged for the life of the installation. It is doubtful that the authors of Item 1 ever anticipated the use of flexible ducts and the higher friction losses associated with such rough-wall ducts. It is also reasonable to assume that all of the sizing criteria in Section 304 are based on smooth-wall ducts. The strength and rigidity of typical flexible ducts would not be equivalent to rigid steel ducts. The exception expresses a concern for fire safety where stud cavities are used as a duct.

Item 2 is intended to prevent blockages and minimize resistance (friction loss) to airflow.

Item 3 prohibits a duct from serving more than one appliance enclosure because the duct and opening sizing criteria are based on a single duct serving a single opening in an appliance location. Multiple openings supplied by a single duct could result in unpredictable airflow and might not create the convective movements of air intended by the combustion air methods.

Item 4 recognizes that where both upper and lower combustion air openings are required, air circulation cannot occur if a single duct serves both combustion air openings. The lower opening functions as an air inlet and the upper opening as an air outlet, thus independent ducts are required (see Commentary Figure 304.11).

Item 5 is intended to prevent obstruction or blockage by insulation materials.

Item 6 is intended to prevent trapping or sloping of the upper combustion air opening in a way that would impede the convective movement of air. The horizontal duct for the upper openings must slope upward as it travels away from the appliance(s) and toward the outdoors.

When a vent or chimney liner is installed within a chimney, an annular space will exist between the vent or liner and the interior chimney walls. According to Item 7, this space cannot be used as a conduit for conveying combustion air because of poor flow characteristics, the tendency for reverse flow to occur as the combustion air is warmed by the liner or vent and the adverse cooling effect on the vent or liner. The exception recognizes that there could be appliances that are designed and listed for the type of installation that this item intends to prohibit.

Item 8 intends to prevent combustion air openings from being obstructed by snow, leaves and vegetation. No part of a combustion air opening is allowed to be less than 12 inches (305 mm) above the finished ground level under such opening.

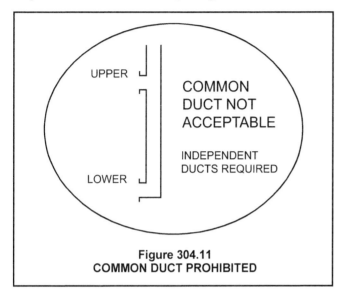

Figure 304.11
COMMON DUCT PROHIBITED

304.12 Protection from fumes and gases. Where corrosive or flammable process fumes or gases, other than products of combustion, are present, means for the disposal of such fumes or gases shall be provided. Such fumes or gases include carbon monoxide, hydrogen sulfide, ammonia, chlorine and halogenated hydrocarbons.

In barbershops, beauty shops and other facilities where chemicals that generate corrosive or flammable products, such as aerosol sprays, are routinely used, nondirect vent-type appliances shall be located in a mechanical room separated or partitioned off from other areas with provisions for *combustion air* and dilution air from the outdoors. *Direct-vent appliances* shall be installed in accordance with the *appliance* manufacturer's instructions.

❖ In many occupancies, the routine use of chemicals contaminates the indoor combustion air. These contaminants can combine with water vapor in the combustion gases to produce acids and other corrosive compounds that can destroy appliance components, vents, chimneys and connectors. For example, it is common to see vents and vent connectors for draft-hood-equipped water heaters in beauty shops that have deteriorated to the point of structural failure. Other examples are swimming pool heaters exposed to chlorine, boilers exposed to laundry and dry cleaning chemicals and boilers exposed to refrigerant leakage. Some contaminants are more toxic to life after they have passed through an appliance combustion chamber. The highly poisonous chemical "phosgene" (carbonyl chloride) can be produced when common refrigerants contaminate combustion air.

Obviously, choosing direct-vent appliances will eliminate this problem in most cases and will not

require isolation enclosures for the appliances. It is possible that some atmospheres are unsuitable for any appliance, direct-vent or otherwise. The code provides an alternative to direct-vent appliances that requires the appliances to be isolated from the contaminated atmosphere by reasonably air-tight enclosures and requires that all combustion air be supplied for the appliance(s) from the outdoors.

SECTION 305 (IFGC)
INSTALLATION

305.1 General. *Equipment* and appliances shall be installed as required by the terms of their approval, in accordance with the conditions of listing, the manufacturer's instructions and this code. Manufacturers' installation instructions shall be available on the job site at the time of inspection. Where a code provision is less restrictive than the conditions of the listing of the *equipment* or *appliance* or the manufacturer's installation instructions, the conditions of the listing and the manufacturer's installation instructions shall apply.

Unlisted appliances *approved* in accordance with Section 301.3 shall be limited to uses recommended by the manufacturer and shall be installed in accordance with the manufacturer's instructions, the provisions of this code and the requirements determined by the code official.

❖ Manufacturers' installation instructions are evaluated by the listing agency verifying compliance with the applicable standard. The listing agency can require that the manufacturer alter, delete or add information to the instructions as necessary to achieve compliance with applicable standards and code requirements. Manufacturers' installation instructions are an enforceable extension of the code and must be in the hands of the code official when an inspection takes place.

When an appliance is tested to obtain a listing and label, the approved agency installs the appliance in accordance with the manufacturer's instructions. The appliance is tested under these conditions; thus, the installation instructions become an integral part of the labeling process.

The listing and labeling process ensures that the appliance and its installation instructions are in compliance with applicable standards. Therefore, an installation in accordance with the manufacturer's instructions is required, except where the code requirements are more stringent. An inspector must carefully and completely read and comprehend the manufacturer's instructions to properly perform an installation inspection.

In some cases, the code will specifically address an installation requirement that is also addressed in the manufacturer's installation instructions. The code requirements may be the same or may exceed the requirements in the manufacturer's installation instructions, or the manufacturer's installation instructions could contain requirements that exceed those in the code. In all such cases, the more restrictive requirements would apply.

In addition to the installation requirements for the appliance or equipment itself, this section also regulates the connections to appliances and equipment by requiring compliance with all other applicable sections of this code and other *International Codes*® (I-Codes®). These connections include, but are not limited to, fuel and electrical supplies, control wiring, hydronic piping, chimneys, vents and ductwork. Some overlap or coincidence of connection requirements may occur in the code and the manufacturer's installation instructions. In the unlikely event that the code was less restrictive (more lenient) than the manufacturer's instructions regarding an installation issue, the manufacturer's instructions must prevail. Where differences or conflicts occur, the requirements that provide the greatest level of safety always apply.

Even if an installation appears to be in compliance with the manufacturer's instructions, the installation cannot be complete or approved until all associated components, connections, and systems that serve the appliance or equipment are also in compliance with the applicable provisions of the code. For example, a gas-fired boiler installation must not be approved if the boiler is connected to a deteriorated, undersized, improper or otherwise unsafe chimney or vent. Likewise, the same installation must not be approved if the existing gas piping is in poor condition, or if the electrical supply circuit is inadequate or unsafe.

In the case of replacement installations, the intent of this section is to require all new work associated with the installation to comply with the code without necessarily requiring full compliance for the existing, unchanged portions of the related ductwork, piping, electrical, venting and similar mechanical systems. In the case of an appliance replacement, such as a furnace, it is the expectation of both the manufacturer's instructions and the code that the new furnace be connected to a code-complying venting system, fuel supply system and electrical circuit, all of which are part of the furnace installation. If the ductwork system was such that the furnace temperature rise across the heat exchanger exceeded the maximum allowable temperature rise, the ductwork system might have to be altered as part of the furnace installation. Existing mechanical systems are accepted on the basis that they are free from hazard although not necessarily compliant with current codes. The code is not retroactive except where specifically stated that it applies to existing systems.

Manufacturer's installation instructions are often updated and changed for various reasons, such as changes in the appliance or equipment design, revisions to the product standards and as a result of field experience related to existing installations. The code official should stay abreast of any changes by reviewing the manufacturer's instructions for every installation.

Equipment and appliances must be installed in accordance with the manufacturer's instructions. The manufacturer's label and installation instructions must be consulted in determining whether or not an appliance or piece of equipment can be installed and operated in a particular hazardous location. Without the manufacturer's installation instructions, an inspector cannot fully perform his or her job, which is why the manufacturer's installation instructions must be available on the job site when the equipment is being inspected.

305.2 Hazardous area. *Equipment* and appliances having an *ignition source* shall not be installed in Group H occupancies or control areas where open use, handling or dispensing of combustible, flammable or explosive materials occurs.

❖ The installation of appliances and equipment having an ignition source within Group H occupancies or control areas where combustible, flammable or explosive materials are used, handled or dispensed is prohibited regardless of compliance with Section 305.3. The presence of an ignition source is too great a risk in such occupancies.

305.3 Elevation of ignition source. *Equipment* and appliances having an *ignition source* shall be elevated such that the source of ignition is not less than 18 inches (457 mm) above the floor in hazardous locations and public garages, private garages, repair garages, motor fuel-dispensing facilities and parking garages. For the purpose of this section, rooms or spaces that are not part of the *living space* of a *dwelling unit* and that communicate directly with a private garage through openings shall be considered to be part of the private garage.

Exception: Elevation of the *ignition source* is not required for appliances that are *listed* as flammable vapor ignition resistant.

❖ To reduce the possibility of ignition of flammable vapors in hazardous areas, the potential sources of ignition must be elevated above the surface supporting the equipment or appliance. It is not the intent of this section for the installer to measure the elevation from the surface of a dedicated appliance stand or other structure built to support the appliance where that stand or structure does not afford room for the storage of flammable liquids. Some flammable and combustible liquids typically associated with hazardous locations give off vapors that are denser than air and tend to collect near the floor. The 18-inch (457 mm) height requirement is intended to reduce the possibility of flammable vapor ignition by keeping the ignition sources elevated above the anticipated level of accumulated vapors. The 18-inch (457 mm) value is a minimum requirement and must be increased when required by the manufacturer's installation instructions.

This section will effectively prohibit the installation of most furnaces, boilers, space heaters and clothes dryers and some water heaters directly on the floor of residential garages.

The accumulation of flammable vapors more than 18 inches (457 mm) deep is unlikely in most ventilated locations; therefore, maintaining all possible sources of ignition at least 18 inches (457 mm) above the floor will substantially reduce the risk of explosion and fire [see Commentary Figures 305.3(1) and 305.3(2)].

In the context of this section, a source of ignition could be a pilot flame, burner, burner igniter or electrical component capable of producing a spark. The term "ignition source" is defined and can be interpreted as meaning an intentional source of ignition, such as for a burner, or an unintentional source of ignition for any flammable vapors that may be present (see the definition of "Ignition source").

An appliance installed in a closet or room that is accessible only from the garage must be considered as part of the garage for application of this section. Even though the room may be separated from the garage by walls and a door, there is no practical means of making the door vapor tight, nor is there any assurance that the door will remain closed during normal use. An appliance room that is accessed only from the outdoors or only from the living space would not be considered as part of the garage (see Section 305.3.1). Rooms such as utility rooms or laundry rooms that communicate with both the garage and the living space and that are considered as part of the living space are not part of the garage. See the definition of "Living space." If a room opens to the garage and that room is not living space, it is part of the garage for the purpose of this section [see Commentary Figure 305.3(3)].

The exception recognizes that new technology exists that allows appliances, such as water heaters, to be tested and listed as being flammable-vapor-ignition resistant and suitable for installation without the 18-inch (457 mm) elevation requirement. This new technology was developed for conventional draft-hood-equipped water heaters, and now the standard for these appliances, ANSI Z21.10.1, mandates that by specified effective dates, all water heaters must comply with the flammable-vapor-ignition criteria in the standard. The technology involves variously configured flame arrestor screens having precisely engineered openings (slots). All combustion air for the appliance is passed through the flame arrestor screen. To pass through the small slots in the screen, the combustion air velocity is increased (Venturi principle) such that it exceeds the velocity at which flames travel through a flammable vapor. If the combustion air contains flammable vapors, the vapors will enter the appliance and be ignited within the combustion chamber. The flames will be contained (trapped) within the combustion chamber because the arrestor screen will not allow flames to exit the combustion

chamber. The result is that vapor ignition is limited to the appliance interior and is not allowed to spread to the atmosphere surrounding the appliance. During such an event, a thermal sensing device shuts down and locks out the appliance. Some appliance manufacturers might require that the appliance be replaced after a flammable-vapor-ignition event. Others require that a sensor be replaced or reset. The appliance standard includes testing requirements to determine whether the arrestor screen is subject to blockage by debris, including lint, plant fibers, dust particles, etc. If the arrestor screen openings become restricted by debris, the appliance will be deprived of combustion air and incomplete combustion and serious appliance malfunction can result. If an appliance shuts down as a result of the arrestor screen being blocked by debris, thus restricting the combustion air supply, typically a thermal device can be reset or replaced to restore operation of the appliance.

If the installation instructions or the label of a water heater equipped with flammable-vapor-ignition-resistant technology state that the appliance must be elevated so that the source of ignition is at least 18 inches (457 mm) above the floor, it must be installed as directed [see Commentary Figures 305.3(1) through 305.3(6)].

For SI: 1 inch = 24.5 mm.

Figure 305.3(1)
GAS-FIRED WATER HEATER
INSTALLATION IN A GARAGE

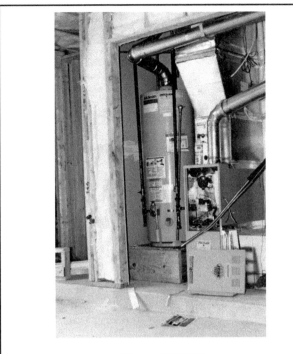

Figure 305.3(2)
SOURCES OF IGNITION
ELEVATED IN PRIVATE GARAGE

305.3.1 (IFGS) Installation in residential garages. In residential garages where appliances are installed in a separate, enclosed space having *access* only from outside of the garage, such appliances shall be permitted to be installed at floor level, provided that the required *combustion air* is taken from the exterior of the garage.

❖ See Commentary Figure 305.3.1 and Section 305.3.

Figure 305.3(4)
EXAMPLE OF FLAMMABLE VAPOR IGNITION-
RESISTANT WATER HEATER TECHNOLOGY
(Figure courtesy of American Water Heater Company)

Figure 305.3(3)
GARAGE SPACE VS. LIVING SPACE

Figure 305.3(5)
EXAMPLE OF FLAMMABLE VAPOR IGNITION-
RESISTANT WATER HEATER TECHNOLOGY
(Photo courtesy of American Water Heater Company)

Figure 305.3(6)
EXAMPLE OF FLAMMABLE VAPOR IGNITION
RESISTANT WATER HEATER TECHNOLOGY
[Photo courtesy of Bradford White Coporation
(Bradford White Defender Safety System™)]

Figure 305.3.1
ELEVATION OF APPLIANCES NOT REQUIRED

305.3.2 Parking garages. Connection of a parking garage with any room in which there is a fuel-fired *appliance* shall be by means of a vestibule providing a two-doorway separation, except that a single door is permitted where the sources of ignition in the *appliance* are elevated in accordance with Section 305.3.

Exception: This section shall not apply to *appliance* installations complying with Section 305.4.

❖ This section addresses parking garages that are covered in Section 406 of the IBC and classified as Group S-2 occupancies. For example, if a boiler room opened into a parking garage, this section requires that the room be separated from the garage by two doors in series. Only a single door is required if the boilers are elevated as required by Section 305.3. The exception recognizes that appliances that are suspended 8 feet (2438 mm) or more above the floor or are installed in accordance with Section 305.3 and NFPA 30A are not likely sources of ignition and therefore, no doors are necessary for isolation. The exception is applicable to appliances installed in a room or anywhere else in a parking garage.

305.4 Public garages. Appliances located in public garages, motor fuel-dispensing facilities, repair garages or other areas frequented by motor vehicles shall be installed not less than 8 feet (2438 mm) above the floor. Where motor vehicles are capable of passing under an appliance, the appliance shall be installed at the clearances required by the appliance manufacturer and not less than 1 foot (305 mm) higher than the tallest vehicle garage door opening.

Exception: The requirements of this section shall not apply where the appliances are protected from motor vehicle impact and installed in accordance with Section 305.3 and NFPA 30A.

❖ Appliances located in motor vehicle areas must be installed to provide protection from vehicle impact. Protection is necessary because the resulting damage from vehicle impact could cause an explosion or fire.

The requirement that appliances be located a minimum of 8 feet (2438 mm) above the floor is intended to provide enough clearance for motor vehicles to pass under the appliance without impact. Although 8 feet (2438 mm) is the minimum height requirement, a clearance of at least 2 feet (610 mm) must always be maintained between the top of the highest vehicle and the bottom of any appliance that is installed where vehicles can pass underneath. If the vehicles that could pass under an appliance are 6 feet (1829 mm) or less in height, the height requirement of 8 feet (2438 mm) will result in a clearance of at least 2 feet (610 mm). If the height of vehicles that could pass under the appliance is greater than 6 feet (1829 mm), the minimum installation height requirement must be

increased to maintain a clearance of not less than 2 feet (610 mm). The height of the largest door or opening that the vehicles must pass through to gain access to the space could be used as a guide to determine the maximum vehicle height in some applications; however, the door height may not be the sole consideration in buildings, such as warehouses, where forklifts or similar elevating vehicles are used. An exception to these requirements occurs when the appliance is installed in accordance with Section 305.3 and protected in accordance with NFPA 30A. Previously, NFPA 88B was referenced, however, this standard was withdrawn as a separate standard and was incorporated into NFPA 30A.

As a means of protecting appliances from motor vehicle impact, the code regulates elevation above the floor as an effective means of protection. Except for suspended unit heaters, the elevation of appliances and equipment above motor vehicles is often impractical, difficult to accomplish and creates access problems. This section allows for an alternative means of protecting appliances located in public garages from vehicle impact.

Although the code does not specify alternative methods of protection, the most apparent method would be to locate the appliance where it could not reasonably be struck by a vehicle. A practical method of protection could be to place a formidable and permanent barrier between the motor vehicles and the appliance. This barrier could include an effectively located vehicle wheel stop that is anchored in place, an elevated platform higher than the vehicle's bumpers or one or more concrete-filled steel pipes.

Where appliances are elevated in accordance with this section, compliance with Section 305.3 is automatically accomplished; however, alternative means of protection may or may not involve elevation of the unit. This section states, in effect, that regardless of the method of protection from vehicular impact, the appliances must be elevated in accordance with either this section or elevated in accordance with Section 305.3. If the installation is in a public repair garage and elevated in accordance with Section 305.3, the installation must also comply with NFPA 30A. NFPA 30A addresses motorized vehicle repair garages and introduces additional criteria and stipulations for nonresidential structures. NFPA 30A addresses appliance elevation, protection and appliance location with respect to operations involving the dispensing or transferring of Class I or II flammable and combustible liquids or liquefied petroleum gas.

In summary, appliances and equipment installed in private and public garages must at minimum comply with Section 305.3 and must be protected from vehicular impact. The area in which they are installed in public garages must also comply with NFPA 30A.

305.5 Private garages. Appliances located in private garages shall be installed with a minimum *clearance* of 6 feet (1829 mm) above the floor.

Exception: The requirements of this section shall not apply where the appliances are protected from motor vehicle impact and installed in accordance with Section 305.3.

❖ Appliances located in a private garage or carport must be protected from vehicle impact (also see Section 304.7). This section is applicable to appliances located in an area where motor vehicles can be operated and includes appliances under which a vehicle can pass and those located anywhere in a vehicle's path where impact is possible. The 6-foot (1829 mm) minimum height requirement is intended to provide adequate clearance above the typical automobile. With the popularity of conversion vans and recreational vehicles, which can be much higher than other automobiles, the 6-foot (1829 mm) minimum installation height above the floor may not provide adequate clearance; additional height might be necessary. The garage door height can be used as a guide in determining the maximum vehicle height [see Commentary Figures 305.5(1) through 305.5(5)].

6'-0" MINIMUM

For SI: 1 foot = 304.8 mm.

For SI: 1 foot = 304.8 mm.

Figure 305.5(1)
SUSPENDED UNIT HEATER

GAS VENT TERMINATIONS

WALLS OF ROOM EXTEND FLOOR TO CEILING

DOOR TO INTERIOR OF GARAGE

GARAGE

FURNACE, SOURCE OF IGNITION MINIMUM OF 18 IN. ABOVE FLOOR

WATER HEATER, SOURCE OF IGNITION MINIMUM OF 18 IN. ABOVE FLOOR

For SI: 1 inch = 25.4 mm.

Figure 305.5(2)
PROTECTED BY WALLS

For SI: 1 inch = 25.4 mm.

Figure 305.5(3)
PROTECTED BY GUARDS

For SI: 1 inch = 25.4 mm.

Figure 305.5(4)
PROTECTED BY A PLATFORM

For SI: 1 inch = 25.4 mm.

Figure 305.5(5)
PROTECTED BY A CURB STOP

305.6 Construction and protection. Boiler rooms and furnace rooms shall be protected as required by the *International Building Code*.

❖ The terms "boiler room" and "furnace room" are defined in the IMC. Section 509 of the IBC addresses boiler and furnace rooms and requires that they be treated as either incidental use areas in Table 509 or as an occupancy group in a mixed occupancy building (see Section 509 of the IBC). In many cases, the IBC requires furnace and boiler rooms to have 1-hour fire-resistance-rated enclosures or automatic fire-extinguishing systems. Note that the gross output of a boiler in Btu/h divided by 33,475 yields boiler horse-power (hp) rating. For example, a boiler having an input rating of 425,000 Btu/h and 80-percent thermal efficiency would have a gross output of 340,000 Btu/h, which is 10.2 hp.

305.7 Clearances from grade. *Equipment* and appliances installed at grade level shall be supported on a level concrete slab or other *approved* material extending not less than 3 inches (76 mm) above adjoining grade or shall be suspended not less than 6 inches (152 mm) above adjoining grade. Such supports shall be installed in accordance with the manufacturer's instructions.

❖ This section addresses outdoor and crawl space installations where an appliance is resting on the earth and not supported by a building. The intent is to maintain level support and protect the appliance from

corrosion and deterioration. The slab should be concrete or a material having comparable strength and longevity. Outdoor appliance installations must be designed to tolerate movement where frost heave will cause a slab to rise and fall.

305.8 Clearances to combustible construction. Heat-producing *equipment* and appliances shall be installed to maintain the required clearances to combustible construction as specified in the listing and manufacturer's instructions. Such clearances shall be reduced only in accordance with Section 308. Clearances to combustibles shall include such considerations as door swing, drawer pull, overhead projections or shelving and window swing. Devices, such as door stops or limits and closers, shall not be used to provide the required clearances.

❖ Section 308 allows clearances to be reduced where protective assemblies are installed to decrease the transfer of heat from the source to the combustible material/assembly. Clearances must be measured from shelving, overhead projections and the possible positions of doors, windows and drawers. Doorstops and closers are easily defeated and subject to failure.

305.9 (IFGS) Parking structures. Appliances installed in enclosed, basement and underground parking structures shall be installed in accordance with NFPA 88A.

❖ NFPA 88A references NFPA 31 for oil-fired appliances and NFPA 211 for venting of appliances.

305.10 (IFGS) Repair garages. Appliances installed in repair garages shall be installed in accordance with NFPA 30A.

❖ The requirements for repair garages are very similar to those for aircraft hangers. The intent is to isolate heating appliances from vehicle fuel vapors that are typically found in garages. Heating appliances are an obvious potential ignition source. It is also important to protect appliances from impact by vehicles in the garage.

305.11 (IFGS) Installation in aircraft hangars. Heaters in aircraft hangars shall be installed in accordance with NFPA 409.

❖ NFPA 409 prohibits open flame heating appliances in aircraft hangers except where such heaters are elevated well above the aircraft. If not elevated, heaters have to be isolated as required in the main paragraph of Section 305.10.

305.12 (IFGS) Avoid strain on gas piping. Appliances shall be supported and connected to the *piping* so as not to exert undue strain on the connections.

❖ Connections under strain can transmit forces to gas controls that can distort valve/control bodies and put strain on burner manifolds and other components. Gas piping must not exert forces on the appliance connection and vice versa. To accomplish this intent, the appliance and the gas piping must be supported independently. "Undue" strain is not specific, but is to be taken to mean any strain that is reasonably capable of threatening the integrity of the appliance connection over any period of time. That being very hard to determine, it would be wise to make the point of connection a neutral point where no forces are transmitted to or from the appliance. For example, where steel gas piping connects to an appliance gas control, it can be determined if the piping is exerting strain on the gas control by simply observing if the piping moves when the union is opened. If the piping moves in some direction, it is exerting a force on the appliance gas control in that same direction.

SECTION 306 (IFGC)
ACCESS AND SERVICE SPACE

[M] 306.1 Access for maintenance and replacement. Appliances, control devices, heat exchangers and HVAC components that utilize energy shall be accessible for inspection, service, repair and replacement without disabling the function of a fire-resistance-rated assembly or removing permanent construction, other appliances, or any other *piping* or ducts not connected to the *appliance* being inspected, serviced, repaired or replaced. A level working space not less than 30 inches (762 mm) deep and 30 inches (762 mm) wide shall be provided in front of the control side to service an *appliance*.

❖ Because equipment and appliances require routine maintenance, repairs and possible replacement, access is required. Additionally, access recommendations or requirements are usually stated in the manufacturer's installation instructions. As a result, the provisions stated here are intended to supplement the manufacturer's installation instructions.

The intent of this section is to ensure access to components that require observation, inspection, adjustment, servicing, repair or replacement. Access is also necessary to conduct operating procedures, such as startup or shutdown.

The code defines access as being able to be reached by a means that first may require the removal of a panel, door or similar obstruction. An appliance or piece of equipment is not accessible if any portion of the structure's permanent finish materials, such as drywall, plaster, paneling, built-in furniture, cabinets or any other similar permanently affixed building component must be removed.

The intent is to ensure access to components such as controls, gauges, burners, filters, blowers and motors that require observation, inspection, adjustment, servicing, repair and replacement.

[M] 306.2 Appliances in rooms. Rooms containing appliances shall be provided with a door and an unobstructed passageway measuring not less than 36 inches (914 mm) wide and 80 inches (2032 mm) high.

> **Exception:** Within a *dwelling unit*, appliances installed in a compartment, alcove, basement or similar space shall be provided with *access* by an opening or door and an unobstructed passageway measuring not less than 24 inches (610 mm) wide and large enough to allow removal of the largest *appliance* in the space, provided that a level service space of not less than 30 inches (762 mm) deep and the height of the *appliance*, but not less than 30 inches (762 mm), is present at the front or service side of the *appliance* with the door open.

❖ Access opening and passageway dimensions are specified to give service personnel reasonable access to appliances and to allow for the passage of system components. Quite often appliances such as furnaces, boilers and water heaters are installed in spaces with little or no forethought about future access for maintenance or replacement.

[M] 306.3 Appliances in attics. Attics containing appliances shall be provided with an opening and unobstructed passageway large enough to allow removal of the largest *appliance*. The passageway shall be not less than 30 inches (762 mm) high and 22 inches (559 mm) wide and not more than 20 feet (6096 mm) in length measured along the centerline of the passageway from the opening to the *appliance*. The passageway shall have continuous solid flooring not less than 24 inches (610 mm) wide. A level service space not less than 30 inches (762 mm) deep and 30 inches (762 mm) wide shall be present at the front or service side of the *appliance*. The clear *access* opening dimensions shall be not less than 20 inches by

30 inches (508 mm by 762 mm) and large enough to allow removal of the largest *appliance*.

Exceptions:

1. The passageway and level service space are not required where the *appliance* is capable of being serviced and removed through the required opening.

2. Where the passageway is not less than 6 feet (1829 mm) high for its entire length, the passageway shall be not greater than 50 feet (15 250 mm) in length.

❖ There is not always sufficient room for installation of equipment and appliances in spaces, such as basements, alcoves, utility rooms and furnace rooms. In an effort to save floor space or simplify an installation, appliances and equipment are often installed in attics or in similar remote locations. Access to appliances and equipment could be difficult because of the lack of a walking surface, such as might occur in an attic or similar space with exposed ceiling joists. The intent of this section is to require a suitable access opening, passageway and work space that will allow reasonably easy access without endangering the service person. The longer the attic passageway, the more the service person will be exposed to extreme tem-

peratures and risk of injury. In the event of a medical emergency, it might be difficult or impossible to escape from an attic space. Note that some appliances might not be listed for attic installation or might otherwise be unsuitable for such conditions (see Commentary Figure 306.3). The access opening must be large enough to allow the appliances to be removed in whole. For example, after disconnecting ductwork plenums and coil cabinets, a furnace or air handler should pass through the access opening without disassembly of the unit itself.

[M] 306.3.1 Electrical requirements. A luminaire controlled by a switch located at the required passageway opening and a receptacle outlet shall be provided at or near the *appliance* location in accordance with NFPA 70.

❖ A lighting outlet and receptacle outlet will encourage and facilitate appliance maintenance. The receptacle will accommodate power tools, drop lights and diagnostic instruments. Also, these provisions will negate the need for extension cords, which can be hazardous to service personnel.

[M] 306.4 Appliances under floors. Under-floor spaces containing appliances shall be provided with an *access* opening

LIGHTING FIXTURE (LUMINAIRE)
(LAMP PROTECTION CAGE DESIRABLE)

RECEPTACLE OUTLET

GAS VENT

UNOBSTRUCTED PASSAGEWAY MINIMUM 30 IN. HIGH AND 22 IN. WIDE

SERVICE SIDE OF APPLIANCE

FURNACE

LIGHT SWITCH

20 MAX. TO FURANCE

24

MINIMUM 30 IN. × 30 IN. FLOOR AREA AT REQUIRED ACCESS SIDE

SOLID FLOORING MINIMUM 24 IN. WIDE

20 IN. × 30 IN. MINIMUM ATTIC OPENING

For SI: 1 inch = 25.4 mm.

**Figure 306.3
ATTIC FURNACE INSTALLATION**

and unobstructed passageway large enough to remove the largest *appliance*. The passageway shall be not less than 30 inches (762 mm) high and 22 inches (559 mm) wide, nor more than 20 feet (6096 mm) in length measured along the centerline of the passageway from the opening to the *appliance*. A level service space not less than 30 inches (762 mm) deep and 30 inches (762 mm) wide shall be present at the front or service side of the *appliance*. If the depth of the passageway or the service space exceeds 12 inches (305 mm) below the adjoining grade, the walls of the passageway shall be lined with concrete or masonry extending 4 inches (102 mm) above the adjoining grade and having sufficient lateral-bearing capacity to resist collapse. The clear *access* opening dimensions shall be not less than 22 inches by 30 inches (559 mm by 762 mm), and large enough to allow removal of the largest *appliance*.

Exceptions:

1. The passageway is not required where the level service space is present when the *access* is open and the *appliance* is capable of being serviced and removed through the required opening.

2. Where the passageway is not less than 6 feet high (1829 mm) for its entire length, the passageway shall not be limited in length.

❖ The intent of this section, which applies to crawl spaces, is the same as for Section 306.3. The more difficult access to appliances and equipment is, the less likely it is that the appliance or equipment will be inspected and serviced. Attic and crawl space installations suffer from the "out-of-sight, out-of-mind" syndrome (see commentary, Section 306.3). Passageways 6 feet (1829 mm) or more in height would allow a person to walk erect or nearly so, thus

allowing quicker egress in the event of an emergency (see Commentary Figure 306.4).

[M] 306.4.1 Electrical requirements. A luminaire controlled by a switch located at the required passageway opening and a receptacle outlet shall be provided at or near the *appliance* location in accordance with NFPA 70.

❖ Appliances located in attics and under-floor spaces are generally not easy to access. The light switch, lighting fixture and receptacle outlet required by this section protect those who service the equipment and appliances. The receptacle will accommodate power tools, drop lights and diagnostic instruments. Also, these provisions will negate the need for extension cords, which can be hazardous to service personnel. NFPA 70 requires ground fault circuit interrupter protection for receptacles located in a crawl space.

[M] 306.5 Equipment and appliances on roofs or elevated structures. Where equipment requiring access or appliances are located on an elevated structure or the roof of a building such that personnel will have to climb higher than 16 feet (4877 mm) above grade to access such equipment or appliances, an interior or exterior means of access shall be provided. Such access shall not require climbing over obstructions greater than 30 inches (762 mm) in height or walking on roofs having a slope greater than 4 units vertical in 12 units horizontal (33-percent slope). Such access shall not require the use of portable ladders.

Permanent ladders installed to provide the required *access* shall comply with the following minimum design criteria:

1. The side railing shall extend above the parapet or roof edge not less than 30 inches (762 mm).

2. Ladders shall have rung spacing not to exceed 14 inches (356 mm) on center. The upper-most rung shall

Figure 306.4
CRAWL SPACE FURNACE INSTALLATION

be not more than 24 inches (610 mm) below the upper edge of the roof hatch, roof or parapet, as applicable.

3. Ladders shall have a toe spacing not less than 6 inches (152 mm) deep.

4. There shall be not less than 18 inches (457 mm) between rails.

5. Rungs shall have a diameter not less than 0.75-inch (19 mm) and be capable of withstanding a 300-pound (136.1 kg) load.

6. Ladders over 30 feet (9144 mm) in height shall be provided with offset sections and landings capable of withstanding 100 pounds per square foot (488.2 kg/m²). Landing dimensions shall be not less than 18 inches (457 mm) and not less than the width of the ladder served. A guard rail shall be provided on all open sides of the landing.

7. Climbing clearance. The distance from the centerline of the rungs to the nearest permanent object on the climbing side of the ladder shall be not less than 30 inches (762 mm) measured perpendicular to the rungs. This distance shall be maintained from the point of ladder access to the bottom of the roof hatch. A minimum clear width of 15 inches (381 mm) shall be provided on both sides of the ladder measured from the midpoint of and parallel with the rungs, except where cages or wells are installed.

8. Landing required. The ladder shall be provided with a clear and unobstructed bottom landing area having a minimum dimension of 30 inches by 30 inches (762 mm by 762 mm) centered in front of the ladder.

9. Ladders shall be protected against corrosion by *approved* means.

10. Access to ladders shall be provided at all times.

Catwalks installed to provide the required *access* shall be not less than 24 inches (610 mm) wide and shall have railings as required for service platforms.

Exception: This section shall not apply to Group R-3 occupancies.

❖ The requirements of this section are intended to prevent portable ladders from being the only means of access to elevated appliances and equipment located above the specified height. These provisions are for the protection of service and inspection personnel. The criteria for constructing and installing permanent ladders and catwalks are intended to make them safe to use. The 16-foot height is always measured from grade. In effect, any building over one story in height will require a permanent means of access to rooftop appliances.

If a roof surface on which an appliance rests is less than 16 feet above grade, but the parapet wall cap is greater than 16 feet above grade, this section will require a permanent means of access because service personnel will have to climb over the parapet if using a portable ladder on the exterior of the building (see Commentary Figure 306.5).

**Figure 306.5
APPLIANCE ACCESS**

[M] 306.5.1 Sloped roofs. Where appliances, *equipment*, fans or other components that require service are installed on a roof having a slope of 3 units vertical in 12 units horizontal (25-percent slope) or greater and having an edge more than 30 inches (762 mm) above grade at such edge, a level platform shall be provided on each side of the *appliance* or *equipment* to which *access* is required for service, repair or maintenance. The platform shall be not less than 30 inches (762 mm) in any dimension and shall be provided with guards. The guards shall extend not less than 42 inches (1067 mm) above the platform, shall be constructed so as to prevent the passage of a 21-inch-diameter (533 mm) sphere and shall comply with the loading requirements for guards specified in the *International Building Code*. *Access* shall not require walking on roofs having a slope greater than 4 units vertical in 12 units horizontal (33-percent slope). Where *access* involves obstructions greater than 30 inches (762 mm) in height, such obstructions shall be provided with ladders installed in accordance with Section 306.5 or stairways installed in accordance with the requirements specified in the *International Building Code* in the path of travel to and from appliances, fans or *equipment* requiring service.

❖ A work space platform with guards will provide protection for service personnel and will facilitate inspection, servicing and repair of appliances and equipment (see Commentary Figure 306.5.1). Working on a sloped surface is difficult and dangerous, and results in tools and materials sliding off roofs, possibly endangering persons below. Because of the additional expense and unattractive appearance of roof platforms, this section may have the effect of discouraging the installation of appliances on sloped roofs. Note that if the roof slope is greater than 4/12, some means such as a catwalk would be required to provide access to the level platform.

[M] 306.5.2 Electrical requirements. A receptacle outlet shall be provided at or near the *appliance* location in accordance with NFPA 70.

❖ A lighting outlet is not required for servicing roof-mounted equipment. A receptacle outlet will encourage and facilitate appliance maintenance. The receptacle will accommodate power tools, drop lights, and diagnostic instruments. Also, these provisions will negate the need for extension cords, which can be hazardous to service personnel. NFPA 70: *National Electrical Code*® (NEC®) requires ground fault circuit interrupter protection for rooftop receptacles.

[M] 306.6 Guards. Guards shall be provided where various components that require service and roof hatch openings are located within 10 feet (3048 mm) of a roof edge or open side of a walking surface and such edge or open side is located more than 30 inches (762 mm) above the floor, roof, or grade below. The guard shall extend not less than 30 inches (762 mm) beyond each end of components that require service. The top of the guard shall be located not less than 42 inches (1067 mm) above the elevated surface adjacent to the guard. The guard shall be constructed so as to prevent the passage of a 21-inch-diameter (533 mm) sphere and shall

Figure 306.5.1
ROOFTOP HVAC UNIT WITH WORK PLATFORM
(Photo courtesy of Reznor/Thomas & Betts Corporation)

comply with the loading requirements for guards specified in the *International Building Code*.

Exception: Guards are not required where permanent fall arrest/restraint anchorage connector devices that comply with ANSI/ASSE Z 359.1 are affixed for use during the entire lifetime of the roof covering. The devices shall be re-evaluated for possible replacement when the entire roof covering is replaced. The devices shall be placed not more than 10 feet (3048 mm) on center along hip and ridge lines and placed not less than 10 feet (3048 mm) from roof edges and the open sides of walking surfaces.

❖ Equipment and appliances require routine inspection, maintenance and repair. Where the units requiring service are located within 10 feet (3048 mm) of a roof edge or open side with a drop greater than 30 inches (762 mm), a possibility exists for the service person to be injured by a fall. The requirements of this section are intended to protect service personnel from the hazard of falls. This section pertains to equipment and appliances regulated by the code that require routine inspection, maintenance and repair and also applies to roof hatch openings.

The guard serves as a warning and as a protective barrier, and is required to be 42 inches (1067 mm) high and constructed to prevent the passage of a 21-inch (533 mm) sphere [see Commentary Figures 306.6(1) and 306.6(2)]. The guard must extend 30 inches (762 mm) beyond each end of the equipment, or appliance or roof hatch opening to provide added protection for the installer or service technician by enlarging the protected area beyond the immediate area of the equipment or appliance.

For SI: 1 inch = 25.4 mm, 1 foot = 304.8 mm.

Figure 306.6(1)
GUARDS

For SI: 1 inch = 25.4 mm, 1 foot = 304.8 mm.

Figure 306.6(2)
ROOF PLAN VIEW

SECTION 307 (IFGC)
CONDENSATE DISPOSAL

307.1 Evaporators and cooling coils. Condensate drainage systems shall be provided for *equipment* and appliances containing evaporators and cooling coils in accordance with the *International Mechanical Code.*

❖ This section recognizes that gas-fired space heating appliances are commonly part of space cooling and conditioning systems and the condensate drainage provisions of the IMC are applicable. Sections 307.2 through 307.5 address condensate relative to combustion of fuel gas, whereas, this section addresses condensate produced by refrigeration and cooling coils.

307.2 Fuel-burning appliances. Liquid combustion by-products of condensing appliances shall be collected and discharged to an *approved* plumbing fixture or disposal area in accordance with the manufacturer's instructions. Condensate *piping* shall be of *approved* corrosion-resistant material and shall be not smaller than the drain connection on the *appliance.* Such *piping* shall maintain a minimum slope in the direction of discharge of not less than one-eighth unit vertical in 12 units horizontal (1-percent slope).

❖ This section contains detailed requirements for the disposal of condensate from appliances such as condensing furnaces, boilers and water heaters. Condensation must be collected from appliances and equipment in accordance with the manufacturer's installation instructions and this section. High-efficiency condensing-type appliances and some mid-efficiency appliances produce water as a combustion byproduct. This condensate is collected at various points in the appliance heat exchangers and venting system and must be disposed of. Because of impurities in the fuel and combustion air, the condensate

can be acidic and thus corrosive to many materials. For example, such condensate can contain hydrochloric and sulfuric acids, although the acid solution would be weak. The slope requirement promotes drainage and prohibits low points (dips) in the piping that would trap water and air that would impede flow.

[M] 307.3 Drain pipe materials and sizes. Components of the condensate disposal system shall be cast iron, galvanized steel, copper, cross-linked polyethylene, polyethylene, ABS, CPVC, PVC or polypropylene pipe or tubing. Components shall be selected for the pressure and temperature rating of the installation. Joints and connections shall be made in accordance with the applicable provisions of Chapter 7 of the *International Plumbing Code* relative to the material type. Condensate waste and drain line size shall be not less than $^{3}/_{4}$-inch (19 mm) internal diameter and shall not decrease in size from the drain pan connection to the place of condensate disposal. Where the drain pipes from more than one unit are manifolded together for condensate drainage, the pipe or tubing shall be sized in accordance with an *approved* method.

❖ Condensate drains must be constructed of a material listed in this section. The materials listed are corrosion resistant. These drains must be straight runs of pipe without sags or dips that would trap liquid. If coiled stock tubing is used, it would be difficult to maintain a uniform slope, and the low points (dips) in the tubing would trap water, causing the drain to be double trapped (air bound). Air-bound drains will not allow flow. When drains are merged, the piping must be properly sized for the aggregate flow.

307.4 Traps. Condensate drains shall be trapped as required by the *equipment* or *appliance* manufacturer.

❖ The appliance or equipment manufacturer determines the need for a trap and often specifies the depth and configuration of the trap. The traps addressed in this section are unrelated to plumbing traps and serve a different purpose. Condensate drain traps do not directly connect in any way to the plumbing drain or the waste and vent system of a building. Condensate drain traps are installed to prevent air and combustion gases from being pushed or pulled through the drain piping. Such airflow can impede condensate flow, causing overflow or abnormal water depth in drain pans and components.

307.5 Auxiliary drain pan. Category IV condensing appliances shall be provided with an auxiliary drain pan where damage to any building component will occur as a result of stoppage in the condensate drainage system. Such pan shall be installed in accordance with the applicable provisions of Section 307 of the *International Mechanical Code.*

> **Exception:** An auxiliary drain pan shall not be required for appliances that automatically shut down operation in the event of a stoppage in the condensate drainage system.

❖ This section treats condensing (Category IV) appliances the same as the IMC treats appliances and equipment having evaporators and cooling coils. Category IV appliances produce condensate and, like cooling equipment and evaporators, can cause structural damage to a building if the condensate drainage

system fails. This section parallels Section 307 of the IMC. Section 307.2.3 of the IMC is applicable, however, only Items 1 and 3 because this section of the code requires an auxiliary pan in all cases. In many cases, this provision will require nothing more than what is already being provided for the cooling coil or evaporator that is part of the gas appliance installation (e.g., the typical gas-fired furnace with cooling coil/evaporator). For example, an up-flow Category IV furnace with an "A" coil on its outlet is installed in an attic, and it is sitting in an auxiliary drain pan. This pan can also serve as protection for the furnace; therefore, no additional installation is required because the furnace happened to be a condensing type (see Commentary Figure 307.4).

The exception recognizes that some condensing appliances are designed to shut down automatically upon failure of the condensate drainage system. For example, the design/test standards for some appliances allow a choice of appliance reaction to a failed condensate drain. The appliance may either continue to function safely with a failed condensate drain or the appliance must shut down upon failure of the condensate drain.

Figure 307.4
AUXILIARY DRAIN PAN REQUIRED (WITH OR WITHOUT COOLING COIL)

307.6 Condensate pumps. Condensate pumps located in uninhabitable spaces, such as attics and crawl spaces, shall be connected to the *appliance* or *equipment* served such that when the pump fails, the *appliance* or *equipment* will be prevented from operating. Pumps shall be installed in accordance with the manufacturer's instructions.

❖ Condensate pumps are often located in attics and crawl spaces and above ceilings where they are not readily observable. If they fail, the condensate overflow can cause structural damage to the building, especially where the overflow will not be noticed immediately. The majority of such pumps are equipped with simple float controls that can be wired in series with the appliance/equipment control circuit. When the pump system fails, the float will rise in the reservoir and open a switch before the condensate starts to overflow the reservoir. These float controls are commonly not connected and, in other cases, the pump might not be equipped with an overflow switch. This new code section requires the installation of condensate pumps that have this overflow shutoff capability and requires that the appliance/equipment served be connected to take advantage of that feature (see Commentary Figure 307.3).

SECTION 308 (IFGS) CLEARANCE REDUCTION

308.1 Scope. This section shall govern the reduction in required clearances to *combustible materials*, including gypsum board, and *combustible assemblies* for chimneys, vents, appliances, devices and equipment. Clearance requirements for air-conditioning equipment and central heating boilers and furnaces shall comply with Sections 308.3 and 308.4.

❖ Heat-producing appliances and mechanical equipment must be installed with the required minimum clearances to combustible materials indicated by their listing label. It is not uncommon to encounter practical or structural difficulties in maintaining clearances. Therefore, clearance reduction methods have been developed to allow, in some cases, reduction of the

minimum prescribed clearance distance while achieving equivalent protection. An important understanding is that all prescribed clearances to combustibles are airspace clearances measured from the heat source to the face of the nearest combustible surface (see Commentary Figure 308.1). When using the assemblies described in Table 308.2, the clearance is measured as described in Note b. In the case of listed appliances and equipment, the required clearances are intended to be clear airspace, and therefore the space is not to be filled with insulation or any other material other than an assembly intended to allow clearance reduction. This is especially important where clearances are required from appliances and equipment that rely on the airspace for convection cooling to maintain their proper operating temperature.

The provisions contained in this section are based on the principles of heat transfer. Mechanical equipment or appliances producing heat can become hot, and many appliances have hot exterior surfaces by design. The heat energy is then radiated to objects surrounding the appliances or equipment. When mechanical equipment and appliances are tested, the minimum clearances are established so that radiant and, to a lesser extent, convective heat transfer do not represent an ignition hazard to adjacent surfaces and objects. This distance is called the "required clearance" to combustible materials. Appliance and equipment labels must specify minimum clearances in all directions.

This section permits the use of materials and systems as radiation shields, decreasing the amount of heat energy transferred to surrounding objects and reducing the required clearances between gas-fired appliances and equipment to combustibles.

Plaster and gypsum by themselves are classified as noncombustible materials. Under continued exposure to heat, however, these materials will gradually decompose as water molecules are driven out of the material. Plaster on wood lath, plasterboard, sheetrock and drywall are all considered to be combustible

For SI: 1 inch = 25.4 mm.

Figure 307.3
CONDENSATE PUMP

materials when the code provisions are applied.

Additionally, gypsum wallboard is faced with paper that has a flame spread index that is measurable in the ASTM E84 test. This alone identifies the need to classify gypsum wallboard as a combustible material for the purpose of requiring a separation from heat-producing equipment and appliances.

The code specifically prohibits the reduction of clearances in certain applications. For examples, see Sections 308.2 through 308.4.

Figure 308.1
CLEARANCE TO COMBUSTIBLES—EXAMPLES OF HOW TO MEASURE

308.2 Reduction table. The allowable *clearance* reduction shall be based on one of the methods specified in Table 308.2 or shall utilize a reduced *clearance* protective assembly *listed* and *labeled* in accordance with UL 1618. Where required clearances are not listed in Table 308.2, the reduced clearances shall be determined by linear interpolation between the distances listed in the table. Reduced clearances shall not be derived by extrapolation below the range of the table. The reduction of the required clearances to combustibles for *listed* and *labeled* appliances and *equipment* shall be in accordance with the requirements of this section, except that such clearances shall not be reduced where reduction is specifically prohibited by the terms of the *appliance* or *equipment* listing [see Figures 308.2(1) through 308.2(3)].

❖ Another option for reducing the required clearances to combustible materials is to use one of the on-site field-constructed methods specified in Table 308.2.

"A" equals the clearance with no protection.
"B" equals the reduced clearance permitted in accordance with Table 308.2. The protection applied to the construction using *combustible material* shall extend far enough in each direction to make "C" equal to "A."

FIGURE 308.2(1)
EXTENT OF PROTECTION NECESSARY TO REDUCE CLEARANCES FROM APPLIANCE OR VENT CONNECTIONS

For SI: 1 inch = 25.4 mm.

FIGURE 308.2(3)
MASONRY CLEARANCE REDUCTION SYSTEM

For SI: 1 inch = 25.4 mm.

FIGURE 308.2(2)
WALL PROTECTOR CLEARANCE REDUCTION SYSTEM

TABLE 308.2. See below.

❖ The column headings of Table 308.2 list required clearances without protection. The numbers to the right of each method indicate the permissible reduced clearance measured from the heat-producing appliances/equipment to the face of the combustible material.

The rationale behind the methods of protection listed in Table 308.2 is based on the ability of the protection to reduce radiant heat transmission from the appliance and equipment to the combustible material so that the temperature rise of the combustible material will remain below the maximum allowed.

Although the materials referred to in Table 308.2 are common construction materials, confusion often arises over what satisfies the requirement for "insula-

tion board" (Item 6 in the table), sometimes referred to as inorganic insulating board, noncombustible mineral board or noncombustible insulating board. These products are not made of carbon-based compounds.

Carbon-based compounds are those found in cellulose (wood), plastics and other materials manufactured from raw materials that once existed in living organisms. Cement board materials must have a specified maximum "C" (conductance) value in addition to being noncombustible.

Note h specifies a maximum thermal conductivity of insulation board. Conductivity is the amount of heat in Btus that will flow each hour through a 1-foot-square slab of material, 1-inch thick with a 1°F temperature difference between both sides and is usually identified by the symbol k.

TABLE 308.2 a through k
REDUCTION OF CLEARANCES WITH SPECIFIED FORMS OF PROTECTION

TYPE OF PROTECTION APPLIED TO AND COVERING ALL SURFACES OF COMBUSTIBLE MATERIAL WITHIN THE DISTANCE SPECIFIED AS THE REQUIRED CLEARANCE WITH NO PROTECTION [see Figures 308.2(1), 308.2(2), and 308.2(3)]	WHERE THE REQUIRED CLEARANCE WITH NO PROTECTION FROM APPLIANCE, VENT CONNECTOR, OR SINGLE-WALL METAL PIPE IS: (inches)									
	36		18		12		9		6	
	Allowable clearances with specified protection (inches)									
	Use Column 1 for clearances above appliance or horizontal connector. Use Column 2 for clearances from appliance, vertical connector and single-wall metal pipe.									
	Above Col. 1	Sides and rear Col. 2	Above Col. 1	Sides and rear Col. 2	Above Col. 1	Sides and rear Col. 2	Above Col. 1	Sides and rear Col. 2	Above Col. 1	Sides and rear Col. 2
1. 3½-inch-thick masonry wall without ventilated airspace	—	24	—	12	—	9	—	6	—	5
2. ½-inch insulation board over 1-inch glass fiber or mineral wool batts	24	18	12	9	9	6	6	5	4	3
3. 0.024-inch (nominal 24 gage) sheet metal over 1-inch glass fiber or mineral wool batts reinforced with wire on rear face with ventilated airspace	18	12	9	6	6	4	5	3	3	3
4. 3½-inch-thick masonry wall with ventilated airspace	—	12	—	6	—	6	—	6	—	6
5. 0.024-inch (nominal 24 gage) sheet metal with ventilated airspace	18	12	9	6	6	4	5	3	3	2
6. ½-inch-thick insulation board with ventilated airspace	18	12	9	6	6	4	5	3	3	3
7. 0.024-inch (nominal 24 gage) sheet metal with ventilated airspace over 0.024-inch (nominal 24 gage) sheet metal with ventilated airspace	18	12	9	6	6	4	5	3	3	3
8. 1-inch glass fiber or mineral wool batts sandwiched between two sheets 0.024-inch (nominal 24 gage) sheet metal with ventilated airspace	18	12	9	6	6	4	5	3	3	3

For SI: 1 inch = 25.4 mm, °C = [(°F - 32)/1.8], 1 pound per cubic foot = 16.02 kg/m³, 1 Btu per inch per square foot per hour per °F = 0.144 W/m² · K.

a. Reduction of clearances from *combustible materials* shall not interfere with combustion air, draft hood clearance and relief, and accessibility of servicing.

b. All clearances shall be measured from the outer surface of the *combustible material* to the nearest point on the surface of the appliance, disregarding any intervening protection applied to the *combustible material*.

c. Spacers and ties shall be of *noncombustible material*. No spacer or tie shall be used directly opposite an appliance or connector.

d. For all clearance reduction systems using a ventilated airspace, adequate provision for air circulation shall be provided as described [see Figures 308.2(2) and 308.2(3)].

e. There shall be not less than 1 inch between clearance reduction systems and combustible walls and ceilings for reduction systems using ventilated airspace.

f. Where a wall protector is mounted on a single flat wall away from corners, it shall have a minimum 1-inch air gap. To provide air circulation, the bottom and top edges, or only the side and top edges, or all edges shall be left open.

g. Mineral wool batts (blanket or board) shall have a minimum density of 8 pounds per cubic foot and a minimum melting point of 1500°F.

h. Insulation material used as part of a clearance reduction system shall have a thermal conductivity of 1.0 Btu per inch per square foot per hour per °F or less.

i. There shall be not less than 1 inch between the appliance and the protector. In no case shall the clearance between the appliance and the combustible surface be reduced below that allowed in this table.

j. All clearances and thicknesses are minimum; larger clearances and thicknesses are acceptable.

k. Listed single-wall connectors shall be installed in accordance with the manufacturer's instructions.

2015 INTERNATIONAL FUEL GAS CODE® COMMENTARY

Thermal conductance (overall) is the time rate of heat flow through a body not taking thickness into account and is usually identified by the symbol C.

Thermal resistance (overall) is the reciprocal of overall thermal conductance and is usually identified by the symbol R.

This translates into a minimum required insulation R-value of 1.0 (square foot × hour × °F)/Btu per inch of insulating material. The methods in Table 308.2 control heat transmission by reflecting heat radiation, retarding thermal conductance and providing convective cooling. Where sheet metal materials or metal plates are specified, the effectiveness of the protection can be enhanced by the reflective surface of the metal. Painting or otherwise covering the surface would reduce the metal's ability to reflect radiant heat and, depending on the color, could increase heat absorption. The airspace between the protected surface and the clearance-reduction assembly allows convection air currents to cool the protection assembly by carrying away heat that has been conducted through the assembly. Where a clearance-reduction assembly must be spaced 1 inch (25 mm) off the wall, the top, bottom and sides of the assembly must remain open as required by Notes d and f to permit unrestricted airflow (convection currents). If the openings were not provided, the air-cooling effect would not take place, and the protection assembly would not be as effective in limiting the temperature rise on the protected surfaces. Ideally, the protection assembly should be open on all sides to provide maximum ventilation. Figure 308.2(2) and Commentary Figure 308.2 show assemblies incorporating airspace.

Spacers must be noncombustible. Spacers should not be placed directly behind the heat source because the location would increase the amount of heat conduction through the spacer, thus creating a "hot spot." Figure 308.2(2) specifically shows a noncombustible spacer arrangement.

The performance of a protective assembly when applied to a horizontal surface, such as a ceiling, will differ substantially from the same assembly placed in a vertical plane. Obviously, temperatures at a ceiling surface will be higher because of natural convection and because the air circulation between the method of protection and the protected ceiling surface will be substantially reduced or nonexistent. It is for these reasons that Table 308.2 is divided into two application groups.

The manufacturer's instructions or label for many appliances will state an absolute minimum clearance, regardless of any clearance reduction method used. Those clearance requirements take precedence over Table 308.2.

The methods in Table 308.2 are intended to be permanent installations properly supported to prevent displacement or deformation. Movement could adversely affect the performance of the protection method, thus posing a potential fire hazard.

The assemblies in Table 308.2 are the product of experience and testing. To achieve predictable and dependable performance, the components of the various assemblies cannot be mixed, matched, combined or otherwise rearranged to comprise new assemblies, and materials cannot be substituted for those prescribed in the table. Any alterations or sub-

Figure 308.2
REDUCED CLEARANCE TO COMBUSTIBLES

stitutions could have an effect on the assembly, and its performance must be tested and approved.

The reduced clearance is measured from the heat source to the combustible material, disregarding any intervening protection assembly.

Factory-built clearance reduction assemblies are available that are listed to UL 1618. Such assemblies may or may not require field assembly of the components.

308.3 Clearances for indoor air-conditioning appliances. *Clearance* requirements for indoor air-conditioning appliances shall comply with Sections 308.3.1 through 308.3.4.

❖ Requirements for clearances to combustibles are emphasized because of the potential fire hazard posed when those clearances are not observed. Maintaining an appropriate distance from the outer surfaces of an appliance or piece of equipment reduces the possibility of ignition of combustible materials.

The minimum clearances to combustibles are specified in the manufacturer's installation instructions for a labeled appliance. Because an approved agency tests appliances in accordance with these instructions, the clearances required are necessary for correct installation and operation of the appliance.

Reduction of the required clearances to combustibles is allowed only when the combustibles are protected by one of the methods outlined in Section 308 and clearance reduction is not prohibited by Section 308, or the appliance and equipment listing (see the commentary for Section 308). The manufacturer's specified minimum clearances and the clearances specified in Section 308 are all airspace clearances, and such spaces cannot be filled with insulation or any other material, even if the material is noncombustible. In some cases, the manufacturer's installation instructions will specify absolute minimum clearances that must not be reduced by any clearance reduction method.

The most common wall covering material, gypsum wallboard, is a combustible finish material for the purpose of the code. As a result, gypsum wallboard, as well as all other combustible wall finishes, must be separated from an appliance or equipment in accordance with the prescribed clearance to combustibles. Clearances to combustibles also apply to furnishings, window treatments and movable items that can be placed within the required clearance range of appliances and equipment.

308.3.1 Appliance clearances. Air-conditioning appliances shall be installed with clearances in accordance with the manufacturer's instructions.

❖ The concept of "rooms large or not large in comparison with the appliance" is no longer part of the code. The code has been simplified by no longer tying rules for clearances to whether or not an appliance is installed in a room having a specific volume. The dated concept of room volume in comparison to appliance volume was meant to determine when the room was considered to be a closet or alcove. If the room

was determined to be a closet or alcove, the code would then prohibit the use of clearance reduction methods and require the appliances to be listed for installation in closets and alcoves because the small room volume would not allow adequate dissipation of heat. This concept had outlived its intent because modern appliances are more efficient, have less jacket heat losses and are listed to current appliance standards. Now, the code simply defers to the appliance manufacturer's instructions and the appliance listing to establish the clearance requirements. In some cases, the installation instructions for an appliance might specifically prohibit the application of a clearance reduction method (see Section 308.2).

308.3.2 Clearance reduction. Air-conditioning appliances shall be permitted to be installed with reduced clearances to *combustible material*, provided that the *combustible material* or *appliance* is protected as described in Table 308.2 and such reduction is allowed by the manufacturer's instructions.

❖ The clearance reduction methods in the code are always allowed unless the manufacturer's instructions specifically prohibit the reduction of clearances or specify different clearance reduction methods (see commentary, Section 308.3.1). Note that some appliances specify clearances that are absolute minimums that cannot be reduced by any means.

308.3.3 Plenum clearances. Where the *furnace plenum* is adjacent to plaster on metal lath or *noncombustible material* attached to *combustible material*, the *clearance* shall be measured to the surface of the plaster or other noncombustible finish where the *clearance* specified is 2 inches (51 mm) or less.

❖ Where the air-conditioning equipment plenum is adjacent to surfaces covered by noncombustible material as described in this section and the required clearance from combustibles is 2 inches (51 mm) or less, the required clearance must be measured to the surface of the adjacent material. Measuring to the combustible material would reduce the clearance between the equipment plenum and adjacent surface. Reducing this clearance would affect the cooling of the equipment and could cause overheating.

308.3.4 Clearance from supply ducts. Supply air ducts connecting to listed central heating furnaces shall have the same minimum clearance to combustibles as required for the furnace supply plenum for a distance of not less than 3 feet (914 mm) from the supply plenum. Clearance is not required beyond the 3-foot (914 mm) distance.

❖ Portions of ducts that are within 3 feet (914 mm) of a supply plenum to which they connect will reach nearly the same temperature as the plenum.

308.4 Central-heating boilers and furnaces. *Clearance* requirements for central-heating boilers and furnaces shall comply with Sections 308.4.1 through 308.4.5. The *clearance* to these appliances shall not interfere with *combustion air*; draft hood *clearance* and relief; and accessibility for servicing.

❖ See the commentary to Section 308.3.

308.4.1 Appliance clearances. Central-heating furnaces and low-pressure boilers shall be installed with clearances in accordance with the manufacturer's instructions.

❖ See the commentary to Section 308.3.1.

308.4.2 Clearance reduction. Central-heating furnaces and low-pressure boilers shall be permitted to be installed with reduced clearances to *combustible material* provided that the *combustible material* or *appliance* is protected as described in Table 308.2 and such reduction is allowed by the manufacturer's instructions.

❖ See the commentary to Section 308.3.2.

308.4.3 Clearance for servicing appliances. Front *clearance* shall be sufficient for servicing the burner and the furnace or boiler.

❖ Because mechanical equipment and appliances require routine maintenance, repairs and possible replacement, access is required. Additionally, access recommendations or requirements are usually stated in the manufacturer's installation instructions.

The intent of this section is to secure access to components that require observation, inspection, adjustment, servicing, repair or replacement, such as controls, gauges, burners, filters, blowers, and motors. Access is also necessary to conduct operating procedures, such as startup or shutdown.

The code defines access as being able to be reached, but first may require the removal of a panel, door or similar obstruction. An appliance or piece of equipment is not accessible if any portion of the structure's permanent finish materials, such as drywall, plaster, paneling, built-in furniture, cabinets or any other similar permanently affixed building component must be removed.

308.4.4 Plenum clearances. Where the *furnace plenum* is adjacent to plaster on metal lath or *noncombustible material* attached to *combustible material*, the *clearance* shall be measured to the surface of the plaster or other noncombustible finish where the *clearance* specified is 2 inches (51 mm) or less.

❖ See the commentary to Section 308.3.3.

308.4.5 Clearance from supply ducts. Supply air ducts connecting to listed central heating furnaces shall have the same minimum clearance to combustibles as required for the furnace supply plenum for a distance of not less than 3 feet (914 mm) from the supply plenum. Clearance is not required beyond the 3-foot (914 mm) distance.

❖ See the commentary to Section 308.3.4.

SECTION 309 (IFGC)
ELECTRICAL

309.1 Grounding. Gas *piping* shall not be used as a grounding electrode.

❖ This section prohibits metal underground gas piping from being used as a grounding electrode. Metal gas piping must be bonded to the electrical service

grounding electrode system, but it cannot serve as a grounding electrode. This is somewhat contradictory and this section has been the center of much confusion relative to the "equal-potential" bonding of gas piping. Underground gas piping cannot be used as an earth electrode because it is not recognized as such in the NEC and because of the perceived risk of potential current flow on piping that conveys flammable gas. There is also the obvious fact that underground piping is typically plastic or is metal with a dielectric fitting isolating the metal piping from above-ground piping. However, bonding underground metal gas piping to the grounding electrode system will cause the piping to act as an additional electrode whether or not this is intended or desired. This is unavoidable unless dielectric fittings are used to electrically isolate the undergound portion of the metal piping from the above-ground portion. Any fault current will take all possible paths to the earth with the intensity dictated by the impedance of the path. The key to remember is that this section does not negate the requirement to bond all metal above-ground gas piping, as required by the NEC and Section 310. See Sections 404.10 and 310.1 for clarification of gas piping bonding means.

309.2 Connections. Electrical connections between appliances and the building wiring, including the grounding of the appliances, shall conform to NFPA 70.

❖ Field-installed power wiring and control wiring for appliances and equipment must be installed in accordance with NEC (NFPA 70).

The power wiring includes all of the wiring, disconnects, overcurrent protection devices, starters and related hardware used to supply electrical power to the appliance or equipment. The control wiring includes all of the wiring, devices and related hardware that connect the main unit to all external controls and accessories, such as temperature and pressure sensors, thermostats, exhausters, equipment contactors, interlock controls and remote damper motors. The internal factory wiring of appliances and equipment is not covered by this section, however, the wiring is covered by the testing and review performed by an approved agency as part of the labeling process.

The mechanical or electrical code official responsible for the inspection of appliances and equipment must be familiar with the applicable sections of NFPA 70.

SECTION 310 (IFGS)
ELECTRICAL BONDING

310.1 Pipe and tubing other than CSST. Each above-ground portion of a gas *piping* system other than corrugated stainless steel tubing (CSST) that is likely to become energized shall be electrically continuous and bonded to an effective ground-fault current path. Gas *piping* other than CSST shall be considered to be bonded where it is connected to appliances that are connected to the *equipment* grounding conductor of the circuit supplying that *appliance*.

❖ This section is consistent with Section 250.104(B) of the 2014 edition of the NEC (NFPA 70). Although difficult to define, gas piping that is likely to become energized is typically considered to be any gas piping that connects to one or more appliances that are electrically powered. The metal components of the appliances are required to be grounded, therefore, the metal gas piping that connects to the appliances should be as well because the piping has the same likelihood of being energized as the appliance components. The term "bonding" simply means to electrically tie metal parts together by means of conductors to create a low impedance path for electricity to flow between them. Metallic objects that are bonded together will have no voltage difference between them and they will all have the same potential with respect to ground.

In the majority of cases, the gas piping system will be connected to at least one, if not several, appliances that use electrical power. The gas connections involve metal-to-metal joints that ensure electrical continuity with the appliance. Those appliances will be connected to an equipment grounding conductor that will serve to bond the gas piping to the grounding electrode system for the building. Where the appliance branch circuits include grounding conductors properly connected to the appliances and one or more such appliances are connected to the gas-piping system, the gas-piping system is automatically bonded as required by this section (see Commentary Figure 310.1). This method of bonding is not allowed as the sole means of bonding for corrugated stainless steel tubing (CSST) gas piping systems and the manufacturer's installation instructions for some gas tubing systems will require additional bonding beyond what is required by this section. This is true for CSST systems and is also mandated by Section 310.1.1 (see Section 310.1.1).

The purpose of the bonding required by this section is to make sure that there is no electrical potential (voltage difference) between gas piping and appliances, between gas piping and any other piping and between any piping and the earth so as to eliminate electrical shock hazards. If all of the metal piping systems and appliances are bonded together, there cannot be a voltage difference between them, therefore, human contact across any such piping systems and appliances will not result in electrical shock. This is referred to as "equal-potential" bonding, the primary purpose of which is to prevent shock. Section 310.1.1 addresses bonding that has an additional purpose.

310.1.1 CSST. Corrugated stainless steel tubing (CSST) gas *piping* systems and piping systems containing one or more segments of CSST shall be bonded to the electrical service grounding electrode system or, where provided, the lightning protection grounding electrode system.

❖ This section addresses gas pipe bonding specific to CSST and does so for the purpose of lessening the possibility of damage to the CSST caused by the electrical energy created by indirect lightning strikes.

Figure 310.1
GAS PIPE BONDING BY APPLIANCE BRANCH CIRCUIT

It is assumed that nothing is capable of fully protecting anything subjected to a direct strike by lightning because of the extremely high energy levels involved and unpredictability of the paths of flow. It is more common for building components to be energized by nearby lightning strikes to the earth, trees, poles, towers, etc. This intense electrical energy can travel through the earth or other objects and energize gas piping systems, among other metal systems and components. This can happen by simple conduction to the piping or by induction through forms of magnetic field coupling. The electrical energy can travel along a pipe or tube and unpredictably jump off and onto another metal pipe, wire, duct or other metal object. This "jumping off" point will experience an arc that has enough energy to burn through a pipe, tube or connector wall. The resulting exit wound caused by the arcing is a perforation that will result in gas leakage and possibly a fire or explosion. This damage can happen to any piping material, including steel pipe, CSST and copper tube, and also gas connectors. Relatively recent experience has shown that CSST is susceptible to this type of lightning damage and this section intends to mitigate the possibility of such damage.

Recall that the CSST piping will already be bonded to some degree by the appliance grounding conductor(s), as explained in the commentary to Section 310.1, assuming that it is connected to appliances that are electrically powered. This bonding, however, is not considered to be adequate to protect CSST, therefore, Section 310.1.1 mandates a more effective bonding means. Directly bonding the CSST piping to the electrical service grounding electrode system has been shown by laboratory experiments to substantially reduce the risk of lightning damage. The minimum size bonding jumper was chosen based on the maximum size of grounding electrode conductor that is required to connect to "made" electrodes such as ground rods and plates as addressed in the NEC (NFPA 70). A number 6 AWG copper bonding jumper provides an electrical path with considerably less impedance than the typical equipment grounding conductor found in an appliance branch circuit. The gas service and electrical service are often in close proximity, thus allowing for a short bonding jumper between the point of delivery and the electrical service. The shorter the bonding jumper, the better it will perform. In cases where the electrical service and gas service are not in close proximity, the bonding jumper should be routed to take the shortest path, noting the length limit of Section 310.1.1.3.

The requirement of this section is also a requirement in the manufacturer's installation instructions for CSST.

As stated in the manufacturer's instructions, the bonding clamp must connect only to rigid steel pipe or a brass CSST fitting and it must never connect directly to the tubing itself. Clamping directly to the corrugated tubing would result in a poor electrical connection and could promote failure from current flow at that location. Bonding clamps must be listed for the application, for example, clamps used outdoors must be a type identified for outdoor use. The typical installation will have Schedule 40 steel pipe extending from the point of delivery (typically the meter outlet) and passing through the outside wall into the building for some distance. The bonding clamp will be attached to this steel pipe either indoors or outdoors, or attached to any other section of steel pipe or any downstream CSST fitting. Note that this section applies to gas piping systems that have any CSST in them. For example, if a piping system is constructed mostly of steel pipe, but contains one or more sections of CSST, such system is treated as if it was constructed entirely of CSST and this code section applies. [see Commentary Figures 310.1.1(1) and 310.1.1(2)].

Some CSST manufacturers have designed and are marketing a type of CSST that has an electrically conductive outer jacket or jacket assembly intended to conduct stray currents and dissipate them in such a manner that arcing is prevented. The code allows these conductive jacketed types, but the bonding provisions of this section still apply unless the code official approves otherwise in accordance with Section 105.2 (see Sections 310.1.1.1 through 310.1.1.5).

310.1.1.1 Point of connection. The bonding jumper shall connect to a metallic pipe, pipe fitting or CSST fitting.

❖ The bonding jumper must connect to a metal pipe or a CSST fitting, but such a connection point can be anywhere in the distribution system. The bonding connection does not have to be between the point of delivery and the first downstream CSST fitting (see Commentary Figure 310.1.1.1).

**Figure 310.1.1.1
POINT OF CONNECTION**

Figure 310.1.1(1)
CSST Bonding

Figure 310.1.1(2)
ALTERNATIVE LOCATION FOR BONDING CLAMP

310.1.1.2 Size and material of jumper. The bonding jumper shall be not smaller than 6 AWG copper wire or equivalent.

❖ See Section 310.1.1.

310.1.1.3 Bonding jumper length. The length of the bonding jumper between the connection to a gas piping system and the connection to a grounding electrode system shall not exceed 75 feet (22 860 mm). Any additional grounding electrodes used shall be bonded to the electrical service grounding electrode system or, where provided, the lightning protection grounding electrode system.

❖ It is a well-known fact that the longer a bonding jumper is, the less effective it becomes because of the increasing impedance to electrical flow on the wire; therefore, the shorter the better for jumper effectiveness. Extensive testing was performed by the CSST industry to determine how well the bonding protects the CSST from indirect lightning strikes and lightning-induced currents. The testing concluded that the bonding was effective in preventing perforations in the CSST under the conditions of the predicted lightning events. The testing also determined that the bonding jumper was functionally adequate up to approximately 100 feet in preventing arcing, thus suggesting the need for a length limit. A length limit of 75 feet was chosen to provide a safety factor and also because it was believed that 75 feet would accommodate the majority of building designs and utility service entrances.

Bonding the CSST to an independent grounding electrode is prohibited; however, the code does not prevent a designer or installer from installing a supplemental grounding electrode ("additional," as stated in the code text) for perceived additional protection. Where such supplemental electrodes are installed, the code requires that they be bonded back to the electrical service grounding electrode system, as this is consistent with NFPA 70 requirements for a common grounding electrode system. The author believes that the code does not intend to allow the length limit to be circumvented by the installation of supplemental electrodes. Where supplemental electrodes are installed by choice, the code implies that the bonding jumpers that connect to the electrical service grounding electrode system are still limited to 75 feet in combined length. An opposing interpretation is that the length of the bonding jumper between the CSST and the supplemental grounding electrode is limited to 75 feet, and the length of the jumper that connects the supplemental grounding electrode back to the electrical service grounding electrode system is limited only by the NEC (NFPA 70). The new code text implies that the more conservative interpretation is intended.

The points of connection to the electrical service grounding electrode system, the methods of connection and the protection of the bonding conductors must be in accordance with NEC (NFPA 70). The devices that are used to connect the bonding jumper on both ends, such as clamps, must be listed for the

application and environment in which they are installed. For example, clamps used outdoors must be listed for exposure to the elements. Some commonly used bonding clamps are suitable only for indoor use, while others are suitable for both indoor and outdoor use (see Commentary Figure 310.1.1.3).

310.1.1.4 Bonding connections. Bonding connections shall be in accordance with NFPA 70.

❖ See the commentary to Section 310.1.1.3.

310.1.1.5 Connection devices. Devices used for making the bonding connections shall be listed for the application in accordance with UL 467.

❖ See the commentary to Section 310.1.1.3.

Bibliography

The following resource materials were used in the preparation of the commentary for this chapter of the code.

IBC-2015, *International Building Code*. Washington, DC: International Code Council, 2014.

IECC-2015, *International Energy Conservation Code*. Washington, DC: International Code Council, 2014.

IMC-2015, *International Mechanical Code*. Washington, DC: International Code Council, 2014.

IPC-2015, *International Plumbing Code*. Washington, DC: International Code Council, 2014.

NFPA 70-2014, *National Electrical Code*. Quincy, MA: National Fire Protection Association, 2013.

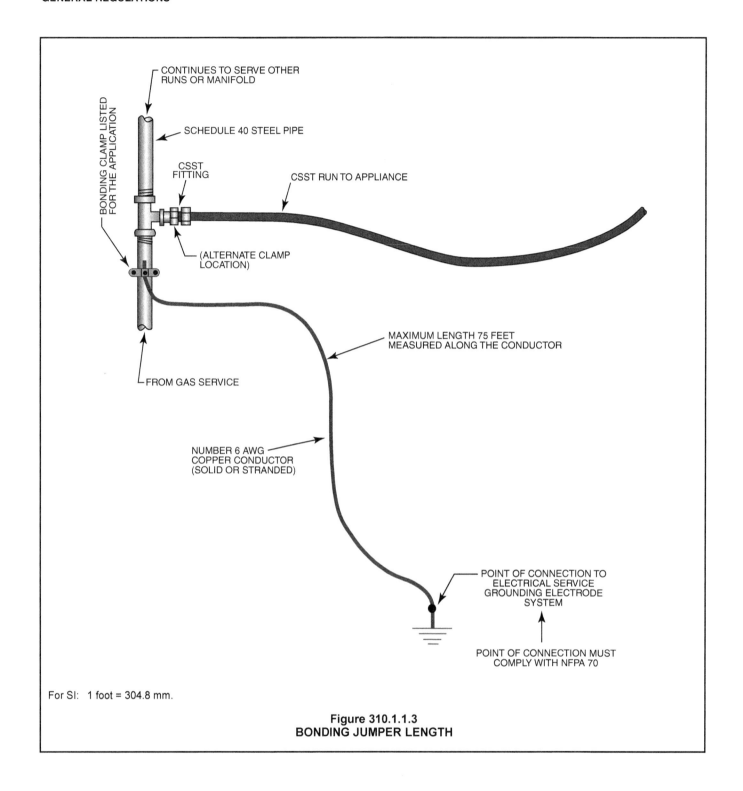

For SI: 1 foot = 304.8 mm.

**Figure 310.1.1.3
BONDING JUMPER LENGTH**

Chapter 4:
Gas Piping Installations

General Comments

The nature of fuel gases makes proper design, installation and selection of materials and devices necessary to minimize the risk to an acceptable level.

The two most commonly used fuel gases are natural gas and liquified petroleum gas (LP or LP-gas). These fuel gases have the following characteristics or properties.

Natural gas: The principal component of natural gas is methane (CH_4). It can also contain small quantities of nitrogen, carbon dioxide, hydrogen, hydrogen sulfide, water vapor, other hydrocarbons and various trace elements. Natural gas is colorless, tasteless and odorless; however, an odorant is added to the gas so that it can be readily detected. Natural gas is lighter than air (specific gravity < 1) and has the tendency to rise when escaping to the atmosphere. Natural gas has a rather narrow flammability range above and below which the gas-to-air mixture ratio is too rich or too lean to support combustion (see Commentary Figure 400). The heating value of natural gas varies depending on the source, but

it averages 1,050 British thermal units (Btu) per cubic foot (89 MJ/m^3).

Liquified petroleum gas: Liquified petroleum gases (LP or LP-gas) include commercial propane (C_3H_8) and commercial butane (C_4H_{10}). LP-gas vapors are heavier than air (specific gravity > 1) and tend to accumulate in low areas and near the floor. The ranges of flammability for LP-gases are narrower than those of natural gas (see Commentary Figure 400). Like natural gas, LP-gases are odorized to make them detectable.

The heating value of propane is approximately 2,500 Btu per cubic foot (93 MJ/m^3) of gas. The heating value of butane is approximately 3,300 Btu per cubic foot (1263 MJ/m^3) of gas.

Purpose

This chapter is intended to regulate the design and installation of fuel-gas distribution piping and systems from the point of delivery of the fuel gas to the appliances and equipment that consume the fuel. The intent is to minimize the hazards associated with the use and distribution of highly flammable/explosive fuel gases.

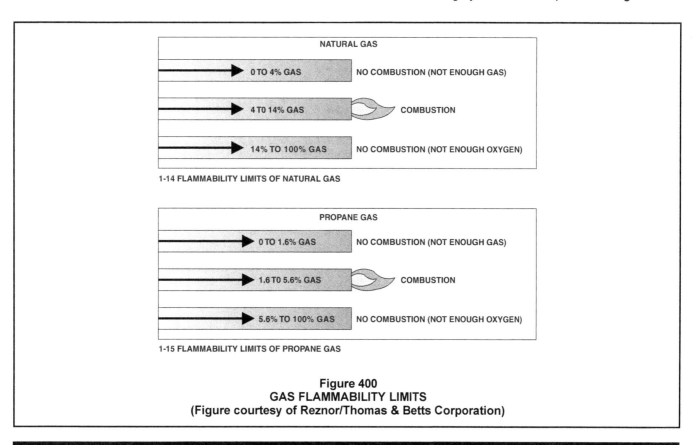

Figure 400
GAS FLAMMABILITY LIMITS
(Figure courtesy of Reznor/Thomas & Betts Corporation)

SECTION 401 (IFGC)
GENERAL

401.1 Scope. This chapter shall govern the design, installation, modification and maintenance of *piping* systems. The applicability of this code to *piping* systems extends from the *point of delivery* to the connections with the *appliance*s and includes the design, materials, components, fabrication, assembly, installation, testing, inspection, operation and maintenance of such *piping* systems.

❖ This section regulates aspects of fuel-gas distribution systems, including design, installation, testing, repair and maintenance. The applicability of this chapter is limited to the consumer side of the public utility company's gas distribution system.

The code governs all piping and system components from the end point of the gas purveyor's (utility company) service line to the supplied appliances and equipment. Typically, the gas utility company service line terminates at the service pressure regulator and meter setting. In other words, the code does not apply to piping and components that are owned by the gas utility company. The construction of utility-owned gas piping is governed by the federal Department of Transportation (DOT) regulations wherever the piping is located. However, in cases where the utility-owned piping runs through, in or on a building, it must do so in a way that does not jeopardize the structural integrity or fire safety of the building. Therefore, the relationship of the utility-owned piping to the building (prohibited locations, penetrations, etc.) is governed by the building portion of this code (see Section 401.1.1). No other provision within this chapter applies to utility-owned gas piping.

Although LP-gas storage systems generally remain under the ownership of the gas supplier, these systems are governed by the code through the reference to NFPA 58 (see commentary, Section 401.2). An LP-gas piping system begins at the outlet side of the first-stage pressure regulator. The piping between the first- and second-stage (final) regulator is governed by NFPA 58, and the piping downstream of the second-stage regulator is governed by this code. The first-stage pressure regulator reduces the storage container pressure to 10 pounds per square inch gauge (psig) (69 kPa) or less.

The termination point of gas utility services varies depending on the particular company's policies, but typically the service terminates at the property line or outdoors, immediately adjacent to the building or structure served by the service [see Commentary Figures 401.1(1) and 401.1(2)].

401.1.1 Utility piping systems located within buildings. Utility service *piping* located within buildings shall be installed in accordance with the structural safety and fire protection provisions of the *International Building Code*.

❖ Piping and components upstream of the point of delivery are designed, installed, tested, owned and maintained by the gas utility company and are regulated by federal DOT regulations, not this code. This code does, however, regulate two aspects of the installation related to protection of the structure from a structural and fire safety standpoint. By reference,

Figure 401.1(1)
SCOPE OF CODES

POINT OF DELIVERY

PIPING AND SERVICE METER ASSEMBLY REGULATED BY FEDERAL DEPT. OF TRANSPORTATION

Figure 401.1(2)
TYPICAL POINT OF DELIVERY FOR NATURAL GAS SERVICE
(Figure courtesy of NORMAC)

the *International Building Code®* (IBC®) would regulate the piping penetrations through structural components and through fire-resistance-rated assemblies. The IBC would also regulate piping support to handle structural loads and seismic forces.

401.2 Liquefied petroleum gas storage. The storage system for liquefied petroleum gas shall be designed and installed in accordance with the *International Fire Code* and NFPA 58.

❖ LP-gas tanks, containers, cylinders, storage vessels and related components must comply with NFPA 58. The storage system includes storage vessels of any type and the vessel appurtenances, such as shutoff valves, pressure gauges, liquid level indicators, pressure regulators, pressure relief valves and filling connections. The first-stage and second-stage pressure regulators and interconnecting piping are also addressed by the standard and could therefore be considered part of the LP-gas storage system. If the piping between the first-stage regulator and the point of delivery [second-stage or 2-psi (13.7 kPa) regulator] is not placed under the control of NFPA 58, it would be unregulated because Section 401.1 states that the code coverage begins at the point of delivery

(see definition of "Point of delivery"). Second-stage 2-psi regulators reduce the outlet pressure of the first-stage regulator to 2 psi (13.7 kPa) and a line pressure regulator downstream of the 2-psi regulator reduces the pressure from 2 psi to a maximum of 14 inches water column (w.c.) (3.5 kPa). Second-stage regulators reduce the outlet pressure of the first-stage regulator to a maximum of 14 inches w.c. NFPA 58 addresses all aspects of LP-gas storage systems such as container design and construction, container appurtenances, safety devices, container location and installation. A referenced standard is enforceable only in the context in which it is used. Any portions of NFPA 58 that extend beyond the scope of the storage system are not applicable because of the context of the reference in this section.

401.3 Modifications to existing systems. In modifying or adding to existing *piping* systems, sizes shall be maintained in accordance with this chapter.

❖ The gas volume demand on any portion of a distribution system cannot be increased beyond the capacity of the piping serving that portion. Modifications and additions to existing piping systems can have a detrimental effect on the existing piping system's ability to serve the connected loads. Commonly, new gas-fired equipment or appliances will be added to existing distribution piping without any regard for the effect of the extra load. To prevent dangerously low gas pressures, existing piping must be able to adequately supply additional loads, or the piping size must be increased. For example, a substantial load increase would occur where a storage (tank)-type water heater is replaced by an instantaneous tankless water heater.

401.4 Additional appliances. Where an additional *appliance* is to be served, the existing *piping* shall be checked to determine if it has adequate capacity for all *appliances* served. If inadequate, the existing system shall be enlarged as required or separate *piping* of adequate capacity shall be provided.

❖ This section parallels Section 401.8 and makes it clear that when any appliances are added to an existing gas piping system, the system must be carefully evaluated to verify that adding an appliance will not create a dangerous low-gas-pressure condition within the gas piping system. The evaluation must be conducted on the completed system as it will be configured after the appliance is added, and if the results indicate that the existing piping system cannot adequately serve the gas demand of the new appliance, the system must be revised accordingly. Undersized (overloaded) piping will cause pressure losses that can result in serious appliance malfunction.

401.5 Identification. For other than steel pipe, exposed *piping* shall be identified by a yellow label marked "Gas" in black letters. The marking shall be spaced at intervals not exceeding 5 feet (1524 mm). The marking shall not be required on pipe located in the same room as the *appliance* served.

❖ The intent of this section is to prevent gas piping from being mistaken for other piping. A case of mistaken identity could lead to a dangerous condition if gas piping were to be cut or opened unintentionally. It is not uncommon for fuel-gas piping systems and various other gas or liquid piping systems in a building to be constructed of the same materials and, thus, have the same appearance. The method of identifying fuel gas piping must be permanent, legible and conspicuous. Steel pipe (typically referred to as "black iron") is exempt from the identification requirements because it is the traditional gas pipe material and is generally recognized as such. Corrugated stainless steel tubing (CSST) is manufactured in compliance with the marking requirements of this section, and some copper tubing is manufactured with a similar yellow plastic jacket [see Commentary Figures 403.5 and 403.5.4(1)]. Steel pipe does not have to be identified, regardless of its color and regardless of any applied paint. Note that steel pipe that has been painted another color is still steel pipe.

401.6 Interconnections. Where two or more meters are installed on the same premises but supply separate consum-

ers, the *piping* systems shall not be interconnected on the *outlet* side of the meters.

❖ If gas distribution systems with independent meters were connected downstream of the meters, it would be impossible to determine the actual consumption of each of the interconnected systems. As the result of building renovations, remodeling and additions, or as the result of intentionally valved cross connections between separately metered systems, it is possible for a consumer to be paying for fuel gas consumed by another consumer. Accurate consumption metering can occur only if each metered distribution system is independent of all other metered distribution systems on the premises. Also, a system thought to be shut off could be backfed from an interconnected system, thereby maintaining pressure in the system that was intended to be shut off.

401.7 Piping meter identification. *Piping* from multiple meter installations shall be marked with an *approved* permanent identification by the installer so that the *piping* system supplied by each meter is readily identifiable.

❖ This provision allows service and emergency personnel to readily locate the shutoff valve that serves each piping system. The identification also helps to prevent interconnections between systems and between spaces when additions and/or alterations are made in the piping systems.

401.8 Minimum sizes. Pipe utilized for the installation, extension and *alteration* of any *piping* system shall be sized to supply the full number of outlets for the intended purpose and shall be sized in accordance with Section 402.

❖ Undersized gas piping systems are not capable of delivering the required volume of fuel at the required pressure. Inadequate gas pressure can cause hazardous operation of appliances. Appliances depend on proper gas pressure to maintain the design Btu (J) input rate, to maintain the required minimum manifold pressure and to produce the required gas velocity in the burners. If the gas pressure to an appliance is too low, the result can be incomplete combustion, burner malfunction and flashback, soot production and appliance malfunction and damage. Gas piping must be sized to maintain the required minimum gas pressures at the appliance inlet.

When designing a gas piping distribution system, all connected appliances and equipment are assumed to operate simultaneously, except as provided for in Section 402.2. In other words, the system is designed for maximum demand. The code contains provisions for sizing gas piping systems using an exact equation or design tables. The tabular method is most often used because it is simpler and less likely to produce errors. Factors that affect the sizing of gas piping systems include the specific gravity of the gas, the length of the pipe and the number of fittings installed, the maximum gas demand, the allowable pressure loss through the system and any diversity factor that is applicable.

The design of gas piping installations should also take into account the possibility of future increases in gas demand (load). The farther the gas meter or point of delivery is from the building, the more extensive will be the disruption of the property in the event that a larger yard line between the meter and the building becomes necessary in the future as a result of increased gas demand within the building, a change in fuel gas type or a change in service pressure.

401.9 Identification. Each length of pipe and tubing and each pipe fitting, utilized in a fuel gas system, shall bear the identification of the manufacturer.

❖ All pipe, tubing and fittings for such must be able to be traced back to their source in case there is an issue with the material or its listing. Fittings of all sizes are easily marked with the symbol or logo of the manufacturer, but this requirement is more difficult for close and shoulder pipe nipples because there is little or no surface area suitable for such a marking. Pipe nipples constructed in the field must be cut and threaded using pipe that meets the code. The pipe is marked, but the field-cut nipples might not be.

401.10 Third-party testing and certification. Piping, tubing and fittings shall comply with the applicable referenced standards, specifications and performance criteria of this code and shall be identified in accordance with Section 401.9. Piping, tubing and fittings shall either be tested by an approved third-party testing agency or certified by an approved *third-party certification agency*.

❖ These materials have always been required to comply with the standards prescribed by the code and this section requires verification of such by a third-party entity. The materials, or in some cases their packaging, must bear proof that a third-party entity has performed the verification. This requirement will make it easier for builders, designers and code officials to determine that materials comply with the appropriate product standards.

SECTION 402 (IFGS)
PIPE SIZING

402.1 General considerations. Piping systems shall be of such size and so installed as to provide a supply of gas sufficient to meet the maximum demand and supply gas to each *appliance* inlet at not less than the minimum supply pressure required by the *appliance*.

❖ This section is a summation of the provisions of Sections 401.8 and 402.5, and is intended to give guidance to the designer (see commentary, Sections 401.8 and 402.5). The pressure loss (drop) in a gas piping system must not exceed the allowable loss described in Section 402.5. Pressure loss must be controlled to make sure that the supply pressure at each appliance inlet is not less than the minimum required by the appliance. Inadequate supply pressure can result in hazardous appliance operation or malfunction.

402.2 Maximum gas demand. The volumetric flow rate of gas to be provided shall be the sum of the maximum input of the *appliance*s served.

The total connected hourly load shall be used as the basis for pipe sizing, assuming that all appliances could be operating at full capacity simultaneously. Where a diversity of load can be established, pipe sizing shall be permitted to be based on such loads.

The volumetric flow rate of gas to be provided shall be adjusted for altitude where the installation is above 2,000 feet (610 m) in elevation.

❖ To determine the demand volume required by an appliance in cubic feet (m³) of gas per hour, the input rating of the appliance in British thermal units per hour (Btu/h) (W) as specified by the appliance manufacturer must be known. This information will be on the appliance nameplate. Note that, depending on the elevation above sea level at which the appliance is installed, the manufacturer of the appliance might also require that the input be derated to compensate for the effect of altitude. If a derating adjustment of input is required, the adjusted input rate would be used in the calculation. If the average heating value of the gas is known [typically about 1,000 Btu/ft³ (37.3 MJ/m³) for natural gas], the volume of gas required per hour can be calculated. The heating value can be obtained from the gas supplier. The input rating of all appliances must be shown on the appliance label. When the input rating of the appliances is unknown at the time of piping system design, the gas demand will have to be estimated based on input from appliance manufacturers, gas utilities or other source. However, when the actual appliance demands are determined upon appliance selection, any approximate calculations must be checked to verify that the piping system is, in fact, adequate to supply the known demand for all appliances. The designer must be aware of the fact that a system designed in accordance with an estimated load might be undersized relative to the actual connected load.

In all cases, the piping must have the capacity to supply the actual connected load of appliances. As also stated in this section and Section 401.8, the load on the piping system must be based on the simultaneous operation of all appliances at full output. The only exception to this would be the demonstration of actual load diversity. For example, the designer could have access to data that demonstrates that the total load for cooking ranges or water heaters in a multiple-family dwelling (apartment building) is less than the sum of all of the appliance demands at full output. The code official would have to be convinced that any such diversity factor had been established. Any known diversity factors would likely come from the serving gas supplier.

The following is an example of how gas demand is calculated for a representative appliance: an appliance installed at sea level has a nameplate (label) input rating of 120,000 Btu/h (35.2 kW). Using natural gas with an average heating value of 1,000 Btu/ft³

(37.3 MJ/m³), the volume of gas required for the appliance is 120,000/1,000 = 120 cubic feet per hour (3.4 m³/h).

402.3 Sizing. Gas *piping* shall be sized in accordance with one of the following:

1. Pipe sizing tables or sizing equations in accordance with Section 402.4.

2. The sizing tables included in a *listed piping* system's manufacturer's installation instructions.

3. Other *approved* engineering methods.

❖ Using the sizing tables is generally simpler and less time consuming than using the equations found in Section 402.4.The gas system pressure, allowable pressure drop, maximum gas demand and the specific gravity of the fuel gas must be known in order to use the tables. The allowable pressure drop is the designer's choice because the intent of the code is to provide the appliance or equipment with the required rate of gas flow at the required minimum pressure. The tables are based on a given pressure drop and cannot be used if the designer wishes to use any other pressure drop. The sizing tables are based on using smooth-bore pipe or tube and can be used with the piping materials listed in Section 403. Pipe materials with higher resistance to flow cannot be sized using these tables. CSST is sized using CSST-specific tables. The pipe sizes in the pipe sizing tables are based on the traditional steel pipe size designation as used in the material standards. More information may be obtained from a variety of pipe design publications, including the *Piping Handbook* by Sabin Crocker and the *ASHRAE Handbook of HVAC Systems and Equipment*. Each of the tables presents different piping material and design parameters and represents the most common system design applications. The tables assume that the system being sized is constructed of a single-type pipe or tubing material throughout; thus, a single table cannot be used for systems constructed of multiple materials. For example, it is common to find hybrid systems composed of both steel pipe and CSST. Such systems would have to be sized from two different tables based on the longest run of the entire system.

The code does not limit sizing methodology to the tables or equations. Manufacturers of gas piping/tubing may provide sizing charts, tables or slide calculators that, in effect, become part of the installation instructions. This is the case for CSST, which is a listed piping system.

Section 402.4 describes a sizing method that uses an exact thermodynamic flow equation to determine the required pipe sizes. Although this is an exact method, the degree of accuracy it affords may not always prove valuable, because pipe sizes are standardized and the pipe diameter calculated may not correspond to an available pipe size. For example, a calculated pipe diameter of 0.80 inch (20.3 mm) will require the selection of the next larger standard pipe size [1-inch (25 mm) pipe].

There are two equations in Section 402.4 to be used, depending on the gas pressure within the piping system.

As stated in Appendix A, Sections A.2.2 and A.2.3, an equivalent length of pipe should be added to the length of any piping run having four or more fittings (see Appendix A, Table A.2.2). Some designers simply add 50 percent of the actual length of piping as an all-inclusive fitting allowance (i.e., actual pipe length × 1.5). This is a conservative estimate and is used because the necessary number and type of fittings are usually not known when the system is being designed.

Steel pipe sizes $^1/_4$ inch and $^3/_8$ inch have been deleted because the use of such sizes is very rare. The pressure of "0.50 or less" that appeared in some table titles has been replaced by "less than 2 psi" because such tables can be used with any pressure less than 2 psi (13.8 kPa), including 0.50 psi (3.5 kPa) and less.

402.4 Sizing tables and equations. Where Tables 402.4(1) through 402.4(37) are used to size *piping* or tubing, the pipe length shall be determined in accordance with Section 402.4.1, 402.4.2 or 402.4.3.

Where Equations 4-1 and 4-2 are used to size *piping* or tubing, the pipe or tubing shall have smooth inside walls and the pipe length shall be determined in accordance with Section 402.4.1, 402.4.2 or 402.4.3.

1. Low-pressure gas equation [Less than $1^1/_2$ pounds per square inch (psi) (10.3 kPa)]:

$$D = \frac{Q^{0.381}}{19.17\left(\frac{\Delta H}{C_r \times L}\right)^{0.206}} \qquad \textbf{(Equation 4-1)}$$

2. High-pressure gas equation [$1^1/_2$ psi (10.3 kPa) and above]:

$$D = \frac{Q^{0.381}}{18.93\left[\frac{(P_1^2 - P_2^2) \times Y}{C_r \times L}\right]^{0.206}} \qquad \textbf{(Equation 4-2)}$$

where:

D = Inside diameter of pipe, inches (mm).

Q = Input rate *appliance*(s), cubic feet per hour at 60°F (16°C) and 30-inch mercury column.

P_1 = Upstream pressure, psia (P_1 + 14.7).

P_2 = Downstream pressure, psia (P_2 + 14.7).

L = Equivalent length of pipe, feet.

ΔH = Pressure drop, inch water column (27.7 inch water column = 1 psi).

❖ The length of pipe or tube to be used with the tables and equations is determined by the sizing method being used. Section 402.4.1 involves only one length whereas Sections 402.4.2 and 402.4.3 involve multiple lengths.

TABLE 402.4
C_r AND Y VALUES FOR NATURAL GAS AND UNDILUTED PROPANE AT STANDARD CONDITIONS

GAS	EQUATION FACTORS	
	C_r	Y
Natural gas	0.6094	0.9992
Undiluted propane	1.2462	0.9910

For SI: 1 cubic foot = 0.028 m^3, 1 foot = 305 mm,
    1-inch water column = 0.2488 kPa,
    1 pound per square inch = 6.895 kPa,
    1 British thermal unit per hour = 0.293 W.

RANGE
70,000 Btu/h

POINT OF DELIVERY

40-GALLON
WATER HEATER
45,000 Btu/h

FURNACE
140,000 Btu/h

For SI: 1 foot = 304.8 mm, 1 Btu/h = 0.29 kW.

Figure 402.4
GAS PIPING LAYOUT

❖ **Example:**

Determine the required pipe sizes for the gas distribution system in Commentary Figure 402.4 using the equation method outlined in this section. The system is designed to operate at less than 0.5 pound per square inch (psig) (3.5 kPa), using steel pipe, Assume the following:

Fuel gas   Natural
Specific gravity   0.60
Heating value   1,000 Btu per cubic foot
Maximum allowable
   pressure drop $^1/_2$ inch of water column
Negligible number of fittings

For SI: 1 inch = 25.4 mm, 1 foot = 304.8 mm, 1 Btu/h = 0.2931 W, 1 Btu/ft^3 = 0.0373 MJ/m^3, 1-inch water column = 0.2488 kPa, 1 cubic foot = 0.02832 m^3, 1 pound per square inch gauge = 6.895 kPa.

To use equations in this section, the gas volume of each appliance must be known.

Gas volume of an appliance = input rating of an appliance (Btu/h)/Heating value of the fuel gas (Btu/ft^3).

APPLIANCE	INPUT RATING (Btu/h)	GAS VOLUME (ft^3/h)
Range	70,000	70
Furnace	140,000	140
Water heater	45,000	45

Using the equation, the minimum pipe diameter (D) can be solved as follows:

$D = Q^{0.381}/19.17 [(\Delta H)/(C_r \times L]^{0.206}$

Starting at the most remote appliance from the point of delivery (Point G - water heater), determine the minimum pipe size for pipe segment D.

$D = (45)^{0.381}/19.17 [(^1/_2)/(0.609 \times 80)]0.206 = 0.574$

Repeat the calculation for each pipe segment. The gas piping system is sized as follows:

PIPE SEGMENT	FLOW RATE (Q)	CALCULATED PIPE DIAMETER (D) (in.)	NOMINAL PIPE SIZE (in.)
A-B	255	1.11	$1^1/_4$
B-C	70	0.679	$^3/_4$
B-D	185	0.983	1
D-E	140	0.884	1
D-G	45	0.574	$^1/_2$

Most gas pipe sizing will be determined by using the sizing tables. In the event that circumstances warrant it, however, pipe sizing may be done by calculation using the provisions of Appendix A or other approved methods. If in a particular design, the piping length, desired pressure loss, system supply pressure, specific gravity or pipe size does not fall within the parameters of a table, an approved alternative sizing procedure must be used.

Instead of using the methods specified in this section, piping sizes may be determined by the use of accurate gas flow, computer or pressure drop charts. Such sizing tools must be acceptable to the code official. Note that calculators, charts and computer programs can be designed to size gas piping for any pressure drop because the intent of the code is to make sure that the supply pressure at the appliance is greater than or equal to the minimum pressure required for proper operation of the appliance (see Section 402.5).

402.4.1 Longest length method. The pipe size of each section of gas *piping* shall be determined using the longest length of *piping* from the *point of delivery* to the most remote *outlet* and the load of the section.

❖ This section provides a step-by-step approach to proper application of the tables using the traditional longest-length method. When using the tables, the maximum pipe length from the point of delivery to the farthest outlet must be determined, including any allowance for the equivalent length of fittings. The designer will use the row in the table that equals the determined length or the next higher row if the determined length is between table values. The system will be designed using only values taken from this row. This is known as the longest-run method: Basing the sizing on the most demanding circuit (longest run) compensates for pressure losses throughout the entire system. Commentary Figure 402.4.1 and the example serve to illustrate the longest length method.

Longest-length method example: Basing the sizing on the most demanding circuit (longest), using Table 402.4(2), size the piping system in Commentary Figure 402.4.1. The longest run is from the meter (point of delivery) to Roof Top Unit #5 and is 210 feet (64 008 mm), therefore, all pipe sizes will be chosen from the 250-foot (76 200 mm) row of the table.

PIPE SECTION	LOAD (MBH)	SIZE (in.)
A	690	2
B	690	2
C	540	2
D	450	2
E	250	$1^1/_2$
F	250	$1^1/_2$
G	150	$1^1/_4$
H	90	1
I	200	$1^1/_4$
J	100	1
K	150	$1^1/_4$

For SI: 1 foot = 304.8 mm, 1 MBH = 293 W.

Figure 402.4.1
STEEL PIPE BRANCH SYSTEM FOR ROOFTOP UNITS

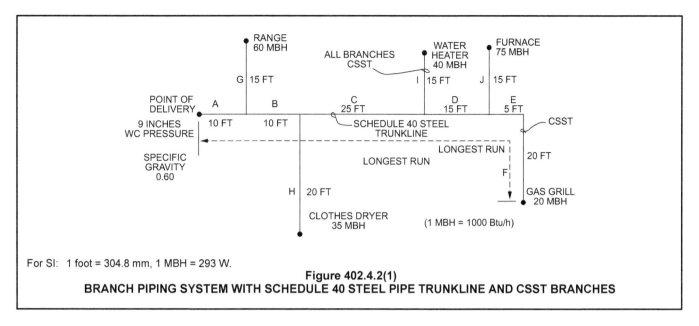

For SI: 1 foot = 304.8 mm, 1 MBH = 293 W.

Figure 402.4.2(1)
BRANCH PIPING SYSTEM WITH SCHEDULE 40 STEEL PIPE TRUNKLINE AND CSST BRANCHES

For SI: 1 foot = 304.8 mm,
         1 MBH = 1,000 Btu/h = 293 W.

Figure 402.4.2(2)
SINGLE PRESSURE CSST MANIFOLD SYSTEM

402.4.2 Branch length method. Pipe shall be sized as follows:

1. Pipe size of each section of the longest pipe run from the *point of delivery* to the most remote *outlet* shall be determined using the longest run of *piping* and the load of the section.

2. The pipe size of each section of branch *piping* not previously sized shall be determined using the length of *piping* from the *point of delivery* to the most remote *outlet* in each branch and the load of the section.

❖ This sizing method is a variation of the longest-length method, and the results can be less conservative. Because less headroom is built into this method, it is especially important to account for the equivalent length of fittings installed in the system [see Commentary Figures 402.4.2(1) and 402.4.2(2) and examples]. This method involves multiple piping lengths within a system for application of the tables or equations, whereas the longest-length method involves only one piping length per system.

402.4.3 Hybrid pressure. The pipe size for each section of higher pressure gas *piping* shall be determined using the longest length of *piping* from the *point of delivery* to the most remote line pressure regulator. The pipe size from the line pressure regulator to each *outlet* shall be determined using the length of *piping* from the regulator to the most remote outlet served by the regulator.

❖ Hybrid pressure piping systems convey gas at different pressures in different parts of the system. One or more pressure regulators are used to reduce the pressure in portions of the system. In some cases, a single gas piping system must deliver widely varying pressure to loads throughout a building. The most common hybrid pressure systems allow small economically sized piping to carry large loads over long distances by conveying gas at high pressures. Near the load being supplied, pressure regulators reduce the higher pressure to suit the load. For example, rooftop units on buildings are typically designed for a natural gas pressure of 14 inches w.c. (3.5 kPa) or less. Running hundreds of feet of pipe across expansive roofs would require large expensive pipe at such low pressure. At 5-psi (34.4 kPa) pressure, small, less expensive pipe can be run over long distances, and regulators at each rooftop unit reduce the 5-psi (34.4 kPa) pressure to less than $^1/_2$ psi (3.4 kPa), which is typically 8 inches w.c. (2 kPa) [see Commentary Figures 402.4.3(1) and (2) and examples.

Branch length method [Commentary Figure 402.4.2(1)] Example 1:

1. In accordance with Section 402.4.2, Item 1, determine the size of Sections A, B, C, D and E (constructed of Schedule 40 steel pipe) based on the load of each section and the longest run length of 85 feet (25 908 mm). Because the system pressure is less than 0.5 psi [(14 inches w.c.) (3.5 kPa), Table 402.4(2) is chosen. Because the longest run length is between

rows in the table, the 90-foot (27 432 mm) row must be chosen.

PIPE SECTION	LOAD (MBH)	SIZE (in.)
A	230	$1^1/_4$
B	170	1
C	135	1
D	95	$^3/_4$
E	20	$^1/_2$

2. In accordance with Section 402.4.2 Item 2, determine the size of Branch Sections F, G, H, I and J (constructed of CSST) based on the load of each section and the length of piping and tubing from the point of delivery to the outlet on that section. Table 402.4(15) is chosen because the branch sections are constructed of CSST and the pressure is less than 0.5 psi [(14 inches w.c.) (3.5 kPa)]. Where a length falls between entries in the table, use the next longer length row.

CSST BRANCH	LOAD (MBH)	LENGTH PIPING AND TUBING (ft)	SIZE EHD
F	20	85	18
G	60	25	19
H	35	40	18
I	40	60	23
J	75	75	30

Branch length method [Commentary Figure 402.4.2(2)] Example 2:

Determine the minimum required size of each piping section using branch length method.

Given:

• Piping arranged in parallel.

• Main section and individual appliance runs constructed of CSST.

• Supply pressure is 6 inches w.c. downstream of meter.

• Natural gas with specific gravity of 0.60.

• Chosen pressure drop 0.5 inch w.c.

• Minimum appliance supply pressure requirements are 5 inches w.c.

• Heating value of gas is 1,000 Btu per ft³.

Sizing Section A:

• Size Section A based on appliance total load and length from meter to furthest appliance.

• Total load = 80 + 36 + 52 + 28 + 25 = 221 cfh.

• Section length = 45 + 20 + 5 = 70 feet.

• Use Table 402.4(15).

Sizing Section B:

- Size Section B based on appliance total load and length from meter to appliance.
- Total load = 80 cfh.
- Section length (furnace) = 45 + 15 = 60 feet.
- Use Table 402.4(15).

Sizing Section C:

- Size Section C based on appliance total load and length from meter to appliance.
- Total load = 36 cfh.
- Section length (water heater) = 45 + 10 = 55.
- Use Table 402.4(15).

Sizing Section D:

- Size Section D based on appliance total load and length from meter to appliance.
- Total load = 52 cfh.
- Section length (range) = 45 + 20 = 65 feet.
- Use Table 402.4(15).

Sizing Section E:

- Size Section E based on appliance total load and length from meter to furthest appliance.
- Total load = 25 + 28 = 53 cfh.
- Section length (dryer/room heater) = 45 + 20 + 5 = 70 feet.
- Use Table 402.4(15).

Sizing Section F:

- Size Section F based on appliance total load and length from meter to furthest appliance.
- Total load = 28 cfh.
- Section length (dryer) = 45 + 20 + 5 = 70 feet.
- Use Table 402.4(15).

Sizing Section G:

- Size Section G based on appliance total load and length from meter to furthest appliance.
- Total load (room heater) = 25 cfh.
- Section length = 45 + 20 + 5 = 70 feet.
- Use Table 402.4(15).

Hybrid pressure [Commentary Figure 402.4.3(1)] Example 1:

Note: This example illustrates hybrid system design, but is not typical for dwellings.

1. Using Table 402.4(7), determine the minimum required size of piping Sections A through D. The longest run of piping from the point of delivery to the most remote regulator is 175 feet.

   - Section A serves a load of 625 MBH and in the 175-foot row of the table, $^{1}/_{2}$-inch pipe is shown to have a capacity of 728 ft^3/hr or 728,000 Btu/h for gas with a heating value of 1,000 Btu/ft^3.
   - Because Sections B, C and D all serve lesser loads than Section A, they too can be $^{1}/_{2}$ inch in size.
   - Because Table 402.4(7) is based upon a pressure drop of 3.5 psi, the available pressure at the inlets of the "pounds-to-inches" regulators under full load condition will be at least 1.5 psig.

2. Determine size of piping Sections E, F and G, using Table 402.4(2) and the longest length of piping from the pounds-to-inches regulator to the most remote outlet. In this case, rooftop unit 1 is the most remote at 65 feet from the regulator.

3. Determine size of piping Sections H, I and J in the same manner described in Item 2 above.

NOTE: MBH = 1,000 Btu/h

Hybrid pressure [Commentary Figure 402.4.3(2) Example 2:

Given:

- All piping is CSST.
- Piping branches run in parallel.
- Individual CSST appliance branch runs supplied by manifold.
- Supply pressure is 2 psi downstream of meter.
- Chosen pressure drop is 1 psi for Section A.
- Natural gas with specific gravity of 0.60.
- Heating value of gas is 1,000 Btu/ft^3.

SUMMARY TABLE FOR BRANCH LENGTH METHOD EXAMPLE 2				
SECTION	LOAD (cfh)	SECTION LENGTH (ft)	SECTION USED WITH TABLE (ft)	REQUIRED SIZE (CSST)
A	221	45	70	
B	80	15	60	
C	36	10	55	
D	52	20	65	
E	53	20	70	
F	28	5	70	
G	25	5	70	

- Supply pressure at manifold is 8 inches w.c.
- Chosen pressure drop is 3 inches w.c. for Sections B through G.
- Appliances need 5 inches w.c. minimum supply pressure.
- Pressure drop across regulator is less than $1/2$ psi at full load of this system.

Sizing Section A:

- Size Section A based on total load and length from meter to regulator.
- Total load = 80 + 36 + 52 + 28 + 25 = 221 cfh.
- Section length = 45 feet.
- Use Table 402.4(18).

Sizing Section B:

- Size Section B based on appliance load and length from regulator to appliance.
- Appliance load = 80 cfh.
- Section length (furnace) = 15 feet.
- Use Table 402.4(16).

Sizing Section C:

- Size Section C based on appliance load and length from regulator to appliance.
- Total load = 36 cfh.
- Section length (water heater) = 10 feet.
- Use Table 402.4(16).

Sizing Section D:

- Size Section D based on appliance load and length from regulator to appliance.
- Total load = 52 cfh.
- Section length (range) = 20 feet.
- Use Table 402.4(16).

Sizing Section E:

- Size Section E based on appliance total load and length from regulator to furthest appliance.

- Total load = 25 + 28 = 53 cfh.
- Section length (dryer/room heater) = 20 + 5 = 25 feet.
- Use Table 402.4(16).

Sizing Section F:

- Size Section F based on appliance total load and length from regulator to furthest appliance.
- Total load (dryer) = 28 cfh.
- Section length = 20 + 5 = 25 feet.
- Use Table 402.4(16).

Sizing Section G:

- Size Section G based on appliance total load and length from regulator to furthest appliance.
- Total load (room heater) = 25 cfh.
- Section length = 20 + 5 = 25 feet.
- Use Table 402.4(16).

402.5 Allowable pressure drop. The design pressure loss in any *piping* system under maximum probable flow conditions, from the *point of delivery* to the inlet connection of the *appliance*, shall be such that the supply pressure at the *appliance* is greater than or equal to the minimum pressure required by the *appliance*.

❖ Between the point of delivery and the load, gas pressure losses will occur because of pipe friction. The design of the system must control these losses to ensure that the pressure at the load (appliance) connection point will be equal to or greater than the minimum required pressure as dictated by the appliance manufacturer. As stated in the commentary to Section 401.8, inadequate gas pressures can result in dangerous appliance operation. This section speaks of "maximum probable flow conditions," which must be taken to mean the total connected load of all appliances operating at full capacity except where a diversity-of-load factor is applied.

SUMMARY TABLE FOR HYBRID PRESSURE EXAMPLE 2

SECTION	LOAD (cfh)	ACTUAL SECTION LENGTH (ft) (including equivalent length of fittings and bends)	SECTION LENGTH (ft) USED WITH TABLE	REQUIRED SIZE (CSST)
A	221	45	45[a]	
B	80	15	15[b]	
C	36	10	10[b]	
D	52	20	20[b]	
E	53	20	25[b]	
F	28	5	25[b]	
G	25	5	25[b]	

a. Distance from meter to regulator is the run length for the main section.

b. Distance from regulator to each appliance is the run length for appliance branch run sections.

For SI: 1 foot = 304.8 mm, 1 MBH = 293 W, 1 inch water column = 249 Pa.

Figure 402.4.3(1)
HYBRID PRESSURE BRANCH SYSTEM

For SI: 1 foot = 304.8 mm,
1 MBH = 293 W, 1 inch water column = 249 Pa.

Figure 402.4.3(2)
HYBRID PRESSURE PARALLEL
DISTRIBUTION SYSTEM

The designer can choose any pressure loss for a piping system if the goal of this section is met. The code does not specify any maximum pressure drop other than what is required by this section. For example, consider an appliance that has a minimum required inlet pressure of 5 inches w.c. (1.25 kPa), and the gas pressure at the point of delivery is 8 inches w.c. (2 kPa). The maximum allowable pressure drop through the piping system would be 3 inches w.c. (0.75 kPa) because the pressure at the appliance connection must be greater than or equal to 5 inches w.c. Some sizing tables, such as Table 402.4(2), are based on very small pressure drops [0.5 inch w.c. (125 Pa)] and, thus, are quite conservative sizing methods that account for the friction losses through a limited number of fittings used in a piping system. The tables are optional sizing methods placed in the code for convenience. The code allows piping systems to be designed for the exact pressure losses allowed by this section. Some tables are based on much larger pressure losses and can be used where the designer wants to keep pipe sizes to their minimum. The code intends to allow the designer to have the option of choosing the system pressure losses, provided that such losses meet the requirement of this section.

402.6 Maximum design operating pressure. The maximum design operating pressure for *piping* systems located inside buildings shall not exceed 5 pounds per square inch gauge (psig) (34 kPa gauge) except where one or more of the following conditions are met:

1. The *piping* system is welded.

2. The *piping* is located in a ventilated chase or otherwise enclosed for protection against accidental gas accumulation.

3. The *piping* is located inside buildings or separate areas of buildings used exclusively for any of the following:

    3.1. Industrial processing or heating.

    3.2. Research.

    3.3. Warehousing.

    3.4. Boiler or mechanical rooms.

4. The *piping* is a temporary installation for buildings under construction.

5. The piping serves appliances or *equipment* used for agricultural purposes.

6. The *piping* system is an LP-gas *piping* system with a design operating pressure greater than 20 psi (137.9 kPa) and complies with NFPA 58.

❖ This section is specific to indoor piping systems. The pressure is limited to 5 psig (34 kPa) to minimize the consequences of a leak. Also, the higher the pressure, the more difficult it is to contain any gas, and thus leakage is more likely. Item 1 recognizes the superior strength and integrity of steel piping with welded joints. Item 2 allows the option of higher pressures where the piping is enclosed to prevent leakage from entering the building. Item 4 allows higher pressures for temporary piping for construction site applications, such as temporary space heating, because of low occupant density, the presence of trained personnel and the short duration of use. Item 4 recognizes the unique nature of temporary piping on construction sites where small piping is desirable, thus making higher pressures necessary. Item 6 recognizes that NFPA 58 allows higher pressures for LG-gas systems in industrial occupancies.

This section applies to all fuel gases, including LP-gas (see Section 402.6.1).

402.6.1 Liquefied petroleum gas systems. LP-gas systems designed to operate below -5°F (-21°C) or with butane or a propane-butane mix shall be designed to either accommodate liquid LP-gas or prevent LP-gas vapor from condensing into a liquid.

❖ In low-temperature environments, LP-gases might condense into the liquid state within the piping system. If the liquid fuel enters system components or appliances, dangerous malfunctions could result. The temperature and pressure of a gas both affect the point at which a gas will change states to a liquid phase. LP-gas can be made to condense to liquid by applying pressure, by cooling the gas and by a combination of both. At atmospheric pressure, 14.7 psia at sea level, propane will condense at minus 44°F (-42°C) and butane will condense at 15°F (-9°C), meaning that LP-liquid would remain as a liquid in an open container at temperatures of minus 44°F and lower. As the liquid warms to temperatures above the condensing temperature, the liquid will begin to boil, producing gas vapor. The resultant vapor pressure will have to be held in a closed container to prevent vaporization of all of the liquid. This is what happens in an LP-gas storage tank. The tank holds both liquid and gaseous LP under pressure that varies with the ambient temperature. Within a piping system, the pressure could be higher than one atmosphere and LP-gas would, therefore, condense at a higher temperature. For example, at 20 psig (138 kPa), propane will condense at -5°F (-21°C). The -5°F threshold requirement of this section was chosen because 20 psi system piping was commonly installed outdoors.

TABLES 402.4(1) through 402.4(37). See pages 4-15 through 4-50.

❖ See the commentary to Section 402.4.

TABLE 402.4(1)
SCHEDULE 40 METALLIC PIPE

Gas	Natural
Inlet Pressure	Less than 2 psi
Pressure Drop	0.3 in. w.c.
Specific Gravity	0.60

PIPE SIZE (inch)

Nominal	1/2	3/4	1	1 1/4	1 1/2	2	2 1/2	3	4	5	6	8	10	12
Actual ID	0.622	0.824	1.049	1.380	1.610	2.067	2.469	3.068	4.026	5.047	6.065	7.981	10.020	11.938
Length (ft)	Capacity in Cubic Feet of Gas Per Hour												
10	131	273	514	1,060	1,580	3,050	4,860	8,580	17,500	31,700	51,300	105,000	191,000	303,000
20	90	188	353	726	1,090	2,090	3,340	5,900	12,000	21,800	35,300	72,400	132,000	208,000
30	72	151	284	583	873	1,680	2,680	4,740	9,660	17,500	28,300	58,200	106,000	167,000
40	62	129	243	499	747	1,440	2,290	4,050	8,270	15,000	24,200	49,800	90,400	143,000
50	55	114	215	442	662	1,280	2,030	3,590	7,330	13,300	21,500	44,100	80,100	127,000
60	50	104	195	400	600	1,160	1,840	3,260	6,640	12,000	19,500	40,000	72,600	115,000
70	46	95	179	368	552	1,060	1,690	3,000	6,110	11,100	17,900	36,800	66,800	106,000
80	42	89	167	343	514	989	1,580	2,790	5,680	10,300	16,700	34,200	62,100	98,400
90	40	83	157	322	482	928	1,480	2,610	5,330	9,650	15,600	32,100	58,300	92,300
100	38	79	148	304	455	877	1,400	2,470	5,040	9,110	14,800	30,300	55,100	87,200
125	33	70	131	269	403	777	1,240	2,190	4,460	8,080	13,100	26,900	48,800	77,300
150	30	63	119	244	366	704	1,120	1,980	4,050	7,320	11,900	24,300	44,200	70,000
175	28	58	109	224	336	648	1,030	1,820	3,720	6,730	10,900	22,400	40,700	64,400
200	26	54	102	209	313	602	960	1,700	3,460	6,260	10,100	20,800	37,900	59,900
250	23	48	90	185	277	534	851	1,500	3,070	5,550	8,990	18,500	33,500	53,100
300	21	43	82	168	251	484	771	1,360	2,780	5,030	8,150	16,700	30,400	48,100
350	19	40	75	154	231	445	709	1,250	2,560	4,630	7,490	15,400	28,000	44,300
400	18	37	70	143	215	414	660	1,170	2,380	4,310	6,970	14,300	26,000	41,200
450	17	35	66	135	202	389	619	1,090	2,230	4,040	6,540	13,400	24,400	38,600
500	16	33	62	127	191	367	585	1,030	2,110	3,820	6,180	12,700	23,100	36,500
550	15	31	59	121	181	349	556	982	2,000	3,620	5,870	12,100	21,900	34,700
600	14	30	56	115	173	333	530	937	1,910	3,460	5,600	11,500	20,900	33,100
650	14	29	54	110	165	318	508	897	1,830	3,310	5,360	11,000	20,000	31,700
700	13	27	52	106	159	306	488	862	1,760	3,180	5,150	10,600	19,200	30,400
750	13	26	50	102	153	295	470	830	1,690	3,060	4,960	10,200	18,500	29,300
800	12	26	48	99	148	285	454	802	1,640	2,960	4,790	9,840	17,900	28,300
850	12	25	46	95	143	275	439	776	1,580	2,860	4,640	9,530	17,300	27,400
900	11	24	45	93	139	267	426	752	1,530	2,780	4,500	9,240	16,800	26,600
950	11	23	44	90	135	259	413	731	1,490	2,700	4,370	8,970	16,300	25,800
1,000	11	23	43	87	131	252	402	711	1,450	2,620	4,250	8,720	15,800	25,100
1,100	10	21	40	83	124	240	382	675	1,380	2,490	4,030	8,290	15,100	23,800
1,200	NA	20	39	79	119	229	364	644	1,310	2,380	3,850	7,910	14,400	22,700
1,300	NA	20	37	76	114	219	349	617	1,260	2,280	3,680	7,570	13,700	21,800
1,400	NA	19	35	73	109	210	335	592	1,210	2,190	3,540	7,270	13,200	20,900
1,500	NA	18	34	70	105	203	323	571	1,160	2,110	3,410	7,010	12,700	20,100
1,600	NA	18	33	68	102	196	312	551	1,120	2,030	3,290	6,770	12,300	19,500
1,700	NA	17	32	66	98	189	302	533	1,090	1,970	3,190	6,550	11,900	18,800
1,800	NA	16	31	64	95	184	293	517	1,050	1,910	3,090	6,350	11,500	18,300
1,900	NA	16	30	62	93	178	284	502	1,020	1,850	3,000	6,170	11,200	17,700
2,000	NA	16	29	60	90	173	276	488	1,000	1,800	2,920	6,000	10,900	17,200

For SI: 1 inch = 25.4 mm, 1 foot = 304.8 mm, 1 pound per square inch = 6.895 kPa, 1-inch water column = 0.2488 kPa, 1 British thermal unit per hour = 0.2931 W, 1 cubic foot per hour = 0.0283 m³/h, 1 degree = 0.01745 rad.

Notes:
1. NA means a flow of less than 10 cfh.
2. All table entries have been rounded to three significant digits.

❖ See the commentary to Section 402.4.

TABLE 402.4(2)
SCHEDULE 40 METALLIC PIPE

Gas	Natural
Inlet Pressure	Less than 2 psi
Pressure Drop	0.5 in. w.c.
Specific Gravity	0.60

	PIPE SIZE (inch)													
Nominal	$^1/_2$	$^3/_4$	1	$1^1/_4$	$1^1/_2$	2	$2^1/_2$	3	4	5	6	8	10	12
Actual ID	0.622	0.824	1.049	1.380	1.610	2.067	2.469	3.068	4.026	5.047	6.065	7.981	10.020	11.938
Length (ft)	Capacity in Cubic Feet of Gas Per Hour													
10	172	360	678	1,390	2,090	4,020	6,400	11,300	23,100	41,800	67,600	139,000	252,000	399,000
20	118	247	466	957	1,430	2,760	4,400	7,780	15,900	28,700	46,500	95,500	173,000	275,000
30	95	199	374	768	1,150	2,220	3,530	6,250	12,700	23,000	37,300	76,700	139,000	220,000
40	81	170	320	657	985	1,900	3,020	5,350	10,900	19,700	31,900	65,600	119,000	189,000
50	72	151	284	583	873	1,680	2,680	4,740	9,660	17,500	28,300	58,200	106,000	167,000
60	65	137	257	528	791	1,520	2,430	4,290	8,760	15,800	25,600	52,700	95,700	152,000
70	60	126	237	486	728	1,400	2,230	3,950	8,050	14,600	23,600	48,500	88,100	139,000
80	56	117	220	452	677	1,300	2,080	3,670	7,490	13,600	22,000	45,100	81,900	130,000
90	52	110	207	424	635	1,220	1,950	3,450	7,030	12,700	20,600	42,300	76,900	122,000
100	50	104	195	400	600	1,160	1,840	3,260	6,640	12,000	19,500	40,000	72,600	115,000
125	44	92	173	355	532	1,020	1,630	2,890	5,890	10,600	17,200	35,400	64,300	102,000
150	40	83	157	322	482	928	1,480	2,610	5,330	9,650	15,600	32,100	58,300	92,300
175	37	77	144	296	443	854	1,360	2,410	4,910	8,880	14,400	29,500	53,600	84,900
200	34	71	134	275	412	794	1,270	2,240	4,560	8,260	13,400	27,500	49,900	79,000
250	30	63	119	244	366	704	1,120	1,980	4,050	7,320	11,900	24,300	44,200	70,000
300	27	57	108	221	331	638	1,020	1,800	3,670	6,630	10,700	22,100	40,100	63,400
350	25	53	99	203	305	587	935	1,650	3,370	6,100	9,880	20,300	36,900	58,400
400	23	49	92	189	283	546	870	1,540	3,140	5,680	9,190	18,900	34,300	54,300
450	22	46	86	177	266	512	816	1,440	2,940	5,330	8,620	17,700	32,200	50,900
500	21	43	82	168	251	484	771	1,360	2,780	5,030	8,150	16,700	30,400	48,100
550	20	41	78	159	239	459	732	1,290	2,640	4,780	7,740	15,900	28,900	45,700
600	19	39	74	152	228	438	699	1,240	2,520	4,560	7,380	15,200	27,500	43,600
650	18	38	71	145	218	420	669	1,180	2,410	4,360	7,070	14,500	26,400	41,800
700	17	36	68	140	209	403	643	1,140	2,320	4,190	6,790	14,000	25,300	40,100
750	17	35	66	135	202	389	619	1,090	2,230	4,040	6,540	13,400	24,400	38,600
800	16	34	63	130	195	375	598	1,060	2,160	3,900	6,320	13,000	23,600	37,300
850	16	33	61	126	189	363	579	1,020	2,090	3,780	6,110	12,600	22,800	36,100
900	15	32	59	122	183	352	561	992	2,020	3,660	5,930	12,200	22,100	35,000
950	15	31	58	118	178	342	545	963	1,960	3,550	5,760	11,800	21,500	34,000
1,000	14	30	56	115	173	333	530	937	1,910	3,460	5,600	11,500	20,900	33,100
1,100	14	28	53	109	164	316	503	890	1,810	3,280	5,320	10,900	19,800	31,400
1,200	13	27	51	104	156	301	480	849	1,730	3,130	5,070	10,400	18,900	30,000
1,300	12	26	49	100	150	289	460	813	1,660	3,000	4,860	9,980	18,100	28,700
1,400	12	25	47	96	144	277	442	781	1,590	2,880	4,670	9,590	17,400	27,600
1,500	11	24	45	93	139	267	426	752	1,530	2,780	4,500	9,240	16,800	26,600
1,600	11	23	44	89	134	258	411	727	1,480	2,680	4,340	8,920	16,200	25,600
1,700	11	22	42	86	130	250	398	703	1,430	2,590	4,200	8,630	15,700	24,800
1,800	10	22	41	84	126	242	386	682	1,390	2,520	4,070	8,370	15,200	24,100
1,900	10	21	40	81	122	235	375	662	1,350	2,440	3,960	8,130	14,800	23,400
2,000	NA	20	39	79	119	229	364	644	1,310	2,380	3,850	7,910	14,400	22,700

For SI: 1 inch = 25.4 mm, 1 foot = 304.8 mm, 1 pound per square inch = 6.895 kPa, 1-inch water column = 0.2488 kPa, 1 British thermal unit per hour = 0.2931 W, 1 cubic foot per hour = 0.0283 m³/h, 1 degree = 0.01745 rad.

Notes:
1. NA means a flow of less than 10 cfh.
2. All table entries have been rounded to three significant digits.

❖ See the commentary to Section 402.4.

TABLE 402.4(3)
SCHEDULE 40 METALLIC PIPE

Gas	Natural
Inlet Pressure	Less than 2 psi
Pressure Drop	3.0 in. w.c.
Specific Gravity	0.60

INTENDED USE: Initial supply pressure of 8.0 inches w.c. or greater									
PIPE SIZE (inch)									
Nominal	1/2	3/4	1	1 1/4	1 1/2	2	2 1/2	3	4
Actual ID	0.622	0.824	1.049	1.380	1.610	2.067	2.469	3.068	4.026
Length (ft)	Capacity in Cubic Feet of Gas Per Hour								
10	454	949	1,790	3,670	5,500	10,600	16,900	29,800	60,800
20	312	652	1,230	2,520	3,780	7,280	11,600	20,500	41,800
30	250	524	986	2,030	3,030	5,840	9,310	16,500	33,600
40	214	448	844	1,730	2,600	5,000	7,970	14,100	28,700
50	190	397	748	1,540	2,300	4,430	7,070	12,500	25,500
60	172	360	678	1,390	2,090	4,020	6,400	11,300	23,100
70	158	331	624	1,280	1,920	3,700	5,890	10,400	21,200
80	147	308	580	1,190	1,790	3,440	5,480	9,690	19,800
90	138	289	544	1,120	1,680	3,230	5,140	9,090	18,500
100	131	273	514	1,060	1,580	3,050	4,860	8,580	17,500
125	116	242	456	936	1,400	2,700	4,300	7,610	15,500
150	105	219	413	848	1,270	2,450	3,900	6,890	14,100
175	96	202	380	780	1,170	2,250	3,590	6,340	12,900
200	90	188	353	726	1,090	2,090	3,340	5,900	12,000
250	80	166	313	643	964	1,860	2,960	5,230	10,700
300	72	151	284	583	873	1,680	2,680	4,740	9,700
350	66	139	261	536	803	1,550	2,470	4,360	8,900
400	62	129	243	499	747	1,440	2,290	4,060	8,300
450	58	121	228	468	701	1,350	2,150	3,800	7,800
500	55	114	215	442	662	1,280	2,030	3,590	7,300
550	52	109	204	420	629	1,210	1,930	3,410	7,000
600	50	104	195	400	600	1,160	1,840	3,260	6,640
650	47	99	187	384	575	1,110	1,760	3,120	6,360
700	46	95	179	368	552	1,060	1,700	3,000	6,110
750	44	92	173	355	532	1,020	1,630	2,890	5,890
800	42	89	167	343	514	989	1,580	2,790	5,680
850	41	86	162	332	497	957	1,530	2,700	5,500
900	40	83	157	322	482	928	1,480	2,620	5,330
950	39	81	152	312	468	901	1,440	2,540	5,180
1,000	38	79	148	304	455	877	1,400	2,470	5,040
1,100	36	75	141	289	432	833	1,330	2,350	4,780
1,200	34	71	134	275	412	794	1,270	2,240	4,560
1,300	33	68	128	264	395	761	1,210	2,140	4,370
1,400	31	65	123	253	379	731	1,170	2,060	4,200
1,500	30	63	119	244	366	704	1,120	1,980	4,050
1,600	29	61	115	236	353	680	1,080	1,920	3,910
1,700	28	59	111	228	342	658	1,050	1,850	3,780
1,800	27	57	108	221	331	638	1,020	1,800	3,670
1,900	27	56	105	215	322	619	987	1,750	3,560
2,000	26	54	102	209	313	602	960	1,700	3,460

For SI: 1 inch = 25.4 mm, 1 foot = 304.8 mm, 1 pound per square inch = 6.895 kPa, 1-inch water column = 0.2488 kPa,
1 British thermal unit per hour = 0.2931 W, 1 cubic foot per hour = 0.0283 m³/h, 1 degree = 0.01745 rad.

Note: All table entries have been rounded to three significant digits.

❖ See the commentary to Section 402.4.

TABLE 402.4(4)
SCHEDULE 40 METALLIC PIPE

Gas	Natural
Inlet Pressure	Less than 2 psi
Pressure Drop	6.0 in. w.c.
Specific Gravity	0.60

INTENDED USE: Initial supply pressure of 11.0 inches w.c. or greater									
PIPE SIZE (inch)									
Nominal	$^1/_2$	$^3/_4$	1	$1^1/_4$	$1^1/_2$	2	$2^1/_2$	3	4
Actual ID	0.622	0.824	1.049	1.380	1.610	2.067	2.469	3.068	4.026
Length (ft)	Capacity in Cubic Feet of Gas Per Hour								
10	660	1,380	2,600	5,340	8,000	15,400	24,600	43,400	88,500
20	454	949	1,790	3,670	5,500	10,600	16,900	29,900	60,800
30	364	762	1,440	2,950	4,420	8,500	13,600	24,000	48,900
40	312	652	1,230	2,520	3,780	7,280	11,600	20,500	41,800
50	276	578	1,090	2,240	3,350	6,450	10,300	18,200	37,100
60	250	524	986	2,030	3,030	5,840	9,310	16,500	33,600
70	230	482	907	1,860	2,790	5,380	8,570	15,100	30,900
80	214	448	844	1,730	2,600	5,000	7,970	14,100	28,700
90	201	420	792	1,630	2,440	4,690	7,480	13,200	27,000
100	190	397	748	1,540	2,300	4,430	7,070	12,500	25,500
125	168	352	663	1,360	2,040	3,930	6,260	11,100	22,600
150	153	319	601	1,230	1,850	3,560	5,670	10,000	20,500
175	140	293	553	1,140	1,700	3,280	5,220	9,230	18,800
200	131	273	514	1,060	1,580	3,050	4,860	8,580	17,500
250	116	242	456	936	1,400	2,700	4,300	7,610	15,500
300	105	219	413	848	1,270	2,450	3,900	6,890	14,100
350	96	202	380	780	1,170	2,250	3,590	6,340	12,900
400	90	188	353	726	1,090	2,090	3,340	5,900	12,000
450	84	176	332	681	1,020	1,970	3,130	5,540	11,300
500	80	166	313	643	964	1,860	2,960	5,230	10,700
550	76	158	297	611	915	1,760	2,810	4,970	10,100
600	72	151	284	583	873	1,680	2,680	4,740	9,660
650	69	144	272	558	836	1,610	2,570	4,540	9,250
700	66	139	261	536	803	1,550	2,470	4,360	8,890
750	64	134	252	516	774	1,490	2,380	4,200	8,560
800	62	129	243	499	747	1,440	2,290	4,060	8,270
850	60	125	235	483	723	1,390	2,220	3,920	8,000
900	58	121	228	468	701	1,350	2,150	3,800	7,760
950	56	118	221	454	681	1,310	2,090	3,700	7,540
1,000	55	114	215	442	662	1,280	2,030	3,590	7,330
1,100	52	109	204	420	629	1,210	1,930	3,410	6,960
1,200	50	104	195	400	600	1,160	1,840	3,260	6,640
1,300	47	99	187	384	575	1,100	1,760	3,120	6,360
1,400	46	95	179	368	552	1,060	1,700	3,000	6,110
1,500	44	92	173	355	532	1,020	1,630	2,890	5,890
1,600	42	89	167	343	514	989	1,580	2,790	5,680
1,700	41	86	162	332	497	957	1,530	2,700	5,500
1,800	40	83	157	322	482	928	1,480	2,620	5,330
1,900	39	81	152	312	468	901	1,440	2,540	5,180
2,000	38	79	148	304	455	877	1,400	2,470	5,040

For SI: 1 inch = 25.4 mm, 1 foot = 304.8 mm, 1 pound per square inch = 6.895 kPa, 1-inch water column = 0.2488 kPa,
   1 British thermal unit per hour = 0.2931 W, 1 cubic foot per hour = 0.0283 m^3/h, 1 degree = 0.01745 rad.

Note: All table entries have been rounded to three significant digits.

❖ See the commentary to Section 402.4.

TABLE 402.4(5)
SCHEDULE 40 METALLIC PIPE

Gas	Natural
Inlet Pressure	2.0 psi
Pressure Drop	1.0 psi
Specific Gravity	0.60

	PIPE SIZE (inch)								
Nominal	$^1/_2$	$^3/_4$	1	$1^1/_4$	$1^1/_2$	2	$2^1/_2$	3	4
Actual ID	0.622	0.824	1.049	1.380	1.610	2.067	2.469	3.068	4.026
Length (ft)	Capacity in Cubic Feet of Gas Per Hour								
10	1,510	3,040	5,560	11,400	17,100	32,900	52,500	92,800	189,000
20	1,070	2,150	3,930	8,070	12,100	23,300	37,100	65,600	134,000
30	869	1,760	3,210	6,590	9,880	19,000	30,300	53,600	109,000
40	753	1,520	2,780	5,710	8,550	16,500	26,300	46,400	94,700
50	673	1,360	2,490	5,110	7,650	14,700	23,500	41,500	84,700
60	615	1,240	2,270	4,660	6,980	13,500	21,400	37,900	77,300
70	569	1,150	2,100	4,320	6,470	12,500	19,900	35,100	71,600
80	532	1,080	1,970	4,040	6,050	11,700	18,600	32,800	67,000
90	502	1,010	1,850	3,810	5,700	11,000	17,500	30,900	63,100
100	462	934	1,710	3,510	5,260	10,100	16,100	28,500	58,200
125	414	836	1,530	3,140	4,700	9,060	14,400	25,500	52,100
150	372	751	1,370	2,820	4,220	8,130	13,000	22,900	46,700
175	344	695	1,270	2,601	3,910	7,530	12,000	21,200	43,300
200	318	642	1,170	2,410	3,610	6,960	11,100	19,600	40,000
250	279	583	1,040	2,140	3,210	6,180	9,850	17,400	35,500
300	253	528	945	1,940	2,910	5,600	8,920	15,800	32,200
350	232	486	869	1,790	2,670	5,150	8,210	14,500	29,600
400	216	452	809	1,660	2,490	4,790	7,640	13,500	27,500
450	203	424	759	1,560	2,330	4,500	7,170	12,700	25,800
500	192	401	717	1,470	2,210	4,250	6,770	12,000	24,400
550	182	381	681	1,400	2,090	4,030	6,430	11,400	23,200
600	174	363	650	1,330	2,000	3,850	6,130	10,800	22,100
650	166	348	622	1,280	1,910	3,680	5,870	10,400	21,200
700	160	334	598	1,230	1,840	3,540	5,640	9,970	20,300
750	154	322	576	1,180	1,770	3,410	5,440	9,610	19,600
800	149	311	556	1,140	1,710	3,290	5,250	9,280	18,900
850	144	301	538	1,100	1,650	3,190	5,080	8,980	18,300
900	139	292	522	1,070	1,600	3,090	4,930	8,710	17,800
950	135	283	507	1,040	1,560	3,000	4,780	8,460	17,200
1,000	132	275	493	1,010	1,520	2,920	4,650	8,220	16,800
1,100	125	262	468	960	1,440	2,770	4,420	7,810	15,900
1,200	119	250	446	917	1,370	2,640	4,220	7,450	15,200
1,300	114	239	427	878	1,320	2,530	4,040	7,140	14,600
1,400	110	230	411	843	1,260	2,430	3,880	6,860	14,000
1,500	106	221	396	812	1,220	2,340	3,740	6,600	13,500
1,600	102	214	382	784	1,180	2,260	3,610	6,380	13,000
1,700	99	207	370	759	1,140	2,190	3,490	6,170	12,600
1,800	96	200	358	736	1,100	2,120	3,390	5,980	12,200
1,900	93	195	348	715	1,070	2,060	3,290	5,810	11,900
2,000	91	189	339	695	1,040	2,010	3,200	5,650	11,500

For SI: 1 inch = 25.4 mm, 1 foot = 304.8 mm, 1 pound per square inch = 6.895 kPa, 1-inch water column = 0.2488 kPa,
 1 British thermal unit per hour = 0.2931 W, 1 cubic foot per hour = 0.0283 m³/h, 1 degree = 0.01745 rad.
Note: All table entries have been rounded to three significant digits.

❖ See the commentary to Section 402.4.

TABLE 402.4(6)
SCHEDULE 40 METALLIC PIPE

Gas	Natural
Inlet Pressure	3.0 psi
Pressure Drop	2.0 psi
Specific Gravity	0.60

PIPE SIZE (inch)									
Nominal	$^1/_2$	$^3/_4$	1	$1^1/_4$	$1^1/_2$	2	$2^1/_2$	3	4
Actual ID	0.622	0.824	1.049	1.380	1.610	2.067	2.469	3.068	4.026
Length (ft)	Capacity in Cubic Feet of Gas Per Hour								
10	2,350	4,920	9,270	19,000	28,500	54,900	87,500	155,000	316,000
20	1,620	3,380	6,370	13,100	19,600	37,700	60,100	106,000	217,000
30	1,300	2,720	5,110	10,500	15,700	30,300	48,300	85,400	174,000
40	1,110	2,320	4,380	8,990	13,500	25,900	41,300	73,100	149,000
50	985	2,060	3,880	7,970	11,900	23,000	36,600	64,800	132,000
60	892	1,870	3,520	7,220	10,800	20,800	33,200	58,700	120,000
70	821	1,720	3,230	6,640	9,950	19,200	30,500	54,000	110,000
80	764	1,600	3,010	6,180	9,260	17,800	28,400	50,200	102,000
90	717	1,500	2,820	5,800	8,680	16,700	26,700	47,100	96,100
100	677	1,420	2,670	5,470	8,200	15,800	25,200	44,500	90,800
125	600	1,250	2,360	4,850	7,270	14,000	22,300	39,500	80,500
150	544	1,140	2,140	4,400	6,590	12,700	20,200	35,700	72,900
175	500	1,050	1,970	4,040	6,060	11,700	18,600	32,900	67,100
200	465	973	1,830	3,760	5,640	10,900	17,300	30,600	62,400
250	412	862	1,620	3,330	5,000	9,620	15,300	27,100	55,300
300	374	781	1,470	3,020	4,530	8,720	13,900	24,600	50,100
350	344	719	1,350	2,780	4,170	8,020	12,800	22,600	46,100
400	320	669	1,260	2,590	3,870	7,460	11,900	21,000	42,900
450	300	627	1,180	2,430	3,640	7,000	11,200	19,700	40,200
500	283	593	1,120	2,290	3,430	6,610	10,500	18,600	38,000
550	269	563	1,060	2,180	3,260	6,280	10,000	17,700	36,100
600	257	537	1,010	2,080	3,110	5,990	9,550	16,900	34,400
650	246	514	969	1,990	2,980	5,740	9,150	16,200	33,000
700	236	494	931	1,910	2,860	5,510	8,790	15,500	31,700
750	228	476	897	1,840	2,760	5,310	8,470	15,000	30,500
800	220	460	866	1,780	2,660	5,130	8,180	14,500	29,500
850	213	445	838	1,720	2,580	4,960	7,910	14,000	28,500
900	206	431	812	1,670	2,500	4,810	7,670	13,600	27,700
950	200	419	789	1,620	2,430	4,670	7,450	13,200	26,900
1,000	195	407	767	1,580	2,360	4,550	7,240	12,800	26,100
1,100	185	387	729	1,500	2,240	4,320	6,890	12,200	24,800
1,200	177	369	695	1,430	2,140	4,120	6,570	11,600	23,700
1,300	169	353	666	1,370	2,050	3,940	6,290	11,100	22,700
1,400	162	340	640	1,310	1,970	3,790	6,040	10,700	21,800
1,500	156	327	616	1,270	1,900	3,650	5,820	10,300	21,000
1,600	151	316	595	1,220	1,830	3,530	5,620	10,000	20,300
1,700	146	306	576	1,180	1,770	3,410	5,440	9,610	19,600
1,800	142	296	558	1,150	1,720	3,310	5,270	9,320	19,000
1,900	138	288	542	1,110	1,670	3,210	5,120	9,050	18,400
2,000	134	280	527	1,080	1,620	3,120	4,980	8,800	18,000

For SI: 1 inch = 25.4 mm, 1 foot = 304.8 mm, 1 pound per square inch = 6.895 kPa, 1-inch water column = 0.2488 kPa,
  1 British thermal unit per hour = 0.2931 W, 1 cubic foot per hour = 0.0283 m³/h, 1 degree = 0.01745 rad.
Note: All table entries have been rounded to three significant digits.

❖ See the commentary to Section 402.4.

TABLE 402.4(7)
SCHEDULE 40 METALLIC PIPE

Gas	Natural
Inlet Pressure	5.0 psi
Pressure Drop	3.5 psi
Specific Gravity	0.60

	PIPE SIZE (inch)								
Nominal	$^1/_2$	$^3/_4$	1	$1^1/_4$	$1^1/_2$	2	$2^1/_2$	3	4
Actual ID	0.622	0.824	1.049	1.380	1.610	2.067	2.469	3.068	4.026
Length (ft)	Capacity in Cubic Feet of Gas Per Hour								
10	3,190	6,430	11,800	24,200	36,200	69,700	111,000	196,000	401,000
20	2,250	4,550	8,320	17,100	25,600	49,300	78,600	139,000	283,000
30	1,840	3,720	6,790	14,000	20,900	40,300	64,200	113,000	231,000
40	1,590	3,220	5,880	12,100	18,100	34,900	55,600	98,200	200,000
50	1,430	2,880	5,260	10,800	16,200	31,200	49,700	87,900	179,000
60	1,300	2,630	4,800	9,860	14,800	28,500	45,400	80,200	164,000
70	1,200	2,430	4,450	9,130	13,700	26,400	42,000	74,300	151,000
80	1,150	2,330	4,260	8,540	12,800	24,700	39,300	69,500	142,000
90	1,060	2,150	3,920	8,050	12,100	23,200	37,000	65,500	134,000
100	979	1,980	3,620	7,430	11,100	21,400	34,200	60,400	123,000
125	876	1,770	3,240	6,640	9,950	19,200	30,600	54,000	110,000
150	786	1,590	2,910	5,960	8,940	17,200	27,400	48,500	98,900
175	728	1,470	2,690	5,520	8,270	15,900	25,400	44,900	91,600
200	673	1,360	2,490	5,100	7,650	14,700	23,500	41,500	84,700
250	558	1,170	2,200	4,510	6,760	13,000	20,800	36,700	74,900
300	506	1,060	1,990	4,090	6,130	11,800	18,800	33,300	67,800
350	465	973	1,830	3,760	5,640	10,900	17,300	30,600	62,400
400	433	905	1,710	3,500	5,250	10,100	16,100	28,500	58,100
450	406	849	1,600	3,290	4,920	9,480	15,100	26,700	54,500
500	384	802	1,510	3,100	4,650	8,950	14,300	25,200	51,500
550	364	762	1,440	2,950	4,420	8,500	13,600	24,000	48,900
600	348	727	1,370	2,810	4,210	8,110	12,900	22,900	46,600
650	333	696	1,310	2,690	4,030	7,770	12,400	21,900	44,600
700	320	669	1,260	2,590	3,880	7,460	11,900	21,000	42,900
750	308	644	1,210	2,490	3,730	7,190	11,500	20,300	41,300
800	298	622	1,170	2,410	3,610	6,940	11,100	19,600	39,900
850	288	602	1,130	2,330	3,490	6,720	10,700	18,900	38,600
900	279	584	1,100	2,260	3,380	6,520	10,400	18,400	37,400
950	271	567	1,070	2,190	3,290	6,330	10,100	17,800	36,400
1,000	264	551	1,040	2,130	3,200	6,150	9,810	17,300	35,400
1,100	250	524	987	2,030	3,030	5,840	9,320	16,500	33,600
1,200	239	500	941	1,930	2,900	5,580	8,890	15,700	32,000
1,300	229	478	901	1,850	2,770	5,340	8,510	15,000	30,700
1,400	220	460	866	1,780	2,660	5,130	8,180	14,500	29,500
1,500	212	443	834	1,710	2,570	4,940	7,880	13,900	28,400
1,600	205	428	806	1,650	2,480	4,770	7,610	13,400	27,400
1,700	198	414	780	1,600	2,400	4,620	7,360	13,000	26,500
1,800	192	401	756	1,550	2,330	4,480	7,140	12,600	25,700
1,900	186	390	734	1,510	2,260	4,350	6,930	12,300	25,000
2,000	181	379	714	1,470	2,200	4,230	6,740	11,900	24,300

For SI: 1 inch = 25.4 mm, 1 foot – 304.8 mm, 1 pound per square inch = 6.895 kPa, 1-inch water column = 0.2488 kPa,
1 British thermal unit per hour = 0.2931 W, 1 cubic foot per hour = 0.0283 m³/h, 1 degree = 0.01745 rad.
Note: All table entries have been rounded to three significant digits.

❖ See the commentary to Section 402.4.

**TABLE 402.4(8)
SEMIRIGID COPPER TUBING**

Gas	Natural
Inlet Pressure	Less than 2 psi
Pressure Drop	0.3 in. w.c.
Specific Gravity	0.60

Nominal	K & L	\(^1/_4\)	\(^3/_8\)	\(^1/_2\)	\(^5/_8\)	\(^3/_4\)	1	\(1^1/_4\)	\(1^1/_2\)	2
	ACR	\(^3/_8\)	\(^1/_2\)	\(^5/_8\)	\(^3/_4\)	\(^7/_8\)	\(1^1/_8\)	\(1^3/_8\)	—	—
Outside		0.375	0.500	0.625	0.750	0.875	1.125	1.375	1.625	2.125
Inside		0.305	0.402	0.527	0.652	0.745	0.995	1.245	1.481	1.959
Length (ft)		\multicolumn{9}{c}{Capacity in Cubic Feet of Gas Per Hour}								
10		20	42	85	148	210	448	806	1,270	2,650
20		14	29	58	102	144	308	554	873	1,820
30		11	23	47	82	116	247	445	701	1,460
40		10	20	40	70	99	211	381	600	1,250
50		NA	17	35	62	88	187	337	532	1,110
60		NA	16	32	56	79	170	306	482	1,000
70		NA	14	29	52	73	156	281	443	924
80		NA	13	27	48	68	145	262	413	859
90		NA	13	26	45	64	136	245	387	806
100		NA	12	24	43	60	129	232	366	761
125		NA	11	22	38	53	114	206	324	675
150		NA	10	20	34	48	103	186	294	612
175		NA	NA	18	31	45	95	171	270	563
200		NA	NA	17	29	41	89	159	251	523
250		NA	NA	15	26	37	78	141	223	464
300		NA	NA	13	23	33	71	128	202	420
350		NA	NA	12	22	31	65	118	186	387
400		NA	NA	11	20	28	61	110	173	360
450		NA	NA	11	19	27	57	103	162	338
500		NA	NA	10	18	25	54	97	153	319
550		NA	NA	NA	17	24	51	92	145	303
600		NA	NA	NA	16	23	49	88	139	289
650		NA	NA	NA	15	22	47	84	133	277
700		NA	NA	NA	15	21	45	81	128	266
750		NA	NA	NA	14	20	43	78	123	256
800		NA	NA	NA	14	20	42	75	119	247
850		NA	NA	NA	13	19	40	73	115	239
900		NA	NA	NA	13	18	39	71	111	232
950		NA	NA	NA	13	18	38	69	108	225
1,000		NA	NA	NA	12	17	37	67	105	219
1,100		NA	NA	NA	12	16	35	63	100	208
1,200		NA	NA	NA	11	16	34	60	95	199
1,300		NA	NA	NA	11	15	32	58	91	190
1,400		NA	NA	NA	10	14	31	56	88	183
1,500		NA	NA	NA	NA	14	30	54	84	176
1,600		NA	NA	NA	NA	13	29	52	82	170
1,700		NA	NA	NA	NA	13	28	50	79	164
1,800		NA	NA	NA	NA	13	27	49	77	159
1,900		NA	NA	NA	NA	12	26	47	74	155
2,000		NA	NA	NA	NA	12	25	46	72	151

For SI: 1 inch = 25.4 mm, 1 foot = 304.8 mm, 1 pound per square inch = 6.895 kPa, 1-inch water column = 0.2488 kPa,
    1 British thermal unit per hour = 0.2931 W, 1 cubic foot per hour = 0.0283 m³/h, 1 degree = 0.01745 rad.

Notes:

1. Table capacities are based on Type K copper tubing inside diameter (shown), which has the smallest inside diameter of the copper tubing products.

2. NA means a flow of less than 10 cfh.

3. All table entries have been rounded to three significant digits.

❖ See the commentary to Section 402.4.

TABLE 402.4(9)
SEMIRIGID COPPER TUBING

Gas	Natural
Inlet Pressure	Less than 2 psi
Pressure Drop	0.5 in. w.c.
Specific Gravity	0.60

Nominal	K & L	¹/₄	³/₈	¹/₂	⁵/₈	³/₄	1	1¹/₄	1¹/₂	2
	ACR	³/₈	¹/₂	⁵/₈	³/₄	⁷/₈	1¹/₈	1³/₈	—	—
Outside		0.375	0.500	0.625	0.750	0.875	1.125	1.375	1.625	2.125
Inside		0.305	0.402	0.527	0.652	0.745	0.995	1.245	1.481	1.959
Length (ft)		Capacity in Cubic Feet of Gas Per Hour								
10		27	55	111	195	276	590	1,060	1,680	3,490
20		18	38	77	134	190	406	730	1,150	2,400
30		15	30	61	107	152	326	586	925	1,930
40		13	26	53	92	131	279	502	791	1,650
50		11	23	47	82	116	247	445	701	1,460
60		10	21	42	74	105	224	403	635	1,320
70		NA	19	39	68	96	206	371	585	1,220
80		NA	18	36	63	90	192	345	544	1,130
90		NA	17	34	59	84	180	324	510	1,060
100		NA	16	32	56	79	170	306	482	1,000
125		NA	14	28	50	70	151	271	427	890
150		NA	13	26	45	64	136	245	387	806
175		NA	12	24	41	59	125	226	356	742
200		NA	11	22	39	55	117	210	331	690
250		NA	NA	20	34	48	103	186	294	612
300		NA	NA	18	31	44	94	169	266	554
350		NA	NA	16	28	40	86	155	245	510
400		NA	NA	15	26	38	80	144	228	474
450		NA	NA	14	25	35	75	135	214	445
500		NA	NA	13	23	33	71	128	202	420
550		NA	NA	13	22	32	68	122	192	399
600		NA	NA	12	21	30	64	116	183	381
650		NA	NA	12	20	29	62	111	175	365
700		NA	NA	11	20	28	59	107	168	350
750		NA	NA	11	19	27	57	103	162	338
800		NA	NA	10	18	26	55	99	156	326
850		NA	NA	10	18	25	53	96	151	315
900		NA	NA	NA	17	24	52	93	147	306
950		NA	NA	NA	17	24	50	90	143	297
1,000		NA	NA	NA	16	23	49	88	139	289
1,100		NA	NA	NA	15	22	46	84	132	274
1,200		NA	NA	NA	15	21	44	80	126	262
1,300		NA	NA	NA	14	20	42	76	120	251
1,400		NA	NA	NA	13	19	41	73	116	241
1,500		NA	NA	NA	13	18	39	71	111	232
1,600		NA	NA	NA	13	18	38	68	108	224
1,700		NA	NA	NA	12	17	37	66	104	217
1,800		NA	NA	NA	12	17	36	64	101	210
1,900		NA	NA	NA	11	16	35	62	98	204
2,000		NA	NA	NA	11	16	34	60	95	199

For SI: 1 inch = 25.4 mm, 1 foot = 304.8 mm, 1 pound per square inch = 6.895 kPa, 1-inch water column = 0.2488 kPa,
1 British thermal unit per hour = 0.2931 W, 1 cubic foot per hour = 0.0283 m³/h, 1 degree = 0.01745 rad.

Notes:

1. Table capacities are based on Type K copper tubing inside diameter (shown), which has the smallest inside diameter of the copper tubing products.
2. NA means a flow of less than 10 cfh.
3. All table entries have been rounded to three significant digits.

❖ See the commentary to Section 402.4.

TABLE 402.4(10)
SEMIRIGID COPPER TUBING

Gas	Natural
Inlet Pressure	Less than 2 psi
Pressure Drop	1.0 in. w.c.
Specific Gravity	0.60

INTENDED USE: SIZING BETWEEN HOUSE LINE REGULATOR AND THE APPLIANCE									
TUBE SIZE (inch)									
Nominal	K & L	$^1/_4$	$^3/_8$	$^1/_2$	$^5/_8$	$^3/_4$	1	$1^1/_4$	$1^1/_2$	2
	ACR	$^3/_8$	$^1/_2$	$^5/_8$	$^3/_4$	$^7/_8$	$1^1/_8$	$1^3/_8$	—	—
Outside		0.375	0.500	0.625	0.750	0.875	1.125	1.375	1.625	2.125
Inside		0.305	0.402	0.527	0.652	0.745	0.995	1.245	1.481	1.959
Length (ft)		Capacity in Cubic Feet of Gas Per Hour								
10		39	80	162	283	402	859	1,550	2,440	5,080
20		27	55	111	195	276	590	1,060	1,680	3,490
30		21	44	89	156	222	474	853	1,350	2,800
40		18	38	77	134	190	406	730	1,150	2,400
50		16	33	68	119	168	359	647	1,020	2,130
60		15	30	61	107	152	326	586	925	1,930
70		13	28	57	99	140	300	539	851	1,770
80		13	26	53	92	131	279	502	791	1,650
90		12	24	49	86	122	262	471	742	1,550
100		11	23	47	82	116	247	445	701	1,460
125		NA	20	41	72	103	219	394	622	1,290
150		NA	18	37	65	93	198	357	563	1,170
175		NA	17	34	60	85	183	329	518	1,080
200		NA	16	32	56	79	170	306	482	1,000
250		NA	14	28	50	70	151	271	427	890
300		NA	13	26	45	64	136	245	387	806
350		NA	12	24	41	59	125	226	356	742
400		NA	11	22	39	55	117	210	331	690
450		NA	10	21	36	51	110	197	311	647
500		NA	NA	20	34	48	103	186	294	612
550		NA	NA	19	32	46	98	177	279	581
600		NA	NA	18	31	44	94	169	266	554
650		NA	NA	17	30	42	90	162	255	531
700		NA	NA	16	28	40	86	155	245	510
750		NA	NA	16	27	39	83	150	236	491
800		NA	NA	15	26	38	80	144	228	474
850		NA	NA	15	26	36	78	140	220	459
900		NA	NA	14	25	35	75	135	214	445
950		NA	NA	14	24	34	73	132	207	432
1,000		NA	NA	13	23	33	71	128	202	420
1,100		NA	NA	13	22	32	68	122	192	399
1,200		NA	NA	12	21	30	64	116	183	381
1,300		NA	NA	12	20	29	62	111	175	365
1,400		NA	NA	11	20	28	59	107	168	350
1,500		NA	NA	11	19	27	57	103	162	338
1,600		NA	NA	10	18	26	55	99	156	326
1,700		NA	NA	10	18	25	53	96	151	315
1,800		NA	NA	NA	17	24	52	93	147	306
1,900		NA	NA	NA	17	24	50	90	143	297
2,000		NA	NA	NA	16	23	49	88	139	289

For SI: 1 inch = 25.4 mm, 1 foot = 304.8 mm, 1 pound per square inch = 6.895 kPa, 1-inch water column = 0.2488 kPa,
   1 British thermal unit per hour = 0.2931 W, 1 cubic foot per hour = 0.0283 m³/h, 1 degree = 0.01745 rad.

Notes:

1. Table capacities are based on Type K copper tubing inside diameter (shown), which has the smallest inside diameter of the copper tubing products.
2. NA means a flow of less than 10 cfh.
3. All table entries have been rounded to three significant digits.

❖ See the commentary to Section 402.4.

TABLE 402.4(11)
SEMIRIGID COPPER TUBING

	Gas	Natural
	Inlet Pressure	Less than 2 psi
	Pressure Drop	17.0 in. w.c.
	Specific Gravity	0.60

Nominal	K & L	¹/₄	³/₈	¹/₂	⁵/₈	³/₄	1	1¹/₄	1¹/₂	2
	ACR	³/₈	¹/₂	⁵/₈	³/₄	⁷/₈	1¹/₈	1³/₈	—	—
Outside		0.375	0.500	0.625	0.750	0.875	1.125	1.375	1.625	2.125
Inside		0.305	0.402	0.527	0.652	0.745	0.995	1.245	1.481	1.959
Length (ft)		\multicolumn{9}{Capacity in Cubic Feet of Gas Per Hour}
10		190	391	796	1,390	1,970	4,220	7,590	12,000	24,900
20		130	269	547	956	1,360	2,900	5,220	8,230	17,100
30		105	216	439	768	1,090	2,330	4,190	6,610	13,800
40		90	185	376	657	932	1,990	3,590	5,650	11,800
50		79	164	333	582	826	1,770	3,180	5,010	10,400
60		72	148	302	528	749	1,600	2,880	4,540	9,460
70		66	137	278	486	689	1,470	2,650	4,180	8,700
80		62	127	258	452	641	1,370	2,460	3,890	8,090
90		58	119	243	424	601	1,280	2,310	3,650	7,590
100		55	113	229	400	568	1,210	2,180	3,440	7,170
125		48	100	203	355	503	1,080	1,940	3,050	6,360
150		44	90	184	321	456	974	1,750	2,770	5,760
175		40	83	169	296	420	896	1,610	2,540	5,300
200		38	77	157	275	390	834	1,500	2,370	4,930
250		33	69	140	244	346	739	1,330	2,100	4,370
300		30	62	126	221	313	670	1,210	1,900	3,960
350		28	57	116	203	288	616	1,110	1,750	3,640
400		26	53	108	189	268	573	1,030	1,630	3,390
450		24	50	102	177	252	538	968	1,530	3,180
500		23	47	96	168	238	508	914	1,440	3,000
550		22	45	91	159	226	482	868	1,370	2,850
600		21	43	87	152	215	460	829	1,310	2,720
650		20	41	83	145	206	441	793	1,250	2,610
700		19	39	80	140	198	423	762	1,200	2,500
750		18	38	77	135	191	408	734	1,160	2,410
800		18	37	74	130	184	394	709	1,120	2,330
850		17	35	72	126	178	381	686	1,080	2,250
900		17	34	70	122	173	370	665	1,050	2,180
950		16	33	68	118	168	359	646	1,020	2,120
1,000		16	32	66	115	163	349	628	991	2,060
1,100		15	31	63	109	155	332	597	941	1,960
1,200		14	29	60	104	148	316	569	898	1,870
1,300		14	28	57	100	142	303	545	860	1,790
1,400		13	27	55	96	136	291	524	826	1,720
1,500		13	26	53	93	131	280	505	796	1,660
1,600		12	25	51	89	127	271	487	768	1,600
1,700		12	24	49	86	123	262	472	744	1,550
1,800		11	24	48	84	119	254	457	721	1,500
1,900		11	23	47	81	115	247	444	700	1,460
2,000		11	22	45	79	112	240	432	681	1,420

For SI: 1 inch = 25.4 mm, 1 foot = 304.8 mm, 1 pound per square inch = 6.895 kPa, 1-inch water column = 0.2488 kPa, 1 British thermal unit per hour = 0.2931 W, 1 cubic foot per hour = 0.0283 m³/h, 1 degree = 0.01745 rad.

Notes:
1. Table capacities are based on Type K copper tubing inside diameter (shown), which has the smallest inside diameter of the copper tubing products.
2. All table entries have been rounded to three significant digits.

❖ See the commentary to Section 402.4.

TABLE 402.4(12)
SEMIRIGID COPPER TUBING

	Gas	Natural
	Inlet Pressure	2.0 psi
	Pressure Drop	1.0 psi
	Specific Gravity	0.60

		TUBE SIZE (inch)								
Nominal	K & L	$^1/_4$	$^3/_8$	$^1/_2$	$^5/_8$	$^3/_4$	1	$1^1/_4$	$1^1/_2$	2
	ACR	$^3/_8$	$^1/_2$	$^5/_8$	$^3/_4$	$^7/_8$	$1^1/_8$	$1^3/_8$	—	—
Outside		0.375	0.500	0.625	0.750	0.875	1.125	1.375	1.625	2.125
Inside		0.305	0.402	0.527	0.652	0.745	0.995	1.245	1.481	1.959
Length (ft)		Capacity in Cubic Feet of Gas Per Hour								
10		245	506	1,030	1,800	2,550	5,450	9,820	15,500	32,200
20		169	348	708	1,240	1,760	3,750	6,750	10,600	22,200
30		135	279	568	993	1,410	3,010	5,420	8,550	17,800
40		116	239	486	850	1,210	2,580	4,640	7,310	15,200
50		103	212	431	754	1,070	2,280	4,110	6,480	13,500
60		93	192	391	683	969	2,070	3,730	5,870	12,200
70		86	177	359	628	891	1,900	3,430	5,400	11,300
80		80	164	334	584	829	1,770	3,190	5,030	10,500
90		75	154	314	548	778	1,660	2,990	4,720	9,820
100		71	146	296	518	735	1,570	2,830	4,450	9,280
125		63	129	263	459	651	1,390	2,500	3,950	8,220
150		57	117	238	416	590	1,260	2,270	3,580	7,450
175		52	108	219	383	543	1,160	2,090	3,290	6,850
200		49	100	204	356	505	1,080	1,940	3,060	6,380
250		43	89	181	315	448	956	1,720	2,710	5,650
300		39	80	164	286	406	866	1,560	2,460	5,120
350		36	74	150	263	373	797	1,430	2,260	4,710
400		33	69	140	245	347	741	1,330	2,100	4,380
450		31	65	131	230	326	696	1,250	1,970	4,110
500		30	61	124	217	308	657	1,180	1,870	3,880
550		28	58	118	206	292	624	1,120	1,770	3,690
600		27	55	112	196	279	595	1,070	1,690	3,520
650		26	53	108	188	267	570	1,030	1,620	3,370
700		25	51	103	181	256	548	986	1,550	3,240
750		24	49	100	174	247	528	950	1,500	3,120
800		23	47	96	168	239	510	917	1,450	3,010
850		22	46	93	163	231	493	888	1,400	2,920
900		22	44	90	158	224	478	861	1,360	2,830
950		21	43	88	153	217	464	836	1,320	2,740
1,000		20	42	85	149	211	452	813	1,280	2,670
1,100		19	40	81	142	201	429	772	1,220	2,540
1,200		18	38	77	135	192	409	737	1,160	2,420
1,300		18	36	74	129	183	392	705	1,110	2,320
1,400		17	35	71	124	176	376	678	1,070	2,230
1,500		16	34	68	120	170	363	653	1,030	2,140
1,600		16	33	66	116	164	350	630	994	2,070
1,700		15	31	64	112	159	339	610	962	2,000
1,800		15	30	62	108	154	329	592	933	1,940
1,900		14	30	60	105	149	319	575	906	1,890
2,000		14	29	59	102	145	310	559	881	1,830

For SI: 1 inch = 25.4 mm, 1 foot = 304.8 mm, 1 pound per square inch = 6.895 kPa, 1-inch water column = 0.2488 kPa,
    1 British thermal unit per hour = 0.2931 W, 1 cubic foot per hour = 0.0283 m³/h, 1 degree = 0.01745 rad.

Notes:

1. Table capacities are based on Type K copper tubing inside diameter (shown), which has the smallest inside diameter of the copper tubing products.

2. All table entries have been rounded to three significant digits.

❖ See the commentary to Section 402.4.

TABLE 402.4(13)
SEMIRIGID COPPER TUBING

Gas	Natural
Inlet Pressure	2.0 psi
Pressure Drop	1.5 psi
Specific Gravity	0.60

INTENDED USE			Pipe sizing between point of delivery and the house line regulator. Total load supplied by a single house line regulator not exceeding 150 cubic feet per hour.

						TUBE SIZE (inch)				
Nominal	K & L	$^1/_4$	$^3/_8$	$^1/_2$	$^5/_8$	$^3/_4$	1	$1^1/_4$	$1^1/_2$	2
	ACR	$^3/_8$	$^1/_2$	$^5/_8$	$^3/_4$	$^7/_8$	$1^1/_8$	$1^3/_8$	—	—
Outside		0.375	0.500	0.625	0.750	0.875	1.125	1.375	1.625	2.125
Inside		0.305	0.402	0.527	0.652	0.745	0.995	1.245	1.481	1.959
Length (ft)			Capacity in Cubic Feet of Gas Per Hour							
10		303	625	1,270	2,220	3,150	6,740	12,100	19,100	39,800
20		208	430	874	1,530	2,170	4,630	8,330	13,100	27,400
30		167	345	702	1,230	1,740	3,720	6,690	10,600	22,000
40		143	295	601	1,050	1,490	3,180	5,730	9,030	18,800
50		127	262	532	931	1,320	2,820	5,080	8,000	16,700
60		115	237	482	843	1,200	2,560	4,600	7,250	15,100
70		106	218	444	776	1,100	2,350	4,230	6,670	13,900
80		98	203	413	722	1,020	2,190	3,940	6,210	12,900
90		92	190	387	677	961	2,050	3,690	5,820	12,100
100		87	180	366	640	907	1,940	3,490	5,500	11,500
125		77	159	324	567	804	1,720	3,090	4,880	10,200
150		70	144	294	514	729	1,560	2,800	4,420	9,200
175		64	133	270	472	670	1,430	2,580	4,060	8,460
200		60	124	252	440	624	1,330	2,400	3,780	7,870
250		53	110	223	390	553	1,180	2,130	3,350	6,980
300		48	99	202	353	501	1,070	1,930	3,040	6,320
350		44	91	186	325	461	984	1,770	2,790	5,820
400		41	85	173	302	429	916	1,650	2,600	5,410
450		39	80	162	283	402	859	1,550	2,440	5,080
500		36	75	153	268	380	811	1,460	2,300	4,800
550		35	72	146	254	361	771	1,390	2,190	4,560
600		33	68	139	243	344	735	1,320	2,090	4,350
650		32	65	133	232	330	704	1,270	2,000	4,160
700		30	63	128	223	317	676	1,220	1,920	4,000
750		29	60	123	215	305	652	1,170	1,850	3,850
800		28	58	119	208	295	629	1,130	1,790	3,720
850		27	57	115	201	285	609	1,100	1,730	3,600
900		27	55	111	195	276	590	1,060	1,680	3,490
950		26	53	108	189	268	573	1,030	1,630	3,390
1,000		25	52	105	184	261	558	1,000	1,580	3,300
1,100		24	49	100	175	248	530	954	1,500	3,130
1,200		23	47	95	167	237	505	910	1,430	2,990
1,300		22	45	91	160	227	484	871	1,370	2,860
1,400		21	43	88	153	218	465	837	1,320	2,750
1,500		20	42	85	148	210	448	806	1,270	2,650
1,600		19	40	82	143	202	432	779	1,230	2,560
1,700		19	39	79	138	196	419	753	1,190	2,470
1,800		18	38	77	134	190	406	731	1,150	2,400
1,900		18	37	74	130	184	394	709	1,120	2,330
2,000		17	36	72	126	179	383	690	1,090	2,270

For SI: 1 inch = 25.4 mm, 1 foot = 304.8 mm, 1 pound per square inch = 6.895 kPa, 1-inch water column = 0.2488 kPa,
   1 British thermal unit per hour = 0.2931 W, 1 cubic foot per hour = 0.0283 m³/h, 1 degree = 0.01745 rad.

Notes:

1. Table capacities are based on Type K copper tubing inside diameter (shown), which has the smallest inside diameter of the copper tubing products.

2. Where this table is used to size the tubing upstream of a line pressure regulator, the pipe or tubing downstream of the line pressure regulator shall be sized using a pressure drop not greater than 1 inch w.c.

3. All table entries have been rounded to three significant digits.

❖ See the commentary to Section 402.4.

TABLE 402.4(14)
SEMIRIGID COPPER TUBING

	Gas	Natural
	Inlet Pressure	5.0 psi
	Pressure Drop	3.5 psi
	Specific Gravity	0.60

Nominal	K & L	1/4	3/8	1/2	5/8	3/4	1	1 1/4	1 1/2	2
	ACR	3/8	1/2	5/8	3/4	7/8	1 1/8	1 3/8	—	—
Outside		0.375	0.500	0.625	0.750	0.875	1.125	1.375	1.625	2.125
Inside		0.305	0.402	0.527	0.652	0.745	0.995	1.245	1.481	1.959
Length (ft)		Capacity in Cubic Feet of Gas Per Hour								
10		511	1,050	2,140	3,750	5,320	11,400	20,400	32,200	67,100
20		351	724	1,470	2,580	3,650	7,800	14,000	22,200	46,100
30		282	582	1,180	2,070	2,930	6,270	11,300	17,800	37,000
40		241	498	1,010	1,770	2,510	5,360	9,660	15,200	31,700
50		214	441	898	1,570	2,230	4,750	8,560	13,500	28,100
60		194	400	813	1,420	2,020	4,310	7,750	12,200	25,500
70		178	368	748	1,310	1,860	3,960	7,130	11,200	23,400
80		166	342	696	1,220	1,730	3,690	6,640	10,500	21,800
90		156	321	653	1,140	1,620	3,460	6,230	9,820	20,400
100		147	303	617	1,080	1,530	3,270	5,880	9,270	19,300
125		130	269	547	955	1,360	2,900	5,210	8,220	17,100
150		118	243	495	866	1,230	2,620	4,720	7,450	15,500
175		109	224	456	796	1,130	2,410	4,350	6,850	14,300
200		101	208	424	741	1,050	2,250	4,040	6,370	13,300
250		90	185	376	657	932	1,990	3,580	5,650	11,800
300		81	167	340	595	844	1,800	3,250	5,120	10,700
350		75	154	313	547	777	1,660	2,990	4,710	9,810
400		69	143	291	509	722	1,540	2,780	4,380	9,120
450		65	134	273	478	678	1,450	2,610	4,110	8,560
500		62	127	258	451	640	1,370	2,460	3,880	8,090
550		58	121	245	429	608	1,300	2,340	3,690	7,680
600		56	115	234	409	580	1,240	2,230	3,520	7,330
650		53	110	224	392	556	1,190	2,140	3,370	7,020
700		51	106	215	376	534	1,140	2,050	3,240	6,740
750		49	102	207	362	514	1,100	1,980	3,120	6,490
800		48	98	200	350	497	1,060	1,910	3,010	6,270
850		46	95	194	339	481	1,030	1,850	2,910	6,070
900		45	92	188	328	466	1,000	1,790	2,820	5,880
950		43	90	182	319	452	967	1,740	2,740	5,710
1,000		42	87	177	310	440	940	1,690	2,670	5,560
1,100		40	83	169	295	418	893	1,610	2,530	5,280
1,200		38	79	161	281	399	852	1,530	2,420	5,040
1,300		37	76	154	269	382	816	1,470	2,320	4,820
1,400		35	73	148	259	367	784	1,410	2,220	4,630
1,500		34	70	143	249	353	755	1,360	2,140	4,460
1,600		33	68	138	241	341	729	1,310	2,070	4,310
1,700		32	65	133	233	330	705	1,270	2,000	4,170
1,800		31	63	129	226	320	684	1,230	1,940	4,040
1,900		30	62	125	219	311	664	1,200	1,890	3,930
2,000		29	60	122	213	302	646	1,160	1,830	3,820

For SI: 1 inch = 25.4 mm, 1 foot = 304.8 mm, 1 pound per square inch = 6.895 kPa, 1-inch water column = 0.2488 kPa, 1 British thermal unit per hour = 0.2931 W, 1 cubic foot per hour = 0.0283 m³/h, 1 degree = 0.01745 rad.

Notes:

1. Table capacities are based on Type K copper tubing inside diameter (shown), which has the smallest inside diameter of the copper tubing products.

2. All table entries have been rounded to three significant digits.

❖ See the commentary to Section 402.4.

TABLE 402.4(15)
CORRUGATED STAINLESS STEEL TUBING (CSST)

Gas	Natural
Inlet Pressure	Less than 2 psi
Pressure Drop	0.5 in. w.c.
Specific Gravity	0.60

	TUBE SIZE (EHD)													
Flow Designation	13	15	18	19	23	25	30	31	37	39	46	48	60	62
Length (ft)	Capacity in Cubic Feet of Gas Per Hour													
5	46	63	115	134	225	270	471	546	895	1,037	1,790	2,070	3,660	4,140
10	32	44	82	95	161	192	330	383	639	746	1,260	1,470	2,600	2,930
15	25	35	66	77	132	157	267	310	524	615	1,030	1,200	2,140	2,400
20	22	31	58	67	116	137	231	269	456	536	888	1,050	1,850	2,080
25	19	27	52	60	104	122	206	240	409	482	793	936	1,660	1,860
30	18	25	47	55	96	112	188	218	374	442	723	856	1,520	1,700
40	15	21	41	47	83	97	162	188	325	386	625	742	1,320	1,470
50	13	19	37	42	75	87	144	168	292	347	559	665	1,180	1,320
60	12	17	34	38	68	80	131	153	267	318	509	608	1,080	1,200
70	11	16	31	36	63	74	121	141	248	295	471	563	1,000	1,110
80	10	15	29	33	60	69	113	132	232	277	440	527	940	1,040
90	10	14	28	32	57	65	107	125	219	262	415	498	887	983
100	9	13	26	30	54	62	101	118	208	249	393	472	843	933
150	7	10	20	23	42	48	78	91	171	205	320	387	691	762
200	6	9	18	21	38	44	71	82	148	179	277	336	600	661
250	5	8	16	19	34	39	63	74	133	161	247	301	538	591
300	5	7	15	17	32	36	57	67	95	148	226	275	492	540

For SI: 1 inch = 25.4 mm, 1 foot = 304.8 mm, 1 pound per square inch = 6.895 kPa, 1-inch water column = 0.2488 kPa,
1 British thermal unit per hour = 0.2931 W, 1 cubic foot per hour = 0.0283 m³/h, 1 degree = 0.01745 rad.

Notes:

1. Table includes losses for four 90-degree bends and two end fittings. Tubing runs with larger numbers of bends or fittings shall be increased by an equivalent length of tubing to the following equation: $L = 1.3n$, where L is additional length (feet) of tubing and n is the number of additional fittings or bends.

2. EHD—Equivalent Hydraulic Diameter, which is a measure of the relative hydraulic efficiency between different tubing sizes. The greater the value of EHD, the greater the gas capacity of the tubing.

3. All table entries have been rounded to three significant digits.

❖ See the commentary to Section 402.4.

TABLE 402.4(16)
CORRUGATED STAINLESS STEEL TUBING (CSST)

Gas	Natural
Inlet Pressure	Less than 2 psi
Pressure Drop	3.0 in. w.c.
Specific Gravity	0.60

INTENDED USE: Initial supply pressure of 8.0 inches w.c. or greater													
TUBE SIZE (EHD)													
Flow Designation	13	15	18	19	23	25	30	31	37	46	48	60	62
Length (ft)	Capacity in Cubic Feet of Gas Per Hour												
5	120	160	277	327	529	649	1,180	1,370	2,140	4,430	5,010	8,800	10,100
10	83	112	197	231	380	462	828	958	1,530	3,200	3,560	6,270	7,160
15	67	90	161	189	313	379	673	778	1,250	2,540	2,910	5,140	5,850
20	57	78	140	164	273	329	580	672	1,090	2,200	2,530	4,460	5,070
25	51	69	125	147	245	295	518	599	978	1,960	2,270	4,000	4,540
30	46	63	115	134	225	270	471	546	895	1,790	2,070	3,660	4,140
40	39	54	100	116	196	234	407	471	778	1,550	1,800	3,180	3,590
50	35	48	89	104	176	210	363	421	698	1,380	1,610	2,850	3,210
60	32	44	82	95	161	192	330	383	639	1,260	1,470	2,600	2,930
70	29	41	76	88	150	178	306	355	593	1,170	1,360	2,420	2,720
80	27	38	71	82	141	167	285	331	555	1,090	1,280	2,260	2,540
90	26	36	67	77	133	157	268	311	524	1,030	1,200	2,140	2,400
100	24	34	63	73	126	149	254	295	498	974	1,140	2,030	2,280
150	19	27	52	60	104	122	206	240	409	793	936	1,660	1,860
200	17	23	45	52	91	106	178	207	355	686	812	1,440	1,610
250	15	21	40	46	82	95	159	184	319	613	728	1,290	1,440
300	13	19	37	42	75	87	144	168	234	559	665	1,180	1,320

For SI: 1 inch = 25.4 mm, 1 foot = 304.8 mm, 1 pound per square inch = 6.895 kPa, 1-inch water column = 0.2488 kPa,
1 British thermal unit per hour = 0.2931 W, 1 cubic foot per hour = 0.0283 m³/h, 1 degree = 0.01745 rad.

Notes:

1. Table includes losses for four 90-degree bends and two end fittings. Tubing runs with larger numbers of bends or fittings shall be increased by an equivalent length of tubing to the following equation: $L = 1.3n$ where L is additional length (feet) of tubing and n is the number of additional fittings or bends.

2. EHD—Equivalent Hydraulic Diameter, which is a measure of the relative hydraulic efficiency between different tubing sizes. The greater the value of EHD, the greater the gas capacity of the tubing.

3. All table entries have been rounded to three significant digits.

❖ See the commentary to Section 402.4.

TABLE 402.4(17)
CORRUGATED STAINLESS STEEL TUBING (CSST)

Gas	Natural
Inlet Pressure	Less than 2 psi
Pressure Drop	6.0 in. w.c.
Specific Gravity	0.60

INTENDED USE: Initial supply pressure of 11.0 inches w.c. or greater

TUBE SIZE (EHD)

Flow Designation	13	15	18	19	23	25	30	31	37	46	48	60	62
Length (ft)	Capacity in Cubic Feet of Gas Per Hour												
5	173	229	389	461	737	911	1,690	1,950	3,000	6,280	7,050	12,400	14,260
10	120	160	277	327	529	649	1,180	1,370	2,140	4,430	5,010	8,800	10,100
15	96	130	227	267	436	532	960	1,110	1,760	3,610	4,100	7,210	8,260
20	83	112	197	231	380	462	828	958	1,530	3,120	3,560	6,270	7,160
25	74	99	176	207	342	414	739	855	1,370	2,790	3,190	5,620	6,400
30	67	90	161	189	313	379	673	778	1,250	2,540	2,910	5,140	5,850
40	57	78	140	164	273	329	580	672	1,090	2,200	2,530	4,460	5,070
50	51	69	125	147	245	295	518	599	978	1,960	2,270	4,000	4,540
60	46	63	115	134	225	270	471	546	895	1,790	2,070	3,660	4,140
70	42	58	106	124	209	250	435	505	830	1,660	1,920	3,390	3,840
80	39	54	100	116	196	234	407	471	778	1,550	1,800	3,180	3,590
90	37	51	94	109	185	221	383	444	735	1,460	1,700	3,000	3,390
100	35	48	89	104	176	210	363	421	698	1,380	1,610	2,850	3,210
150	28	39	73	85	145	172	294	342	573	1,130	1,320	2,340	2,630
200	24	34	63	73	126	149	254	295	498	974	1,140	2,030	2,280
250	21	30	57	66	114	134	226	263	447	870	1,020	1,820	2,040
300	19	27	52	60	104	122	206	240	409	793	936	1,660	1,860

For SI: 1 inch = 25.4 mm, 1 foot = 304.8 mm, 1 pound per square inch = 6.895 kPa, 1-inch water column = 0.2488 kPa,
1 British thermal unit per hour = 0.2931 W, 1 cubic foot per hour = 0.0283 m³/h, 1 degree = 0.01745 rad.

Notes:

1. Table includes losses for four 90-degree bends and two end fittings. Tubing runs with larger numbers of bends or fittings shall be increased by an equivalent length of tubing to the following equation: $L = 1.3n$ where L is additional length (feet) of tubing and n is the number of additional fittings or bends.

2. EHD—Equivalent Hydraulic Diameter, which is a measure of the relative hydraulic efficiency between different tubing sizes. The greater the value of EHD, the greater the gas capacity of the tubing.

3. All table entries have been rounded to three significant digits.

❖ See the commentary to Section 402.4.

TABLE 402.4(18)
CORRUGATED STAINLESS STEEL TUBING (CSST)

Gas	Natural
Inlet Pressure	2 psi
Pressure Drop	1.0 psi
Specific Gravity	0.60

TUBE SIZE (EHD)													
Flow Designation	13	15	18	19	23	25	30	31	37	39	46	48	60	62
Length (ft)	Capacity in Cubic Feet of Gas Per Hour													
10	270	353	587	700	1,100	1,370	2,590	2,990	4,510	5,037	9,600	10,700	18,600	21,600
25	166	220	374	444	709	876	1,620	1,870	2,890	3,258	6,040	6,780	11,900	13,700
30	151	200	342	405	650	801	1,480	1,700	2,640	2,987	5,510	6,200	10,900	12,500
40	129	172	297	351	567	696	1,270	1,470	2,300	2,605	4,760	5,380	9,440	10,900
50	115	154	266	314	510	624	1,140	1,310	2,060	2,343	4,260	4,820	8,470	9,720
75	93	124	218	257	420	512	922	1,070	1,690	1,932	3,470	3,950	6,940	7,940
80	89	120	211	249	407	496	892	1,030	1,640	1,874	3,360	3,820	6,730	7,690
100	79	107	189	222	366	445	795	920	1,470	1,685	3,000	3,420	6,030	6,880
150	64	87	155	182	302	364	646	748	1,210	1,389	2,440	2,800	4,940	5,620
200	55	75	135	157	263	317	557	645	1,050	1,212	2,110	2,430	4,290	4,870
250	49	67	121	141	236	284	497	576	941	1,090	1,890	2,180	3,850	4,360
300	44	61	110	129	217	260	453	525	862	999	1,720	1,990	3,520	3,980
400	38	52	96	111	189	225	390	453	749	871	1,490	1,730	3,060	3,450
500	34	46	86	100	170	202	348	404	552	783	1,330	1,550	2,740	3,090

For SI: 1 inch = 25.4 mm, 1 foot = 304.8 mm, 1 pound per square inch = 6.895 kPa, 1-inch water column = 0.2488 kPa,
1 British thermal unit per hour = 0.2931 W, 1 cubic foot per hour = 0.0283 m^3/h, 1 degree = 0.01745 rad.

Notes:

1. Table does not include effect of pressure drop across the line regulator. Where regulator loss exceeds $^3/_4$ psi, DO NOT USE THIS TABLE. Consult with the regulator manufacturer for pressure drops and capacity factors. Pressure drops across a regulator may vary with flow rate.

2. CAUTION: Capacities shown in the table might exceed maximum capacity for a selected regulator. Consult with the regulator or tubing manufacturer for guidance.

3. Table includes losses for four 90-degree bends and two end fittings. Tubing runs with larger numbers of bends or fittings shall be increased by an equivalent length of tubing to the following equation: $L = 1.3n$ where L is additional length (feet) of tubing and n is the number of additional fittings or bends.

4. EHD—Equivalent Hydraulic Diameter, which is a measure of the relative hydraulic efficiency between different tubing sizes. The greater the value of EHD, the greater the gas capacity of the tubing.

5. All table entries have been rounded to three significant digits.

❖ See the commentary to Section 402.4.

TABLE 402.4(19)
CORRUGATED STAINLESS STEEL TUBING (CSST)

Gas	Natural
Inlet Pressure	5.0 psi
Pressure Drop	3.5 psi
Specific Gravity	0.60

	TUBE SIZE (EHD)													
Flow Designation	13	15	18	19	23	25	30	31	37	39	46	48	60	62
Length (ft)	Capacity in Cubic Feet of Gas Per Hour													
10	523	674	1,080	1,300	2,000	2,530	4,920	5,660	8,300	9,140	18,100	19,800	34,400	40,400
25	322	420	691	827	1,290	1,620	3,080	3,540	5,310	5,911	11,400	12,600	22,000	25,600
30	292	382	632	755	1,180	1,480	2,800	3,230	4,860	5,420	10,400	11,500	20,100	23,400
40	251	329	549	654	1,030	1,280	2,420	2,790	4,230	4,727	8,970	10,000	17,400	20,200
50	223	293	492	586	926	1,150	2,160	2,490	3,790	4,251	8,020	8,930	15,600	18,100
75	180	238	403	479	763	944	1,750	2,020	3,110	3,506	6,530	7,320	12,800	14,800
80	174	230	391	463	740	915	1,690	1,960	3,020	3,400	6,320	7,090	12,400	14,300
100	154	205	350	415	665	820	1,510	1,740	2,710	3,057	5,650	6,350	11,100	12,800
150	124	166	287	339	548	672	1,230	1,420	2,220	2,521	4,600	5,200	9,130	10,500
200	107	143	249	294	478	584	1,060	1,220	1,930	2,199	3,980	4,510	7,930	9,090
250	95	128	223	263	430	524	945	1,090	1,730	1,977	3,550	4,040	7,110	8,140
300	86	116	204	240	394	479	860	995	1,590	1,813	3,240	3,690	6,500	7,430
400	74	100	177	208	343	416	742	858	1,380	1,581	2,800	3,210	5,650	6,440
500	66	89	159	186	309	373	662	766	1,040	1,422	2,500	2,870	5,060	5,760

For SI: 1 inch = 25.4 mm, 1 foot = 304.8 mm, 1 pound per square inch = 6.895 kPa, 1-inch water column = 0.2488 kPa,
  1 British thermal unit per hour = 0.2931 W, 1 cubic foot per hour = 0.0283 m³/h, 1 degree = 0.01745 rad.

Notes:

1. Table does not include effect of pressure drop across the line regulator. Where regulator loss exceeds $^3/_4$ psi, DO NOT USE THIS TABLE. Consult with the regulator manufacturer for pressure drops and capacity factors. Pressure drops across a regulator may vary with flow rate.

2. CAUTION: Capacities shown in the table might exceed maximum capacity for a selected regulator. Consult with the regulator or tubing manufacturer for guidance.

3. Table includes losses for four 90-degree bends and two end fittings. Tubing runs with larger numbers of bends or fittings shall be increased by an equivalent length of tubing to the following equation: $L = 1.3n$ where L is additional length (feet) of tubing and n is the number of additional fittings or bends.

4. EHD—Equivalent Hydraulic Diameter, which is a measure of the relative hydraulic efficiency between different tubing sizes. The greater the value of EHD, the greater the gas capacity of the tubing.

5. All table entries have been rounded to three significant digits.

❖ See the commentary to Section 402.4.

TABLE 402.4(20)
POLYETHYLENE PLASTIC PIPE

Gas	Natural
Inlet Pressure	Less than 2 psi
Pressure Drop	0.3 in. w.c.
Specific Gravity	0.60

PIPE SIZE (inch)								
Nominal OD	$^1/_2$	$^3/_4$	1	$1^1/_4$	$1^1/_2$	2	3	4
Designation	SDR 9	SDR 11	SDR 11	SDR 10	SDR 11	SDR 11	SDR 11	SDR 11
Actual ID	0.660	0.860	1.077	1.328	1.554	1.943	2.864	3.682
Length (ft)	Capacity in Cubic Feet of Gas per Hour							
10	153	305	551	955	1,440	2,590	7,170	13,900
20	105	210	379	656	991	1,780	4,920	9,520
30	84	169	304	527	796	1,430	3,950	7,640
40	72	144	260	451	681	1,220	3,380	6,540
50	64	128	231	400	604	1,080	3,000	5,800
60	58	116	209	362	547	983	2,720	5,250
70	53	107	192	333	503	904	2,500	4,830
80	50	99	179	310	468	841	2,330	4,500
90	46	93	168	291	439	789	2,180	4,220
100	44	88	159	275	415	745	2,060	3,990
125	39	78	141	243	368	661	1,830	3,530
150	35	71	127	221	333	598	1,660	3,200
175	32	65	117	203	306	551	1,520	2,940
200	30	60	109	189	285	512	1,420	2,740
250	27	54	97	167	253	454	1,260	2,430
300	24	48	88	152	229	411	1,140	2,200
350	22	45	81	139	211	378	1,050	2,020
400	21	42	75	130	196	352	974	1,880
450	19	39	70	122	184	330	914	1,770
500	18	37	66	115	174	312	863	1,670

For SI: 1 inch = 25.4 mm, 1 foot = 304.8 mm, 1 pound per square inch = 6.895 kPa, 1-inch water column = 0.2488 kPa,
1 British thermal unit per hour = 0.2931 W, 1 cubic foot per hour = 0.0283 m³/h, 1 degree = 0.01745 rad.
Note: All table entries have been rounded to three significant digits.

❖ See the commentary to Section 402.4.

**TABLE 402.4(21)
POLYETHYLENE PLASTIC PIPE**

Gas	Natural
Inlet Pressure	Less than 2 psi
Pressure Drop	0.5 in. w.c.
Specific Gravity	0.60

	PIPE SIZE (inch)							
Nominal OD	$^1/_2$	$^3/_4$	1	$1^1/_4$	$1^1/_2$	2	3	4
Designation	SDR 9	SDR 11	SDR 11	SDR 10	SDR 11	SDR 11	SDR 11	SDR 11
Actual ID	0.660	0.860	1.077	1.328	1.554	1.943	2.864	3.682
Length (ft)	Capacity in Cubic Feet of Gas per Hour							
10	201	403	726	1,260	1,900	3,410	9,450	18,260
20	138	277	499	865	1,310	2,350	6,490	12,550
30	111	222	401	695	1,050	1,880	5,210	10,080
40	95	190	343	594	898	1,610	4,460	8,630
50	84	169	304	527	796	1,430	3,950	7,640
60	76	153	276	477	721	1,300	3,580	6,930
70	70	140	254	439	663	1,190	3,300	6,370
80	65	131	236	409	617	1,110	3,070	5,930
90	61	123	221	383	579	1,040	2,880	5,560
100	58	116	209	362	547	983	2,720	5,250
125	51	103	185	321	485	871	2,410	4,660
150	46	93	168	291	439	789	2,180	4,220
175	43	86	154	268	404	726	2,010	3,880
200	40	80	144	249	376	675	1,870	3,610
250	35	71	127	221	333	598	1,660	3,200
300	32	64	115	200	302	542	1,500	2,900
350	29	59	106	184	278	499	1,380	2,670
400	27	55	99	171	258	464	1,280	2,480
450	26	51	93	160	242	435	1,200	2,330
500	24	48	88	152	229	411	1,140	2,200

For SI: 1 inch = 25.4 mm, 1 foot = 304.8 mm, 1 pound per square inch = 6.895 kPa, 1-inch water column = 0.2488 kPa,
1 British thermal unit per hour = 0.2931 W, 1 cubic foot per hour = 0.0283 m³/h, 1 degree = 0.01745 rad.
Note: All table entries have been rounded to three significant digits.

❖ See the commentary to Section 402.4.

TABLE 402.4(22)
POLYETHYLENE PLASTIC PIPE

Gas	Natural
Inlet Pressure	2.0 psi
Pressure Drop	1.0 psi
Specific Gravity	0.60

PIPE SIZE (inch)								
Nominal OD	¹/₂	³/₄	1	1¹/₄	1¹/₂	2	3	4
Designation	SDR 9	SDR 11	SDR 11	SDR 10	SDR 11	SDR 11	SDR 11	SDR 11
Actual ID	0.660	0.860	1.077	1.328	1.554	1.943	2.864	3.682
Length (ft)	Capacity in Cubic Feet of Gas per Hour							
10	1,860	3,720	6,710	11,600	17,600	31,600	87,300	169,000
20	1,280	2,560	4,610	7,990	12,100	21,700	60,000	116,000
30	1,030	2,050	3,710	6,420	9,690	17,400	48,200	93,200
40	878	1,760	3,170	5,490	8,300	14,900	41,200	79,700
50	778	1,560	2,810	4,870	7,350	13,200	36,600	70,700
60	705	1,410	2,550	4,410	6,660	12,000	33,100	64,000
70	649	1,300	2,340	4,060	6,130	11,000	30,500	58,900
80	603	1,210	2,180	3,780	5,700	10,200	28,300	54,800
90	566	1,130	2,050	3,540	5,350	9,610	26,600	51,400
100	535	1,070	1,930	3,350	5,050	9,080	25,100	48,600
125	474	949	1,710	2,970	4,480	8,050	22,300	43,000
150	429	860	1,550	2,690	4,060	7,290	20,200	39,000
175	395	791	1,430	2,470	3,730	6,710	18,600	35,900
200	368	736	1,330	2,300	3,470	6,240	17,300	33,400
250	326	652	1,180	2,040	3,080	5,530	15,300	29,600
300	295	591	1,070	1,850	2,790	5,010	13,900	26,800
350	272	544	981	1,700	2,570	4,610	12,800	24,700
400	253	506	913	1,580	2,390	4,290	11,900	22,900
450	237	475	856	1,480	2,240	4,020	11,100	21,500
500	224	448	809	1,400	2,120	3,800	10,500	20,300
550	213	426	768	1,330	2,010	3,610	9,990	19,300
600	203	406	733	1,270	1,920	3,440	9,530	18,400
650	194	389	702	1,220	1,840	3,300	9,130	17,600
700	187	374	674	1,170	1,760	3,170	8,770	16,900
750	180	360	649	1,130	1,700	3,050	8,450	16,300
800	174	348	627	1,090	1,640	2,950	8,160	15,800
850	168	336	607	1,050	1,590	2,850	7,890	15,300
900	163	326	588	1,020	1,540	2,770	7,650	14,800
950	158	317	572	990	1,500	2,690	7,430	14,400
1,000	154	308	556	963	1,450	2,610	7,230	14,000
1,100	146	293	528	915	1,380	2,480	6,870	13,300
1,200	139	279	504	873	1,320	2,370	6,550	12,700
1,300	134	267	482	836	1,260	2,270	6,270	12,100
1,400	128	257	463	803	1,210	2,180	6,030	11,600
1,500	124	247	446	773	1,170	2,100	5,810	11,200
1,600	119	239	431	747	1,130	2,030	5,610	10,800
1,700	115	231	417	723	1,090	1,960	5,430	10,500
1,800	112	224	404	701	1,060	1,900	5,260	10,200
1,900	109	218	393	680	1,030	1,850	5,110	9,900
2,000	106	212	382	662	1,000	1,800	4,970	9,600

For SI: 1 inch = 25.4 mm, 1 foot = 304.8 mm, 1 pound per square inch = 6.895 kPa, 1-inch water column = 0.2488 kPa,
1 British thermal unit per hour = 0.2931 W, 1 cubic foot per hour = 0.0283 m³/h, 1 degree = 0.01745 rad.

Note: All table entries have been rounded to three significant digits.

❖ See the commentary to Section 402.4.

TABLE 402.4(23)
POLYETHYLENE PLASTIC TUBING

Gas	Natural
Inlet Pressure	Less than 2.0 psi
Pressure Drop	0.3 in. w.c.
Specific Gravity	0.60

	PLASTIC TUBING SIZE (CTS) (inch)	
Nominal OD	$^1/_2$	$^3/_4$
Designation	SDR 7	SDR 11
Actual ID	0.445	0.927
Length (ft)	Capacity in Cubic Feet of Gas per Hour	
10	54	372
20	37	256
30	30	205
40	26	176
50	23	156
60	21	141
70	19	130
80	18	121
90	17	113
100	16	107
125	14	95
150	13	86
175	12	79
200	11	74
225	10	69
250	NA	65
275	NA	62
300	NA	59
350	NA	54
400	NA	51
450	NA	47
500	NA	45

For SI: 1 inch = 25.4 mm, 1 foot = 304.8 mm,
1 pound per square inch = 6.895 kPa,
1-inch water column = 0.2488 kPa,
1 British thermal unit per hour = 0.2931 W,
1 cubic foot per hour = 0.0283 m³/h, 1 degree = 0.01745 rad.

Notes:
1. NA means a flow of less than 10 cfh.
2. All table entries have been rounded to three significant digits.

❖ See the commentary to Section 402.4.

TABLE 402.4(24)
POLYETHYLENE PLASTIC TUBING

Gas	Natural
Inlet Pressure	Less than 2.0 psi
Pressure Drop	0.5 in. w.c.
Specific Gravity	0.60

	PLASTIC TUBING SIZE (CTS) (inch)	
Nominal OD	$^1/_2$	$^3/_4$
Designation	SDR 7	SDR 11
Actual ID	0.445	0.927
Length (ft)	Capacity in Cubic Feet of Gas per Hour	
10	72	490
20	49	337
30	39	271
40	34	232
50	30	205
60	27	186
70	25	171
80	23	159
90	22	149
100	21	141
125	18	125
150	17	113
175	15	104
200	14	97
225	13	91
250	12	86
275	11	82
300	11	78
350	10	72
400	NA	67
450	NA	63
500	NA	59

For SI: 1 inch = 25.4 mm, 1 foot = 304.8 mm,
1 pound per square inch = 6.895 kPa,
1-inch water column = 0.2488 kPa,
1 British thermal unit per hour = 0.2931 W,
1 cubic foot per hour = 0.0283 m³/h, 1 degree = 0.01745 rad.

Notes:
1. NA means a flow of less than 10 cfh.
2. All table entries have been rounded to three significant digits.

❖ See the commentary to Section 402.4.

TABLE 402.4(25)
SCHEDULE 40 METALLIC PIPE

Gas	Undiluted Propane
Inlet Pressure	10.0 psi
Pressure Drop	1.0 psi
Specific Gravity	1.50

INTENDED USE	Pipe sizing between first stage (high-pressure regulator) and second stage (low-pressure regulator).								
	PIPE SIZE (inch)								
Nominal	$^1/_2$	$^3/_4$	1	$1^1/_4$	$1^1/_2$	2	$2^1/_2$	3	4
Actual ID	0.622	0.824	1.049	1.380	1.610	2.067	2.469	3.068	4.026
Length (ft)	Capacity in Thousands of Btu per Hour								
10	3,320	6,950	13,100	26,900	40,300	77,600	124,000	219,000	446,000
20	2,280	4,780	9,000	18,500	27,700	53,300	85,000	150,000	306,000
30	1,830	3,840	7,220	14,800	22,200	42,800	68,200	121,000	246,000
40	1,570	3,280	6,180	12,700	19,000	36,600	58,400	103,000	211,000
50	1,390	2,910	5,480	11,300	16,900	32,500	51,700	91,500	187,000
60	1,260	2,640	4,970	10,200	15,300	29,400	46,900	82,900	169,000
70	1,160	2,430	4,570	9,380	14,100	27,100	43,100	76,300	156,000
80	1,080	2,260	4,250	8,730	13,100	25,200	40,100	70,900	145,000
90	1,010	2,120	3,990	8,190	12,300	23,600	37,700	66,600	136,000
100	956	2,000	3,770	7,730	11,600	22,300	35,600	62,900	128,000
125	848	1,770	3,340	6,850	10,300	19,800	31,500	55,700	114,000
150	768	1,610	3,020	6,210	9,300	17,900	28,600	50,500	103,000
175	706	1,480	2,780	5,710	8,560	16,500	26,300	46,500	94,700
200	657	1,370	2,590	5,320	7,960	15,300	24,400	43,200	88,100
250	582	1,220	2,290	4,710	7,060	13,600	21,700	38,300	78,100
300	528	1,100	2,080	4,270	6,400	12,300	19,600	34,700	70,800
350	486	1,020	1,910	3,930	5,880	11,300	18,100	31,900	65,100
400	452	945	1,780	3,650	5,470	10,500	16,800	29,700	60,600
450	424	886	1,670	3,430	5,140	9,890	15,800	27,900	56,800
500	400	837	1,580	3,240	4,850	9,340	14,900	26,300	53,700
550	380	795	1,500	3,070	4,610	8,870	14,100	25,000	51,000
600	363	759	1,430	2,930	4,400	8,460	13,500	23,900	48,600
650	347	726	1,370	2,810	4,210	8,110	12,900	22,800	46,600
700	334	698	1,310	2,700	4,040	7,790	12,400	21,900	44,800
750	321	672	1,270	2,600	3,900	7,500	12,000	21,100	43,100
800	310	649	1,220	2,510	3,760	7,240	11,500	20,400	41,600
850	300	628	1,180	2,430	3,640	7,010	11,200	19,800	40,300
900	291	609	1,150	2,360	3,530	6,800	10,800	19,200	39,100
950	283	592	1,110	2,290	3,430	6,600	10,500	18,600	37,900
1,000	275	575	1,080	2,230	3,330	6,420	10,200	18,100	36,900
1,100	261	546	1,030	2,110	3,170	6,100	9,720	17,200	35,000
1,200	249	521	982	2,020	3,020	5,820	9,270	16,400	33,400
1,300	239	499	940	1,930	2,890	5,570	8,880	15,700	32,000
1,400	229	480	903	1,850	2,780	5,350	8,530	15,100	30,800
1,500	221	462	870	1,790	2,680	5,160	8,220	14,500	29,600
1,600	213	446	840	1,730	2,590	4,980	7,940	14,000	28,600
1,700	206	432	813	1,670	2,500	4,820	7,680	13,600	27,700
1,800	200	419	789	1,620	2,430	4,670	7,450	13,200	26,900
1,900	194	407	766	1,570	2,360	4,540	7,230	12,800	26,100
2,000	189	395	745	1,530	2,290	4,410	7,030	12,400	25,400

For SI: 1 inch = 25.4 mm, 1 foot = 304.8 mm, 1 pound per square inch = 6.895 kPa, 1-inch water column = 0.2488 kPa,
1 British thermal unit per hour = 0.2931 W, 1 cubic foot per hour = 0.0283 m^3/h, 1 degree = 0.01745 rad.

Note: All table entries have been rounded to three significant digits.

❖ See the commentary to Section 402.4.

TABLE 402.4(26)
SCHEDULE 40 METALLIC PIPE

Gas	Undiluted Propane
Inlet Pressure	10.0 psi
Pressure Drop	3.0 psi
Specific Gravity	1.50

INTENDED USE	Pipe sizing between first stage (high-pressure regulator) and second stage (low-pressure regulator).								
	PIPE SIZE (inch)								
Nominal	1/2	3/4	1	1 1/4	1 1/2	2	2 1/2	3	4
Actual ID	0.622	0.824	1.049	1.380	1.610	2.067	2.469	3.068	4.026
Length (ft)	Capacity in Thousands of Btu per Hour								
10	5,890	12,300	23,200	47,600	71,300	137,000	219,000	387,000	789,000
20	4,050	8,460	15,900	32,700	49,000	94,400	150,000	266,000	543,000
30	3,250	6,790	12,800	26,300	39,400	75,800	121,000	214,000	436,000
40	2,780	5,810	11,000	22,500	33,700	64,900	103,000	183,000	373,000
50	2,460	5,150	9,710	19,900	29,900	57,500	91,600	162,000	330,000
60	2,230	4,670	8,790	18,100	27,100	52,100	83,000	147,000	299,000
70	2,050	4,300	8,090	16,600	24,900	47,900	76,400	135,000	275,000
80	1,910	4,000	7,530	15,500	23,200	44,600	71,100	126,000	256,000
90	1,790	3,750	7,060	14,500	21,700	41,800	66,700	118,000	240,000
100	1,690	3,540	6,670	13,700	20,500	39,500	63,000	111,000	227,000
125	1,500	3,140	5,910	12,100	18,200	35,000	55,800	98,700	201,000
150	1,360	2,840	5,360	11,000	16,500	31,700	50,600	89,400	182,000
175	1,250	2,620	4,930	10,100	15,200	29,200	46,500	82,300	167,800
200	1,160	2,430	4,580	9,410	14,100	27,200	43,300	76,500	156,100
250	1,030	2,160	4,060	8,340	12,500	24,100	38,400	67,800	138,400
300	935	1,950	3,680	7,560	11,300	21,800	34,800	61,500	125,400
350	860	1,800	3,390	6,950	10,400	20,100	32,000	56,500	115,300
400	800	1,670	3,150	6,470	9,690	18,700	29,800	52,600	107,300
450	751	1,570	2,960	6,070	9,090	17,500	27,900	49,400	100,700
500	709	1,480	2,790	5,730	8,590	16,500	26,400	46,600	95,100
550	673	1,410	2,650	5,450	8,160	15,700	25,000	44,300	90,300
600	642	1,340	2,530	5,200	7,780	15,000	23,900	42,200	86,200
650	615	1,290	2,420	4,980	7,450	14,400	22,900	40,500	82,500
700	591	1,240	2,330	4,780	7,160	13,800	22,000	38,900	79,300
750	569	1,190	2,240	4,600	6,900	13,300	21,200	37,400	76,400
800	550	1,150	2,170	4,450	6,660	12,800	20,500	36,200	73,700
850	532	1,110	2,100	4,300	6,450	12,400	19,800	35,000	71,400
900	516	1,080	2,030	4,170	6,250	12,000	19,200	33,900	69,200
950	501	1,050	1,970	4,050	6,070	11,700	18,600	32,900	67,200
1,000	487	1,020	1,920	3,940	5,900	11,400	18,100	32,000	65,400
1,100	463	968	1,820	3,740	5,610	10,800	17,200	30,400	62,100
1,200	442	923	1,740	3,570	5,350	10,300	16,400	29,000	59,200
1,300	423	884	1,670	3,420	5,120	9,870	15,700	27,800	56,700
1,400	406	849	1,600	3,280	4,920	9,480	15,100	26,700	54,500
1,500	391	818	1,540	3,160	4,740	9,130	14,600	25,700	52,500
1,600	378	790	1,490	3,060	4,580	8,820	14,100	24,800	50,700
1,700	366	765	1,440	2,960	4,430	8,530	13,600	24,000	49,000
1,800	355	741	1,400	2,870	4,300	8,270	13,200	23,300	47,600
1,900	344	720	1,360	2,780	4,170	8,040	12,800	22,600	46,200
2,000	335	700	1,320	2,710	4,060	7,820	12,500	22,000	44,900

For SI: 1 inch = 25.4 mm, 1 foot = 304.8 mm, 1 pound per square inch = 6.895 kPa, 1-inch water column = 0.2488 kPa,
1 British thermal unit per hour = 0.2931 W, 1 cubic foot per hour = 0.0283 m³/h, 1 degree = 0.01745 rad.
Note: All table entries have been rounded to three significant digits.

❖ See the commentary to Section 402.4.

TABLE 402.4(27)
SCHEDULE 40 METALLIC PIPE

Gas	Undiluted Propane
Inlet Pressure	2.0 psi
Pressure Drop	1.0 psi
Specific Gravity	1.50

INTENDED USE	Pipe sizing between 2 psig service and line pressure regulator.								
	PIPE SIZE (inch)								
Nominal	$^1/_2$	$^3/_4$	1	$1^1/_4$	$1^1/_2$	2	$2^1/_2$	3	4
Actual ID	0.622	0.824	1.049	1.380	1.610	2.067	2.469	3.068	4.026
Length (ft)	Capacity in Thousands of Btu per Hour								
10	2,680	5,590	10,500	21,600	32,400	62,400	99,500	176,000	359,000
20	1,840	3,850	7,240	14,900	22,300	42,900	68,400	121,000	247,000
30	1,480	3,090	5,820	11,900	17,900	34,500	54,900	97,100	198,000
40	1,260	2,640	4,980	10,200	15,300	29,500	47,000	83,100	170,000
50	1,120	2,340	4,410	9,060	13,600	26,100	41,700	73,700	150,000
60	1,010	2,120	4,000	8,210	12,300	23,700	37,700	66,700	136,000
70	934	1,950	3,680	7,550	11,300	21,800	34,700	61,400	125,000
80	869	1,820	3,420	7,020	10,500	20,300	32,300	57,100	116,000
90	815	1,700	3,210	6,590	9,880	19,000	30,300	53,600	109,000
100	770	1,610	3,030	6,230	9,330	18,000	28,600	50,600	103,000
125	682	1,430	2,690	5,520	8,270	15,900	25,400	44,900	91,500
150	618	1,290	2,440	5,000	7,490	14,400	23,000	40,700	82,900
175	569	1,190	2,240	4,600	6,890	13,300	21,200	37,400	76,300
200	529	1,110	2,080	4,280	6,410	12,300	19,700	34,800	71,000
250	469	981	1,850	3,790	5,680	10,900	17,400	30,800	62,900
300	425	889	1,670	3,440	5,150	9,920	15,800	27,900	57,000
350	391	817	1,540	3,160	4,740	9,120	14,500	25,700	52,400
400	364	760	1,430	2,940	4,410	8,490	13,500	23,900	48,800
450	341	714	1,340	2,760	4,130	7,960	12,700	22,400	45,800
500	322	674	1,270	2,610	3,910	7,520	12,000	21,200	43,200
550	306	640	1,210	2,480	3,710	7,140	11,400	20,100	41,100
600	292	611	1,150	2,360	3,540	6,820	10,900	19,200	39,200
650	280	585	1,100	2,260	3,390	6,530	10,400	18,400	37,500
700	269	562	1,060	2,170	3,260	6,270	9,990	17,700	36,000
750	259	541	1,020	2,090	3,140	6,040	9,630	17,000	34,700
800	250	523	985	2,020	3,030	5,830	9,300	16,400	33,500
850	242	506	953	1,960	2,930	5,640	9,000	15,900	32,400
900	235	490	924	1,900	2,840	5,470	8,720	15,400	31,500
950	228	476	897	1,840	2,760	5,310	8,470	15,000	30,500
1,000	222	463	873	1,790	2,680	5,170	8,240	14,600	29,700
1,100	210	440	829	1,700	2,550	4,910	7,830	13,800	28,200
1,200	201	420	791	1,620	2,430	4,680	7,470	13,200	26,900
1,300	192	402	757	1,550	2,330	4,490	7,150	12,600	25,800
1,400	185	386	727	1,490	2,240	4,310	6,870	12,100	24,800
1,500	178	372	701	1,440	2,160	4,150	6,620	11,700	23,900
1,600	172	359	677	1,390	2,080	4,010	6,390	11,300	23,000
1,700	166	348	655	1,340	2,010	3,880	6,180	10,900	22,300
1,800	161	337	635	1,300	1,950	3,760	6,000	10,600	21,600
1,900	157	327	617	1,270	1,900	3,650	5,820	10,300	21,000
2,000	152	318	600	1,230	1,840	3,550	5,660	10,000	20,400

For SI: 1 inch = 25.4 mm, 1 foot = 304.8 mm, 1 pound per square inch = 6.895 kPa, 1-inch water column = 0.2488 kPa,
1 British thermal unit per hour = 0.2931 W, 1 cubic foot per hour = 0.0283 m³/h, 1 degree = 0.01745 rad.
Note: All table entries have been rounded to three significant digits.

❖ See the commentary to Section 402.4.

TABLE 402.4(28)
SCHEDULE 40 METALLIC PIPE

Gas	Undiluted Propane
Inlet Pressure	11.0 in. w.c.
Pressure Drop	0.5 in. w.c.
Specific Gravity	1.50

INTENDED USE	Pipe sizing between single- or second-stage (low pressure) regulator and appliance.								
	PIPE SIZE (inch)								
Nominal	1/2	3/4	1	1 1/4	1 1/2	2	2 1/2	3	4
Actual ID	0.622	0.824	1.049	1.380	1.610	2.067	2.469	3.068	4.026
Length (ft)	Capacity in Thousands of Btu per Hour								
10	291	608	1,150	2,350	3,520	6,790	10,800	19,100	39,000
20	200	418	787	1,620	2,420	4,660	7,430	13,100	26,800
30	160	336	632	1,300	1,940	3,750	5,970	10,600	21,500
40	137	287	541	1,110	1,660	3,210	5,110	9,030	18,400
50	122	255	480	985	1,480	2,840	4,530	8,000	16,300
60	110	231	434	892	1,340	2,570	4,100	7,250	14,800
80	101	212	400	821	1,230	2,370	3,770	6,670	13,600
100	94	197	372	763	1,140	2,200	3,510	6,210	12,700
125	89	185	349	716	1,070	2,070	3,290	5,820	11,900
150	84	175	330	677	1,010	1,950	3,110	5,500	11,200
175	74	155	292	600	899	1,730	2,760	4,880	9,950
200	67	140	265	543	814	1,570	2,500	4,420	9,010
250	62	129	243	500	749	1,440	2,300	4,060	8,290
300	58	120	227	465	697	1,340	2,140	3,780	7,710
350	51	107	201	412	618	1,190	1,900	3,350	6,840
400	46	97	182	373	560	1,080	1,720	3,040	6,190
450	42	89	167	344	515	991	1,580	2,790	5,700
500	40	83	156	320	479	922	1,470	2,600	5,300
550	37	78	146	300	449	865	1,380	2,440	4,970
600	35	73	138	283	424	817	1,300	2,300	4,700
650	33	70	131	269	403	776	1,240	2,190	4,460
700	32	66	125	257	385	741	1,180	2,090	4,260
750	30	64	120	246	368	709	1,130	2,000	4,080
800	29	61	115	236	354	681	1,090	1,920	3,920
850	28	59	111	227	341	656	1,050	1,850	3,770
900	27	57	107	220	329	634	1,010	1,790	3,640
950	26	55	104	213	319	613	978	1,730	3,530
1,000	25	53	100	206	309	595	948	1,680	3,420
1,100	25	52	97	200	300	578	921	1,630	3,320
1,200	24	50	95	195	292	562	895	1,580	3,230
1,300	23	48	90	185	277	534	850	1,500	3,070
1,400	22	46	86	176	264	509	811	1,430	2,930
1,500	21	44	82	169	253	487	777	1,370	2,800
1,600	20	42	79	162	243	468	746	1,320	2,690
1,700	19	40	76	156	234	451	719	1,270	2,590
1,800	19	39	74	151	226	436	694	1,230	2,500
1,900	18	38	71	146	219	422	672	1,190	2,420
2,000	18	37	69	142	212	409	652	1,150	2,350

For SI: 1 inch = 25.4 mm, 1 foot = 304.8 mm, 1 pound per square inch = 6.895 kPa, 1-inch water column = 0.2488 kPa,
1 British thermal unit per hour = 0.2931 W, 1 cubic foot per hour = 0.0283 m³/h, 1 degree = 0.01745 rad.
Note: All table entries have been rounded to three significant digits.

❖ See the commentary to Section 402.4.

TABLE 402.4(29)
SEMIRIGID COPPER TUBING

Gas	Undiluted Propane
Inlet Pressure	10.0 psi
Pressure Drop	1.0 psi
Specific Gravity	1.50

INTENDED USE		Sizing between first stage (high-pressure regulator) and second stage (low-pressure regulator).								
		TUBE SIZE (in.)								
Nominal	K & L	$^1/_4$	$^3/_8$	$^1/_2$	$^5/_8$	$^3/_4$	1	$1^1/_4$	$1^1/_2$	2
	ACR	$^3/_8$	$^1/_2$	$^5/_8$	$^3/_4$	$^7/_8$	$1^1/_8$	$1^3/_8$	—	—
Outside		0.375	0.500	0.625	0.750	0.875	1.125	1.375	1.625	2.125
Inside		0.305	0.402	0.527	0.652	0.745	0.995	1.245	1.481	1.959
Length (ft)		Capacity in Thousands of Btu per Hour								
10		513	1,060	2,150	3,760	5,330	11,400	20,500	32,300	67,400
20		352	727	1,480	2,580	3,670	7,830	14,100	22,200	46,300
30		283	584	1,190	2,080	2,940	6,290	11,300	17,900	37,200
40		242	500	1,020	1,780	2,520	5,380	9,690	15,300	31,800
50		215	443	901	1,570	2,230	4,770	8,590	13,500	28,200
60		194	401	816	1,430	2,020	4,320	7,780	12,300	25,600
70		179	369	751	1,310	1,860	3,980	7,160	11,300	23,500
80		166	343	699	1,220	1,730	3,700	6,660	10,500	21,900
90		156	322	655	1,150	1,630	3,470	6,250	9,850	20,500
100		147	304	619	1,080	1,540	3,280	5,900	9,310	19,400
125		131	270	549	959	1,360	2,910	5,230	8,250	17,200
150		118	244	497	869	1,230	2,630	4,740	7,470	15,600
175		109	225	457	799	1,130	2,420	4,360	6,880	14,300
200		101	209	426	744	1,060	2,250	4,060	6,400	13,300
250		90	185	377	659	935	2,000	3,600	5,670	11,800
300		81	168	342	597	847	1,810	3,260	5,140	10,700
350		75	155	314	549	779	1,660	3,000	4,730	9,840
400		70	144	292	511	725	1,550	2,790	4,400	9,160
450		65	135	274	480	680	1,450	2,620	4,130	8,590
500		62	127	259	453	643	1,370	2,470	3,900	8,120
550		59	121	246	430	610	1,300	2,350	3,700	7,710
600		56	115	235	410	582	1,240	2,240	3,530	7,350
650		54	111	225	393	558	1,190	2,140	3,380	7,040
700		51	106	216	378	536	1,140	2,060	3,250	6,770
750		50	102	208	364	516	1,100	1,980	3,130	6,520
800		48	99	201	351	498	1,060	1,920	3,020	6,290
850		46	96	195	340	482	1,030	1,850	2,920	6,090
900		45	93	189	330	468	1,000	1,800	2,840	5,910
950		44	90	183	320	454	970	1,750	2,750	5,730
1,000		42	88	178	311	442	944	1,700	2,680	5,580
1,100		40	83	169	296	420	896	1,610	2,540	5,300
1,200		38	79	161	282	400	855	1,540	2,430	5,050
1,300		37	76	155	270	383	819	1,470	2,320	4,840
1,400		35	73	148	260	368	787	1,420	2,230	4,650
1,500		34	70	143	250	355	758	1,360	2,150	4,480
1,600		33	68	138	241	343	732	1,320	2,080	4,330
1,700		32	66	134	234	331	708	1,270	2,010	4,190
1,800		31	64	130	227	321	687	1,240	1,950	4,060
1,900		30	62	126	220	312	667	1,200	1,890	3,940
2,000		29	60	122	214	304	648	1,170	1,840	3,830

For SI: 1 inch = 25.4 mm, 1 foot = 304.8 mm, 1 pound per square inch = 6.895 kPa, 1-inch water column = 0.2488 kPa,
1 British thermal unit per hour = 0.2931 W, 1 cubic foot per hour = 0.0283 m³/h, 1 degree = 0.01745 rad.

Notes:

1. Table capacities are based on Type K copper tubing inside diameter (shown), which has the smallest inside diameter of the copper tubing products.
2. All table entries have been rounded to three significant digits.

❖ See the commentary to Section 402.4.

**TABLE 402.4(30)
SEMIRIGID COPPER TUBING**

Gas	Undiluted Propane
Inlet Pressure	11.0 in. w.c.
Pressure Drop	0.5 in. w.c.
Specific Gravity	1.50

INTENDED USE		Sizing between single or second stage (low-pressure regulator) and appliance.								
		TUBE SIZE (inch)								
Nominal	K & L	$^1/_4$	$^3/_8$	$^1/_2$	$^5/_8$	$^3/_4$	1	$1^1/_4$	$1^1/_2$	2
	ACR	$^3/_8$	$^1/_2$	$^5/_8$	$^3/_4$	$^7/_8$	$1^1/_8$	$1^3/_8$	—	—
Outside		0.375	0.500	0.625	0.750	0.875	1.125	1.375	1.625	2.125
Inside		0.305	0.402	0.527	0.652	0.745	0.995	1.245	1.481	1.959
Length (ft)		Capacity in Thousands of Btu per Hour								
10		45	93	188	329	467	997	1,800	2,830	5,890
20		31	64	129	226	321	685	1,230	1,950	4,050
30		25	51	104	182	258	550	991	1,560	3,250
40		21	44	89	155	220	471	848	1,340	2,780
50		19	39	79	138	195	417	752	1,180	2,470
60		17	35	71	125	177	378	681	1,070	2,240
70		16	32	66	115	163	348	626	988	2,060
80		15	30	61	107	152	324	583	919	1,910
90		14	28	57	100	142	304	547	862	1,800
100		13	27	54	95	134	287	517	814	1,700
125		11	24	48	84	119	254	458	722	1,500
150		10	21	44	76	108	230	415	654	1,360
175		NA	20	40	70	99	212	382	602	1,250
200		NA	18	37	65	92	197	355	560	1,170
250		NA	16	33	58	82	175	315	496	1,030
300		NA	15	30	52	74	158	285	449	936
350		NA	14	28	48	68	146	262	414	861
400		NA	13	26	45	63	136	244	385	801
450		NA	12	24	42	60	127	229	361	752
500		NA	11	23	40	56	120	216	341	710
550		NA	11	22	38	53	114	205	324	674
600		NA	10	21	36	51	109	196	309	643
650		NA	NA	20	34	49	104	188	296	616
700		NA	NA	19	33	47	100	180	284	592
750		NA	NA	18	32	45	96	174	274	570
800		NA	NA	18	31	44	93	168	264	551
850		NA	NA	17	30	42	90	162	256	533
900		NA	NA	17	29	41	87	157	248	517
950		NA	NA	16	28	40	85	153	241	502
1,000		NA	NA	16	27	39	83	149	234	488
1,100		NA	NA	15	26	37	78	141	223	464
1,200		NA	NA	14	25	35	75	135	212	442
1,300		NA	NA	14	24	34	72	129	203	423
1,400		NA	NA	13	23	32	69	124	195	407
1,500		NA	NA	13	22	31	66	119	188	392
1,600		NA	NA	12	21	30	64	115	182	378
1,700		NA	NA	12	20	29	62	112	176	366
1,800		NA	NA	11	20	28	60	108	170	355
1,900		NA	NA	11	19	27	58	105	166	345
2,000		NA	NA	11	19	27	57	102	161	335

For SI: 1 inch = 25.4 mm, 1 foot = 304.8 mm, 1 pound per square inch = 6.895 kPa, 1-inch water column = 0.2488 kPa, 1 British thermal unit per hour = 0.2931 W, 1 cubic foot per hour = 0.0283 m³/h, 1 degree = 0.01745 rad.

Notes:
1. Table capacities are based on Type K copper tubing inside diameter (shown), which has the smallest inside diameter of the copper tubing products.
2. NA means a flow of less than 10,000 Btu/hr.
3. All table entries have been rounded to three significant digits.

❖ See the commentary to Section 402.4.

TABLE 402.4(31)
SEMIRIGID COPPER TUBING

Gas	Undiluted Propane
Inlet Pressure	2.0 psi
Pressure Drop	1.0 psi
Specific Gravity	1.50

INTENDED USE		Tube sizing between 2 psig service and line pressure regulator.								
		TUBE SIZE (inch)								
Nominal	K & L	$1/4$	$3/8$	$1/2$	$5/8$	$3/4$	1	$1^1/4$	$1^1/2$	2
	ACR	$3/8$	$1/2$	$5/8$	$3/4$	$7/8$	$1^1/8$	$1^3/8$	—	—
Outside		0.375	0.500	0.625	0.750	0.875	1.125	1.375	1.625	2.125
Inside		0.305	0.402	0.527	0.652	0.745	0.995	1.245	1.481	1.959
Length (ft)		Capacity in Thousands of Btu per Hour								
10		413	852	1,730	3,030	4,300	9,170	16,500	26,000	54,200
20		284	585	1,190	2,080	2,950	6,310	11,400	17,900	37,300
30		228	470	956	1,670	2,370	5,060	9,120	14,400	29,900
40		195	402	818	1,430	2,030	4,330	7,800	12,300	25,600
50		173	356	725	1,270	1,800	3,840	6,920	10,900	22,700
60		157	323	657	1,150	1,630	3,480	6,270	9,880	20,600
70		144	297	605	1,060	1,500	3,200	5,760	9,090	18,900
80		134	276	562	983	1,390	2,980	5,360	8,450	17,600
90		126	259	528	922	1,310	2,790	5,030	7,930	16,500
100		119	245	498	871	1,240	2,640	4,750	7,490	15,600
125		105	217	442	772	1,100	2,340	4,210	6,640	13,800
150		95	197	400	700	992	2,120	3,820	6,020	12,500
175		88	181	368	644	913	1,950	3,510	5,540	11,500
200		82	168	343	599	849	1,810	3,270	5,150	10,700
250		72	149	304	531	753	1,610	2,900	4,560	9,510
300		66	135	275	481	682	1,460	2,620	4,140	8,610
350		60	124	253	442	628	1,340	2,410	3,800	7,920
400		56	116	235	411	584	1,250	2,250	3,540	7,370
450		53	109	221	386	548	1,170	2,110	3,320	6,920
500		50	103	209	365	517	1,110	1,990	3,140	6,530
550		47	97	198	346	491	1,050	1,890	2,980	6,210
600		45	93	189	330	469	1,000	1,800	2,840	5,920
650		43	89	181	316	449	959	1,730	2,720	5,670
700		41	86	174	304	431	921	1,660	2,620	5,450
750		40	82	168	293	415	888	1,600	2,520	5,250
800		39	80	162	283	401	857	1,540	2,430	5,070
850		37	77	157	274	388	829	1,490	2,350	4,900
900		36	75	152	265	376	804	1,450	2,280	4,750
950		35	72	147	258	366	781	1,410	2,220	4,620
1,000		34	71	143	251	356	760	1,370	2,160	4,490
1,100		32	67	136	238	338	721	1,300	2,050	4,270
1,200		31	64	130	227	322	688	1,240	1,950	4,070
1,300		30	61	124	217	309	659	1,190	1,870	3,900
1,400		28	59	120	209	296	633	1,140	1,800	3,740
1,500		27	57	115	201	286	610	1,100	1,730	3,610
1,600		26	55	111	194	276	589	1,060	1,670	3,480
1,700		26	53	108	188	267	570	1,030	1,620	3,370
1,800		25	51	104	182	259	553	1,000	1,570	3,270
1,900		24	50	101	177	251	537	966	1,520	3,170
2,000		23	48	99	172	244	522	940	1,480	3,090

For SI: 1 inch = 25.4 mm, 1 foot = 304.8 mm, 1 pound per square inch = 6.895 kPa, 1-inch water column = 0.2488 kPa,
1 British thermal unit per hour = 0.2931 W, 1 cubic foot per hour = 0.0283 m^3/h, 1 degree = 0.01745 rad.

Notes:
1. Table capacities are based on Type K copper tubing inside diameter (shown), which has the smallest inside diameter of the copper tubing products.
2. All table entries have been rounded to three significant digits.

❖ See the commentary to Section 402.4.

TABLE 402.4(32)
CORRUGATED STAINLESS STEEL TUBING (CSST)

Gas	Undiluted Propane
Inlet Pressure	11.0 in. w.c.
Pressure Drop	0.5 in. w.c.
Specific Gravity	1.50

INTENDED USE: SIZING BETWEEN SINGLE OR SECOND STAGE (Low Pressure) REGULATOR AND THE APPLIANCE SHUTOFF VALVE													
TUBE SIZE (EHD)													
Flow Designation	13	15	18	19	23	25	30	31	37	39	46	48	60	62
Length (ft)	Capacity in Thousands of Btu per Hour												
5	72	99	181	211	355	426	744	863	1,420	1,638	2,830	3,270	5,780	6,550
10	50	69	129	150	254	303	521	605	971	1,179	1,990	2,320	4,110	4,640
15	39	55	104	121	208	248	422	490	775	972	1,620	1,900	3,370	3,790
20	34	49	91	106	183	216	365	425	661	847	1,400	1,650	2,930	3,290
25	30	42	82	94	164	192	325	379	583	762	1,250	1,480	2,630	2,940
30	28	39	74	87	151	177	297	344	528	698	1,140	1,350	2,400	2,680
40	23	33	64	74	131	153	256	297	449	610	988	1,170	2,090	2,330
50	20	30	58	66	118	137	227	265	397	548	884	1,050	1,870	2,080
60	19	26	53	60	107	126	207	241	359	502	805	961	1,710	1,900
70	17	25	49	57	99	117	191	222	330	466	745	890	1,590	1,760
80	15	23	45	52	94	109	178	208	307	438	696	833	1,490	1,650
90	15	22	44	50	90	102	169	197	286	414	656	787	1,400	1,550
100	14	20	41	47	85	98	159	186	270	393	621	746	1,330	1,480
150	11	15	31	36	66	75	123	143	217	324	506	611	1,090	1,210
200	9	14	28	33	60	69	112	129	183	283	438	531	948	1,050
250	8	12	25	30	53	61	99	117	163	254	390	476	850	934
300	8	11	23	26	50	57	90	107	147	234	357	434	777	854

For SI: 1 inch = 25.4 mm, 1 foot = 304.8 mm, 1 pound per square inch = 6.895 kPa, 1-inch water column = 0.2488 kPa,
    1 British thermal unit per hour = 0.2931 W, 1 cubic foot per hour = 0.0283 m³/h, 1 degree = 0.01745 rad.

Notes:

1. Table includes losses for four 90-degree bends and two end fittings. Tubing runs with larger numbers of bends or fittings shall be increased by an equivalent length of tubing to the following equation: $L = 1.3n$ where L is additional length (feet) of tubing and n is the number of additional fittings or bends.

2. EHD—Equivalent Hydraulic Diameter, which is a measure of the relative hydraulic efficiency between different tubing sizes. The greater the value of EHD, the greater the gas capacity of the tubing.

3. All table entries have been rounded to three significant digits.

❖ See the commentary to Section 402.4.

TABLE 402.4(33)
CORRUGATED STAINLESS STEEL TUBING (CSST)

Gas	Undiluted Propane
Inlet Pressure	2.0 psi
Pressure Drop	1.0 psi
Specific Gravity	1.50

INTENDED USE: SIZING BETWEEN 2 PSI SERVICE AND THE LINE PRESSURE REGULATOR													
TUBE SIZE (EHD)													
Flow Designation	13	15	18	19	23	25	30	31	37	39	46	48	60	62
Length (ft)	Capacity in Thousands of Btu per Hour												
10	426	558	927	1,110	1,740	2,170	4,100	4,720	7,130	7,958	15,200	16,800	29,400	34,200
25	262	347	591	701	1,120	1,380	2,560	2,950	4,560	5,147	9,550	10,700	18,800	21,700
30	238	316	540	640	1,030	1,270	2,330	2,690	4,180	4,719	8,710	9,790	17,200	19,800
40	203	271	469	554	896	1,100	2,010	2,320	3,630	4,116	7,530	8,500	14,900	17,200
50	181	243	420	496	806	986	1,790	2,070	3,260	3,702	6,730	7,610	13,400	15,400
75	147	196	344	406	663	809	1,460	1,690	2,680	3,053	5,480	6,230	11,000	12,600
80	140	189	333	393	643	768	1,410	1,630	2,590	2,961	5,300	6,040	10,600	12,200
100	124	169	298	350	578	703	1,260	1,450	2,330	2,662	4,740	5,410	9,530	10,900
150	101	137	245	287	477	575	1,020	1,180	1,910	2,195	3,860	4,430	7,810	8,890
200	86	118	213	248	415	501	880	1,020	1,660	1,915	3,340	3,840	6,780	7,710
250	77	105	191	222	373	448	785	910	1,490	1,722	2,980	3,440	6,080	6,900
300	69	96	173	203	343	411	716	829	1,360	1,578	2,720	3,150	5,560	6,300
400	60	82	151	175	298	355	616	716	1,160	1,376	2,350	2,730	4,830	5,460
500	53	72	135	158	268	319	550	638	1,030	1,237	2,100	2,450	4,330	4,880

For SI: 1 inch = 25.4 mm, 1 foot = 304.8 mm, 1 pound per square inch = 6.895kPa, 1-inch water column = 0.2488 kPa,
1 British thermal unit per hour = 0.293 1 W, 1 cubic foot per hour = 0.0283 m³/h, 1 degree = 0.01745 rad.

Notes:
1. Table does not include effect of pressure drop across the line regulator. Where regulator loss exceeds $^1/_2$ psi (based on 13 in. w.c. outlet pressure), DO NOT USE THIS TABLE. Consult with the regulator manufacturer for pressure drops and capacity factors. Pressure drops across a regulator may vary with flow rate.
2. CAUTION: Capacities shown in the table might exceed maximum capacity for a selected regulator. Consult with the regulator or tubing manufacturer for guidance.
3. Table includes losses for four 90-degree bends and two end fittings. Tubing runs with larger numbers of bends or fittings shall be increased by an equivalent length of tubing to the following equation: $L = 1.3n$ where L is additional length (feet) of tubing and n is the number of additional fittings or bends.
4. EHD—Equivalent Hydraulic Diameter, which is a measure of the relative hydraulic efficiency between different tubing sizes. The greater the value of EHD, the greater the gas capacity of the tubing.
5. All table entries have been rounded to three significant digits.

❖ See the commentary to Section 402.4.

TABLE 402.4(34)
CORRUGATED STAINLESS STEEL TUBING (CSST)

Gas	Undiluted Propane
Inlet Pressure	5.0 psi
Pressure Drop	3.5 psi
Specific Gravity	1.50

TUBE SIZE (EHD)														
Flow Designation	13	15	18	19	23	25	30	31	37	39	46	48	60	62
Length (ft)	Capacity in Thousands of Btu per Hour													
10	826	1,070	1,710	2,060	3,150	4,000	7,830	8,950	13,100	14,441	28,600	31,200	54,400	63,800
25	509	664	1,090	1,310	2,040	2,550	4,860	5,600	8,400	9,339	18,000	19,900	34,700	40,400
30	461	603	999	1,190	1,870	2,340	4,430	5,100	7,680	8,564	16,400	18,200	31,700	36,900
40	396	520	867	1,030	1,630	2,030	3,820	4,400	6,680	7,469	14,200	15,800	27,600	32,000
50	352	463	777	926	1,460	1,820	3,410	3,930	5,990	6,717	12,700	14,100	24,700	28,600
75	284	376	637	757	1,210	1,490	2,770	3,190	4,920	5,539	10,300	11,600	20,300	23,400
80	275	363	618	731	1,170	1,450	2,680	3,090	4,770	5,372	9,990	11,200	19,600	22,700
100	243	324	553	656	1,050	1,300	2,390	2,760	4,280	4,830	8,930	10,000	17,600	20,300
150	196	262	453	535	866	1,060	1,940	2,240	3,510	3,983	7,270	8,210	14,400	16,600
200	169	226	393	464	755	923	1,680	1,930	3,050	3,474	6,290	7,130	12,500	14,400
250	150	202	352	415	679	828	1,490	1,730	2,740	3,124	5,620	6,390	11,200	12,900
300	136	183	322	379	622	757	1,360	1,570	2,510	2,865	5,120	5,840	10,300	11,700
400	117	158	279	328	542	657	1,170	1,360	2,180	2,498	4,430	5,070	8,920	10,200
500	104	140	251	294	488	589	1,050	1,210	1,950	2,247	3,960	4,540	8,000	9,110

For SI: 1 inch = 25.4 mm, 1 foot = 304.8 mm, 1 pound per square inch = 6.895 kPa, 1-inch water column = 0.2488 kPa,
　　1 British thermal unit per hour = 0.2931 W, 1 cubic foot per hour = 0.0283 m³/h, 1 degree = 0.01745 rad.

Notes:

1　Table does not include effect of pressure drop across line regulator. Where regulator loss exceeds 1 psi, DO NOT USE THIS TABLE. Consult with the regulator manufacturer for pressure drops and capacity factors. Pressure drop across regulator may vary with the flow rate.

2. CAUTION: Capacities shown in the table might exceed maximum capacity of selected regulator. Consult with the tubing manufacturer for guidance.

3. Table includes losses for four 90-degree bends and two end fittings. Tubing runs with larger numbers of bends or fittings shall be increased by an equivalent length of tubing to the following equation: $L = 1.3n$ where L is additional length (feet) of tubing and n is the number of additional fittings or bends.

4. EHD—Equivalent Hydraulic Diameter, which is a measure of the relative hydraulic efficiency between different tubing sizes. The greater the value of EHD, the greater the gas capacity of the tubing.

5. All table entries have been rounded to three significant digits.

❖ See the commentary to Section 402.4.

TABLE 402.4(35)
POLYETHYLENE PLASTIC PIPE

Gas	Undiluted Propane
Inlet Pressure	11.0 in. w.c.
Pressure Drop	0.5 in. w.c.
Specific Gravity	1.50

INTENDED USE	PE pipe sizing between integral two-stage regulator at tank or second stage (low-pressure regulator) and building.							
	PIPE SIZE (inch)							
Nominal OD	$^1/_2$	$^3/_4$	1	$1^1/_4$	$1^1/_2$	2	3	4
Designation	SDR 9	SDR 11	SDR 11	SDR 10	SDR 11	SDR 11	SDR 11	SDR 11
Actual ID	0.660	0.860	1.077	1.328	1.554	1.943	2.864	3.682
Length (ft)	**Capacity in Thousands of Btu per Hour**							
10	340	680	1,230	2,130	3,210	5,770	16,000	30,900
20	233	468	844	1,460	2,210	3,970	11,000	21,200
30	187	375	677	1,170	1,770	3,180	8,810	17,000
40	160	321	580	1,000	1,520	2,730	7,540	14,600
50	142	285	514	890	1,340	2,420	6,680	12,900
60	129	258	466	807	1,220	2,190	6,050	11,700
70	119	237	428	742	1,120	2,010	5,570	10,800
80	110	221	398	690	1,040	1,870	5,180	10,000
90	103	207	374	648	978	1,760	4,860	9,400
100	98	196	353	612	924	1,660	4,590	8,900
125	87	173	313	542	819	1,470	4,070	7,900
150	78	157	284	491	742	1,330	3,690	7,130
175	72	145	261	452	683	1,230	3,390	6,560
200	67	135	243	420	635	1,140	3,160	6,100
250	60	119	215	373	563	1,010	2,800	5,410
300	54	108	195	338	510	916	2,530	4,900
350	50	99	179	311	469	843	2,330	4,510
400	46	92	167	289	436	784	2,170	4,190
450	43	87	157	271	409	736	2,040	3,930
500	41	82	148	256	387	695	1,920	3,720

For SI: 1 inch = 25.4 mm, 1 foot = 304.8 mm, 1 pound per square inch = 6.895 kPa, 1-inch water column = 0.2488 kPa,
   1 British thermal unit per hour = 0.2931 W, 1 cubic foot per hour = 0.0283 m³/h, 1 degree = 0.01745 rad.

Note: All table entries have been rounded to three significant digits.

❖ See the commentary to Section 402.4.

TABLE 402.4(36)
POLYETHYLENE PLASTIC PIPE

Gas	Undiluted Propane
Inlet Pressure	2.0 psi
Pressure Drop	1.0 psi
Specific Gravity	1.50

INTENDED USE	PE pipe sizing between 2 psig service regulator and line pressure regulator.							
	PIPE SIZE (inch)							
Nominal OD	$^1/_2$	$^3/_4$	1	$1^1/_4$	$1^1/_2$	2	3	4
Designation	SDR 9	SDR 11	SDR 11	SDR 10	SDR 11	SDR 11	SDR 11	SDR 11
Actual ID	0.660	0.860	1.077	1.328	1.554	1.943	2.864	3.682
Length (ft)	**Capacity in Thousands of Btu per Hour**							
10	3,130	6,260	11,300	19,600	29,500	53,100	147,000	284,000
20	2,150	4,300	7,760	13,400	20,300	36,500	101,000	195,000
30	1,730	3,450	6,230	10,800	16,300	29,300	81,100	157,000
40	1,480	2,960	5,330	9,240	14,000	25,100	69,400	134,100
50	1,310	2,620	4,730	8,190	12,400	22,200	61,500	119,000
60	1,190	2,370	4,280	7,420	11,200	20,100	55,700	108,000
70	1,090	2,180	3,940	6,830	10,300	18,500	51,300	99,100
80	1,010	2,030	3,670	6,350	9,590	17,200	47,700	92,200
90	952	1,910	3,440	5,960	9,000	16,200	44,700	86,500
100	899	1,800	3,250	5,630	8,500	15,300	42,300	81,700
125	797	1,600	2,880	4,990	7,530	13,500	37,500	72,400
150	722	1,450	2,610	4,520	6,830	12,300	33,900	65,600
175	664	1,330	2,400	4,160	6,280	11,300	31,200	60,300
200	618	1,240	2,230	3,870	5,840	10,500	29,000	56,100
250	548	1,100	1,980	3,430	5,180	9,300	25,700	49,800
300	496	994	1,790	3,110	4,690	8,430	23,300	45,100
350	457	914	1,650	2,860	4,320	7,760	21,500	41,500
400	425	851	1,530	2,660	4,020	7,220	12,000	38,600
450	399	798	1,440	2,500	3,770	6,770	18,700	36,200
500	377	754	1,360	2,360	3,560	6,390	17,700	34,200
550	358	716	1,290	2,240	3,380	6,070	16,800	32,500
600	341	683	1,230	2,140	3,220	5,790	16,000	31,000
650	327	654	1,180	2,040	3,090	5,550	15,400	29,700
700	314	628	1,130	1,960	2,970	5,330	14,700	28,500
750	302	605	1,090	1,890	2,860	5,140	14,200	27,500
800	292	585	1,050	1,830	2,760	4,960	13,700	26,500
850	283	566	1,020	1,770	2,670	4,800	13,300	25,700
900	274	549	990	1,710	2,590	4,650	12,900	24,900
950	266	533	961	1,670	2,520	4,520	12,500	24,200
1,000	259	518	935	1,620	2,450	4,400	12,200	23,500
1,100	246	492	888	1,540	2,320	4,170	11,500	22,300
1,200	234	470	847	1,470	2,220	3,980	11,000	21,300
1,300	225	450	811	1,410	2,120	3,810	10,600	20,400
1,400	216	432	779	1,350	2,040	3,660	10,100	19,600
1,500	208	416	751	1,300	1,960	3,530	9,760	18,900
1,600	201	402	725	1,260	1,900	3,410	9,430	18,200
1,700	194	389	702	1,220	1,840	3,300	9,130	17,600
1,800	188	377	680	1,180	1,780	3,200	8,850	17,100
1,900	183	366	661	1,140	1,730	3,110	8,590	16,600
2,000	178	356	643	1,110	1,680	3,020	8,360	16,200

For SI: 1 inch = 25.4 mm, 1 foot = 304.8 mm, 1 pound per square inch = 6.895 kPa, 1-inch water column = 0.2488 kPa,
  1 British thermal unit per hour = 0.2931 W, 1 cubic foot per hour = 0.0283 m³/h, 1 degree = 0.01745 rad.
Note: All table entries have been rounded to three significant digits.

❖ See the commentary to Section 402.4.

TABLE 402.4(37)
POLYETHYLENE PLASTIC TUBING

Gas	Undiluted Propane
Inlet Pressure	11.0 in. w.c.
Pressure Drop	0.5 in. w.c.
Specific Gravity	1.50

INTENDED USE	PE pipe sizing between integral two-stage regulator at tank or second stage (low-pressure regulator) and building.	
Plastic Tubing Size (CTS) (inch)		
Nominal OD	$^1/_2$	1
Designation	SDR 7	SDR 11
Actual ID	0.445	0.927
Length (ft)	**Capacity in Cubic Feet of Gas per Hour**	
10	121	828
20	83	569
30	67	457
40	57	391
50	51	347
60	46	314
70	42	289
80	39	269
90	37	252
100	35	238
125	31	211
150	28	191
175	26	176
200	24	164
225	22	154
250	21	145
275	20	138
300	19	132
350	18	121
400	16	113
450	15	106
500	15	100

For SI: 1 inch = 25.4 mm, 1 foot = 304.8 mm,
     1 pound per square inch = 6.895 kPa,
     1-inch water column = 0.2488 kPa,
     1 British thermal unit per hour = 0.293 1 W,
     1 cubic foot per hour = 0.0283 m³/h, 1 degree = 0.0 1745 rad.
Note: All table entries have been rounded to three significant digits.

❖ See the commentary to Section 402.4.

SECTION 403 (IFGS)
PIPING MATERIALS

403.1 General. Materials used for *piping* systems shall comply with the requirements of this chapter or shall be *approved*.

❖ This section dictates what materials and components can be used to construct gas distribution systems and also specifies the allowable applications of those materials.

403.2 Used materials. Pipe, fittings, valves and other materials shall not be used again except where they are free of foreign materials and have been ascertained to be adequate for the service intended.

❖ Ideally, gas pipe installations should be constructed of new materials. However, this section recognizes that there are occasions when used piping materials may be perfectly acceptable for an installation if they meet a number of very strict criteria that intend to reduce the potential for poor performance. To judge equivalency of reused piping materials, compliance with approval criteria for new materials should be used.

403.3 Other materials. Material not covered by the standards specifications listed herein shall be investigated and tested to determine that it is safe and suitable for the proposed service, and, in addition, shall be recommended for that service by the manufacturer and shall be *approved* by the code official.

❖ This section echoes the intent of Section 105.2.

403.4 Metallic pipe. Metallic pipe shall comply with Sections 403.4.1 through 403.4.4.

❖ This section addresses the traditional metal pipes used in piping installations.

403.4.1 Cast iron. Cast-iron pipe shall not be used.

❖ Because cast-iron pipe is brittle compared to the malleable metals, it is more likely to fail under stress and therefore is not suitable for conveying fuel gases.

403.4.2 Steel. Steel and wrought-iron pipe shall be not less than standard weight (Schedule 40) and shall comply with one of the following standards:

1. ASME B36.10, 10M.

2. ASTM A53/A53M.

3. ASTM A106.

❖ Steel pipe must be Schedule 40 or heavier, must comply with one of the listed standards, and can be black iron or galvanized. Contrary to popular belief, natural gas does not adversely react with the zinc coating on galvanized pipe.

403.4.3 Copper and copper alloy. Copper and copper alloy pipe shall not be used if the gas contains more than an average of 0.3 grains of hydrogen sulfide per 100 standard cubic feet of gas (0.7 milligrams per 100 liters). Threaded copper, copper alloy and aluminum-alloy pipe shall not be used with gases corrosive to such materials.

❖ The intent of this section is to prohibit the use of a piping and tubing material where it would be subject to corrosion from the fuel gas. Gas piping materials must be chemically compatible with the fuel gas that will be conveyed in the piping. For example, natural gas contains trace amounts of hydrogen sulfide and, depending on the concentration of the sulfur compound, can be corrosive to copper and brass pipe and tubing. The chemical reaction can create a precipitant on the pipe interior walls. The precipitant can flake off and be carried into appliance gas controls, causing hazardous operation. This is one of the reasons why the code requires sediment traps at connections to appliances. The code considers natural gas to be corrosive if it contains more than an average of 0.3 grain (0.0194 g) of hydrogen sulfide per 100 cubic feet (2.83 m³) of gas. Most pipeline-quality gas does not exceed this limit; however, the gas supplier must be consulted to obtain the chemical composition of the gas that will be conveyed by copper pipe or tubing.

403.4.4 Aluminum. Aluminum-alloy pipe shall comply with ASTM B241 except that the use of alloy 5456 is prohibited. Aluminum-alloy pipe shall be marked at each end of each length indicating compliance. Aluminum-alloy pipe shall be coated to protect against external corrosion where it is in contact with masonry, plaster or insulation, or is subject to repeated wettings by such liquids as water, detergents or sewage. Aluminum-alloy pipe shall not be used in exterior locations or underground.

❖ Aluminum pipe is not commonly used for gas service. Because aluminum is a fairly reactive metal, it corrodes readily when exposed to a variety of materials and the environment.

403.5 Metallic tubing. Seamless copper, aluminum alloy and steel tubing shall not be used with gases corrosive to such materials.

❖ Aluminum and steel tubing are used in the manufacturing of appliances but are rarely used for gas distribution. Copper tubing is widely used in gas distribution systems and is used as an alternative to CSST in parallel distribution manifold systems. As with copper pipe, copper tubing cannot be used with gases that will cause it to corrode. Commentary Figure 403.5 illustrates a type of sheathed copper tubing

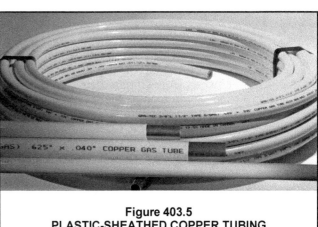

Figure 403.5
PLASTIC-SHEATHED COPPER TUBING
(Photo courtesy of KAMCO Products Ltd.)

that is designed for gas distribution. The corrosiveness of the gas can be determined by contacting the gas supplier. Depending upon the chemical composition of the gas, copper can form scale deposits on the interior pipe wall. Such scale can find its way into gas controls, regulators and valves, causing malfunctions.

403.5.1 Steel tubing. Steel tubing shall comply with ASTM A254.

❖ Steel tubing is used only in industrial applications and in the construction of appliances.

403.5.2 Copper and copper alloy tubing. Copper tubing shall comply with Standard Type K or L of ASTM B88 or ASTM B280.

Copper and copper alloy tubing shall not be used if the gas contains more than an average of 0.3 grains of hydrogen sulfide per 100 standard cubic feet of gas (0.7 milligrams per 100 liters).

❖ ASTM B88 applies to copper water tube, and Types K and L refer to wall thickness. Type M tubing has the thinnest wall and is not allowed for gas service.

403.5.3 Aluminum tubing. Aluminum-alloy tubing shall comply with ASTM B210 or ASTM B241. Aluminum-alloy tubing shall be coated to protect against external corrosion where it is in contact with masonry, plaster or insulation, or is subject to repeated wettings by such liquids as water, detergent or sewage.

Aluminum-alloy tubing shall not be used in exterior locations or underground.

❖ See the commentary to Section 403.4.4.

403.5.4 Corrugated stainless steel tubing. Corrugated stainless steel tubing shall be *listed* in accordance with ANSI LC 1/CSA 6.26.

❖ ANSI LC-1 contains material, performance criteria and installation requirements for CSST gas distribution systems. CSST systems are used as an alternative to more traditional gas piping systems for the distribution of natural gas within buildings.

Corrugated stainless steel tubing is a semirigid stainless steel tubing with a plastic jacket and is available in coil lengths varying from 100 to 250 feet (30 480 to 76 200 mm) and in the following standard sizes: $^3/_8$, $^1/_2$, $^3/_4$, 1, $1^1/_4$, $1^1/_2$ and 2 inch (larger sizes may be available) [see Commentary Figure 403.5.4(1)]. Note that CSST sizes are designated by equivalent hydraulic diameters (EHD) not standard tube sizes in inches. Any standard size markings in inches provided by the manufacturer are approximations only. Special proprietary fittings are required for connection of the tubing [see Commentary Figures 403.5.4(2) through 403.5.4(4)]. CSST can be used in low-pressure systems [up to 0.5 psig (14 inches of water column) (3.45 kPa)] and in higher pressure systems operating at pressures up to 5 psig (34.5 kPa). Commentary Figure 403.5.4(5) shows a typical manifold installation using CSST and shows a multiport manifold with a pressure regulator and shutoff valve

installed. To prevent damage or puncture by screws and nails, specialized steel shield plates and protective flexible conduit are used to protect CSST systems that pass through or near structural members. Such protection must employ the parts specified by the CSST manufacturer as opposed to standard off-the-shelf protection devices. [see Commentary Figures 403.5.4(1) and 403.5.4(6)].

CSST systems are usually designed as parallel distribution (manifold) systems, whereas traditional gas distribution systems are typically designed as

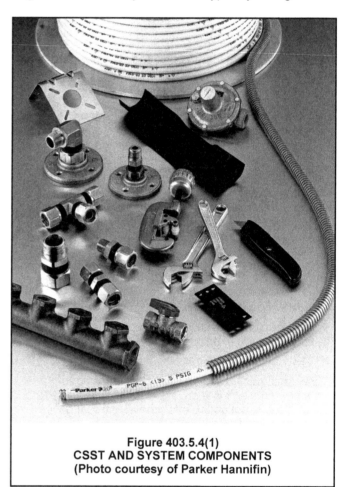

Figure 403.5.4(1)
CSST AND SYSTEM COMPONENTS
(Photo courtesy of Parker Hannifin)

Figure 403.5.4(2)
CSST FITTING
(Figure courtesy of Parker Hannifin)

branch (series) systems [see Commentary Figure 403.5.4(7)].

Medium-pressure CSST systems, typically 2 psig (13.8 kPa) maximum, are designed to allow the use of small diameter tubing for long runs. Significant distribution pressure losses can be compensated for by providing sufficiently high supply pressures. Medium-pressure distribution systems require the installation of pressure regulators, commonly called pounds-to-inches regulators, to reduce the supply pressure to the pressure required for delivery to the gas-fired appliances [see commentary, Section 410 and Commentary Figure 403.5.4(8)]. Note that systems with pressures over 2 psi (13.8 kPa) that supply appliances designed for a maximum inlet pressure of 14 inches w.c. (3.5 kPa) must have over pressure protection devices (OPD) in addition to the pounds-to-inches line pressure regulators. Such OPD devices are factory-installed on pressure regulators shipped for use in systems having pressures greater than 2 psi (13.8 kPa).

Section 404.3 prohibits tubing fittings in concealed locations but this does not apply to CSST fittings. In accordance with the manufacturer's installation instructions and the material listing, CSST fittings can be concealed in the building construction. The tubing,

Figure 403.5.4(5)
MULTIPORT MANIFOLD INSTALLATION USING CSST
(Photo courtesy of Gastite Division/Titeflex Corporation)

Figure 403.5.4(3)
CSST FITTING
(Figure courtesy of Gastite Division/Titeflex Corporation)

Figure 403.5.4(4)
CSST FITTING
(Photo courtesy of Gastite Division/Titeflex Corporation)

Figure 403.5.4(6)
TOP PLATE TUBING PROTECTION SHIELD
(Photo courtesy of Gastite Division/Titeflex Corporation)

fittings, shields and outlet terminals are all part of a system and must be installed together as a system.

The manufacturer's installation instructions are to be considered as a complete system installation manual, which also includes the proprietary sizing criteria. Except for those tables that are specific to CSST, the sizing tables in Section 402 do not apply to and cannot be used for CSST systems.

Installation of CSST is not allowed underground (directly buried) unless it is of a type specially designed and listed for underground applications [see Commentary Figure 403.5.4(9)]. Some manufacturers' installation instructions allow CSST to be "indirectly" buried by installing the tubing within a nonmetallic, water-tight conduit, such as PVC, PE or ABS plastic pipe. Other manufacturers may offer

For SI: 1 pound per square inch = 6.895 kPa.

Figure 403.5.4(7)
LOW-PRESSURE GAS PIPING SYSTEM USING CSST

For SI: 1 pound per square inch = 6.895 kPa.

Figure 403.5.4(8)
MEDIUM-PRESSURE GAS PIPING SYSTEM USING CSST

CSST with an integral casing and sealing fittings specifically designed for underground installation. See Sections 404.6 and 404.14 regarding piping buried in slabs and beneath buildings.

CSST is allowed by the manufacturers' installation instructions to connect directly to fixed-in-place, nonportable, nonmovable appliances, such as furnaces, boilers, gas fireplaces and water heaters. CSST is not intended for direct connection to movable appliances such as clothes dryers or ranges, nor is it intended to connect to appliances suspended on chains or rods, such as radiant heaters and unit heaters. Also, connecting directly to log lighters installed in solid-fuel-burning fireplaces is not allowed. Direct connection is allowed for gas fireplace appliances in accordance with the CSST and fireplace manufacturers' instructions. At range and clothes dryer locations, the CSST must terminate at a CSST termination fitting assembly, and a gas appliance connector must be used to connect the appliance to the termination fitting assembly. CSST must not be exposed to impact, vibration or repeated movement. "Drops" to fixed-in-place, nonmovable appliances must be well supported and routed to avoid being disturbed. In other words "flying runs" of CSST are prohibited. CSST, like copper tubing, is not flexible, but rather is semi-rigid, allowing it to be formed into shape as it is installed. CSST, like any metallic tubing, must not be repeatedly bent in the same location because this can result in kinking, work hardening and structural failure.

Generally, the sale of CSST materials is restricted and only individuals who have been specially trained and certified are allowed to purchase and install CSST systems.

Traditionally, the utility company (gas supplier) service regulator delivers natural gas to the building at a pressure of approximately 6 to 10 inches w.c. (1.49 to 2.49 kPa). Where medium-pressure [2 psig (13.8 kPa)] systems are used, the service regulator is set to deliver gas at a pressure of 2 psig (13.8 kPa). A medium-pressure gas distribution system must be properly designed to deliver gas to the connected appliances at pressures not exceeding the maximum inlet (supply) pressure for which the appliance is designed. A hazardous condition could develop if gas is supplied to appliances at pressures greater than the design inlet pressure of the appliances. See Section 416 of the code.

ANSI LC-1/CSA6.26 addresses CSST systems and contains installation requirements. Installers and inspectors must have copies of the standard and the

Figure 403.5.4(9)
CSST MANUFACTURED FOR UNDERGROUND INSTALLATION
(Photo courtesy of Omega Flex)

CSST manufacturer's installation instructions to properly install and inspect CSST systems. Manufacturers' instructions contain detailed requirements for protection hardware designed to prevent damage to installed CSST systems. As with most tubing systems, the most common installation errors involve abuse of the CSST tubing, lack of support and improperly installed puncture protection devices or the omission of protection devices. CSST manufacturers publish design and installation guides that are very helpful and should be on the bookshelves of inspectors, installers and designers [see Commentary Figures 403.5.4(10) and 403.5.4(11)]. Some CSST manufacturers offer pre-assembled manifolds in a rough-in cabinet to simplify installation [see Commentary Figure 403.5.4(12)]. CSST transitions to rigid pipe appliance connections and listed appliance connectors must be made with fittings that prevent stress and rotational forces from being applied to the CSST. Commentary Figures 403.5.4(13) through 403.5.4(15) illustrate typical fitting assemblies used to make the transition from CSST to appliance connections or connectors. It is a code violation to construct appliance outlets with a standard pipe elbow and rigid pipe nipple stubbed out of a wall or up through a floor because any rotational forces applied by tools to the outlet nipple will cause the CSST to rotate behind the

For SI: 1 inch = 25.4 mm.

Figure 403.5.4(11)
EXCERPT FROM CSST MANUFACTURER'S
INSTALLATION INSTRUCTIONS
(Figure courtesy of Parker Hannifin)

For SI: 1 inch = 25.4 mm.

Figure 403.5.4(10)
EXCERPT FROM CSST MANUFACTURER'S
INSTALLATION INSTRUCTIONS
(Figure courtesy of Parker Hannifin)

Figure 403.5.4(12)
MANIFOLD FOR 2 PSI PARALLEL
DISTRIBUTION SYSTEM
(Photo courtesy of Parker Hannifin)

Figure 403.5.4(13)
"STUB-OUT" TERMINATION TRANSITION FITTING
(Photo courtesy of Gastite Division/Titeflex
Corporation)

Figure 403.5.4(15)
TERMINATION TRANSITION FITTING
(Photo courtesy of Gastite Division/Titeflex
Corporation)

Figure 403.5.4(14)
"VALVE BOX" TERMINATION TRANSITION FITTING
(Photo courtesy of Gastite Division/Titeflex
Corporation)

wall or below the floor. The same is true for tee-handle shutoff valves concealed in a wall. The proper terminal fittings and brackets must be used to prevent forces from being transferred to the CSST.

403.6 Plastic pipe, tubing and fittings. Polyethylene plastic pipe, tubing and fittings used to supply fuel gas shall conform to ASTM D2513. Such pipe shall be marked "Gas" and "ASTM D2513."

Plastic pipe, tubing and fittings, other than polyethylene, shall be identified and conform to the 2008 edition of ASTM D2513. Such pipe shall be marked "Gas" and "ASTM D2513."

Polyvinyl chloride (PVC) and chlorinated polyvinyl chloride (CPVC) plastic pipe, tubing and fittings shall not be used to supply fuel gas.

❖ The installation of plastic pipe and tubing is limited to areas that are both outside the building (outdoors) and underground because of the potential hazard

associated with the use of a material that has lower resistance to physical damage and heat compared to metallic pipe. Plastic pipe and tubing must never be installed in or under any building. Plastic pipe and tubing are widely used for underground gas distribution systems because of their ease of installation and inherent resistance to corrosion (See Section 404.17.1). The scope of ASTM D2513 was recently changed to limit its application to PE pipe, therefore, an older edition is referenced for other types of plastic. This section specifically prohibits the use of PVC and CPVC pipe, tubing and fittings for conveying fuel gas.

The code now references the 2013 edition of ASTM D2513, which has been revised to address polyethylene (PE) plastic pipe, tubing and fittings only, whereas the 2008 edition addressed all plastic materials. The code had to maintain a reference to the 2008 edition of the standard in order to address plastics other than PE, such as polyamide (nylon). It was determined that polyamide pipe is currently used to supply fuel gas; however, PVC and CPVC are not. Further, it was decided that because of the brittle nature of PVC and CPVC, especially at low temperatures, these materials are not suitable for conveying fuel gas. Rather than being silent, the code now prohibits what the marketplace has failed to embrace as a viable material for fuel gas (see Commentary Figure 403.6).

403.6.1 Anodeless risers. Plastic pipe, tubing and anodeless risers shall comply with the following:

1. Factory-assembled anodeless risers shall be recommended by the manufacturer for the gas used and shall be leak tested by the manufacturer in accordance with written procedures.

2. Service head adapters and field-assembled anodeless risers incorporating service head adapters shall be recommended by the manufacturer for the gas used, and shall be designed and certified to meet the requirements of Category I of ASTM D2513, and U.S. Department of Transportation, Code of Federal Regulations, Title 49, Part 192.281(e). The manufacturer shall provide the

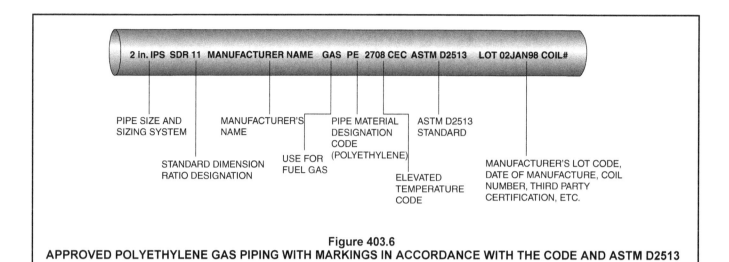

Figure 403.6
APPROVED POLYETHYLENE GAS PIPING WITH MARKINGS IN ACCORDANCE WITH THE CODE AND ASTM D2513

user with qualified installation instructions as prescribed by the U.S. Department of Transportation, Code of Federal Regulations, Title 49, Part 192.283(b).

❖ This section is an exception to the absolute prohibition of plastic gas piping above ground. Plastic pipe is allowed to rise from underground within a steel riser conduit constructed as an anodeless riser assembly. Such risers are coated to resist corrosion and do not use a sacrificial anode for cathodic corrosion protection, hence the name "anodeless riser." The steel riser protects the PE plastic pipe from physical damage and terminates the plastic pipe with a threaded steel pipe transition fitting. A sweeping 90-degree (1.57 rad) bend is placed on the underground end of rigid risers to prevent them from rotating in the ground when threaded connections are made at the terminal end. The plastic pipe could be damaged by rotational and bending stresses transmitted by the riser; therefore, a horizontal portion of steel piping is installed at the bottom of the riser to limit such movement. These risers can eliminate the need for an underground plastic-to-metal transition fitting and simplify installation. They are used extensively by utility companies for service laterals and can be used for other installations such as supply runs to out-buildings, pool heaters, gas grills and yard lights.

Service head adapters are also a type of terminal transition fitting that allow plastic pipe to be pulled through a retired steel gas service lateral and terminated at the point where the concentric piping passes through the building foundation wall. The adapter makes the transition from plastic pipe to steel pipe and seals the end of the retired steel service lateral [see Commentary Figures 202.(1), 403.6.1(1) 403.6.1(2) and 403.6.1(3)].

403.6.2 LP-gas systems. The use of plastic pipe, tubing and fittings in undiluted liquefied petroleum gas *piping* systems shall be in accordance with NFPA 58.

❖ NFPA 58 limits plastic piping for LP-gas service to only polyethylene pipe and tubing. The PE pipe and tubing must be marked as being in compliance with

Figure 403.6.1(1)
SERVICE HEAD ADAPTERS
(Photo courtesy of Perfection Corporation)

ASTM D2513 and is allowed for LP-gas service only if so recommended by the material manufacturer.

403.6.3 Regulator vent piping. Plastic pipe and fittings used to connect regulator vents to remote vent terminations shall be PVC conforming to ANSI/UL 651. PVC vent *piping* shall not be installed indoors.

❖ The referenced standard is for Schedule 40 and 80 PVC electrical conduit. This material was chosen because it is required to be sunlight-resistant for use above ground and the typical application for regulator vent piping is above ground, outdoors where it would be subject to the degrading effects of ultraviolet radiation in sunlight. Consistent with other provisions in the code, plastic pipe that does or could convey fuel gas is prohibited indoors. Regulator vent outlets are piped to maintain separation between vent outlets and sources of ignition and air intake openings in building envelopes.

Figure 403.6.1(2)
SERVICE HEAD ADAPTERS
(Figure courtesy of NORMAC)

403.7 Workmanship and defects. Pipe, tubing and fittings shall be clear and free from cutting burrs and defects in structure or threading, and shall be thoroughly brushed, and chip and scale blown.

Defects in pipe, tubing and fittings shall not be repaired. Defective pipe, tubing and fittings shall be replaced.

❖ This section stresses that proper installation procedures are necessary to secure a safe installation. In the context of this section, good workmanship is an enforceable requirement. Burrs result from the cutting of pipe or tube and cause flow restriction. Pipe scale, dirt, construction site debris, metal chips from cutting and threading operations and excess thread compound are potentially harmful contaminants that might cause damage to valves, regulators and appliances. Damaged and/or improperly cut threads can cause leakage or joint failure. Defective materials must be replaced because attempts to repair them are always makeshift at best, and any cost savings would not be worth the risk.

403.8 Protective coating. Where in contact with material or atmosphere exerting a corrosive action, metallic *piping* and fittings coated with a corrosion-resistant material shall be used. External or internal coatings or linings used on *piping* or components shall not be considered as adding strength.

❖ Protective coatings in the form of wrappings, tapes, enamels, epoxies, sleeves, casings and factory-applied coverings are a common method of isolating metallic piping materials from the atmosphere or from the earth in which they are buried. Application of these materials under strictly controlled factory conditions improves the quality control process and reduces the likelihood of coating defects. Corrosion can be caused by weather exposure, burial in the soil

Figure 403.6.1(3)
FLEXIBLE ANODELESS RISER
(Photo courtesy of NORMAC)

or construction materials that are in contact with the pipe. Cinders (also known as dross) are the waste residue from steel production and carry the waste products skimmed from the molten metal during the smelting process. Cinders are also the waste residue of coal-fired appliances and equipment. These materials are extremely corrosive to metallic piping and may not be used in any form (backfill or cinder blocks, for example) where in direct contact with the pipe. Protection is usually provided by a factory-applied coating or by field wrapping the pipe with a protective covering, such as a coal-tar-based or plastic wrapping. Galvanized piping in contact with soil would require additional coating or protection. The zinc coating does not usually provide long-term protection in underground applications. Where possible, a piping material is chosen that is not subject to the type of corrosion of the application.

Corrosion can also be caused by galvanic action that takes place where dissimilar metals are joined in a current-carrying medium, such as soil or water. For example, if steel and copper pipe are joined in a medium that conducts electrical current, the steel pipe will corrode at an accelerated rate because of the electrochemical process between the dissimilar metals. In this case, the soil acts as an electrolyte, the steel will be a sacrificial anode, and the copper will be a cathode. Because a cell is created, current will flow through the metal junction and the soil, resulting in the gradual deterioration of the steel. To protect against galvanic corrosion, dielectric fittings and couplings are used to join the piping, thus breaking the circuit of the cell.

This section is commonly interpreted as requiring priming and painting of black steel pipe that is exposed to the weather as the minimum form of protection.

403.9 Metallic pipe threads. Metallic pipe and fitting threads shall be taper pipe threads and shall comply with ASME B1.20.1.

❖ Threads must be taper cut in accordance with the stated standard. The standard regulates tapered and straight threads, but this section specifies tapered threads. Tapered pipe threads, when made up, form a metal-to-metal (interference fit) seal. Pipe-joint compound or polytetrafluoroethylene (PTFE) tape is to be applied to male threads only to decrease the possibility of tape fragments or compound entering the piping system. Such debris in piping systems can block orifices, restrict flow or interfere with the operation of controls. The primary purpose of pipe-thread compounds is to act as a lubricant to allow proper tightening and to achieve a metal-to-metal seal. They also fill in small imperfections on the threaded surfaces. This section prohibits the use of straight tapped fittings such as couplings because the threaded joint will not provide the same seal as a joint made with taper-tapped fittings. Steel taper-threaded couplings are easily identified because they look exactly like the run of a tee fitting (see Commentary Figure 403.9). The common designation "NPT" is misinterpreted as "nominal pipe threads" and actually stands for "national standard, pipe, tapered."

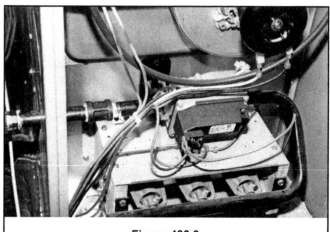

Figure 403.9
TAPERED THREAD COUPLING IN CONNECTION TO FURNACE

403.9.1 Damaged threads. Pipe with threads that are stripped, chipped, corroded or otherwise damaged shall not be used. Where a weld opens during the operation of cutting or threading, that portion of the pipe shall not be used.

❖ Damaged or improperly cut threads and defective welded seam pipe must be eliminated from any gas piping system because of the potential for leaks.

403.9.2 Number of threads. Field threading of metallic pipe shall be in accordance with Table 403.9.2.

TABLE 403.9.2
SPECIFICATIONS FOR THREADING METALLIC PIPE

IRON PIPE SIZE (inches)	APPROXIMATE LENGTH OF THREADED PORTION (inches)	APPROXIMATE NUMBER OF THREADS TO BE CUT
$^1/_2$	$^3/_4$	10
$^3/_4$	$^3/_4$	10
1	$^7/_8$	10
$1^1/_4$	1	11
$1^1/_2$	1	11
2	1	11
$2^1/_2$	$1^1/_2$	12
3	$1^1/_2$	12
4	$1^5/_8$	13

For SI: 1 inch = 25.4 mm.

❖ The table sets forth the required number of threads per unit length of pipe based on the pipe size. These specifications are intended to be both maximums and minimums because both undercutting and overcutting can result in faulty joints. Proper threading of pipe requires skill, practice, and well-maintained tools.

403.9.3 Thread joint compounds. Thread joint compounds shall be resistant to the action of liquefied petroleum gas or to any other chemical constituents of the gases to be conducted through the *piping*.

❖ Joint compounds in both paste and tape forms are commonly misapplied and used for the wrong application. Some compounds react chemically with the gas being conveyed, which could result in leakage. The label on the compound container will specify the applications for which the compound is suitable. Thread compounds act as a lubricant for the threads during assembly and also act as a sealant for the life of the joint. Care must be taken to keep all compounds out of the piping system interior because the contamination can cause damage to components and appliances. Pipe-joint compound or tape is limited to application on the male threads only to decrease the possibility of tape fragments or compound entering the piping system. Such debris in piping systems can block orifices, restrict flow or interfere with the operation of safety controls. It is important to leave the first one or two threads on the end of the male threads bare to help prevent the compound from entering the piping system.

Once made from lead compounds and linseed oil, joint compound formulations now commonly contain PTFE (Teflon). The primary purpose of pipe-thread compounds is to act as a lubricant to allow proper tightening and to achieve a metal-to-metal seal. They also fill in small imperfections on the threaded surfaces. Pipe-thread compounds and tapes must be compatible with both the piping material and the contents of the piping.

LP-gases can be a solvent for some pipe joint compound formulations. Therefore, a type must be chosen that will not react with the gas.

403.10 Metallic piping joints and fittings. The type of *piping* joint used shall be suitable for the pressure-temperature conditions and shall be selected giving consideration to joint tightness and mechanical strength under the service conditions. The joint shall be able to sustain the maximum end force caused by the internal pressure and any additional forces caused by temperature expansion or contraction, vibration, fatigue or the weight of the pipe and its contents.

❖ This section is in performance language and requires the designer and/or installer to choose the type of joining means that is appropriate for the application. Besides being leak tight, the joints must be able to withstand the conditions of service. Many joining methods are addressed in the code, but not all of them are suitable for all applications. For example, PE pipe for LP-gas use is limited to a maximum pressure of 30 psig (206.8 kPa), and most tubing materials would be unsuitable for uses involving equipment vibration that is transferred to the tubing. Also, some fittings might not be suitable for exposure to extreme changes in temperature.

403.10.1 Pipe joints. Pipe joints shall be threaded, flanged, brazed or welded. Where nonferrous pipe is brazed, the brazing materials shall have a melting point in excess of 1,000°F

(538°C). Brazing alloys shall not contain more than 0.05-percent phosphorus.

❖ Joints in piping materials can be made by four methods: threaded (screwed), brazed, welded and flanged. Flanged joints must also involve threading or welding as the means of attaching the flanges to the pipe. The limit on phosphorus content is related to corrosion that can occur from the reaction of phosphorus with a trace contaminant (sulphur) in the gas (see commentary, Section 403.10.2).

403.10.2 Tubing joints. Tubing joints shall be made with *approved* gas tubing fittings, brazed with a material having a melting point in excess of 1,000°F (538°C) or made with press-connect fittings complying with ANSI LC-4. Brazing alloys shall not contain more than 0.05-percent phosphorus.

❖ Tubing joints can be made by three methods: brazing, tubing fittings and press-connect fittings. Tubing fittings include flare, compression and CSST. Press-connect fittings require special tools that make an irreversible joint (see Commentary Figure 403.10.2).

Brazing is the act of producing a brazed joint. Brazing is often referred to as silver soldering. Silver soldering is more accurately described as silver brazing and employs high-silver-bearing alloys primarily composed of silver, copper and zinc. Silver soldering (brazing) typically requires temperatures in excess of 1,000°F (538°C), and such solders are classified as "hard" solders (see the commentary for "Brazed joint").

Confusion has always been present with respect to the distinction between "silver solder" and "silver-bearing" solder. Silver solders are unique and can be further subdivided into soft and hard categories, which are determined by the percentages of silver and the other component elements of the particular alloy. The distinction is that silver-bearing solders [melting point below 600°F (316°C)] are used in soft-soldered joints, and silver solders [melting point higher than 1,000°F (538°C)] are used in silver-brazed joints.

Brazed joints are considered to have superior strength and stress resistance and, because of the

Figure 403.10.2
PRESS-CONNECT COPPER FITTING

high melting point, are less likely to fail when exposed to fire.

Brazing with a filler metal conforming to AWS A5.8 produces a strong joint that will perform under extreme service conditions. The surfaces to be brazed must be cleaned free of oxides and impurities. Flux should be applied as soon as possible after the surfaces have been cleaned. Flux helps to remove residual traces of oxides, to promote wetting and to protect the surfaces from oxidation during heating. Care should be taken to prevent flux from entering the piping system during the brazing operation because flux that remains may corrode the pipe or contaminate the system.

Air should be removed from the pipe being brazed by purging the piping with a nonflammable gas such as carbon dioxide or nitrogen. Purging the system has several benefits, such as preventing oxidation from occurring on the inside of the pipe. Mechanical joints are usually proprietary joints that are developed and marketed by individual manufacturers. Many types of mechanical joints use a sleeve or ferrule that is compressed around the circumference of the pipe or tube. Mechanical joints must be specifically designed for and compatible with the type of pipe or tube to be joined.

403.10.3 Flared joints. Flared joints shall be used only in systems constructed from nonferrous pipe and tubing where experience or tests have demonstrated that the joint is suitable for the conditions and where provisions are made in the design to prevent separation of the joints.

❖ Flared joints are typically used with copper and aluminum tubing and are prohibited for steel tubing. Flared joints require the use of specialized tools. Because the pipe end is expanded in a flared joint, only annealed and bending tempered (drawn) copper tubing may be flared. Commonly used flaring tools use a screw yoke and block assembly or an expander tool that is driven into the tube with a hammer. The flared tubing end is compressed between a fitting seat and a threaded nut to form a metal-to-metal seal.

403.10.4 Metallic fittings. Metallic fittings shall comply with the following:

1. Threaded fittings in sizes larger than 4 inches (102 mm) shall not be used.

2. Fittings used with steel or wrought-iron pipe shall be steel, copper alloy, malleable iron or cast iron.

3. Fittings used with copper or copper alloy pipe shall be copper or copper alloy.

4. Fittings used with aluminum-alloy pipe shall be of aluminum alloy.

5. Cast-iron fittings:

   5.1. Flanges shall be permitted.

   5.2. Bushings shall not be used.

   5.3. Fittings shall not be used in systems containing flammable gas-air mixtures.

   5.4. Fittings in sizes 4 inches (102 mm) and larger shall not be used indoors except where *approved*.

   5.5. Fittings in sizes 6 inches (152 mm) and larger shall not be used except where *approved*.

6. Aluminum-alloy fittings. Threads shall not form the joint seal.

7. Zinc aluminum-alloy fittings. Fittings shall not be used in systems containing flammable gas-air mixtures.

8. Special fittings. Fittings such as couplings, proprietary-type joints, saddle tees, gland-type compression fittings and flared, flareless and compression-type tubing fittings shall be: used within the fitting manufacturer's pressure-temperature recommendations; used within the service conditions anticipated with respect to vibration, fatigue, thermal expansion and contraction; and shall be approved.

9. Where pipe fittings are drilled and tapped in the field, the operation shall be in accordance with all of the following:

   9.1. The operation shall be performed on systems having operating pressures of 5 psi (34.5 kPa) or less.

   9.2. The operation shall be performed by the gas supplier or the gas supplier's designated representative.

   9.3. The drilling and tapping operation shall be performed in accordance with written procedures prepared by the gas supplier.

   9.4. The fittings shall be located outdoors.

   9.5. The tapped fitting assembly shall be inspected and proven to be free of leakage.

❖ Item 1 limits threaded fittings to a maximum size of 4 inches (102 mm) because larger pipe threads are impractical, difficult to cut and difficult to make leak tight.

Item 2 allows use of cast iron fittings with steel, but cast iron pipe is prohibited. Malleable iron (steel) fittings are used almost exclusively.

Items 3 and 4 require fittings for nonferrous pipes to be of a material consistent with the pipe material.

Item 5 places limitations on cast-iron fittings because of the brittle nature of cast iron. Bushings have been known to split from overtightening. Fittings sizes 4 inches (102 mm) and up are allowed only outdoors and fittings 6 inches (152 mm) and up are not allowed anywhere.

Item 8 speaks of special fittings of the mechanical type. None of the joining methods addressed in the code use a nonmetallic or elastomeric sealing element except those used outdoors only.

Item 9 addresses a practice that is allowed only for gas utility (gas supplier) companies. Occasionally, an installer will drill a hole in a Schedule 40 or heavier pipe fitting and tap threads in the hole for the purpose of making a test connection or branch connection. The pipe fittings, however, were never designed to be drilled and tapped, and the thickness of the fitting wall does not allow for even one complete thread in Schedule 40 fittings. The resulting pipe connection or plugged opening would not have sufficient threads to make a joint that is both strong and free of leakage. Lacking the required number of complete threads, such joints would rely solely on the thread-sealant paste or tape to make a seal. Normal threaded joints rely on an interference fit of several threads and metal-to-metal sealing, while depending on pipe joint pastes and tapes only as thread lubricants and for sealing tiny imperfections in the threads. Field drilling and tapping of outdoor fittings on the outlet side of the meter is performed by some gas suppliers (utilities) for a unique purpose. If the gas supplier needs to replace a customer's gas meter, and it is not convenient to shut off the gas supply to the customer in order to replace the meter, the gas supplier may use a special apparatus to tap into the customer's piping near the meter for the purpose of temporarily supplying gas to the system at the same time the meter is isolated, removed and replaced. The gas is temporarily supplied from tanks of compressed gas on the utility's service trucks. After replacing the meter, the tapped opening is plugged. The five conditions in the new text limit this practice to gas-supplier personnel and authorized contractors employed by the gas supplier. The practice is limited to outdoor fittings, and the entire operation must be performed with specialized tools and equipment in accordance with the prescribed procedures. In cases where a leak-free plugged assembly cannot be achieved, the gas supplier will cover the affected area of the fitting with specialized devices or leak-repair methods used in the gas industry. The drilling and tapping of pipe fittings is prohibited in all cases except where performed in accordance with all five conditions that are intended to mitigate the risk [see Commentary Figures 403.10.4(1) and 403.10.4(2)].

403.11 Plastic pipe, joints and fittings. Plastic pipe, tubing and fittings shall be joined in accordance with the manufacturer's instructions. Such joint shall comply with the following:

1. The joint shall be designed and installed so that the longitudinal pull-out resistance of the joint will be at least equal to the tensile strength of the plastic *piping* material.

2. Heat-fusion joints shall be made in accordance with qualified procedures that have been established and proven by test to produce gas-tight joints at least as strong as the pipe or tubing being joined. Joints shall be made with the joining method recommended by the pipe manufacturer. Heat fusion fittings shall be marked "ASTM D2513."

3. Where compression-type mechanical joints are used, the gasket material in the fitting shall be compatible with the plastic *piping* and with the gas distributed by the system. An internal tubular rigid stiffener shall be used in conjunction with the fitting. The stiffener shall

be flush with the end of the pipe or tubing and shall extend at least to the outside end of the compression fitting when installed. The stiffener shall be free of rough or sharp edges and shall not be a force fit in the plastic. Split tubular stiffeners shall not be used.

4. Plastic *piping* joints and fittings for use in liquefied petroleum gas *piping* systems shall be in accordance with NFPA 58.

❖ This section contains general requirements for the proper joining of plastic gas pipe materials. Although plastic pipe materials offer many advantages such as ease of installation, corrosion resistance, lighter weight and lower cost, these materials require differ-

Figure 403.10.4(1)
METER SETTING

Figure 403.10.4(2)
ELBOW WITH DRILLED AND TAPPED HOLE

ent handling and installation methods than steel or other metal pipe or tubing. In all cases, the pipe material manufacturer's installation instructions must be strictly adhered to in order to reduce the likelihood of creating improper joints, which could lead to gas leakage or piping or joint failure. Dissimilar types of plastic materials may not be joined by heat fusion methods. Because of the resulting reduction in wall thickness, plastic pipe and tubing must not be field-threaded as a joining method. All approved joints must be designed to account for the forces acting upon the piping system, such as expansion, contraction and external loads imposed by burial in underground installations. The longitudinal pull-out resistance of the joint should be approximately the same as the tensile strength of the materials being joined.

Heat-fusion joints for plastic pipe are analogous to the welding of steel pipe. Heat fusion must be performed in accordance with the pipe manufacturer's instructions. The process involves heating the pipe and fittings with a special iron. When the parts to be joined reach their melting points, they are assembled and allowed to fuse. Electrical fusion fittings are available and employ an integral heating element that is connected to a power supply unit that allows the fitting to be heated at the required rate and duration.

Mechanical compression joints employ an elastomeric seal that must be compatible with not only the piping material but also the gas that is conveyed in the system. The manufacturer's instructions normally require an internal stiffener insert used in conjunction with the fitting to provide additional support. The stiffener insert prevents the pipe from deforming under the compression force exerted by the seal and compression nut assembly.

Category 1 fittings in accordance with ASTM D2513 are designed to provide a gas-tight seal and resist axial (pull-out) forces equal to the yield strength of the piping material [see Commentary Figures 403.11(1) through 403.11(4)].

403.12 Flanges. Flanges and flange gaskets shall comply with Sections 403.12.1 through 403.12.7.

❖ Flanges serve the same function as a union and are used for large piping at locations where it is necessary to disassemble and reassemble piping connections to valves, meters, regulators and other sections of piping. A flange consists of two mating fittings that bolt together to form a butt joint. A gasket is compressed between the flanges as the through-bolts are tightened (see Commentary Figure 403.12).

403.12.1 Cast iron. Cast-iron flanges shall be in accordance with ASME B 16.1.

❖ See the commentary to Section 403.12.

403.12.2 Steel. Steel flanges shall be in accordance with ASME B 16.5 or ASME B 16.47.

❖ See the commentary to Section 403.12.

403.12.3 Nonferrous. Nonferrous flanges shall be in accordance with ASME B 16.24.

❖ See the commentary to Section 403.12.

403.12.4 Ductile iron. Ductile-iron flanges shall be in accordance with ASME B 16.42.

❖ See the commentary to Section 403.12.

403.12.5 Raised face. Raised face flanges shall not be joined to flat faced cast-iron, ductile-iron or nonferrous material flanges.

❖ See the commentary to Section 403.12.

403.12.6 Flange facings. Standard facings shall be permitted for use under this code. Where 150-pound (1034 kPa) pressure-rated steel flanges are bolted to Class 125 cast-iron flanges, the raised face on the steel flange shall be removed.

❖ Because cast-iron fittings are brittle, a joint failure could occur from bolt tightening if the two flange surfaces are not in full contact with each other and the intervening gasket. In this application, Section 403.13 requires gaskets to cover the entire surface area of the flanges. Any partial raised face on a flange would create a fulcrum over which the cast-iron flange could be broken.

403.12.7 Lapped flanges. Lapped flanges shall be used only above ground or in exposed locations accessible for inspection.

❖ Lapped flanges are held on the pipe by flaring the end of the pipe. Such joints are rarely used compared to gasket-type flanges that are threaded or welded onto the pipe.

403.13 Flange gaskets. Material for gaskets shall be capable of withstanding the design temperature and pressure of the *piping* system, and the chemical constituents of the gas being conducted, without change to its chemical and physical properties. The effects of fire exposure to the joint shall be considered in choosing material. Acceptable materials include metal (plain or corrugated), composition, aluminum "O" rings, spiral wound metal gaskets, rubber-faced phenolic and elastomeric. Where a flanged joint is opened, the gasket shall be replaced. Full-face flange gaskets shall be used with all nonsteel flanges.

❖ Similar to a ground joint union, a flanged joint creates a seal by compressing a softer metal between the two flange fitting surfaces. This author assumes that asbestos gaskets are no longer made. Once compressed, the gasket in a disassembled joint is spent and must be discarded (see commentary, Section 403.12.1).

403.13.1 Metallic gaskets. Metallic flange gaskets shall be in accordance with ASME B 16.20.

❖ See the commentary to Section 403.13.

403.13.2 Nonmetallic gaskets. Nonmetallic flange gaskets shall be in accordance with ASME B 16.21.

❖ See the commentary to Section 403.13.

SECTION 404 (IFGC)
PIPING SYSTEM INSTALLATION

404.1 Installation of materials. Materials used shall be installed in strict accordance with the standards under which the materials are accepted and approved. In the absence of such installation procedures, the manufacturer's instructions shall be followed. Where the requirements of referenced standards or manufacturer's instructions do not conform to minimum provisions of this code, the provisions of this code shall apply.

❖ Piping materials and related system components must be installed in accordance with the standards to which they are listed. Improper installation can result in substandard performance, product failures and hazardous conditions. In some cases, the product standard will contain installation requirements such as found in ANSI LC-1 for CSST. In other cases the product standards might not contain any installation requirements, therefore, the manufacturer of the product supplies the only installation requirements. The manufacturer could also have installation

Figure 403.11(1)
MECHANICAL JOINTS
(Photo courtesy of Perfection Coproation)

Figure 403.11(2)
MECHANICAL JOINT
(Figure courtesy of Perfection Corporation)

Figure 403.11(3)
PLASTIC PIPE TRANSITION FITTING
(Figure courtesy of NORMAC)

Figure 403.11(4)
PLASTIC PIPE TRANSITION FITTING
(Photo courtesy of NORMAC)

requirements that are in addition to or in excess of those in the standard or vice versa. If the installation requirements in the product standards or manufacturer's instructions are somehow less stringent that a provision of this code, this code must prevail.

404.2 CSST. CSST piping systems shall be installed in accordance with the terms of their approval, the conditions of listing, the manufacturer's instructions and this code.

❖ The CSST product standard dictates what the manufacturer's instructions must include at a minimum. This section was added because Section 403.5.4 only required that CSST be listed and failed to clarify that the CSST system must also comply with the standard, the manufacturer's instructions, this code and any conditions established in the approval of such systems.

404.3 Prohibited locations. *Piping* shall not be installed in or through a ducted supply, return or exhaust, or a clothes chute, chimney or gas vent, dumbwaiter or elevator shaft. *Piping* installed downstream of the *point of delivery* shall not extend through any townhouse unit other than the unit served by such *piping*.

❖ An air supply duct or areas that create a shaft that provides a path for the gas to travel are considered to be locations where the problem of a gas leak could be compounded by spreading the gas or any resultant fire throughout the building. Also, in some locations, the gas piping could be subject to deterioration because of temperature or corrosive conditions such as in air ducts or chimneys.

Locating fuel gas piping in certain areas or atmospheres may cause corrosion of the pipe, which in

Figure 403.12
FLANGED JOINTS
(Photo courtesy of Washington Gas)

turn may cause a leak in the piping system. Within a supply air duct, the conditioned air may cause moisture to condense on or within the pipe, thereby causing corrosion of the pipe. Fuel gas piping located within a chimney or a vent will be subjected to high temperatures and the corrosive effects of flue gases.

Additionally, piping located in a dumbwaiter or elevator shaft may be subject to an additional hazard resulting in mechanical damage to the pipe. Damage from a dumbwaiter or elevator impacting the fuel gas piping may not only cause a leak that would allow the gas to escape and travel up the shaft but may also put any occupants within an elevator in immediate danger. The elevator codes also intend to prohibit nonessential piping within elevator shafts.

Gas piping is not prohibited in concealed spaces used to convey environmental air; for example, above ceiling and below floor plenums, stud cavities and joist spaces. Gas piping is often installed in return air plenums above a suspended ceiling with no history of problems. The piping system is assumed not to leak in a ceiling air plenum or any other location, for that matter. In fact, when viewed logically, it would be more hazardous for a leak to occur in a location where the air is static and an explosive gas/air mixture could be created. Return, supply and exhaust air ducts are not considered to be a plenum. These ducts are supply, return and exhaust air conduits constructed as components of a ductwork system. A furnace plenum is part of a ductwork system and would be treated as an air duct (see the definition of "Furnace plenum").

This section now prohibits what was once a common practice, that is, the practice of placing a bank of gas meters on one end of a townhouse complex and running the load side gas line from each meter through the attic or crawl spaces of the townhouse units to reach the units served. For example, the gas supply line for the unit farthest away from the meter bank will pass through all of the other units to reach

the unit it serves. This is a concern because of what might be done to the piping within the units through which it passes and because of the potential for one of the units to be destroyed by fire thereby damaging the gas piping that passes through the fire walls/separation walls between units. Townhouses are considered to be separate buildings in the *International Residential Code®* (IRC®) and are constructed to prevent fire from spreading beyond the unit of origin. This is similar to the provision in the IRC that prevents electrical service entrance conductors from passing through dwelling units other than the unit served (see Commentary Figure 404.3).

404.4 Piping in solid partitions and walls. Concealed *piping* shall not be located in solid partitions and solid walls, unless installed in a chase or casing.

❖ As with the alternative installation requirements in Section 404.8, this section allows installation of gas piping within solid walls or partitions only if the piping is installed within a chase or casing to protect the pipe from stress and from corrosive effects of wall materials such as concrete.

404.5 Fittings in concealed locations. Fittings installed in concealed locations shall be limited to the following types:

1. Threaded elbows, tees and couplings.

2. Brazed fittings.

3. Welded fittings.

4. Fittings listed to ANSI LC-1/CSA 6.26 or ANSI LC-4.

❖ A concealed location is a location that requires the removal of permanent construction in order to gain access (see the definition of "Concealed location"). The space above a dropped ceiling having readily removable lay-in panels or other locations that have removable access panels are not considered to be concealed locations for the purposes of this section. Concealed locations include wall, floor and ceiling cavities bounded by permanent finish materials, such as gypsum board, masonry or paneling. This section lists what is allowed to be concealed; therefore, by omission, all other types of fittings are not allowed in concealed locations. Note that threaded plugs and caps are not listed, but the author is not aware of any reason to prohibit them. The *National Fuel Gas Code* 2015 edition, ANSI Z223.1, has been revised to specifically allow threaded plugs and caps in concealed locations. Unions and mechanical joint tubing fittings are not permitted in concealed locations because they are more likely to loosen and leak than other joining means and fittings. Tubing fittings include flare, compression and similar proprietary-type fittings, all of which are mechanical joints.

Joints for CSST systems are mechanical tubing joints; however, these joints use specialized proprietary fittings that are listed for concealment as part of the CSST system. Such fittings undergo stringent testing to determine their ability to remain leak tight.

Right and left couplings are used in some parts of the world and are unknown in others. Right and left

Figure 404.3
PROHIBITED TOWNHOUSE INSTALLATION

couplings are used instead of a union fitting. The coupling is made with right-hand threads on one end and left-hand threads on the other, and the corresponding male pipe threads can be used to connect pipe and fittings within a section of existing piping. The code does not specifically prohibit right and left couplings, as it refers to only "couplings" generically; however, if the coupling was concealed, people would not be aware of such a fitting in a piping system and would not realize that they were loosening a joint as they applied the traditional tightening torque to an exposed section of piping. Piping should always be backed up with wrenches to prevent rotational forces from extending beyond the work piece. When the code refers to couplings, those fittings are tapered-thread fittings (see Section 403.9). Bushings have been known to split after assembly and are not trusted in concealed locations.

Brazing is a commonly used method of joining copper tubing and is as reliable as welded joints for steel pipe.

This section does not prohibit NPT threaded (screwed) elbow, tee and coupling fittings from being concealed; however, it does prohibit threaded unions and bushings from being concealed.

404.6 Underground penetrations prohibited. Gas *piping* shall not penetrate building foundation walls at any point below grade. Gas *piping* shall enter and exit a building at a point above grade and the annular space between the pipe and the wall shall be sealed.

❖ The preferred and required method for bringing outdoor gas piping into or out of a building is to penetrate the exterior wall at a point above grade. In other words, gas piping that comes into or leaves a building should not provide a path for underground gas leakage to enter the interior of a building. This section does not apply to utility company owned piping upstream of the point of delivery (see Section 101.2).

The penetration of the pipe and any protective sleeve must be adequately sealed to prevent the possibility of water or insect entry into the building. A gas pipe is not permitted to enter or exit a building below grade because any gas leakage in the underground piping will be channeled into the building through the surrounding soil and the wall penetration. Gas leakage can travel in the annular space that exists between the outside wall of the pipe and the soil that surrounds it. With an above-ground wall penetration, any outdoor piping leakage will vent to the outdoors with very little possibility of it finding its way into the building. Underground penetrations of building foundations are very difficult to seal against gas migration and the best way to eliminate the possibility of gas entering the building is to simply require the piping to enter and exit above grade. There have been cases of explosions that were believed to be caused by gas entering basements and crawl spaces through underground wall penetrations (see Commentary Figure 404.6). Note that this section creates a conflict with Sections 404.8.1 and 404.14.1 in that gas piping is no longer allowed to enter or exit a building at any point below grade, regardless of any vented conduit that may contain the piping. Sections 404.8.1 and 404.14.1 still have other applications that do not involve piping entering or exiting a building below grade.

Figure 404.6
ABOVE GRADE ENTRANCE/EXIT

404.7 Protection against physical damage. Where *piping* will be concealed within light-frame construction assemblies, the *piping* shall be protected against penetration by fasteners in accordance with Sections 404.7.1 through 404.7.3.

Exception: Black steel piping and galvanized steel piping shall not be required to be protected.

❖ The section on protection of piping has been completely rewritten to address more than just bored holes and notches in structural members. It now addresses piping parallel to framing members and piping within framing members. The new text requires that the protection extend well beyond the edge of members that are bored or notched.

Piping and tubing other than Schedule 40 steel pipe must be protected from penetration by nails and screws where the pipe or tubing is less than $1^1/_2$ inches from the face of the member where membranes will be attached. This protection is necessary whether the pipe or tube is perpendicular or parallel to the framing member. If a pipe or tube is run inside of the "C" channel of a $3^1/_2$-inch metal stud, it will almost certainly be penetrated by screws unless the pipe or tube is $^1/_2$ inch or less in diameter and located

dead center in the stud channel. Where pipes and tubing are attached to and run parallel with the side of a framing member, penetration by a nail or screw is likely if the installer misses the member with the nail or screw and the pipe or tube is less than $1^1/_2$ inches from either face of the stud. Extending the protection shield 4 inches beyond the edges of the framing member is intended to protect against fasteners that miss the member or that exit the member on an angle. To avoid having protection plates run parallel with a member, the installer could simply place the pipe or tube on "standoffs" such that the pipe/tube is not less than $1^1/_2$ inches from the nearest edge of the face of the member. As always, a designer's careful planning of the routing of gas piping or tubing can avoid the need for installation of protection plates by simply relying on a distance of at least $1^1/_2$ inches between the pipe/tubing and the fastener face of the member. Pipes and tubes are sometimes penetrated by screws or nails, and a leak does not develop immediately. It may take years for the fastener to corrode enough for a leak to eventually develop.

This section pertains to piping and tubing that will be concealed within wood or steel light-frame construction assemblies, which is the same scope as the original text (see Commentary Figure 404.7).

This section is intended to minimize the possibility that nails or screws will be driven into the gas pipe or tube. Because nails and screws sometimes miss the stud, rafter, joist or sole, or top plates, the shield plates must extend parallel to the pipe or tube not less than 4 inches (102 mm) beyond the member on each side or not less than 4 inches (102 mm) above or below sole or top wall plates, respectively. Commentary Figures 403.5.4(6) and 404.7 show typical shield plates. Black and galvanized steel pipe (Schedule 40) each have wall thicknesses greater than the required thickness for the shield plates, which makes these piping materials inherently resistant to nail and screw penetrations. This section does not apply to CSST tubing, because CSST systems have their own protection requirements dictated by the manufacturer's instructions and ANSI LC-1/CSA 6.26. The protection requirements for CSST systems are, overall, more stringent than those of this section. Before studs, rafters, joists or other structural members are drilled or notched, the building code should be consulted.

404.7.1 Piping through bored holes or notches. Where *piping* is installed through holes or notches in framing members and the *piping* is located less than $1^1/_2$ inches (38 mm) from the framing member face to which wall, ceiling or floor membranes will be attached, the pipe shall be protected by shield plates that cover the width of the pipe and the framing member and that extend not less than 4 inches (51 mm) to each side of the framing member. Where the framing member that the *piping* passes through is a bottom plate, bottom track, top plate or top track, the shield plates shall cover the framing member and extend not less than 4 inches (51 mm) above the

For SI: 1 inch = 25.4 mm.

Figure 404.7
PIPE PROTECTION

bottom framing member and not less than 4 inches (51 mm) below the top framing member.

❖ See the commentary to Section 404.7.

404.7.2 Piping installed in other locations. Where the *piping* is located within a framing member and is less than $1^1/_2$ inches (38 mm) from the framing member face to which wall, ceiling or floor membranes will be attached, the *piping* shall be protected by shield plates that cover the width and length of the *piping*. Where the *piping* is located outside of a framing member and is located less than $1^1/_2$ inches (38 mm) from the nearest edge of the face of the framing member to which the membrane will be attached, the *piping* shall be protected by shield plates that cover the width and length of the *piping*.

❖ This section addresses piping that is run along the side of framing members and that is run inside the channel of metal framing members. Where the piping is within $1^1/_2$ inches of the face of the framing member where fasteners will penetrate the member, there is the danger that fasteners will miss the target or be driven at an angle such that the piping is penetrated. Piping is sometimes run within the "C" channel of metal studs, which practically begs for the pipe/tube to be penetrated by fasteners that attach wall sheathing to the framing members (see commentary, Section 404.7).

404.7.3 Shield plates. Shield plates shall be of steel material having a thickness of not less than 0.0575 inch (1.463 mm) (No. 16 gage).

❖ See the commentary to Section 404.7.

404.8 Piping in solid floors. *Piping* in solid floors shall be laid in channels in the floor and covered in a manner that will allow *access* to the *piping* with a minimum amount of damage to the building. Where such *piping* is subject to exposure to excessive moisture or corrosive substances, the *piping*

shall be protected in an *approved* manner. As an alternative to installation in channels, the *piping* shall be installed in a conduit of Schedule 40 steel, wrought iron, PVC or ABS pipe in accordance with Section 404.8.1 or 404.8.2.

❖ Piping must not be installed in any solid concrete or masonry floor construction. The potential for pipe damage from slab settlement, cracking or the corrosive action of the floor material makes it imperative that one of the installation methods in this section be used. This section does not intend to allow any direct encasement of gas piping in solid concrete or masonry floor systems.

Gas piping installed within a solid floor system must be safeguarded by installation in a sealed casing or in a floor channel with a removable cover for pipe access (see Commentary Figure 404.8). Either of these methods should provide reasonable protection of the pipe from the effects of settling, cracking and being in contact with corrosive materials.

Because casings constructed of metal could corrode where installed within a concrete slab on grade, consideration should be given to corrosion protection for the steel casing, or an alternative material should be chosen.

Figure 404.8
PIPING IN SOLID FLOORS

404.8.1 Conduit with one end terminating outdoors. The conduit shall extend into an occupiable portion of the building and, at the point where the conduit terminates in the building, the space between the conduit and the gas *piping* shall be sealed to prevent the possible entrance of any gas leakage. The conduit shall extend not less than 2 inches (51 mm) beyond the point where the pipe emerges from the floor. If the end sealing is capable of withstanding the full pressure of the gas pipe, the conduit shall be designed for the same pressure as the pipe. Such conduit shall extend not less than 4 inches (102 mm) outside the building, shall be vented above grade to the outdoors and shall be installed so as prevent the entrance of water and insects.

❖ This method cannot be used where piping enters the building from the outdoors or exits the building to the outdoors. The main application of this section would be for interior piping that must enter a solid floor and exit a solid floor all within the interior of the building. The intent is to direct any gas leakage to the outdoors by means of the secondary containment conduit and the outdoor vent terminal. Note that this section does not prohibit NPT threaded (screwed) elbow, tee, coupling, cap and plug fittings from being concealed. It does prohibit threaded unions, right and left couplings, bushings and swing joints from being concealed.

404.8.2 Conduit with both ends terminating indoors. Where the conduit originates and terminates within the same building, the conduit shall originate and terminate in an accessible portion of the building and shall not be sealed. The conduit shall extend not less than 2 inches (51 mm) beyond the point where the pipe emerges from the floor.

❖ This method allows piping to run between any two points within the building without a vent run to the outdoors and without sealing the ends of the conduit. This method is an alternative to and less complicated than the method of Section 404.6.1. The minimum rise of 2 inches (51 mm) above the floor is intended to prevent liquids and debris from falling into the conduit. The ends of the conduit are not to be sealed so as to allow any leakage to be noticed by the occupants, the same as leakage would be noticed from piping located anywhere else in the structure. If the conduit ends were sealed, there would be no evidence of a piping defect, no matter how severe the defect, possibly leading to a major leak in the event of conduit or end seal failure (see Commentary Figure 404.8.2).

404.9 Above-ground outdoor piping. *Piping* installed outdoors shall be elevated not less than $3^1/_2$ inches (152 mm) above ground and where installed across roof surfaces, shall be elevated not less than $3^1/_2$ inches (152 mm) above the roof surface. *Piping* installed above ground, outdoors, and installed across the surface of roofs shall be securely supported and located where it will be protected from physical damage. Where passing through an outside wall, the *piping* shall be protected against corrosion by coating or wrapping with an inert material. Where *piping* is encased in a protective pipe sleeve, the annular space between the *piping* and the sleeve shall be sealed.

❖ Gas piping in any location must be properly supported and protected from physical damage. Protection from damage is especially important where piping is run outdoors near grade or across roof surfaces (see Sections 407 and 415). Piping passing through an outside wall must be protected where the material of the wall could corrode or abrade the piping (concrete, masonry or stucco, for example). To help protect piping from corrosion resulting from

exposure to moisture, the piping must always be located at least $3^1/_2$ inches (89 mm) above the earth and roof surfaces. The distance of $3^1/_2$ inches (89 mm) was chosen because it corresponds to the size of nominal 4-inch by 4-inch (102 mm by 102 mm) pressure-treated lumber which is commonly used to support piping run across roof surfaces.

Where piping is supported on 4-inch by 4-inch (102 mm by 102 mm) wood blocks or similar supports, the piping must be attached to the blocking with suitable clamps or straps. It is common to see wood blocking or piping that has been displaced because the piping was not properly fastened to the blocking. Section 415.1 dictates the minimum required spacing for blocking.

404.10 Isolation. Metallic *piping* and metallic tubing that conveys fuel gas from an LP-gas storage container shall be provided with an *approved* dielectric fitting to electrically isolate the underground portion of the pipe or tube from the above-ground portion that enters a building. Such dielectric fitting shall be installed above ground, outdoors.

❖ Such isolation is common practice for natural gas utility service risers. The required isolation fitting is intended to isolate the piping/tubing in the building from piping extending underground. This section does not require isolation between the LP-storage tank and the piping/tubing running from it to below

Figure 404.8.2
PIPE UNDER SLAB BETWEEN INTERIOR POINTS

ground. In other words, in the typical installation, piping/tubing enters the soil near the tank and rises out of the soil near the building served and the isolation is required where the piping/tubing rises to enter the building. This section is related to Sections 309 and 310. There has been concern raised about bonding metal gas piping/tubing that extends underground and this section is intended to alleviate such concern (see Commentary Figure 404.10).

Figure 404.10
LP-GAS PIPING ISOLATION

404.11 Protection against corrosion. Metallic pipe or tubing exposed to corrosive action, such as soil condition or moisture, shall be protected in an *approved* manner. Zinc coatings (galvanizing) shall not be deemed adequate protection for gas *piping* underground. Where dissimilar metals are joined underground, an insulating coupling or fitting shall be used. *Piping* shall not be laid in contact with cinders.

❖ See the commentary to Section 403.8.

404.11.1 Prohibited use. Uncoated threaded or socket-welded joints shall not be used in *piping* in contact with soil or where internal or external crevice corrosion is known to occur.

❖ Pits, crevices and other surface imperfections can create concentrated corrosion cells where the corrosion effect is amplified. Pipe or fitting failure could occur more quickly than if the corrosion was uniformly distributed over the entire surface of the pipe or fitting (see commentary, Section 404.11.2).

404.11.2 Protective coatings and wrapping. Pipe protective coatings and wrappings shall be *approved* for the application and shall be factory applied.

Exception: Where installed in accordance with the manufacturer's instructions, field application of coatings and wrappings shall be permitted for pipe nipples, fittings and locations where the factory coating or wrapping has been damaged or necessarily removed at joints.

❖ Protective coatings in the form of wrappings, tapes, enamels, epoxies, sleeves, casings and factory-

applied coverings are a common method of isolating metallic piping materials from the atmosphere or from the earth in which they are buried. Corrosion can be caused by soil or construction materials that are in contact with the pipe. Protection is usually provided by a factory-applied coating or by field wrapping the pipe with a protective covering such as a coal-tar-based or plastic wrapping. Where possible, a piping material is chosen that is not subject to the type of corrosion typical of the application. Application of these materials under strictly controlled factory conditions improves the quality control process and reduces the likelihood of coating defects. In the event that coatings or wrappings must be applied in the field, the exception states that they must be installed in accordance with the manufacturer's instructions. Coatings and wrappings should be applied only by persons trained and experienced in such work in order to improve reliability of the coating/wrapping process. Field application is permitted only in cases where the factory-applied material may have been damaged during transit or installation of the pipe, where the applied coating was removed for pipe welding or threading of the pipe or for short sections (nipples) of pipe used in the installation. Corrosion can be concentrated on the pipe or fitting where there are flaws in the coating or wrapping; therefore, improperly applied coatings and wrappings can cause pipe or fitting failures to occur faster than failures would occur in unprotected metals. It is evident why plastic piping is used extensively in underground installations.

404.12 Minimum burial depth. Underground *piping* systems shall be installed a minimum depth of 12 inches (305 mm) below grade, except as provided for in Section 404.12.1.

❖ The depth of 12 inches (305 mm) is considered to be sufficient to avoid possible harm to the pipe from the use of hand tools such as spades and shovels. However, if the piping is located in an area subject to surface loads such as vehicular traffic, the 12-inch (305 mm) depth may not be sufficient to protect the piping from those loads (see Commentary Figure 404.12).

404.12.1 Individual outside appliances. Individual lines to outdoor lights, grills and other *appliances* shall be installed not less than 8 inches (203 mm) below finished grade, provided that such installation is *approved* and is installed in locations not susceptible to physical damage.

❖ Gas piping may be installed within 8 inches (203 mm) of the ground surface where it serves individual outdoor appliances and is not likely to be subjected to damage such as might occur from vehicular traffic, gardening, future excavation, etc. Each individual installation must be reviewed and approved by the code official. The intent is to allow shallow installations for small distribution lines that serve outdoor gas lights, cooking appliances, pool heaters and similar loads only where the piping is unlikely to be disturbed.

For SI: 1 inch = 25.4 mm.

Figure 404.12
MINIMUM BURIAL DEPTH

404.13 Trenches. The trench shall be graded so that the pipe has a firm, substantially continuous bearing on the bottom of the trench.

❖ Where trenches have nonuniform depth or peaks and valleys in the bottom, the piping could lack continuous support and could be subjected to stresses from the backfill and surface loads.

404.14 Piping underground beneath buildings. *Piping* installed underground beneath buildings is prohibited except where the *piping* is encased in a conduit of wrought iron, plastic pipe, steel pipe or other *approved* conduit material designed to withstand the superimposed loads. The conduit shall be protected from corrosion in accordance with Section 404.11 and shall be installed in accordance with Section 404.14.1 or 404.14.2.

❖ This section prohibits the installation of gas piping beneath buildings to reduce the potential for an inaccessible pipe failure caused by settling of the structure. The prohibition also reduces the potential for corrosion-caused failure of piping embedded in soil or fill. See Section 404.6 for gas pipe installations in floor slabs. Where underground installation under a building is unavoidable, the piping must be encased in another pipe (conduit) to act as a secondary containment system or as a protective shield from the fill under the floor (see Commentary Figure 404.8.2).

404.14.1 Conduit with one end terminating outdoors. The conduit shall extend into an occupiable portion of the building and, at the point where the conduit terminates in the building, the space between the conduit and the gas *piping* shall be sealed to prevent the possible entrance of any gas leakage. The conduit shall extend not less than 2 inches (51 mm) beyond the point where the pipe emerges from the floor. Where the end sealing is capable of withstanding the full pressure of the gas pipe, the conduit shall be designed for the same pressure as the pipe. Such conduit shall extend not less than 4 inches (102 mm) outside of the building, shall be vented above grade to the outdoors and shall be installed so as to prevent the entrance of water and insects.

❖ Note the limiting effect that Section 404.6 has on the application of this section (see commentary, Section 404.8.1).

404.14.2 Conduit with both ends terminating indoors. Where the conduit originates and terminates within the same building, the conduit shall originate and terminate in an accessible portion of the building and shall not be sealed. The conduit shall extend not less than 2 inches (51 mm) beyond the point where the pipe emerges from the floor.

❖ See the commentary to Section 404.8.2.

404.15 Outlet closures. Gas *outlets* that do not connect to *appliances* shall be capped gas tight.

   Exception: *Listed* and *labeled* flush-mounted-type quick-disconnect devices and *listed* and *labeled* gas convenience outlets shall be installed in accordance with the manufacturer's instructions.

❖ Unused fuel gas outlets must be capped or plugged gas tight, regardless of whether a shutoff valve is installed at the outlet. A closed valve alone is not dependable and poses an unnecessary risk of leakage, accidental opening and vandalism. The exception recognizes that listed gas outlet devices are built with an inherent safety feature that will automatically shut off the gas flow if the mating connector is disengaged or that will not allow the appliance connector to be disengaged until the integral shutoff valve is manually closed.

404.16 Location of outlets. The unthreaded portion of *piping outlets* shall extend not less than 1 inch (25 mm) through finished ceilings and walls and where extending through floors or outdoor patios and slabs, shall be not less than 2 inches (51 mm) above them. The outlet fitting or *piping* shall be securely supported. *Outlets* shall not be placed behind doors. *Outlets* shall be located in the room or space where the *appliance* is installed.

> **Exception:** *Listed* and *labeled* flush-mounted-type quick-disconnect devices and *listed* and *labeled* gas convenience *outlets* shall be installed in accordance with the manufacturer's instructions.

❖ This section regulates the location, installation and termination of gas piping system outlets to reduce the likelihood of physical damage and to provide clearances for the use of tools. When making connections to piping outlets, "back-up" wrenches are used to prevent piping from rotating, loosening or being damaged. Sufficient pipe length is necessary to allow the application of tools. Gas outlet devices that have been tested and labeled for installation methods other than those addressed in this section must be installed in accordance with the terms of their testing and listing as contained in the manufacturer's installation instructions. Gas outlets for CSST systems must be installed with the termination fitting designed specifically for that purpose and provided by the CSST manufacturer.

404.17 Plastic pipe. The installation of plastic pipe shall comply with Sections 404.17.1 through 404.17.3.

❖ This section places restrictions on the location and operating pressures for plastic pipe in addition to stating installation requirements specific to plastic pipe.

404.17.1 Limitations. Plastic pipe shall be installed outdoors underground only. Plastic pipe shall not be used within or under any building or slab or be operated at pressures greater than 100 psig (689 kPa) for natural gas or 30 psig (207 kPa) for LP-gas.

> **Exceptions:**
> 1. Plastic pipe shall be permitted to terminate above ground outside of buildings where installed in premanufactured anodeless risers or service head adapter risers that are installed in accordance with the manufacturer's instructions.

> 2. Plastic pipe shall be permitted to terminate with a wall head adapter within buildings where the plastic pipe is inserted in a *piping* material for fuel gas use in buildings.

> 3. Plastic pipe shall be permitted under outdoor patio, walkway and driveway slabs provided that the burial depth complies with Section 404.12.

❖ Because of the potential hazard associated with the use of a material that has lower resistance to physical damage and heat as compared to metallic pipe, installation of plastic pipe and tubing is limited to areas that are both outside of the building and underground. Plastic pipe and tubing are widely used for underground gas distribution systems because of their ease of installation and inherent resistance to corrosion. PE and polyamide (PA) SDR 11 pipe are the only allowable plastic pipes for use with LP-gas, and the code user is directed to the referenced standard, NFPA 58, which requires that these piping materials comply with ASTM D2513. Exception 1 makes it clear that if an equivalent level of physical protection is installed, plastic pipe may terminate above ground outside of the building. It is common practice for gas utility companies to install premanufactured riser assemblies at their meter settings. Such risers typically consist of a steel pipe with a corrosion-resistant coating in which a length of PE pipe is preinstalled for coupling to the service lateral. A 90-degree (1.57 rad) sweeping bend is built into the underground end of the riser assembly. Plastic piping is not allowed in or under a concrete slab except as permitted by Exception 3. It is not the intent of this section to prevent an outdoor slab from being placed over customer-owned (private) underground gas lines, such as in the case of walkways, driveways and open patios. The prohibition on piping within a slab is applicable to all slabs without exception. Exceptions 1 and 2 relate to Section 403.6.1; see the commentary for that section (see Commentary Figure 404.17.1).

404.17.2 Connections. Connections made outdoors and underground between metallic and plastic *piping* shall be made only with transition fittings conforming to ASTM D2513 Category I or ASTM F1973.

❖ Because installation of plastic piping is allowed only where both outside and underground, the same requirement holds for joining plastic and metallic piping. Mechanical compression joints employ an elastomeric seal that must be compatible not only with the piping material but also with the gas that is conveyed in the system. The manufacturer's instructions normally require use of an internal stiffener insert in conjunction with the fitting for additional support. The stiffener insert prevents the pipe from deforming under the compression force exerted by the seal and compression nut assembly.

Category 1 fittings are designed to make a gas-tight seal and resist axial (pull-out) forces equal to the

Figure 404.17.1
PIPING UNDER SLABS

yield strength of the piping material [see Commentary Figure 403.11(3)].

404.17.3 Tracer. A yellow insulated copper tracer wire or other *approved* conductor shall be installed adjacent to underground nonmetallic *piping*. *Access* shall be provided to the tracer wire or the tracer wire shall terminate above ground at each end of the nonmetallic *piping*. The tracer wire size shall be not less than 18 AWG and the insulation type shall be suitable for direct burial.

❖ To avoid piping damage and a hazardous condition, the location of underground piping must be known prior to excavation work in the vicinity of such piping.

In the past, gas utility companies, installers and contractors relied on metal detectors to locate gas piping underground. With the advent of nonmetallic (plastic) piping, this is no longer possible. A tracer wire provides a means for locating nonmetallic gas piping without having to excavate, thereby avoiding the possibility of damaging the pipe. An electrical current can be passed through the wire to allow a metal detector to locate the piping. Insulated wire is required because copper wire or other approved conductors may be susceptible to corrosion in some soils.

Yellow is the standardized color used to identify gas piping. Specialized tracer systems made of proprietary materials are available. Tapes that contain a conducting strip and also serve as a visual warning are also available.

404.18 Pipe cleaning. The use of a flammable or combustible gas to clean or remove debris from a *piping* system shall be prohibited.

❖ Such cleaning operations were once used on industrial sites using high-pressure gas. Regardless of the location, fuel gas must never be used as the means to flush debris from gas piping because the discharge of gas from the piping poses a serious threat of explosion.

404.19 Prohibited devices. A device shall not be placed inside the *piping* or fittings that will reduce the cross-sectional area or otherwise obstruct the free flow of gas.

Exceptions:

1. Approved gas filters.

2. An approved fitting or device where the gas piping system has been sized to accommodate the pressure drop of the fitting or device.

❖ Devices such as filters and flow-measuring instruments could create considerable pressure drop and are prohibited except where the piping system is designed to tolerate such losses. An unaccounted for restriction in the piping could make an otherwise properly designed system fail to supply the required pressure. This section is not intended to apply to shutoff valves and excess flow valve devices because it specifically addresses devices installed in the interior of piping as opposed to devices installed "in" piping systems. The second exception addresses devices such as excess flow valves (EFV) that have internal components that can be viewed as being placed inside of pipe or fittings. An EFV will act as a restriction in the piping and the resultant pressure loss from this restriction must be considered to determine if such loss is tolerable and that the load will be supplied with the required minimum inlet pressure.

404.20 Testing of piping. Before any system of *piping* is put in service or concealed, it shall be tested to ensure that it is gas tight. Testing, inspection and purging of *piping* systems shall comply with Section 406.

❖ A pressure test is required after every installation, alteration, addition or repair to the fuel-gas piping system. The location of a leak may be difficult to determine, especially if it is concealed in the building construction. If a leak is found, the leaking component must be repaired or replaced before the system is concealed or put into operation. Section 406 specifies testing pressures based on the type of system, the design working pressure or other parameters. The piping system must sustain the test pressure for the test duration period without exhibiting any sign of leakage.

SECTION 405 (IFGS)
PIPING BENDS AND CHANGES IN DIRECTION

405.1 General. Changes in direction of pipe shall be permitted to be made by the use of fittings, factory bends or field bends.

❖ This section permits making changes in direction of gas piping through the use of pipe fittings compatible with the piping material or, if recommended by the pipe manufacturer, by bending the pipe. If bending is the chosen method, it must be accomplished in accordance with Sections 405.2 and 405.3. In practice, steel gas piping is rarely bent to accomplish changes in direction because pipe failure at a seam is a possibility for steel pipe that is bent.

405.2 Metallic pipe. Metallic pipe bends shall comply with the following:

1. Bends shall be made only with bending tools and procedures intended for that purpose.

2. All bends shall be smooth and free from buckling, cracks or other evidence of mechanical damage.

3. The longitudinal weld of the pipe shall be near the neutral axis of the bend.

4. Pipe shall not be bent through an arc of more than 90 degrees (1.6 rad).

5. The inside radius of a bend shall be not less than six times the outside diameter of the pipe.

❖ Pipe is bent to accomplish changes in direction without the use of fittings. Pipe must be bent with the appropriate bending tools and materials. Some materials are not tempered or intended to be bent; therefore, consideration must be given to choosing the proper materials. Bending tools are designed to make bends without damaging the pipe. Pipe intended to be bent has a minimum bend radius which must be observed to avoid damage to the pipe. For example, bending welded seam pipe with the seam located outside of the neutral axis of the bend may result in a split seam because of the stresses induced from the bend. In other words, the welded seam must not be oriented in a bend that would place it in tension or compression. If pipe is to be bent, it should be stated in the pipe specifications that the pipe is suitable for bending. Rigid gas piping is not commonly bent because of the perceived risk of pipe stress failures at the bend and because bending can be more labor intensive than using fittings.

405.3 Plastic pipe. Plastic pipe bends shall comply with the following:

1. The pipe shall not be damaged and the internal diameter of the pipe shall not be effectively reduced.

2. Joints shall not be located in pipe bends.

3. The radius of the inner curve of such bends shall be not less than 25 times the inside diameter of the pipe.

4. Where the *piping* manufacturer specifies the use of special bending tools or procedures, such tools or procedures shall be used.

❖ Because of the material characteristics of plastic pipe, the code defers to the manufacturer's installation and bending instructions to achieve the performance level contemplated by this section. For example, coiled plastic pipe and tubing have limitations regarding bending beyond or against the natural curvature of the coil. A joint in a bend would be subject to bending stresses and pipe/fitting misalignments that could cause joint structural failure and/or leakage. Thus, a 1-inch (25.4 mm) ID plastic pipe would have a minimum bending radius of 25 inches (635 mm) measured to the inner curve of the bend.

405.4 Elbows. Factory-made welding elbows or transverse segments cut therefrom shall have an arc length measured along the crotch of not less than 1 inch (25 mm) in pipe sizes 2 inches (51 mm) and larger.

❖ Welding elbows of various angles up to 90 degrees (1.57 rad) are commonly used to make changes in direction in steel piping systems. The minimum length of the arc on the inside radius of the fitting must be 1 inch to provide enough metal and room to make a proper weld.

SECTION 406 (IFGS)
INSPECTION, TESTING AND PURGING

406.1 General. Prior to acceptance and initial operation, all *piping* installations shall be visually inspected and pressure tested to determine that the materials, design, fabrication and installation practices comply with the requirements of this code.

❖ Before any gas piping system is put to use, it must be inspected, tested and approved (see Sections 406.1.1 through 406.6.4). Pressure testing is distinct from leakage checking as covered in Section 406.6.

406.1.1 Inspections. Inspection shall consist of visual examination during or after manufacture, fabrication, assembly or pressure tests.

❖ The inspection of piping installations is intended to be a visual observation of the system and the testing procedure. The designer may require that welded joints in, for example, very high pressure applications be examined by a method that is capable of discovering internal defects that are not detectable by visual observation.

406.1.2 Repairs and additions. In the event repairs or additions are made after the pressure test, the affected *piping* shall be tested.

Minor repairs and additions are not required to be pressure tested provided that the work is inspected and connections are tested with a noncorrosive leak-detecting fluid or other *approved* leak-detecting methods.

❖ Some time after the initial test of the piping, a repair could be made or a branch could be added or extended. The repaired, added or extended portion of the piping system needs to be tested without requiring the unchanged portions to be tested again. Minor work on an existing system is allowed without pressure testing the minor work if the work is visually inspected and tested by a leak-detecting method. Leak-detecting methods include bubble test fluids and electronic sensors. Some bubble test fluids could be corrosive to some piping; therefore, a noncorrosive fluid designed for this purpose should always be used. Methods other than bubble test fluids must be acceptable to the code official.

406.1.3 New branches. Where new branches are installed to new *appliance*s, only the newly installed branches shall be required to be pressure tested. Connections between the new *piping* and the existing *piping* shall be tested with a noncorrosive leak-detecting fluid or other *approved* leak-detecting methods.

❖ This section applies to all new branches regardless of where in the piping system they connect. Often, because it is convenient or because the existing piping is fully loaded, a new piping branch will be run

from the point of delivery to serve a new appliance installation. The new branch is typically taken from a tee fitting installed immediately downstream of the meter. The entire new branch piping must be tested, including the tee fitting in the existing piping from which the new branch is supplied. The point of connection to the existing piping can be tested by a means other than pressure testing, although all other piping in the new branch must be isolated from the existing piping and pressure tested as required for new work. Section 406.1.4 requires that the new branch piping be disconnected from the existing piping during pressure testing, except where an assembly consisting of two closed valves in series with an intermediate open valved port is installed to isolate the new piping from the existing piping (see Commentary Figure 406.1.4).

406.1.4 Section testing. A *piping* system shall be permitted to be tested as a complete unit or in sections. Under no circumstances shall a valve in a line be used as a bulkhead between gas in one section of the *piping* system and test medium in an adjacent section, except where a double block and bleed valve system is installed. A valve shall not be subjected to the test pressure unless it can be determined that the valve, including the valve-closing mechanism, is designed to safely withstand the test pressure.

❖ Depending on the progression of a job, it may be desirable to test portions of a system as they are completed. It is also possible that portions of a system will be put in service before the entire system is completed. To prevent a test medium from leaking into piping containing fuel gas or vice versa, portions of piping under test must be isolated from portions that are in service. A single valve cannot be depended on to establish this isolation because all valves have some "leak-through" allowance, and also the valve could be accidentally opened. To eliminate this risk, two valves in series must be used, and a valved open port (tell-tale) must be installed between the valves. This arrangement is called a double block and bleed valve system. The open port will release any gases that have passed through either isolation valve and also will serve to indicate any leakage or accidental opening of an isolation valve (see Commentary Figure 406.1.4).

406.1.5 Regulators and valve assemblies. Regulator and valve assemblies fabricated independently of the *piping* system in which they are to be installed shall be permitted to be tested with inert gas or air at the time of fabrication.

❖ Assemblies of valves, regulators and/or over-pressure devices might be factory or shop assembled and may be tested independently of the piping system in which they will be installed.

406.1.6 Pipe clearing. Prior to testing, the interior of the pipe shall be cleared of all foreign material.

❖ Piping will contain debris from pipe storage, handling and assembly including soil, construction site trash, pipe joint compound, welding debris, threading and reaming scrap and even tools. Clearing the piping of

Figure 406.1.4
PIPING ISOLATION ARRANGEMENT FOR TESTING

foreign materials will prevent such materials from entering system controls, devices and appliances and will also reduce the chances of injury to personnel from ejected debris and objects (see Section 404.18).

406.2 Test medium. The test medium shall be air, nitrogen, carbon dioxide or an inert gas. Oxygen shall not be used.

❖ The gas that is forced into the piping system for pressure testing must be chemically unreactive to prevent undesirable reactions with the piping system components and the fuel to be conveyed in the piping. Oxygen is not inert, and any residual amount could form an explosive mixture with the fuel gas. The typical test gas is air, which is also not inert but is readily available and free. Nitrogen and carbon dioxide are considered to be inert for the purpose of testing piping systems and are inexpensive compared to most truly inert gases.

406.3 Test preparation. Pipe joints, including welds, shall be left exposed for examination during the test.

    Exception: Covered or concealed pipe end joints that have been previously tested in accordance with this code.

❖ This section is consistent with Section 107.1 and requires piping joints to remain exposed during testing to allow locations of defects. This section speaks only of joints, but Section 107.1 requires that the piping and joints be left exposed until they are inspected and approved.

406.3.1 Expansion joints. Expansion joints shall be provided with temporary restraints, if required, for the additional thrust load under test.

❖ If the piping system uses expansion joints to accommodate piping movement, those joints might need to be restrained to resist the additional thrust loading caused by the movement of the pressurized test gas. Generally, testing does not produce noticeable dynamic thrust forces because of the slow rate at which the test gas is put in and taken out of the piping system. The thrust force produced by the static pressure of the test gas is of concern. Expansion joints

are rarely used because expansion loops and offsets perform the same function.

406.3.2 Appliance and equipment isolation. *Appliances* and *equipment* that are not to be included in the test shall be either disconnected from the *piping* or isolated by blanks, blind flanges or caps. Flanged joints at which blinds are inserted to blank off other *equipment* during the test shall not be required to be tested.

❖ Some equipment can be damaged by test pressure and is therefore disconnected from the piping system or is otherwise isolated by blank-offs placed between the flanges.

406.3.3 Appliance and equipment disconnection. Where the *piping* system is connected to *appliances* or *equipment* designed for operating pressures of less than the test pressure, such *appliances* or *equipment* shall be isolated from the *piping* system by disconnecting them and capping the outlet(s).

❖ This requirement is often overlooked in the field, which could lead to appliance damage and serious consequences. Appliances are typically designed for a maximum inlet gas pressure of $^1/_2$ psig (3.5 kPa) or less, above which damage to controls and regulators could occur. The test pressure will be at least 3 psig (20.7 kPa) and oftentimes much higher in accordance with general practice or local traditions or requirements. Installers routinely pressurize piping systems to pressures between 50 and 100 psig (345 and 690 kPa), often at the insistence of the local authority. It is obvious that there is a good potential to expose appliances to extreme pressures during a test, which is why the code requires the precaution of disconnecting the appliances from the supply piping. Disconnection is normally accomplished by opening a union or disconnecting the appliance connector. The appliance shutoff valve must not be used as a means of appliance isolation because the valve can leak through and/or can be accidentally opened or left open (see Section 406.3.4 and Commentary Figure 406.3.3).

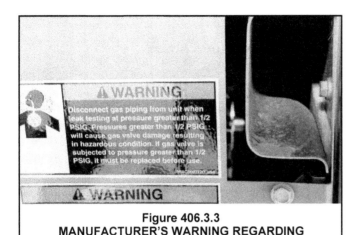

Figure 406.3.3
MANUFACTURER'S WARNING REGARDING
PRESSURE TESTING OF GAS PIPING

406.3.4 Valve isolation. Where the *piping* system is connected to *appliance*s or *equipment* designed for operating pressures equal to or greater than the test pressure, such *appliance*s or *equipment* shall be isolated from the *piping* system by closing the individual *appliance* or *equipment* shutoff valve(s).

❖ If the appliances and other system components are not subject to damage by the test pressure, the equipment shutoff valve can serve to isolate the equipment during the test. Unlike Section 406.3.3, this section does not require disconnection of the appliance/equipment. This section would probably be applicable only in an industrial occupancy. This author believes that no residential or commercial appliance can be safely subjected to even the minimum test pressure of 3 psi; therefore, it is always prudent to disconnect all appliances prior to applying test pressure.

406.3.5 Testing precautions. Testing of *piping* systems shall be performed in a manner that protects the safety of employees and the public during the test.

❖ Compressed gases store energy in direct proportion to the pressure. This stored energy can be destructive and dangerous to personnel if released quickly because of joint, component or pipe failure or an intentional action. Debris will be found in all piping systems after installation. This debris can be harmful to appliances and system components and can act as projectiles during testing and purging operations. Piping is typically purged before testing to flush out foreign materials. Experience has shown that using fuel gas to flush out foreign materials in piping systems can be a dangerous practice.

406.4 Test pressure measurement. Test pressure shall be measured with a manometer or with a pressure-measuring device designed and calibrated to read, record or indicate a pressure loss caused by leakage during the pressure test period. The source of pressure shall be isolated before the pressure tests are made. Mechanical gauges used to measure test pressures shall have a range such that the highest end of the scale is not greater than five times the test pressure.

❖ Test pressures must be measured with a manometer or other pressure-measuring device that is designed and calibrated to indicate a pressure loss caused by leakage during the pressure test period. Pressure-measuring devices must be graduated so that small variations in test pressure will be readily detectable. Small pressure variations might go undetected in instruments having broader ranges. Gauges should have scale increments that are small enough to detect small changes in pressure. Mechanical gauges generally achieve their best accuracy near mid-range of their scale. All test instruments should have been calibrated recently and their accuracy verified before starting the test. Although no piping system can be absolutely leak free, the intent of the code is to make the system free of leaks that are measurable with the testing procedures and instruments prescribed in the code and attainable in the field. Spring-type mechanical pressure gauges must have a high-end scale reading of no greater than five times the test pressure. If the test pressure was 3 psig (20.7 kPa), for example, the gauge would have to have a maximum range of 0 to 15 psig (0 to 104 kPa). Such gauges are a wise choice because the test reading will be higher in the range of the gauge where the gauge accuracy is better. Also, the gauge can have much smaller increments than the typical 100 psig (690 kPa) gauge, thus allowing greater resolution and discernability in detecting a leak [see Commentary Figure 406.4(1)]. All spring-type mechanical gauges are susceptible to inaccuracies caused by conditions of use and mishandling and should be calibrated regularly. Manometers are very accurate and do not need to be calibrated. In all cases, the intent of this section is to require an instrument that is capable of indicating the very small pressure changes that occur as a result of a detectable leak. Test instrument sensitivity is especially important considering the short duration of the test.

The source of pressure in the piping system under test must be disconnected from the system to prevent the source from invalidating the test. For example, the compressed air tank or inert gas cylinder could be adding gas to the piping system at the same rate as a leak is letting gas escape; the test would then be meaningless.

Commentary Figures 406.4(2), 406.4(3) and 406.4(4) show examples of test instruments that can be considerably more accurate than mechanical spring gauges. Electronic instruments can provide high-resolution readings, making very small leaks as well as regular leaks detectable within short time periods. Commentary Figures 406.4(3) and 406.4(4) show instruments that are designed to give accurate pressure readings and give positive indication of a leak. The liquid manometer-type instrument, shown in

Commentary Figure 406.4(3), functions as a manometer that readily indicates test pressure and leakage. A separate feature of the instrument allows small leaks to be recognized quickly. While the system is under test pressure, an integral valve can be closed that will isolate the liquid reservoir from the piping system, thereby trapping the test pressure in the manometer's liquid reservoir. A small diameter vertical dip tube, having its open end submerged in the liquid, rises above the liquid level in the reservoir. The top end of the dip tube is connected to the piping system under test. If there is no leakage, the pressures at the upper end and lower (submerged) end of the dip tube are equal and liquid will not rise. If leakage occurs, the pressure in the piping system will fall below the pressure trapped in the reservoir at the start of the test procedure and liquid will be forced up into the small diameter dip tube, indicating the leak. The use of a small diameter dip tube allows a small change in liquid level in the reservoir to be seen as a comparatively large movement of liquid in the dip tube.

406.4.1 Test pressure. The test pressure to be used shall be not less than $1^1/_2$ times the proposed maximum working pressure, but not less than 3 psig (20 kPa gauge), irrespective of design pressure. Where the test pressure exceeds 125 psig (862 kPa gauge), the test pressure shall not exceed a value that produces a hoop stress in the *piping* greater than 50 percent of the specified minimum yield strength of the pipe.

❖ The minimum test pressure will never be less than 3 psig (20 kPa). The majority of residential and small commercial occupancies are served by piping systems with a pressure of less than $^1/_2$ psig [(14 inches water column) (3.5 kPa)], and for such systems the test pressure must be 3 psig (20 kPa) or higher. The test pressure must also be not less than 1.5 times the pressure at which the piping system is designed to operate. If the piping system is designed to operate at 5 psig (34.5 kPa), the test pressure must be not less than 1.5 x 5 = 7.5 psig (51.7 kPa). In very high pressure applications, the test pressure might approach the structural limits of the piping material, and, therefore, the test pressure must be limited. Hoop stress is the result of internal pressure that tends to expand the pipe walls outward along the circumference of the pipe.

406.4.2 Test duration. Test duration shall be not less than $^1/_2$ hour for each 500 cubic feet (14 m³) of pipe volume or fraction thereof. When testing a system having a volume less than 10 cubic feet (0.28 m³) or a system in a single-family dwelling, the test duration shall be not less than 10 minutes. The duration of the test shall not be required to exceed 24 hours.

❖ As the piping system becomes larger, the test duration must be longer so that any leak can be detected. For example, a leakage rate of 1 cubic foot per hour (472 cm³/min) would produce a more easily detectable pressure drop in a small 10 cubic foot (0.3 m³) volume system than it would in a large 500 cubic foot (14.2 m³) volume system. It would require a longer

Figure 406.4(1)
TYPICAL TEST SET-UP FOR AIR PRESSURE TESTING

Figure 406.4(2)
ELECTRONIC PRESSURE GAUGE
(Photo courtesy of Dwyer Instruments, Inc.)

test period to detect the leak in the large volume system because the leakage of 1 cubic foot represents a much smaller fraction of the total gas volume in the system. If a system has a volume of 501 cubic feet, the minimum test duration would be 1 hour. To put this in perspective, 500 cubic feet of piping system internal volume would equate to 83,333 feet (25 400 m) of 1-inch (25 mm) Schedule 40 steel pipe or 9,747 feet (2971 m) of 3-inch (76 mm) Schedule 40 steel pipe.

Because 10 cubic feet (0.3 m³) equates to 1,667 feet (508.1 m) of 1-inch (25 mm) pipe, it is obvious that single-family dwellings fall in the 10 cubic foot or less volume category. A 10-minute test period is not much time to detect a pressure drop, which makes it even more important to use a pressure instrument that is capable of indicating very small drops in pressure (see commentary, Section 406.4).

Figure 406.4(3)
LEAK DETECTOR INSTRUMENT
(Photo courtesy of KUHLMAN Instrument Company)

Figure 406.4(4)
ELECTRONIC LEAK DETECTOR INSTRUMENT
(Photo courtesy of KUHLMAN Instrument Company)

406.5 Detection of leaks and defects. The *piping* system shall withstand the test pressure specified without showing any evidence of leakage or other defects.

Any reduction of test pressures as indicated by pressure gauges shall be deemed to indicate the presence of a leak unless such reduction can be readily attributed to some other cause.

❖ Any pressure drop, no matter how small, is considered a failure of the test, except where the drop can be shown to result from a change in temperature or other cause. This would be practically impossible to demonstrate in the field, especially for short test durations. A true leak should show up as a continuous drop in test pressure. If the rate of pressure drop slows as the test progresses, it may be caused by cooling of the test medium, which causes it to contract. For example, warm air coming from a compressor will contract when introduced into a cold piping system, thereby causing a test instrument to indicate a drop in pressure. To help eliminate any guesswork, the temperature of the test medium should be allowed to stabilize within the piping system before the test begins.

406.5.1 Detection methods. The leakage shall be located by means of an *approved* gas detector, a noncorrosive leak detection fluid or other *approved* leak detection methods.

❖ Once the pressure measuring instrument indicates leakage, the leak or leaks must be located for repair or replacement. Electronic sensors and bubble fluids are used to find leaks that are not readily found by human senses. Many people who have worked with and/or inspected gas piping installations can recall instances of personal injury, fires and close calls that were caused by searching for leaks with an open flame. Open flames must never be used for leak detection because of the explosion danger and because open flames will not detect small leaks.

406.5.2 Corrections. Where leakage or other defects are located, the affected portion of the *piping* system shall be repaired or replaced and retested.

❖ Where a leak is detected by a method named in Section 406.5.1, the defect must be corrected. Corrections include tightening of fittings or threaded joints and replacing defective pipe, tubing or fittings.

406.6 Piping system and equipment leakage check. Leakage checking of systems and *equipment* shall be in accordance with Sections 406.6.1 through 406.6.4.

❖ Leakage checking is different from pressure testing. Sections 406.1 through 406.5 address the pressure testing of newly installed or altered piping systems for the purpose of locating defects in the installation or materials. This section addresses leakage checks that are intended to discover open outlets, defective appliance connections and defects that have developed since the initial installation (see Section 406.6.2).

406.6.1 Test gases. Leak checks using fuel gas shall be permitted in *piping* systems that have been pressure tested in accordance with Section 406.

❖ Piping systems are pressure tested with air or an inert gas, and leakage checking is done with the fuel gas at whatever pressure the system is designed to operate (see Section 406.6.2).

406.6.2 Before turning gas on. During the process of turning gas on into a system of new gas *piping*, the entire system shall be inspected to determine that there are no open fittings or ends and that all valves at unused outlets are closed and plugged or capped.

❖ This section requires that a precautionary measure be taken to prevent fuel gas from escaping from the piping system when the gas is turned on in the system for the first time. Because the piping system has never been used, it is not to be fully trusted as being gas-tight. Because anything could have been done to the system since it was pressure tested, an additional visual check is justified before the gas is turned on. The condition of appliances may also be unknown, therefore, it may be wise to make sure that all appliance shutoff valves are closed until the appliances are ready to be put in service. Often, parties other than those who installed and tested the piping system are involved with connecting to the outlets in the system. This makes it likely that what was gas tight at the initial pressure testing is no longer gas tight as a result of incomplete or improper connections at the system outlets. Consistent with Section 404.15, all unused outlets must be capped or plugged.

406.6.3 Leak check. Immediately after the gas is turned on into a new system or into a system that has been initially restored after an interruption of service, the *piping* system shall be checked for leakage. Where leakage is indicated, the gas supply shall be shut off until the necessary repairs have been made.

❖ This section is a logical progression from the previous section. After the visual check for the status of outlets and equipment shutoff valves, the gas is introduced into the system, and the system is checked for leakage. This checking is typically accomplished by examining all the outlets, observing a gas meter for indication of flow, monitoring the system pressure for any drop or any combination of these. Because all outlet valves are closed and unused outlets are capped, any gas flow or pressure drop indicates leakage, and the gas must be turned off. This section would also allow leakage checking with the appliance shutoff valves open and the appliances shut down to prevent their operation. Checking in this manner would verify that the appliance gas controls were not faulty. A leakage check is also required after gas service is restored.

406.6.4 Placing appliances and equipment in operation. *Appliances* and *equipment* shall not be placed in operation until after the *piping* system has been checked for leakage in accordance with Section 406.6.3, the *piping* system has been

purged in accordance with Section 406.7 and the connections to the appliances have been checked for leakage.

❖ Gas appliances must not be put into service until the gas piping system that supplies them has been checked for leakage and purged of air and the appliance connections have been checked for leakage (see Section 406.7.2).

406.7 Purging. The purging of piping shall be in accordance with Sections 406.7.1 through 406.7.3.

❖ Purging is intended to prevent a flammable gas/air mixture from being created in the piping. A flammable mixture could be a fire/explosion hazard in the piping, room or space, or in an appliance combustion chamber. There have been documented accidents involving explosions caused by failure to take the necessary precautions during piping purging operations. It has been purported that some accidents were caused by a condition referred to as "odor fade." It is a fact that steel pipe can absorb the chemical additive in fuel gas that gives the gas (natural or LP) its distinctive odor. This is evidenced by the fact that steel gas piping will smell like gas for a long time after it is no longer exposed to gas. Odorant fade appears to happen when gas is introduced into a new piping system that has never before contained gas. There can be other causes of odorant fade. Once a piping system has been conditioned (seasoned) by exposure to gas, its ability to absorb the odorant diminishes with time. If personnel purge piping into an enclosed area and depend on their sense of smell to detect when gas has reached the purge discharge outlet, they could be fooled if the gas has lost its odor and a serious accident could result from such a dangerous practice. Also, it is known that not all humans have the same ability to detect the odorants in fuel gases. The sense of smell should never be relied upon to detect the presence of fuel gas. Using one's sense of smell to detect fuel gas is subjective and unreliable at best, therefore this section prescribes methods that do not depend on the sense of smell. Combustible gas indicators and detectors are commonly used to detect gas during purging operations. See the commentary to Sections 406.7.1 through 406.7.3.

406.7.1 Piping systems required to be purged outdoors. The purging of piping systems shall be in accordance with the provisions of Sections 406.7.1.1 through 406.7.1.4 where the *piping* system meets either of the following:

1. The design operating gas pressure is greater than 2 psig (13.79 kPa).

2. The piping being purged contains one or more sections of pipe or tubing meeting the size and length criteria of Table 406.7.1.1.

❖ This section requires that piping systems meeting either or both of the stated criteria comply with the listed subsections. Note that Section 406.7.1.3 requires that the piping be purged only to the outdoors. The two criteria (and those of Section 406.7.2) were chosen to distinguish the piping systems typically found in industrial, large commercial and large multiple-family residential buildings from those piping systems typically found in small commercial and small residential buildings. The gas piping systems typically installed in industrial, large commercial and large multiple-family occupancies have larger internal pipe volumes and/or higher flow rates that justify special procedures to avoid creating explosive gas/air mixtures within buildings and within the piping. Combustible gas/air mixtures can be the source of explosions in the building or within the piping itself. Because there is no correlation between the occupancy groups established by the IBC and the size and nature of gas piping systems installed in buildings, occupancy groups could not be used to distinguish between the larger, more complex piping systems and the smaller less complex systems. In other words, the operating pressure and size of piping systems are much more dependent upon the type of appliances to be served than dependent upon the occupancy classification of the building. Thus, the criteria of this section and Section 406.7.2 represent the best attempt to establish a threshold for requiring piping systems to be purged to the outdoors. Note that the gas venting and gas displacement requirements of Sections 406.7.1.1 and 406.7.1.2 also apply to piping systems meeting the criteria or criterion of this section.

Criterion 2 is met if any portion of the piping system is larger than 2 inches nominal in size and also longer than 50 feet for $2^1/_2$-inch pipe, 30 feet for 3- and $3^1/_2$-inch pipe, 15 feet for 4- and 5-inch pipe, 10 feet for 6-inch pipe and zero feet for 8-inch and larger pipe. In other words, if any portion of the piping system meets the criteria of Table 406.7.1.1, the piping system meets Criterion 2 of this section. For example, if a piping system is constructed of $1^1/_2$- and 2-inch piping, but contains one section of 3-inch pipe that is over 30 feet in length, that system meets Criterion 2; however, if that same section of 3-inch pipe is 30 feet or less in length, the system does not comply with Criterion 2 (i.e., 30 feet is not >30 feet).

If the operating pressure of the system is greater than 2 psig, Sections 406.7.1.1 through 406.7.1.4 apply regardless of the sizes of pipe in the system.

406.7.1.1 Removal from service. Where existing gas piping is opened, the section that is opened shall be isolated from the gas supply and the line pressure vented in accordance with Section 406.7.1.3. Where gas *piping* meeting the criteria of Table 406.7.1.1 is removed from service, the residual fuel gas in the *piping* shall be displaced with an inert gas.

❖ Where a piping system, to which this section applies, is to be opened for any reason, it must be isolated from the gas supply and then the pressure in the piping system must be relieved by venting the gas to the outdoors in accordance with Section 406.7.1.3. If the piping being opened or retired meets the criteria of Table 406.7.1.1, the unpressurized gas in the piping must be displaced with an inert gas (i.e., a gas that will not support combustion such as nitrogen or argon). The displacement with inert gas will prevent

an explosive mixture from forming inside of the pipe by eliminating any residual gas that could mix with the ambient air. The larger and longer the pipe, the greater the amount of residual gas in the pipe and the greater the potential for the formation of an explosive mixture.

TABLE 406.7.1.1
SIZE AND LENGTH OF PIPING

NOMINAL PIPE SIZE (inches)[a]	LENGTH OF PIPING (feet)
$\geq 2^1/_2 < 3$	> 50
$\geq 3 < 4$	> 30
$\geq 4 < 6$	> 15
$\geq 6 < 8$	> 10
≥ 8	Any length

For SI: 1 inch = 25.4 mm, 1 foot = 304.8 mm.

a. CSST EHD size of 62 is equivalent to nominal 2-inch pipe or tubing size.

406.7.1.2 Placing in operation. Where gas *piping* containing air and meeting the criteria of Table 406.7.1.1 is placed in operation, the air in the *piping* shall first be displaced with an inert gas. The inert gas shall then be displaced with fuel gas in accordance with Section 406.7.1.3.

❖ Where new piping or piping that has been left open for some time is to be put into service, the air that exists in such piping must be removed to prevent an explosive gas/air mixture from forming inside of the piping. This is accomplished by first displacing the air with a gas that does not support combustion (inert gas) and then introducing fuel gas to displace the inert gas. This procedure prevents fuel gas from ever coming into contact with air inside of the piping. As the size and length of the piping increases, so does the volume of air in it and so does the potential for forming explosive mixtures within the piping. To help determine when sufficient inert gas has been introduced during the process of displacing air with an inert gas, the oxygen levels in the piping system could be monitored.

406.7.1.3 Outdoor discharge of purged gases. The open end of a *piping* system being pressure vented or purged shall discharge directly to an outdoor location. Purging operations shall comply with all of the following requirements:

1. The point of discharge shall be controlled with a shut-off valve.

2. The point of discharge shall be located not less than 10 feet (3048 mm) from sources of ignition, not less than 10 feet (3048 mm) from building openings and not less than 25 feet (7620 mm) from mechanical air intake openings.

3. During discharge, the open point of discharge shall be continuously attended and monitored with a combustible gas indicator that complies with Section 406.7.1.4.

4. Purging operations introducing fuel gas shall be stopped when 90 percent fuel gas by volume is detected within the pipe.

5. Persons not involved in the purging operations shall be evacuated from all areas within 10 feet (3048 mm) of the point of discharge.

❖ This section applies to piping systems meeting the criteria of Section 406.7.1 and mandates that the purged gases discharge directly to the outdoors. Discharging purged gases to the building interior is prohibited by this section. Purging to the outdoors can be accomplished by running temporary piping or hoses to the outdoors from the piping being purged or it could be accomplished by installing permanent piping to the outdoors in locations where purging will be repeated regularly. It could also be accomplished by using existing piping that extends to the outdoors for some other purpose, but that is suitable for and able to be reconfigured for the purpose of gas pipe purging. Other configurations are possible. The concept has been proposed by gas industry pundits and some day, the code might contain provisions for purging air from piping by means of evacuation of the air with vacuum pumps and then breaking the vacuum with fuel gas such that the piping will contain nearly 100 percent fuel gas by volume. Purging by vacuum evacuation is only conceptual at this point in time.

Item 1 requires a means to control the purge gas discharge in the event of an emergency. Item 2 intends to prevent purged gases from reaching an ignition source and from entering a building. Item 3 requires that the point of discharge be constantly attended by one or more individuals who are monitoring the discharge gases with an instrument meeting the requirements of Section 406.7.1.4. Item 4 requires purging operations that are introducing fuel gas into the piping to be stopped immediately after determining that the concentration of fuel gas within the purge gas discharge pipe is 90 percent of the purge gas volume. This item intends to limit the volume of purged gases that discharge to the outdoor atmosphere. When the content of the piping is 90% fuel gas by volume, an explosive mixture is not possible within the piping interior and the piping content upstream of the discharge point is expected to be 90 percent fuel gas or higher. Item 5 intends to protect personnel and passersby from the potential hazard of purge gas discharge and also intends to prevent an individual from bringing an ignition source near the point of discharge.

406.7.1.4 Combustible gas indicator. Combustible gas indicators shall be listed and shall be calibrated in accordance with the manufacturer's instructions. Combustible gas indicators shall numerically display a volume scale from zero percent to 100 percent in 1 percent or smaller increments.

❖ Note that a combustible gas indicator is a different instrument than the instrument addressed in Section 406.7.2.2. A CGI is required to be used by Item 3 of Section 406.7.1.3. This instrument must be capable of indicating the percentage of fuel gas in the purge gas discharge and do so by a number scale of 0 to 100 percent. The display resolution requirement means that the instrument must be able to display

percentages across the entire scale with not more than 1 percent between successive numbers. The increments on the 0 to 100 percent scale must be 1 percent or smaller.

In other words, the percentage display must be able to read out 1 percent, 2 percent, 3 percent, 4 percent, 5 percent, etc., all the way up to 100 percent. These instruments need to be calibrated routinely because the nature of the sensors used in these instruments is such that they drift over time and with use.

406.7.2 Piping systems allowed to be purged indoors or outdoors. The purging of piping systems shall be in accordance with the provisions of Section 406.7.2.1 where the piping system meets both of the following:

1. The design operating gas pressure is 2 psig (13.79 kPa) or less.

2. The piping being purged is constructed entirely from pipe or tubing not meeting the size and length criteria of Table 406.7.1.1.

❖ This section addresses purging procedures that allow the purged gases to discharge to the outdoors or to the indoors. The Criteria Items 1 and 2 represent the division between the larger, more complex piping systems and the smaller, less complex piping systems (see commentary, Section 406.7.1). In this section, the piping system must meet both criteria, meaning that the operating pressure of the system must be 2 psig or less and the piping must not meet the criteria of Table 406.7.1.1. Not meeting the criteria of Table 406.7.1.1 means that the piping system is either constructed entirely of 2-inch or smaller nominal size pipe, or the piping is 50 feet or less in length for $2^1/_2$-inch pipe, 30 feet or less in length for 3-inch pipe, 15 feet or less in length for 4-inch pipe, 10 feet or less in length for 6-inch pipe and zero feet in length for 8-inch and longer pipe. For example, if a piping system is constructed of 2-inch and smaller pipe but contains one section of 4-inch pipe 15 feet long, that system does not meet the criteria of Table 406.7.1.1 and thus meets Criterion 2 of this section (i.e., 15 feet is shorter than >15 feet). If **both** criteria of this section are met, the piping system is allowed to be purged in accordance with Section 406.7.2.1; however, if **both** criteria are not met, purging must comply with Sections 406.7.1.1 through 406.7.1.4.

Other examples: A piping system that operates at 2 psig and is constructed of all 2-inch pipe meets both Criterion 1 and 2. A piping system that operates at 2 psig and is constructed of 2-inch and smaller pipe with one 55-foot section of $2^1/_2$-inch pipe meets Criterion 1 but does not meet Criterion 2 because the 55-foot section of $2^1/_2$-inch falls within the size and length criteria of the table.

406.7.2.1 Purging procedure. The piping system shall be purged in accordance with one or more of the following:

1. The piping shall be purged with fuel gas and shall discharge to the outdoors.

2. The piping shall be purged with fuel gas and shall discharge to the indoors or outdoors through an appliance burner not located in a combustion chamber. Such burner shall be provided with a continuous source of ignition.

3. The piping shall be purged with fuel gas and shall discharge to the indoors or outdoors through a burner that has a continuous source of ignition and that is designed for such purpose.

4. The piping shall be purged with fuel gas that is discharged to the indoors or outdoors, and the point of discharge shall be monitored with a listed combustible gas detector in accordance with Section 406.7.2.2. Purging shall be stopped when fuel gas is detected.

5. The *piping* shall be purged by the gas supplier in accordance with written procedures.

❖ The purging provisions of this section are applicable only where the piping system meets both criteria of Section 406.7.2. These provisions are intended for the piping systems typically found in light commercial and small residential occupancies (see commentary, Section 406.7.1). Item 1 allows the simple procedure of purging air in a pipe by displacing it with fuel gas and discharging the purged gases to the outdoors. Item 2 allows a procedure used by some installers and gas suppliers in which air in the piping is displaced with fuel gas and the purged gases are discharged through an open appliance burner, typically on a range or cooktop. Any such open burner must have a continuous ignition source that is either integral with the burner or that is temporarily provided by the personnel performing the purging operation. If the purged gases were discharged into an enclosed combustion chamber such as in a furnace, boiler or water heater, an explosive gas/air mixture could form in the presence of an ignition source resulting in a hazard. Item 3 is basically the same procedure as specified in Item 2, except that the burner is not an appliance burner. Some installers and gas suppliers utilize special purge burners designed for this procedure. Item 4 allows the purged gases to discharge to the indoors or to the outdoors provided that the point of discharge is continuously monitored by the personnel performing the purging operation, using a combustible gas detector (CGD) that complies with Section 406.7.2.2. The purging operation must be immediately stopped when the CGD indicates the presence of fuel gas at the point of discharge. Item 5 allows personnel employed by the gas supplier to perform purging operations in accordance with detailed procedures that are in writing and originating from the gas supplier. This item assumes that the written procedures are based on practices that have been proven by experience to be safe.

406.7.2.2 Combustible gas detector. Combustible gas detectors shall be listed and shall be calibrated or tested in accordance with the manufacturer's instructions. Combustible gas

detectors shall be capable of indicating the presence of fuel gas.

❖ A CGD is different than the CGI required by Section 406.7.1.4; however, both of these instruments need to be calibrated or tested routinely because the nature of the sensors used in these instruments is such that they drift over time and with use.

406.7.3 Purging appliances and equipment. After the *piping* system has been placed in operation, appliances and *equipment* shall be purged before being placed into operation.

❖ This section simply requires that after the gas piping has been properly purged and charged with fuel gas, the appliances served be purged of air before startup.

SECTION 407 (IFGC)
PIPING SUPPORT

407.1 General. *Piping* shall be provided with support in accordance with Section 407.2.

❖ Often, piping support is inadequate, poorly installed or overlooked entirely, despite the fact that support is critical in preventing stresses and strains in piping, fittings, joints, connectors and appliance gas trains (controls) (see Section 407.2).

407.2 Design and installation. *Piping* shall be supported with metal pipe hooks, metal pipe straps, metal bands, metal brackets, metal hangers or building structural components, suitable for the size of *piping*, of adequate strength and quality, and located at intervals so as to prevent or damp out excessive vibration. *Piping* shall be anchored to prevent undue strains on connected *appliance*s and shall not be supported by other *piping*. Pipe hangers and supports shall conform to the requirements of MSS SP-58 and shall be spaced in accordance with Section 415. Supports, hangers and anchors shall be installed so as not to interfere with the free expansion and contraction of the *piping* between anchors. All parts of the supporting *equipment* shall be designed and installed so that they will not be disengaged by movement of the supported *piping*.

❖ As with all piping systems, the support of the fuel gas system is as important as any other part of the overall design. Proper supports are necessary to maintain piping alignment and slope, to support the weight of the pipe and its contents, to control movement and to resist dynamic loads, such as thrust. Inadequate support can cause piping to fail under its own weight, resulting in fire, explosion or property damage. Building design must take into consideration the structural loads created by the support of piping systems.

Hangers or supports must not react with or be detrimental to the pipe they support. Hangers or supports for metallic pipe must be of a material that is compatible with the pipe to prevent any corrosive action. For example, copper, copper-clad or specially coated hangers are required if the piping system is constructed of copper tubing. Hangers and supports must be constructed of noncombustible materials to prevent premature failure in a fire and to prevent deformation of the supports such as creep, sag and elongation. This section prohibits combustible hooks, straps, bands, brackets and hangers. Plastic hooks, straps, hangers, etc., that are used with plumbing piping are prohibited for use with gas piping. Building structural components are listed as a means of support because piping often rests on wall plates, beams, joists and similar structural members where such members are capable of carrying the additional load of the piping. Gas piping must not depend on other piping for support; thus, it must be supported from the structure independently. Gas controls for appliances can be stressed and damaged by unsupported piping drops that bear on or produce bending moments in the controls. During installation, a simple check for proper support at an appliance connected with rigid piping is to open the union and observe any movement of the piping. If the piping drops or moves in any direction, it is evident that stress is being applied to the appliance controls. Eliminating such stress is one of the advantages of using semirigid gas connectors (see Section 415).

SECTION 408 (IFGC)
DRIPS AND SLOPED PIPING

408.1 Slopes. *Piping* for other than dry gas conditions shall be sloped not less than $^1/_4$ inch in 15 feet (6.3 mm in 4572 mm) to prevent traps.

❖ The required minimum slope (0.14 percent) is intended for gas known to contain enough water vapor to cause condensate to form inside the piping. Such conditions are rarely, if ever, encountered today.

408.2 Drips. Where wet gas exists, a drip shall be provided at any point in the line of pipe where condensate could collect. A drip shall be provided at the outlet of the meter and shall be installed so as to constitute a trap wherein an accumulation of condensate will shut off the flow of gas before the condensate will run back into the meter.

❖ Drips, often referred to as "drip legs," are distinct from sediment traps. Modern gas supplies and distribution systems are typically dry; thus, drips would be required only when recommended by the gas supplier.

408.3 Location of drips. Drips shall be provided with ready access to permit cleaning or emptying. A drip shall not be located where the condensate is subject to freezing.

❖ A drip is not the same as a sediment trap; they are required for different reasons (see Section 408.4).

408.4 Sediment trap. Where a sediment trap is not incorporated as part of the appliance, a sediment trap shall be installed downstream of the appliance shutoff valve as close to the inlet of the appliance as practical. The sediment trap shall be either a tee fitting having a capped nipple of any length installed vertically in the bottommost opening of the tee as illustrated in Figure 408.4 or other device approved as an effective sediment trap. Illuminating appliances, ranges,

clothes dryers, decorative vented appliances for installation in vented fireplaces, gas fireplaces and outdoor grills need not be so equipped.

❖ In addition to the code requirement, most appliance manufacturers require the installation of a sediment trap (dirt leg) to protect the appliance from debris in the gas. Note that a drip leg is not the same as a sediment trap (see Section 408.2). Sediment traps are necessary to protect appliance gas controls from the dirt, soil, pipe chips, pipe joint tapes and compounds and construction site debris that enter the piping during storage, handling, installation and repairs. Hazardous appliance operation could result from foreign matter entering gas controls and burners. Despite the fact that utilities supply clean gas, foreign matter can enter the piping prior to and during installation both on the utility side of the system and on the customer side.

Sediment traps are designed to cause the gas flow to change direction 90 degrees (1.57 rad) at the sediment collection point, thus causing the solid or liquid contaminants to drop out of the gas flow [see Commentary Figure 408.4(1)]. The nipple and cap should not be placed in the branch opening of a tee fitting because this would not create a change in direction of flow and would allow debris to simply pass/jump over the capped nipple collection point. Commentary

TO GAS SUPPLY IF BRANCH CONNECTS TO APPLIANCE OR TO APPLIANCE IF BRANCH CONNECTS TO GAS SUPPLY

TEE

NIPPLE OF ANY LENGTH

CAP

FIGURE 408.4
METHOD OF INSTALLING A TEE FITTING SEDIMENT TRAP

Figure 408.4(2) illustrates a relatively ineffective sediment trap; however, such configurations are not expressly prohibited by the wording of this section. The text speaks of the bottommost opening of the tee without prohibiting such opening from being the branch opening. Note that Figure 408.4 illustrates the intent to have the nipple and cap in the run of the tee, thereby suggesting that it is not the intent to have it in the branch of the tee. The code does not specify a minimum length for the capped nipple; therefore, it could be from a close nipple on up. Three to 6 inches is the customary length. The capped nipple must be in a vertical plane to allow the sediments to fall in by gravity. The sediment trap must be as close to the appliance inlet as practical to be able to capture sediment from all of the piping upstream of the appliance connection. The sediment trap must be downstream of the appliance shutoff valve to allow the trap to be serviced after closing the upstream shutoff valve. Manufactured sediment traps are available that have the configuration of a straight section of pipe and are equipped with cleanout openings. Although it would be wise to install sediment traps at all appliance connections, they are not mandated by code for gas lights, ranges, clothes dryers and outdoor grills. These appliances are also susceptible to harm from debris in gas, especially ranges and clothes dryers, and the appliance manufacturer may require sediment traps where the code does not. The code's logic is that these exempt appliances are manually operated rather than automatically operated; therefore, the user would be in attendance and aware of a problem.

SECTION 409 (IFGC) SHUTOFF VALVES

409.1 General. *Piping* systems shall be provided with shutoff valves in accordance with this section.

❖ This section addresses shutoff valves for meters, buildings, tenant spaces, appliances and fireplaces.

409.1.1 Valve approval. Shutoff valves shall be of an *approved* type; shall be constructed of materials compatible with the *piping*; and shall comply with the standard that is applicable for the pressure and application, in accordance with Table 409.1.1.

❖ The code official must approve the type of shutoff valve used. For example, he or she might not approve a valve that requires a wrench to operate. Shutoff valves must be tested to the specified standard and so identified. Ball valves are used almost exclusively today because of their low pressure losses, ease of operation and reliability [see Commentary Figures 409.1.1(1) and 409.1.1(2)].

409.1.2 Prohibited locations. Shutoff valves shall be prohibited in concealed locations and *furnace plenums*.

❖ This section does not prohibit shutoff valves in the typical spaces above ceilings and below floors used for conveying environmental air. See the definition of "Concealed location." A valve in a concealed location would be of no value and a potential leak source. The concern for valves in furnace plenums is temperature extremes and lack of access.

409.1.3 Access to shutoff valves. Shutoff valves shall be located in places so as to provide *access* for operation and shall be installed so as to be protected from damage.

❖ As expressed in the previous section, a shutoff valve has no utility if it is installed where it cannot be accessed. Location is especially important for valves that are intentionally installed for emergency use. Outdoor valves are particularly subject to damage.

Figure 408.4(1)
EFFECTIVE SEDIMENT TRAP

Figure 408.4(2)
LESS EFFECTIVE OR INEFFECTIVE SEDIMENT TRAP CONFIGURATION

Figure 409.1.1(1)
TYPICAL BALL-TYPE GAS SHUTOFF VALVE
(Photo courtesy of Dormont Manufacturing Company)

Figure 409.1.1(2)
BALL-TYPE GAS SHUTOFF VALVE
DESIGNED TO MATE WITH A CONNECTOR
(Photo courtesy of Dormont Manufacturing Company)

409.2 Meter valve. Every meter shall be equipped with a shutoff valve located on the supply side of the meter.

❖ This valve is typically installed and owned by the gas supplier because it is upstream of the point of delivery for utility-owned meters. These valves require a wrench to operate and are capable of being locked off.

409.3 Shutoff valves for multiple-house line systems. Where a single meter is used to supply gas to more than one building or tenant, a separate shutoff valve shall be provided for each building or tenant.

❖ A separate shutoff valve for each building or tenant space will allow isolation of any building or space in the event of an emergency or when work must be performed on the piping system.

409.3.1 Multiple tenant buildings. In multiple tenant buildings, where a common *piping* system is installed to supply other than one- and two-family dwellings, shutoff valves shall be provided for each tenant. Each tenant shall have access to the shutoff valve serving that tenant's space.

❖ Each tenant space and dwelling unit in buildings having more than two dwellings must have its own shutoff valve to allow isolation from the common piping system. The tenant spaces and dwelling units may or may not have separate meters (see Section 409.3). The shutoff valves must be within the tenant space or dwelling unit, or the valves must be located in a common space that can be accessed by all of the occupants.

409.3.2 Individual buildings. In a common system serving more than one building, shutoff valves shall be installed outdoors at each building.

❖ This section requires the shutoff valve for each building to be located outside of the building, thus allowing quick access in an emergency.

409.3.3 Identification of shutoff valves. Each house line shutoff valve shall be plainly marked with an identification tag attached by the installer so that the *piping* systems supplied by such valves are readily identified.

❖ The term "house line" is used to describe the main supply pipe to a building or a dwelling unit. Utilities refer to the house line as the piping on the downstream side of the utility-owned meter. In the case of multiple house lines supplying different spaces in a building, the required shutoff valves must be permanently tagged to indicate which space is supplied by each valve. Without tags, it might not be apparent which space each valve controls. For both service

TABLE 409.1.1
MANUAL GAS VALVE STANDARDS

VALVE STANDARDS	APPLIANCE SHUTOFF VALVE APPLICATION UP TO $^1/_2$ psig PRESSURE	OTHER VALVE APPLICATIONS			
		UP TO $^1/_2$ psig PRESSURE	UP TO 2 psig PRESSURE	UP TO 5 psig PRESSURE	UP TO 125 psig PRESSURE
ANSI Z21.15	X	—	—	—	—
ASME B16.44	X	X	X[a]	X[b]	—
ASME B16.33	X	X	X	X	X

For SI: 1 pound per square inch gauge = 6.895 kPa.
a. If labeled 2G.
b. If labeled 5G.

❖ Table 409.1.1 lists two valve application categories: appliance shutoff application and all other applications. The application of a valve must be consistent with the listing of the valve. For example, a valve used ahead of an MP regulator (see Section 409.4) in a 2-psi system must comply with ASME B16.44 and be marked "2G." It could also comply with ASME B16.33 without being marked "2G."

work and emergency action, the load the valve serves must be identified. The tags should be designed and attached to be permanent and are typically brass medallions attached with small chains.

409.4 MP regulator valves. A *listed* shutoff valve shall be installed immediately ahead of each MP regulator.

❖ See the definition of "Regulator, Medium-pressure." This regulator (also known as "pounds-to-inches" regulator) is placed between the service regulator and the appliance regulator and is typically associated with manifold parallel distribution systems using CSST or copper tubing. A shutoff valve ahead of the MP regulator will allow the regulator to be conveniently serviced, adjusted or replaced. The valve also serves to isolate the higher pressure system from the lower pressure system. For purposes of sizing, the piping downstream of a medium-pressure (MP) regulator is considered as a separate system, with the MP regulator being the source of supply for that system.

409.5 Appliance shutoff valve. Each *appliance* shall be provided with a shutoff valve in accordance with Section 409.5.1, 409.5.2 or 409.5.3.

❖ Each appliance, regardless of type, size or location, must have its own dedicated shutoff valve to isolate the appliance from its gas supply. It is not the intent of this section to allow a single shutoff valve to control more than one appliance. Appliance shutoff valves are not considered to be emergency valves, but rather, are considered to be service valves that allow the appliance to be serviced, repaired, replaced or shut down (see the definition of "Valve, Appliance shutoff"). A shutoff valve is a valve other than the valve that is integral with the appliance gas control. The intent is that the separate shutoff valve be located ahead of all appliance control valves. The location requirements of this section are intended to make the valve conspicuous and easy to operate.

Note that the three subsections to this section are distinct choices and the shutoff valves have to meet one of these choices.

409.5.1 Located within same room. The shutoff valve shall be located in the same room as the *appliance*. The shutoff valve shall be within 6 feet (1829 mm) of the *appliance*, and shall be installed upstream of the union, connector or quick disconnect device it serves. Such shutoff valves shall be provided with *access*. *Appliance* shutoff valves located in the firebox of a *fireplace* shall be installed in accordance with the *appliance* manufacturer's instructions.

❖ This section contains the traditional appliance shutoff valve requirements such as within 6 feet (1820 mm) of and in the same room as the appliance served. This section does not require the appliance shutoff valve to be provided with "ready access." Because such a valve is not an emergency valve, there is no reason to require ready access, and simply providing access is sufficient. Common practice has been to install the appliance shutoff valve behind ranges and clothes dryers and in the control compartment of decorative gas fireplace appliances. This section allows this practice because the valves are accessible. To avoid the use of an exposed valve or concealed "key-handle" valve, it is accepted practice to locate the shutoff valve for gas fireplace appliances in the control compartment in the bottom of the unit, behind the control access panel (see Commentary Figure 409.5.1).

The code does not indicate how the 6-foot limit is to be measured, but traditionally, it is measured along the piping such that no more than 6 feet (1829 mm) of piping extends between the outlet of the shutoff valve and the inlet of the appliance.

This section works with the definition of "Piping system" and Section 402.3 to allow the piping downstream of the shutoff valve to be considered and

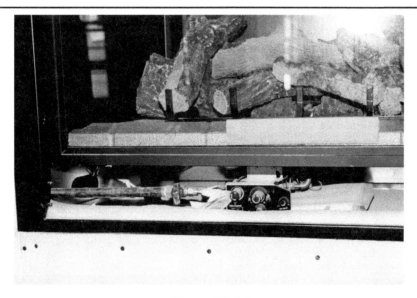

Figure 409.5.1
SHUTOFF VALVE IN CONTROL COMPARTMENT OF GAS FIREPLACE

sized as a connector. The code does not state that the required shutoff valve has to be located external to an appliance cabinet, enclosure or housing; therefore, the valve could be located behind an appliance access door/panel.

The last sentence of this section addresses appliances, such as decorative gas log sets, that are installed in solid fuel-burning fireplaces. The manual shutoff may be in the firebox of the fireplace if it is allowed by the appliance manufacturer's instructions and the valve is located as directed by those instructions. There is typically an area adjacent to the appliance controls or inlet connection that is shielded from radiant heat, and this "cool zone" is where the manual shutoff valve must be located. A shutoff valve must never be located in the firebox if solid fuel is burned.

409.5.2 Vented decorative appliances and room heaters. Shutoff valves for vented decorative appliances, room heaters and decorative appliances for installation in vented *fireplaces* shall be permitted to be installed in an area remote from the appliances where such valves are provided with ready *access*. Such valves shall be permanently identified and shall not serve another *appliance*. The *piping* from the shutoff valve to within 6 feet (1829 mm) of the *appliance* shall be designed, sized and installed in accordance with Sections 401 through 408.

❖ This section allows locating shutoff valves for the stated appliances remote from the appliance if the valves are permanently tagged as to their function, serve only the single appliance and have ready access. The shutoff valve can be located anywhere, indoors or outdoors, as long as it is provided with ready access and is permanently identified. Ready access is required because the shutoff valve will be more difficult to locate, being remote from the appliance served. This section is an alternative to using concealed "key-handle-type" valves located in the wall or floor adjacent to the appliance. The code intends for such valves to be permanently identified by means of a durable tag suitable for the environment, such as brass medallions on chains. The reference to room heaters in this section is not specific as to the type, thus, it would include both vented and unvented room heaters.

The shutoff valve for a log lighter is also the appliance operating valve; therefore, that valve cannot be located remotely.

Because the remote shutoff valve allowed by this section could be any distance (more than 6 feet) from the appliance, the piping from the shutoff valve outlet to within 6 feet of the appliance inlet must be treated and sized as gas piping, not as a connector. Recall that the definition of "Piping system" says that the piping system ends at the appliance shutoff valves, therefore, the piping between the shutoff valve and the appliance becomes a connector, no matter how long it is. This section accounts for the potential pressure loss through lengthy connectors by requiring them to be sized as required for the piping system, except for the last 6 feet before the appliance [see the exception to Section 411.1.3.1 and Commentary Figure 411.1.3.1(2)].

409.5.3 Located at manifold. Where the *appliance* shutoff valve is installed at a manifold, such shutoff valve shall be located within 50 feet (15 240 mm) of the *appliance* served and shall be readily accessible and permanently identified. The *piping* from the manifold to within 6 feet (1829 mm) of the *appliance* shall be designed, sized and installed in accordance with Sections 401 through 408.

❖ Similar to Section 409.5.2, this section allows remote appliance shutoff valves, but, unlike Section 409.5.2, this section specifies the remote location and imposes a distance restriction. Manifolds are part of parallel distribution systems commonly installed with CSST and copper tubing gas distribution systems. A manifold can be a manufactured fitting with multiple branch openings or it can be an assembly of close connected tee fittings. The individual "run-outs" of tubing to each appliance all connect to this manifold and the appliance shutoff valves can be installed at the manifold in a group configuration. Such valves must have ready access and must be labeled as to their function, the same as required in the previous section.

The code does not specify how the 50-foot distance limitation is to be measured. Taken literally, it would be measured in a straight line through walls, floors and ceilings. Clearly, such a measurement could be difficult to perform in the field and should be approached liberally, keeping in mind the intent to simply have the valve within a reasonable distance.

Because the remote shutoff valve allowed by this section could be any distance (up to 50 feet) from the appliance, the piping from the shutoff valve outlet to within 6 feet of the appliance inlet must be treated and sized as gas piping, not as a connector. Recall that the definition of "Piping system" says that the piping system ends at the appliance shutoff valves, therefore, the piping between the shutoff valve and the appliance becomes a connector, no matter how long it is. This section accounts for the potential pressure loss through lengthy connectors by requiring them to be sized as required for the piping system, except for the last 6 feet before the appliance [see the exception to Section 411.1.3.1 and Commentary Figure 411.1.3.1(2)].

409.6 Shutoff valve for laboratories. Where provided with two or more fuel gas outlets, including table-, bench- and hood-mounted outlets, each laboratory space in educational, research, commercial and industrial occupancies shall be provided with a single dedicated shutoff valve through which all such gas outlets shall be supplied. The dedicated shutoff valve shall be readily accessible, located within the laboratory space served, located adjacent to the egress door from the space and shall be identified by approved signage stating "Gas Shutoff."

❖ Laboratories are typically equipped with gas cocks that are mounted on tables, work benches or in exhaust hoods. These gas cocks are designed with tapered and barbed outlets for connection to elastomeric tubing that supplies Bunsen burners. Bunsen burners are used extensively in chemical laboratory work for testing, research, experimentation, production and teaching purposes. These gas cocks are open-ended and have handles much like that of a water faucet. A typical laboratory could have many such gas cocks (outlets) spread throughout the lab and there is a concern for making sure that all such gas cocks are turned off when not in use. Having a master shutoff valve upstream (ahead) of all gas cocks allows the person responsible for the laboratory to shut off the gas supply to all gas cocks simultaneously, thereby making certain that there are no open or partially open gas cocks leaking gas anywhere in the laboratory space. The master shutoff valve could also be used in an emergency situation. Locating the master shutoff valve near the egress door encourages the practice of habitually closing the valve upon exiting the lab. The master shutoff valve must be readily accessible and well marked. Typically, the valve is recessed in a wall, painted yellow and labeled by conspicuous signage.

SECTION 410 (IFGC)
FLOW CONTROLS

410.1 Pressure regulators. A line pressure regulator shall be installed where the *appliance* is designed to operate at a lower pressure than the supply pressure. Line gas pressure regulators shall be *listed* as complying with ANSI Z21.80. *Access* shall be provided to pressure regulators. Pressure regulators shall be protected from physical damage. Regulators installed on the exterior of the building shall be *approved* for outdoor installation.

❖ All gas supply systems employ regulators at some point in the distribution system, and the type and location of these devices depends on the design pressures of the system and utilization equipment. A line pressure regulator reduces the pressure from the service delivery pressure to the pressure of the distribution system. Additional pressure regulators could be required at other points in the system or at the appliance to reduce the pressure even further [see Commentary Figure 410.1(1)]. With the increased use of medium-pressure gas distribution systems, systems are designed with one or more regulators in addition to the service regulators. Appliances and gas utilization equipment are also equipped with factory-installed regulators that control burner input (manifold) pressure.

Pressure regulators are extremely important devices that are relied on to prevent overpressurization of appliances and equipment. Typical gas appliance controls are not designed to tolerate a pressure of more than 14 inches of water column (3.5 kPa). The standards that regulate the design and function of appliance gas controls require that the controls be tested for resistance to permanent damage resulting from exposure to excessive gas supply pressure.

Each function of the gas controls must be tested for exposure to $2^1/_2$-psi (17 kPa) pressure for a period of 1 minute, after which the control must function normally. Excess pressures could lead to severe appliance malfunction. Depending on the application, gas pressure regulators may contain relief mechanisms that are designed to vent gas from the downstream side in the event that the downstream (load) side pressure exceeds a predetermined maximum. This feature is a safeguard intended to prevent the overpressurization of the piping and utilization equipment supplied by the regulator. Regulators with relief devices must be installed outdoors, or the relief vent must be piped to the outdoors to prevent gas from being discharged into the building.

Access to pressure regulators is required for the maintenance, adjustment and servicing of the devices. The definition of "Access" appears in Section 202. Protecting regulators from physical damage essentially means that regulators must be installed in locations that are not prone to impact from equipment or vehicles, or they must be shielded from impact. Because of the range of environmental conditions encountered outside the building, regulators must be evaluated and approved for installation outdoors. Regulators exposed to the weather must be protected from moisture and precipitation. Commentary Figure 410.1(2) shows a service regulator with discharge piping connected to the integral relief valve port. This regulator establishes the pressure at the point of delivery. Commentary Figure 410.1(3) illustrates the basic operating principle of gas pressure regulators. Outlet pressure acts upon the flexible dia-

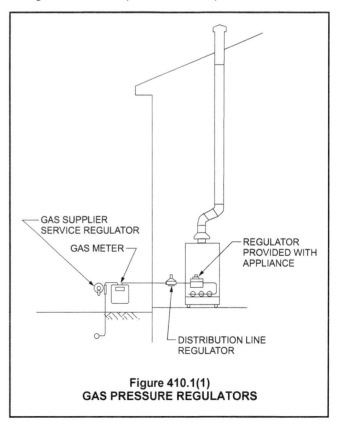

Figure 410.1(1)
GAS PRESSURE REGULATORS

phragm and adjustment spring, causing the valve to close on a rise in outlet pressure and open on a fall in outlet pressure. The vent allows the diaphragm to move freely by admitting or emitting air above the diaphragm. Atmospheric pressure and the adjustment spring force act together to establish the outlet pressure setting of the regulator [see Commentary Figure 410.1(4)].

Figure 410.1(2)
SERVICE REGULATOR
(Photo courtesy of Washington Gas)

Figure 410.1(3)
GAS PRESSURE REGULATOR

410.2 MP regulators. MP pressure regulators shall comply with the following:

1. The MP regulator shall be *approved* and shall be suitable for the inlet and outlet gas pressures for the application.

2. The MP regulator shall maintain a reduced outlet pressure under lock-up (no-flow) conditions.

3. The capacity of the MP regulator, determined by published ratings of its manufacturer, shall be adequate to supply the *appliances* served.

4. The MP pressure regulator shall be provided with *access*. Where located indoors, the regulator shall be vented to the outdoors or shall be equipped with a leak-limiting device, in either case complying with Section 410.3.

5. A tee fitting with one opening capped or plugged shall be installed between the MP regulator and its upstream shutoff valve. Such tee fitting shall be positioned to allow connection of a pressure-measuring instrument and to serve as a sediment trap.

6. A tee fitting with one opening capped or plugged shall be installed not less than 10 pipe diameters downstream of the MP regulator outlet. Such tee fitting shall be positioned to allow connection of a pressure-measuring instrument.

7. Where connected to rigid piping, a union shall be installed within 1 foot (304 mm) of either side of the MP regulator.

❖ This section states the minimum requirements for MP gas regulators installed in 2-psi (13.8 kPa) gas piping systems or portions of systems. These requirements intend to keep the regulators accessible for service and repair, protect them from physical damage, require that they are suitable for the use intended, are properly sized and installed, and, if installed indoors, are properly vented to the outdoors (see commentary, Section 410.3). Item 2 requires a "lockup"-type regulator capable of holding the approximate outlet pressure setpoint indefinitely under no-flow conditions. Section 409.4 requires a gas shutoff valve immediately upstream of the MP regulator to allow isolation of the regulator during testing, repair or replacement of the regulator. Items 5 and 6 require a tee fitting upstream and downstream of the MP regulator to facilitate the connection of pressure-measuring instruments for calibrating the regulator. These fittings should have a $^1/_8$-inch threaded tapping for connection of common gauges and manometers. Item 5 also requires a sediment trap on the inlet side of the regulator. Because a lot of faith is placed in the integrity of this regulator, it deserves the protection from debris afforded by a simple tee, nipple and cap arrangement; see Section 408.4.

Where a line pressure regulator serves appliances rated for $^1/_2$ psi (3.5 kPa) or less inlet pressure and the regulator's inlet pressure is over 2 psi (13.8 kPa), a downstream overpressure protection device must be installed in accordance with ANSI Z21.80, the standard for line pressure regulators [see Commentary Figures 410.2(1) and 410.2(2)]. The OPD is designed to limit the downstream (outlet) pressure to 2 psi (13.8 kPa). MP regulators are line pressure regulators that serve to reduce pressures that are above 0.5 psi and less than or equal to 5 psi to some lower pressure. They are typically installed in 2-psi and 5-psi gas distribution systems that serve appliances having a maximum input pressure of 0.5 psi (14 inches water column). If such regulators are installed with steel piping on the inlet and outlet side, the regulators cannot be removed or isolated without disassembly of the piping system for some distance or cutting the piping. To facilitate removal or isolation of the regulator, a union fitting must be placed near the inlet or outlet side of the regulator. This is simply a common-sense provision [see Commentary Figure 410.2(3)].

410.3 Venting of regulators. Pressure regulators that require a vent shall be vented directly to the outdoors. The vent shall be designed to prevent the entry of insects, water and foreign objects.

> **Exception:** A vent to the outdoors is not required for regulators equipped with and *labeled* for utilization with an *approved* vent-limiting device installed in accordance with the manufacturer's instructions.

❖ All regulators use synthetic or natural rubber diaphragms, one side of which is vented to the atmosphere. The devices must vent one side of the diaphragm to the atmosphere to allow unimpeded movement of the diaphragm within its enclosure. Thus, these devices, in effect, see atmospheric pressure as a reference point for controlling the fuel-gas pressure. Regulators cannot operate without the vent. If the diaphragm were to rupture or perforate, gas would be vented to the exterior of the device. For this reason, regulators installed indoors must be vented to the outdoors or must comply with the exception to this section. See the definition of "Vent

piping, breather." The vent termination outdoors must be protected from blockage and the entry of foreign objects, insects and water, which could cause regulator failure and a severe hazard [see Commentary Figures 410.3(1) and 410.3(2)]. If a regulator vent is blocked by ice, water, insect nests, debris or other foreign material, the air above the diaphragm will be trapped and the regulator port valve could be held in the open position thus disabling the regulator and causing overpressurization of the piping system.

The exception permits installation of regulators indoors without a vent to the outdoors where the regulator is equipped with and labeled for use with a factory-installed vent-limiting device. Vent-limiting devices, as the name implies, are designed to allow the vent to "breathe" under normal operating conditions. In the event of diaphragm failure, the vent-limiting devices will limit gas escapement to a rate near 1 cubic foot per hour (0.0283 m³/h) for natural gas at 2-psi (13.8 kPa) pressure. A vent-limiting device is a ball check valve with a fixed orifice and is designed to allow air inhalation and escapement while restricting gas leakage to a small rate. Even small leakage rates can be hazardous if inadequate ventilation exists to prevent buildup of explosive concentrations. Although a standard is not referenced in this section, ANSI Z21.18 is appropriate for evaluating such vent-limiting devices. The standard limits the volume of gas that a vent-limiting device is allowed to discharge into a room or space. Vent-limiting devices should be installed in locations that are ventilated sufficiently to dissipate any gas discharged from the device.

For SI: 1 inch water column = 249 Pa,
       1 pound per square inch gauge = 6.895 kPa.

Figure 410.1(4)
SYSTEM PRESSURES

For SI: 1 inch = 25.4 mm, 1 psig = 6.895 kPa,
1 inch water column = 0.249 kPa.

**Figure 410.2(1)
SYSTEM PRESSURES**

**Figure 410.2(2)
PRESSURE REGULATOR/OVERPRESSURE
PROTECTION DEVICE ASSEMBLIES
(Photo courtesy of MAXITROL Company)**

410.3.1 Vent piping. Vent *piping* for relief vents and breather vents shall be constructed of materials allowed for gas *piping* in accordance with Section 403. Vent *piping* shall be not smaller than the vent connection on the pressure-regulating device. Vent *piping* serving relief vents and combination relief and breather vents shall be run independently to the outdoors and shall serve only a single device vent. Vent *pip-*

ing serving only breather vents is permitted to be connected in a manifold arrangement where sized in accordance with an *approved* design that minimizes backpressure in the event of diaphragm rupture. Regulator vent *piping* shall not exceed the length specified in the regulator manufacturer's instructions.

❖ See the definitions of "Relief vents" and "Vent piping, breather." Relief vents must not be combined with any other relief vents or breather vents because of the potential to restrict relief capacity and/or cause regulator malfunction. Only breather vents are allowed to be manifolded together and only where in compliance with this section.

In all cases, the minimum size of vent piping is the size of the vent connection on the device served. Although not stated in this section, there is always a limit on how far any size vent piping can be run. Vent piping may have to be increased in size to compensate for long vent runs. The longer the vent piping, the greater the flow resistance and flow resistance must be limited so that the airflow necessary for unimpeded movement of the regulator diaphragm is not restricted. The regulator manufacturer's installation instructions may include vent piping length limitations and vent piping lengths must be in accordance with such instructions. Manifold piping for breather vents must be engineered based on the predicted rise in internal vent pressure in the event of a diaphragm failure. If regulator vents are blocked or the

For SI: 1 foot = 304.8 mm.

Figure 410.2(3)
AT LEAST ONE UNION, INSTALLED UPSTREAM OR DOWNSTREAM,
IS REQUIRED FOR AN MP REGULATOR CONNECTED TO RIGID PIPING

airflow is restricted in any way, including by undersized or excessively long piping runs, the result will be impaired regulator operation, causing retarded regulator response time, underpressure during appliance start-ups and overpressurization of the downstream system, all of which can be hazardous.

This section also requires regulators to be vented individually to the outdoors to prevent a potentially hazardous situation that may arise if more than one regulator is served by a single vent. The hazard could occur if the diaphragm of one regulator failed and the gas leakage into the vent caused an increase of pressure within the vent, which would then act on the other regulator diaphragms, causing all of the other regulators using the common vent to operate at a higher discharge pressure. This section permits use of engineered vent manifold systems to vent more than one regulator if the manifold is designed and sized to allow unimpeded air movement and to minimize backpressure in the event of a diaphragm failure.

Figure 410.3(1)
REGULATOR WITH VENT PROTECTION
(Photo courtesy of Washington Gas)

410.4 Excess flow valves. Where automatic *excess flow valves* are installed, they shall be listed for the application and shall be sized and installed in accordance with the manufacturer's instructions.

❖ The code does not mandate the installation of EFVs, rather, it provides requirements for installations where such installations are chosen. EFVs must be listed for the application in which they are to be used. The manufacturer's instructions provide sizing and installation requirements that must adhered to. Some devices have to be installed in certain positions to work properly. Sizing is very important because the EFV will cause a gas pressure drop and if this drop is not accounted for in the system, the appliance might be supplied with inadequate pressure causing a hazardous condition (see commentary, Section 402.5). Proper sizing for EFVs is also important to ensure that they will function as intended. An undersized EFV might trip unnecessarily resulting in appliance inoperation and an oversized EFV might not trip at all. See Commentary Figure 410.4 and commentary to the definition of "Excess flow valve (EFV)."

410.5 Flashback arrestor check valve. Where fuel gas is used with oxygen in any hot work operation, a listed protective device that serves as a combination flashback arrestor and backflow check valve shall be installed at an approved location on both the fuel gas and oxygen supply lines. Where the pressure of the piped fuel gas supply is insufficient to ensure such safe operation, approved equipment shall be installed between the gas meter and the appliance that increases pressure to the level required for such safe operation.

❖ This section relates to commercial and industrial installations that use fuel gas and oxygen for welding and cutting operations. The devices required by this section are designed to prevent the flow of one gas into the piping for another gas caused by differences in line pressures. The result of any such backflow would be potentially hazardous because an explosive mixture could be created in either piping system and such mixture could be delivered to a burner head.

Figure 410.3(2)
PRESSURE REGULATOR
WITH VENT-LIMITING DEVICE
(Photo courtesy of MAXITROL Company)

Figure 410.4
EXAMPLE OF EXCESS FLOW VALVE

SECTION 411 (IFGC)
APPLIANCE AND MANUFACTURED
HOME CONNECTIONS

411.1 Connecting appliances. Except as required by Section 411.1.1, *appliances* shall be connected to the *piping* system by one of the following:

1. Rigid metallic pipe and fittings.

2. Corrugated stainless steel tubing (CSST) where installed in accordance with the manufacturer's instructions.

3. Semirigid metallic tubing and metallic fittings. Lengths shall not exceed 6 feet (1829 mm) and shall be located entirely in the same room as the *appliance*. Semirigid metallic tubing shall not enter a motor-operated *appliance* through an unprotected knockout opening.

4. *Listed* and *labeled appliance* connectors in compliance with ANSI Z21.24 and installed in accordance with the manufacturer's instructions and located entirely in the same room as the *appliance*.

5. *Listed* and *labeled* quick-disconnect devices used in conjunction with *listed* and *labeled appliance* connectors.

6. *Listed* and *labeled* convenience outlets used in conjunction with *listed* and *labeled appliance* connectors.

7. Listed and *labeled* outdoor *appliance* connectors in compliance with ANSI Z21.75/CSA 6.27 and installed in accordance with the manufacturer's instructions.

8. Listed outdoor gas hose connectors in compliance with ANSI Z21.54 used to connect portable outdoor appliances. The gas hose connection shall be made only in the outdoor area where the appliance is used, and shall be to the gas *piping* supply at an appliance shutoff valve, a listed quick-disconnect device or listed gas convenience outlet.

❖ Connectors are used to connect the appliance or equipment to the gas distribution system outlet. The pipe sizing methods in this chapter size distribution piping up to but not beyond the appliance shutoff valve outlet (Note exceptions in Sections 409.5 and 411.1.3.1) and this section sizes the connection between the shutoff valve outlet and the appliance [see definition of "Piping system" and Commentary Figure 411.1(1)]. The choice of a connection type must take into consideration such factors as appliance movement, vibration, ambient conditions and susceptibility to physical damage. Between the shutoff valve and the appliance, a union fitting or similar arrangement must allow a means of disconnecting the piping. Flared or ground-joint connections that are part of an approved, labeled semirigid ("flexible") connector can serve as a union [see Commentary Figures 411.1(2) and 411.1(3)].

Item 2 addresses CSST connected directly to appliances. Although CSST is allowed to directly connect to specific appliances (fixed/nonmoveable), CSST is not to be used as a substitute for a connector where one is required. CSST manufacturers' instructions specify where CSST can directly connect to appliances and where it cannot [see Commentary Figures 411.1(4) and 411.1(5)].

Item 4 refers to the commonly used, so called "flexible" connectors typically made of corrugated stainless steel and previously made of brass. Appliance connectors are listed and labeled with the label typically being wrapped around the connector tubing.

Most appliance connectors manufactured today have attached plastic labels that state the installation and sizing instructions. These connectors are not designed for repeated movement that causes bending of the metal and should not be reused after the initial installation. Reuse is typically prohibited by the manufacturers' instructions. Repeated bending and/or vibration can cause metal fatigue, stress cracking and gas leakage [see Commentary Figures 411.1(6) and 411.1(7)].

The piping system is sized up to the outlet of the appliance shutoff valve. The connector occurs between the shutoff valve outlet and the appliance and is sized by the connector manufacturer or is based on the appliance input rating, gas inlet size and length of the connector [see Commentary Figure 411.1(1)]. For example, a boiler is supplied by branch piping required to be 1 inch (25 mm) based on the sizing methods in Section 402. The branch piping is 1 inch (25 mm) up to the boiler shutoff valve, but the boiler inlet opening is only $^1/_2$ inch (13 mm). The piping (connector) downstream of the shutoff valve is sized as a connector because it is 6 feet (1829 mm) or less in length (see commentary, Sections 409.5.1, 409.5.2, 409.5.3 and 411.1.3.1).

Item 7 addresses connections for outdoor appliances [see Commentary Figure 411.1(7)].

Because portable appliances can be moved, Items 1 through 7 of this section are not suited for connecting portable appliances to the gas distribution piping system. Most appliance connectors are not designed to be used with any appliance that can be readily moved. All appliance connectors must be used within the parameters of their listings. For outdoor portable appliances, new Item 8 is the only apparent option. Outdoor gas hose connectors have to be resistant to mechanical damage, possible heat exposure and the harmful effects of exposure to the weather. Connectors listed to ANSI Z21.54 are evaluated and tested for the particularly harsh environment of outdoor use. The gas hose connector must be located entirely outdoors and must be connected to the gas piping system at a point outdoors. The point of connection to the gas distribution system piping must be through a listed device that allows the hose to be readily manually disconnected or through an appliance shutoff valve. Quick-disconnect devices have safety features, such as thermal shutoffs, that will close the valve when exposed to high temperatures and interlocking systems that will not allow the hose to be removed until the manual gas valve is closed.

The intent of new Item 8 is to address portable outdoor appliance connections and to mandate that such connectors be listed to a specific safety standard [see Commentary Figure 411.1(8)].

411.1.1 Commercial cooking appliances. Commercial cooking appliances installed on casters and appliances that are moved for cleaning and sanitation purposes shall be connected to the *piping* system with an appliance connector listed as complying with ANSI Z21.69. The commercial cooking appliance connector installation shall be configured in accordance with the manufacturer's instructions. Movement of appliances with casters shall be limited by a restraining device installed in accordance with the connector and appliance manufacturer's instructions.

❖ This section is referred to in Section 411.1 and mandates that a specific type of connector be used for commercial cooking appliances that are routinely moved [see Commentary Figures 411.1.1(1) through 411.1.1(4)]. Connectors listed to ANSI Z21.69 are designed to tolerate the movement anticipated when appliances are moved for cleaning. This section also allows such appliances to be connected with rigid pipe and semirigid metal tubing; however, it would be impractical and somewhat risky to connect such appliances with rigid pipe or metal tubing because employees not skilled in piping installation would be required to break unions or tubing fittings and disconnect and reconnect appliances to the rigid piping or tubing every time the appliances were moved.

411.1.2 Protection against damage. Connectors and tubing shall be installed so as to be protected against physical damage.

❖ Appliance connectors, although sound, are not constructed or tested to withstand the same rigors of service as the gas piping system without physical damage. Therefore, they must be installed to minimize the potential for damage. For example, "flexible" and semirigid tubing connectors should not be used where subject to excessive vibration or impact by occupants, vehicles, animals, doors, stored materials, etc. Impact, repeated movement and vibration can cause connector failure.

It is not the intent of this section to require that appliance connectors be hidden from view behind or under an appliance to consider them as protected from damage. For typical range and clothes dryer installations, the connectors are physically protected by the appliances themselves. However, for water heaters, furnaces, boilers, unit heaters, infrared heaters, etc., the connector will be located in the open and thus will need protection from damage by proper location, placement, elevation or guards. Protection from damage is commonly afforded by locating the connector out of harm's way, such as between a wall and an appliance or 8 feet (2438 mm) or more above the floor.

Figure 411.1(1)
PIPING SYSTEM SIZING AND APPLIANCE CONNECTION AND CONNECTOR SIZING

Figure 411.1(2)
QUICK-DISCONNECT DEVICE FOR USE WITH COMMERCIAL APPLIANCE CONNECTORS
(Photo courtesy of Dormont Manufacturing Company)

Figure 411.1(3)
QUICK-DISCONNECT DEVICE FOR USE WITH COMMERCIAL APPLIANCE CONNECTORS
(Photo courtesy of Dormont Manufacturing Company)

Figure 411.1(4)
CSST DIRECTLY CONNECTED TO NONMOVEABLE APPLIANCE

Figure 411.1(5)
CSST TERMINATED AT APPLIANCE LOCATION

**Figure 411.1(6)
SEMIRIGID (FLEXIBLE) APPLIANCE CONNECTOR
(Photo courtesy of Dormont Manufacturing Company)**

**Figure 411.1(7)
OUTDOOR MOBILE HOME CONNECTOR
(Photo courtesy of Dormont Manufacturing Company)**

PORTABLE OUTDOOR NATURAL GAS FIRE PIT OR PATIO HEATER

OUTDOOR GAS HOSE CONNECTOR LISTED TO ANSI Z21.54

CONNECTS TO NATURAL GAS SUPPLY AT AN APPLIANCE SHUTOFF VALVE, A LISTED QUICK-DISCONNECT DEVICE, OR LISTED GAS CONVENIENCE OUTLET.

**Figure 411.1(8)
NATURAL GAS HOSE AND CONNECTOR FOR PORTABLE OUTDOOR GAS APPLIANCES
(Figure courtesy of Dormont Manufacturing Company)**

411.1.3 Connector installation. *Appliance* fuel connectors shall be installed in accordance with the manufacturer's instructions and Sections 411.1.3.1 through 411.1.3.4.

❖ Flexible (semirigid) appliance fuel connectors are primarily intended for use with cooking ranges, clothes dryers and similar appliances where the gas connection is located behind the appliance and some degree of flexibility is necessary to facilitate the hook-up; however, flexible connectors may be used with any appliance where they are not subject to damage (see commentary, Section 411.1.1). Rigid connections are not practical for movable appliances. Flexible connectors are typically constructed of stainless steel and are labeled with tags or metal rings placed over

the tubing. The manufacturer's instructions dictate the installation requirements that are intended to protect the connector from damage and prevent leakage. Most flexible connectors are intended to be installed only once and are not designed for repeated bending and forming. For other than commercial flexible connections that are designed for repeated movement, a new connector should be used each time an appliance is reinstalled. Repeated flexing or bending can cause metal fatigue and connector failure. Flexible connectors must be protected from physical damage and are not allowed to pass through any walls, floors, ceilings or appliance housings except as allowed by Section 411.1.3.3.

Figure 411.1.1(1)
COMMERCIAL COOKING APPLIANCE
CONNECTOR WITH TRAVEL RESTRAINT
(Photo courtesy of Dormont Manufacturing Company)

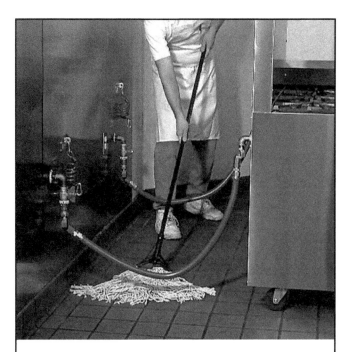

Figure 411.1.1(3)
COMMERCIAL COOKING
APPLIANCE FUEL CONNECTORS
(Photo courtesy of Dormont Manufacturing Company)

Figure 411.1.1(2)
COMMERCIAL COOKING APPLIANCE
CONNECTOR WITH TRAVEL RESTRAINT
(Photo courtesy of Dormont Manufacturing Company)

Figure 411.1.1(4)
COMMERCIAL COOKING
APPLIANCE FUEL CONNECTORS
(Photo courtesy of Dormont Manufacturing Company)

411.1.3.1 Maximum length. Connectors shall have an overall length not to exceed 6 feet (1829 mm). Measurement shall be made along the centerline of the connector. Only one connector shall be used for each *appliance*.

Exception: Rigid metallic *piping* used to connect an *appliance* to the *piping* system shall be permitted to have a total length greater than 6 feet (1829 mm), provided that the connecting pipe is sized as part of the *piping* system in accordance with Section 402 and the location of the *appliance* shutoff valve complies with Section 409.5.

❖ Flexible connectors must bear the label of an approved agency and are limited to a maximum of 6 feet (914 mm) in length. Six-foot lengths have traditionally been used for range and dryer installations to allow the connector to be looped or otherwise arranged to allow limited movement of the appliance without stressing the connector. The safety record for such connectors has been good, therefore, the use of 6-foot-length connectors is not limited to specific appliances. Flexible connectors are evaluated in accordance with ANSI Z21.24 or ANSI Z21.69. Multiple connectors must not be connected together to circumvent the maximum length requirement [see Commentary Figures 411.1.3.1(1) and 411.1.3.1(2)].

The exception allows rigid metallic piping (steel) to exceed 6 feet (1829 mm) in length for connection to any appliance if the shutoff valve is located in accordance with Section 409.5 and the connector is sized as gas piping. If the shutoff must be within 6 feet (1829 mm) of the appliance, the connector length is also limited to 6 feet (1829 mm). If the shutoff is remote in accordance with Section 409.5.2 or 409.5.3, the connector length is unlimited if sized as gas piping (see Sections 409.5.1 through 409.5.3).

For SI: 1 foot = 304.8 mm.

Figure 411.1.3.1(1)
APPLIANCE FUEL CONNECTOR

For SI: 1 foot = 304.8 mm.

Figure 411.1.3.1(2)
APPLIANCE FUEL CONNECTOR

411.1.3.2 Minimum size. Connectors shall have the capacity for the total demand of the connected *appliance*.

❖ Connectors must be chosen on the basis that they will provide the total gas demand of the appliance served at the required minimum pressure, as indicated in the manufacturer's instructions. Listed connectors provide capacity information on the tag attached to the connector.

411.1.3.3 Prohibited locations and penetrations. Connectors shall not be concealed within, or extended through, walls, floors, partitions, ceilings or *appliance* housings.

Exceptions:

1. Connectors constructed of materials allowed for *piping* systems in accordance with Section 403 shall be permitted to pass through walls, floors, partitions and ceilings where installed in accordance with Section 409.5.2 or 409.5.3.

2. Rigid steel pipe connectors shall be permitted to extend through openings in *appliance* housings.

3. *Fireplace* inserts that are factory equipped with grommets, sleeves or other means of protection in accordance with the listing of the *appliance*.

4. Semirigid tubing and *listed* connectors shall be permitted to extend through an opening in an *appliance* housing, cabinet or casing where the tubing or connector is protected against damage.

❖ This section requires connectors to be located entirely in the same room as the appliance served to protect them from damage and because the connector begins at the appliance shutoff valve which must also be in the same room as the appliance served if installed in accordance with Section 409.5.1. Exception 1 applies to piping system materials, not semirigid (flexible) appliance connectors, and recognizes that Sections 409.5.2 and 409.5.3 both permit remote shutoff valves, meaning that in many cases, the piping between the shutoff valve and the appliance will necessarily have to pass through walls, floors, etc., in order to reach the appliance served. For example, a gas fireplace on the first floor has its shutoff valve in the basement and the piping (connector) must pass through the floor to connect to the fireplace. With respect to passing through appliance housings, this section is intended to address listed so-called "flexible" appliance connectors and metal tubing used as a connector. As addressed in Exception 2, rigid steel piping has always been allowed to pass through appliance housings and is often required by appliance installation instructions. CSST installed upstream of the appliance shutoff valve is not considered to be a connector; rather, it is part of the piping system. CSST installed downstream of an appliance shutoff valve, like all other piping materials, is treated as a connector, however, CSST cannot be substituted for a listed flexible appliance connector where one is required, such as for moveable appliances

(see commentary for Section 411.1) (see Commentary Figure 411.1.3.3).

Exception 2 recognizes that rigid steel pipe is not likely to be damaged by passing through an appliance housing/cabinet opening designed for that purpose.

Recall that all piping downstream from a shutoff valve is connector piping by definition of piping system.

Exception 3 allows any connector of any material to pass through fireplace insert appliance housings where means integral with the appliance provide adequate protection from abrasion, cutting and vibration.

Exception 4 allows listed appliance connectors and tubing used as a connector to pass through any appliance housing/cabinet/casing provided that the connector is adequately protected from abrasion, cutting and vibration.

Figure 411.1.3.3
CONNECTOR PIPING DOWNSTREAM
OF SHUTOFF VALVE

411.1.3.4 Shutoff valve. A shutoff valve not less than the nominal size of the connector shall be installed ahead of the connector in accordance with Section 409.5.

❖ Connectors are installed downstream of the shutoff valve and that valve must be located in the same room as the appliance except as allowed by Sections 409.5.2 and 409.5.3. All of the connectors addressed in Section 411.1 must be sized for the connected load.

411.1.4 Movable appliances. Where appliances are equipped with casters or are otherwise subject to periodic movement or relocation for purposes such as routine cleaning and maintenance, such appliances shall be connected to the supply system *piping* by means of an *appliance* connector listed as complying with ANSI Z21.69 or by means of Item 1 of Section 411.1. Such flexible connectors shall be installed and protected against physical damage in accordance with the manufacturer's instructions.

❖ Appliances, such as cooking appliances, commonly use heavy-duty, commercial-type flexible gas connectors. These connectors are designed to tolerate repeated movement to allow appliances on wheels or casters to be moved for cleaning operations or relocation.

Such connectors are commonly equipped with quick-disconnect couplings for ease of disconnection and reconnection and restraint cables to limit the amount of appliance movement, thereby preventing stress and damage to the connector.

Flexible connectors designed for specific applications are also used for the connection of equipment and appliances that produce significant vibrations. For example, gas-burning combustion engines in generator sets and HVAC equipment are connected with specially designed vibration dampening flexible connectors. Some pulse-combustion heating appliances employ flexible connectors to control the transmission of noise to gas distribution piping.

411.1.5 (IFGS) Connection of gas engine-powered air conditioners. Internal combustion engines shall not be rigidly connected to the gas supply *piping*.

❖ Internal combustion engines produce significant vibrations and will move on their mounting during operation. Such movement could weaken and damage gas connections and/or gas controls. Some type of "flexible" connector designed for this purpose must be used to allow for movement.

411.1.6 Unions. A union fitting shall be provided for *appliances* connected by rigid metallic pipe. Such unions shall be accessible and located within 6 feet (1829 mm) of the *appliance*.

❖ Although unions have always been installed in rigid pipe connections to appliances by necessity, previous code editions did not specifically require them. Regardless of where the shutoff valve is located, the union must be within 6 feet of the appliance. Note that this section applies only to rigid piping because tubing systems and listed appliance connectors are joined with fittings that can serve as a union. A union is a special type of threaded coupling fitting consisting of three mating components and it is used "inline" to join two sections of piping that could not otherwise be threaded together. Unions also allow an appliance to be easily disconnected for repairs and during gas piping system testing (see Section 404.5).

411.2 Manufactured home connections. Manufactured homes shall be connected to the distribution *piping* system by one of the following materials:

1. Metallic pipe in accordance with Section 403.4.

2. Metallic tubing in accordance with Section 403.5.

3. *Listed* and *labeled* connectors in compliance with ANSI Z21.75/CSA 6.27 and installed in accordance with the manufacturer's instructions.

❖ Manufactured homes are defined in the National Fuel Gas Code, Z223.1, as having a permanent chassis and for use with or without a permanent foundation (see Commentary Figure 411.1(7)].

411.3 Suspended low-intensity infrared tube heaters. Suspended low-intensity infrared tube heaters shall be connected to the building *piping* system with a connector *listed* for the application complying with ANSI Z21.24/CGA 6.10. The connector shall be installed as specified by the tube heater manufacturer's instructions.

❖ Such heaters are used in factories, garages, shops and recreation/hobby spaces where the desire is to partially condition the space by heating the occupants rather than the ambient air in the space. Although uncommon, radiant tube heaters could be found within dwelling accessory structures and garages. This type of appliance will move each time it cycles on and off because of linear expansion and contraction of the radiant tube heating element. This movement has been known to cause stress failures of the appliance connectors that serve these appliances. The appliance manufacturer's installation instructions will provide very specific details on how the appliance connector is to be installed, configured and oriented. Commentary Figures 411.3(1) and 411.3(2) are examples of one manufacturer's installation instructions. The instructions will vary from manufacturer to manufacturer, but the theme and intent are the same; that is, to orient and configure the connector such that it will flex with the movement of the appliance without causing stress in the connector.

PIPE AND CONNECTOR IN THE SAME VERTICAL PLANE PARALLEL WITH THE LONG DIMENSION OF THE HEATER

APPLIANCE CONNECTOR LISTED TO Z21.24

12"

BURNER/CONTROL BOX

HEATER RADIANT TUBE

←MOVEMENT→

SIDE VIEW

Figure 411.3(1)
EXAMPLE OF ONE MANUFACTURER'S
INSTALLATION INSTRUCTIONS

Figure 411.3(2)
END VIEW OF FIGURE 411.3(1)

(Text inside the figure:)
ALTERNATE LOCATIONS 1-4 FOR SUPPLY PIPING

GAS SUPPLY DROP

NOTE: IN ALL LOCATIONS, THE HORIZONTAL SECTION OF PIPE CONNECTED TO THE BRANCH OF THE TEE, THE APPLIANCE CONNECTOR AND THE CONNECTION TO THE APPLIANCE ARE ALL IN THE SAME PLANE PARALLEL WITH THE HEATER TUBE

BURNER/CONTROL BOX

SECTION 412 (IFGC)
LIQUEFIED PETROLEUM GAS MOTOR
VEHICLE FUEL-DISPENSING FACILITIES

[F] 412.1 General. Motor fuel-dispensing facilities for LP-gas fuel shall be in accordance with this section and the *International Fire Code*. The operation of LP-gas motor fuel-dispensing facilities shall be regulated by the *International Fire Code*.

❖ Section 2308 of the *International Fire Code®* (IFC®) regulates LP-gas service stations and references Chapter 38 of the IFC.

[F] 412.2 Storage and dispensing. Storage vessels and *equipment* used for the storage or dispensing of LP-gas shall be *approved* or *listed* in accordance with Sections 412.3 and 412.4.

❖ Sections 412.3 and 412.4 dictate which components must be listed and which components must be approved by the code official.

[F] 412.3 Approved equipment. Containers; pressure-relief devices, including pressure-relief valves; and pressure regulators and *piping* used for LP-gas shall be *approved*.

❖ Listing of the specified components could be the basis for approval, but listing is not mandated. Any such components must be designed for the application and recommended for the application by the manufacturer.

[F] 412.4 Listed equipment. Hoses, hose connections, vehicle fuel connections, dispensers, LP-gas pumps and electrical *equipment* used for LP-gas shall be listed.

❖ The specified components must be listed and labeled as complying with the relevant product standards. As with all listed products, the testing/listing agency will apply its seal or mark to the product.

[F] 412.5 Attendants. Motor vehicle fueling operations shall be conducted by qualified attendants or in accordance with Section 412.9 by persons trained in the proper handling of LP-gas.

❖ The general public is not allowed to fuel vehicles at LP-gas service stations. The fueling operation must be performed by a service station attendant or trained personnel such as the employees of the company that owns and operates the fuel-dispensing system (see Section 412.8).

[F] 412.6 Location. The point of transfer for LP-gas dispensing operations shall be separated from buildings and other exposures in accordance with the following:

1. Not less than 25 feet (7620 mm) from buildings where the exterior wall is not part of a fire-resistance-rated assembly having a rating of 1 hour or greater.

2. Not less than 25 feet (7620 mm) from combustible overhangs on buildings, measured from a vertical line dropped from the face of the overhang at a point nearest the point of transfer.

3. Not less than 25 feet (7620 mm) from the lot line of property that can be built upon.

4. Not less than 25 feet (7620 mm) from the centerline of the nearest mainline railroad track.

5. Not less than 10 feet (3048 mm) from public streets, highways, thoroughfares, sidewalks and driveways.

6. Not less than 10 feet (3048 mm) from buildings where the exterior wall is part of a fire-resistance-rated assembly having a rating of 1 hour or greater.

> **Exception:** The point of transfer for LP-gas dispensing operations need not be separated from canopies that are constructed in accordance with the *International Building Code* and that provide weather protection for the dispensing equipment.

Liquefied petroleum gas containers shall be located in accordance with the *International Fire Code*. Liquefied petroleum gas storage and dispensing *equipment* shall be located outdoors and in accordance with the *International Fire Code*.

❖ This section specifies a minimum separation distance from buildings, property lines, streets, driveways, sidewalks and railroads to protect them from fire and/or explosion that could occur during a dispensing operation (see Section 2307 and Chapter 61 of the IFC).

[F] 412.7 Additional requirements for LP-gas dispensers and equipment. LP-gas dispensers and related equipment shall comply with the following provisions:

1. Pumps shall be fixed in place and shall be designed to allow control of the flow and to prevent leakage and accidental discharge.

2. Dispensing devices installed within 10 feet (3048 mm) of where vehicle traffic occurs shall be protected against physical damage by mounting on a concrete island 6 inches (152 mm) or more in height, or shall be protected in accordance with Section 312 of the *International Fire Code*.

3. Dispensing devices shall be securely fastened to their mounting surface in accordance with the dispenser manufacturer's instructions.

❖ See Section 2307 of the commentary to the IFC. LP-gas dispensers are the equivalent of gasoline pumps (dispensers). Dispensers are the interface between the fuel supply and the vehicles that receive the fuel and involve operation by personnel.

[F] 412.8 Installation of dispensing devices and equipment. The installation and operation of LP-gas dispensing systems shall be in accordance with this section and the *International Fire Code*. Liquefied petroleum gas dispensers and dispensing stations shall be installed in accordance with manufacturers' specifications and their listing.

❖ Section 412.4 requires that dispensers be listed (see Section 2207 and Chapter 38 of the IFC).

[F] 412.8.1 Product control valves. The dispenser system piping shall be protected from uncontrolled discharge in accordance with the following:

1. Where mounted on a concrete base, a means shall be provided and installed within $^1/_2$ inch (12.7 mm) of the top of the concrete base that will prevent flow from the supply piping in the event that the dispenser is displaced from its mounting.

2. A manual shutoff valve and an excess flow-control check valve shall be located in the liquid line between the pump and the dispenser inlet where the dispensing device is installed at a remote location and is not part of a complete storage and dispensing unit mounted on a common base.

3. An excess flow-control check valve or an emergency shutoff valve shall be installed in or on the dispenser at the point where the dispenser hose is connected to the liquid piping.

4. A *listed* automatic-closing-type hose nozzle valve with or without a latch-open device shall be provided on island-type dispensers.

❖ Item 1 is similar to gasoline pump installations where the fuel supply is automatically cut off in the event that a vehicle impacts the pump/dispenser and displaces it. An excess flow-control check valve senses abnormal flow resulting from a damaged pipe or piece of equipment and then closes to stop the leakage.

[F] 412.8.2 Hoses. Hoses and *piping* for the dispensing of LP-gas shall be provided with hydrostatic relief valves. The hose length shall not exceed 18 feet (5486 mm). An *approved* method shall be provided to protect the hose against mechanical damage.

❖ Hoses convey liquid fuel; therefore, tremendous hydrostatic pressures could result from pumping-control failures and expansion of the liquid resulting from temperature changes. Hoses are limited in length and must be protected to help prevent hose failure and the subsequent fuel leakage.

[F] 412.8.3 Vehicle impact protection. Where installed within 10 feet (3048 mm) of vehicle traffic, LP-gas storage containers, pumps and dispensers shall be protected in accordance with Section 2307.5(2) of the *International Fire Code*.

❖ See Section 2307 of the commentary to the IFC. Item 2 of Section 2307.5 of the IFC requires concrete islands for the dispensers, pumps and containers or requires barriers such as guard posts as prescribed in Section 312 of the IFC.

[F] 412.8.4 Breakaway protection. Dispenser hoses shall be equipped with a *listed* emergency breakaway device designed to retain liquid on both sides of the breakaway point. Where hoses are attached to hose-retrieving mechanisms, the emergency breakaway device shall be located such that the breakaway device activates to protect the dispenser from displacement.

❖ Breakaway protection is needed where vehicles could drive away while still connected to the fuel dispenser, thereby damaging the equipment and causing fuel leakage. The breakaway device is designed to close off both ends of the hose that connect to it so as to trap the liquid fuel in the hose sections that were separated at the device.

[F] 412.9 Public fueling of motor vehicles. Self-service LP-gas dispensing systems, including key, code and card lock dispensing systems, shall be limited to the filling of permanently mounted containers providing fuel to the LP-gas-powered vehicle.

The requirements for self-service LP-gas dispensing systems shall be in accordance with the following:

1. The arrangement and operation of the transfer of product into a vehicle shall be in accordance with this section and Chapter 61 of the *International Fire Code*.

2. The system shall be provided with an emergency shutoff switch located within 100 feet (30 480 mm) of, but not less than 20 feet (6096 mm) from, dispensers.

3. The *owner* of the LP-gas motor fuel-dispensing facility or the owner's designee shall provide for the safe operation of the system and the training of users.

4. The dispenser and hose-end valve shall release not more than 4 cubic centimeters of liquid to the atmosphere upon breaking of the connection with the fill valve on the vehicle.

5. Fire extinguishers shall be provided in accordance with Section 2305.4 of the *International Fire Code*.

6. Warning signs shall be provided in accordance with Section 2305.6 of the *International Fire Code*.

7. The area around the dispenser shall be maintained in accordance with Section 2305.7 of the *International Fire Code*.

❖ There are increased risks associated with self-service dispensers because the users might be uninformed and untrained, might improperly use the equipment and might take risks. The dispenser must be designed to deliver fuel only to onboard vehicle fuel tanks as opposed to any other type of separate container.

SECTION 413 (IFGC)
COMPRESSED NATURAL GAS MOTOR
VEHICLE FUEL-DISPENSING FACILITIES

[F] 413.1 General. Motor fuel-dispensing facilities for CNG fuel shall be in accordance with this section and the *International Fire Code*. The operation of CNG motor fuel-dispensing facilities shall be regulated by the *International Fire Code*.

❖ Section 2308 of the IFC regulates CNG service stations (see Commentary Figure 413.1).

[F] 413.2 General. Storage vessels and *equipment* used for the storage, compression or dispensing of CNG shall be *approved* or *listed* in accordance with Sections 413.2.1 through 413.2.3.

❖ This section dictates which components must be listed and which components must be approved by the code official.

[F] 413.2.1 Approved equipment. Containers; compressors; pressure-relief devices, including pressure-relief valves; and pressure regulators and *piping* used for CNG shall be *approved*.

❖ Listing of the specified components could be the basis for approval, but listing is not mandated. Any such components must be designed for the application and recommended for the application by the manufacturer.

[F] 413.2.2 Listed equipment. Hoses, hose connections, dispensers, gas detection systems and electrical *equipment* used for CNG shall be *listed*. Vehicle fueling connections shall be *listed* and *labeled*.

❖ The specified components must be listed and labeled as complying with the relevant product standards. As

with all listed products, the testing/listing agency will apply its seal or mark to the product.

[F] 413.2.3 General. Residential fueling *appliances* shall be *listed*. The capacity of a residential fueling *appliance* shall not exceed 5 standard cubic feet per minute (0.14 standard cubic meter/min) of natural gas.

❖ This provision is consistent with NFPA 52, 2006 Edition. This section provides installation requirements for listed residential compressed natural gas (CNG) fueling appliances and would allow their installation indoors. These appliances do not consume natural gas but are designed to compress natural gas from the fuel gas supply piping for delivery to CNG vehicles. These fueling appliances are electrically driven, do not have built-in storage, and are designed to provide a slow refill [maximum of 5 cubic feet per minute (2.4 L/s)] of a vehicle's onboard CNG vessel.

[F] 413.3 Location of dispensing operations and equipment. Compression, storage and dispensing *equipment* shall be located above ground outside.

Exceptions:

1. Compression, storage or dispensing *equipment* is allowed in buildings of noncombustible construction, as set forth in the *International Building Code*, which are unenclosed for three-quarters or more of the perimeter.

2. Compression, storage and dispensing *equipment* is allowed to be located indoors or in vaults in accordance with the *International Fire Code*.

3. Residential fueling *appliances* and *equipment* shall be allowed to be installed indoors in accordance with the *equipment* manufacturer's instructions and Section 413.4.3.

❖ Because of the potential for leakage, the compression, storage and dispensing equipment must be located either outdoors or in a noncombustible, substantially open building except as allowed by Section 2208, Chapter 30 of the IFC, and Item 3 of this section (see Section 413.4.3).

[F] 413.3.1 Location on property. In addition to the fuel-dispensing requirements of the *International Fire Code*, compression, storage and dispensing *equipment* not located in vaults complying with the *International Fire Code* and other than residential fueling appliances shall not be installed:

1. Beneath power lines.

2. Less than 10 feet (3048 mm) from the nearest building or property that could be built on, public street, sidewalk or source of ignition.

   Exception: Dispensing *equipment* need not be separated from canopies that provide weather protection for the dispensing *equipment* and are constructed in accordance with the *International Building Code*.

3. Less than 25 feet (7620 mm) from the nearest rail of any railroad track.

4. Less than 50 feet (15 240 mm) from the nearest rail of any railroad main track or any railroad or transit line where power for train propulsion is provided by an outside electrical source, such as third rail or overhead catenary.

5. Less than 50 feet (15 240 mm) from the vertical plane below the nearest overhead wire of a trolley bus line.

❖ This section lists the limitations on location of CNG dispensing stations with respect to power lines, buildings, property lines, railroad tracks, streets, sidewalks, sources of ignition, electric trains and trolleys.

[F] 413.4 Residential fueling appliance installation. Residential fueling *appliance*s shall be installed in accordance with Sections 413.4.1 through 413.4.3.

❖ Residential fueling appliances take fuel gas from distribution piping and compress it to pressures as high as 3,600 psi (24.8 MPa) and deliver it to the motor vehicle's on-board storage vessel. Because natural gas will not liquefy at normal ambient temperatures, the gas must be stored in its gaseous state and high pressures are necessary to increase the amount of storage to practical volumes. For example, a pressure vessel with a volume of 1 cubic foot (0.028 m³) will hold 289 cubic feet (8.2 m³) of natural gas when the vessel pressure is 3,600 psi (24.8 MPa) at 70°F (21°C).

[F] 413.4.1 Gas connections. Residential fueling appliances shall be connected to the premises' gas *piping* system without causing damage to the *piping* system or the connection to the internal *appliance* apparatus.

❖ This requirement is applicable to all appliances regulated by the code. The appliance manufacturer's instructions will dictate the method of connecting to the distribution piping of the building.

[F] 413.4.2 Outdoor installation. Residential fueling *appliance*s located outdoors shall be installed on a firm, noncombustible base.

❖ As with all appliances, a stable and substantial base (pad) must be provided so that the appliance will be properly supported, protected from settling and remain level in accordance with the manufacturer's instructions.

[F] 413.4.3 Indoor installation. Where located indoors, residential fueling *appliance*s shall be vented to the outdoors. A gas detector set to operate at one-fifth of the lower limit of flammability of natural gas shall be installed in the room or space containing the *appliance*. The detector shall be located within 6 inches (152 mm) of the highest point in the room or space. The detector shall stop the operation of the *appliance* and activate an audible or a visual alarm.

❖ Indoor spaces/rooms that house fueling appliances must be ventilated to allow any gas leakage to escape to the outdoors. The required gas detector will stop the appliance operation and activate an alarm before the gas/air mixture in the space/room reaches its threshold of flammability.

  The detector location is based on the fact that natural gas (typical specific gravity is 0.60) is lighter than air and will rise to the highest part of the room or space.

[F] 413.5 Private fueling of motor vehicles. Self-service CNG-dispensing systems, including key, code and card lock dispensing systems, shall be limited to the filling of permanently mounted fuel containers on CNG-powered vehicles.

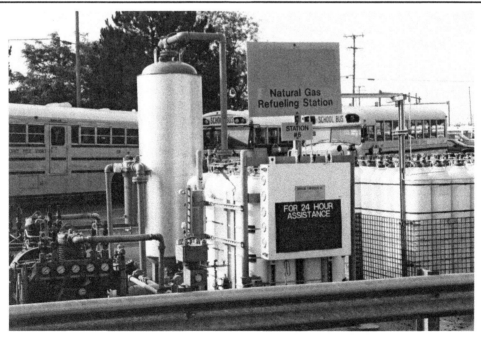

Figure 413.1
CNG FUEL-DISPENSING STATION

In addition to the requirements in the *International Fire Code*, the owner of a self-service CNG-dispensing facility shall ensure the safe operation of the system and the training of users.

❖ Unlike LP-gas self-service dispensing stations, CNG self-service dispensing systems can be open to the public. The owner of the system must provide for the training of any users of the system

[F] 413.6 Pressure regulators. Pressure regulators shall be designed, installed or protected so their operation will not be affected by the elements (freezing rain, sleet, snow, ice, mud or debris). This protection is allowed to be integral with the regulator.

❖ Pressure regulator failure could result in dangerous overpressure and the opening of relief valves; therefore, they must be dependable. The regulator vent is susceptible to blockage by debris and ice.

[F] 413.7 Valves. *Piping* to *equipment* shall be provided with a remote manual shutoff valve. Such valve shall be provided with ready access.

❖ Shutoff valves allow isolation of components for service, repair, replacement and emergency shutdown (see the definition of "Ready access").

[F] 413.8 Emergency shutdown control. An emergency shutdown device shall be located within 75 feet (22 860 mm) of, but not less than 25 feet (7620 mm) from, dispensers and shall also be provided in the compressor area. Upon activation, the emergency shutdown system shall automatically shut off the power supply to the compressor and close valves between the main gas supply and the compressor and between the storage containers and dispensers.

❖ CNG systems take natural gas from the utility supply line, compress it to very high pressures and store the compressed gas in vessels from which the dispensers draw the gas for transfer to the vehicle onboard containers. The emergency shutdown device must be located not farther than 75 feet (22.9 m) from and not closer than 25 feet (7.6 m) to the dispensers. An additional shutdown device must be located near the compressors. The compressors may be remote from the dispensers. The shutdown device must kill the power to the compressors, must actuate automatic valves that isolate the gas supply from the compressors and isolate the storage vessels from the dispenser, thus limiting accidental gas discharge.

[F] 413.9 Discharge of CNG from motor vehicle fuel storage containers. The discharge of CNG from motor vehicle fuel cylinders for the purposes of maintenance, cylinder certification, calibration of dispensers or other activities shall be in accordance with this section. The discharge of CNG from motor vehicle fuel cylinders shall be accomplished through a closed transfer system or an *approved* method of atmospheric venting in accordance with Section 413.9.1 or 413.9.2.

❖ Gas is intentionally discharged from vehicle containers for several reasons including vehicle repairs, certification of the container integrity and container replacement. The gas must be discharged in a safe

manner as dictated by either Section 413.9.1 or 413.9.2. The vehicle storage container is referred to as a "vessel" and as a "cylinder" in the text to follow.

[F] 413.9.1 Closed transfer system. A documented procedure that explains the logical sequence for discharging the cylinder shall be provided to the code official for review and approval. The procedure shall include what actions the operator will take in the event of a low-pressure or high-pressure natural gas release during the discharging activity. A drawing illustrating the arrangement of *piping*, regulators and *equipment* settings shall be provided to the code official for review and approval. The drawing shall illustrate the *piping* and regulator arrangement and shall be shown in spatial relation to the location of the compressor, storage vessels and emergency shutdown devices.

❖ A closed transfer system uses the same basic components as a dispensing system and withdraws the gas from the vehicle container and stores it in vessels. Drawings and a description of the sequence of operation of the transfer system must be submitted to the code official for approval.

[F] 413.9.2 Atmospheric venting. Atmospheric venting of motor vehicle fuel cylinders shall be in accordance with Sections 413.9.2.1 through 413.9.2.6.

❖ The six subsections that follow list the conditions and requirements under which gas may be discharged to the atmosphere. Such discharge should be avoided where practical because:

1. There is inherent hazard in doing so;

2. Methane is an air contaminant with a high global warming potential; and

3. Natural resources should never be wasted.

A closed transfer system is the preferable way to remove gas from vehicle containers.

[F] 413.9.2.1 Plans and specifications. A drawing illustrating the location of the vessel support, *piping*, the method of grounding and bonding, and other requirements specified herein shall be provided to the code official for review and approval.

❖ Plans of the proposed atmospheric venting apparatus and piping system must be reviewed and approved by the code official.

[F] 413.9.2.2 Cylinder stability. A method of rigidly supporting the vessel during the venting of CNG shall be provided. The selected method shall provide not less than two points of support and shall prevent horizontal and lateral movement of the vessel. The system shall be designed to prevent movement of the vessel based on the highest gas-release velocity through valve orifices at the vessel's rated pressure and volume. The structure or appurtenance shall be constructed of *noncombustible materials*.

❖ Vehicle CNG containers (vessels) can contain extremely high pressures, which if released quickly, can produce large thrust forces that would propel the container like a rocket. Natural gas does not liquify at normal ambient temperatures; therefore, to hold the required amount of fuel on board the vehicle, the gaseous fuel must be compressed to extreme pressures of up to 3,600 psia (24.8 MPa). Such pressures are unnecessary for LP-gas because it can be stored as a liquid at normal ambient temperatures.

[F] 413.9.2.3 Separation. The structure or appurtenance used for stabilizing the cylinder shall be separated from the site *equipment*, features and exposures and shall be located in accordance with Table 413.9.2.3.

❖ This section and Table 413.9.2.3 require that the location of the cylinder to be discharged (vessel) be separated from buildings, building openings, lot lines, public ways, vehicles and other CNG equipment. This precaution is reasonable considering the potential hazard of working with highly pressurized containers of a flammable gas.

[F] TABLE 413.9.2.3
SEPARATION DISTANCE FOR ATMOSPHERIC VENTING OF CNG

EQUIPMENT OR FEATURE	MINIMUM SEPARATION (feet)
Buildings	25
Building openings	25
Lot lines	15
Public ways	15
Vehicles	25
CNG compressor and storage vessels	25
CNG dispensers	25

For SI: 1 foot = 304.8 mm.

❖ See the commentary to Section 413.9.2.3.

[F] 413.9.2.4 Grounding and bonding. The structure or appurtenance used for supporting the cylinder shall be grounded in accordance with NFPA 70. The cylinder valve shall be bonded prior to the commencement of venting operations.

❖ "Grounding" means to intentionally connect to the earth. This could be accomplished by installing a conductor between the cylinder support and the building grounding electrode system. Bonding means to join metallic parts to form a continuous electrical pathway. Grounding and bonding required by this section are intended to control sparking that could result from current flow produced by voltage differentials across parts of the venting set-up and the building components. Grounding the venting set-up and attaching a bonding jumper to the cylinder valve will put all such components and the building components at the same voltage potential, thereby reducing the possibility of sparks that could ignite flammable vapors and/or harm the cylinder assembly. The provisions of this section also help prevent the buildup of static electrical charges that could be a source of ignition.

[F] 413.9.2.5 Vent tube. A vent tube that will divert the gas flow to the atmosphere shall be installed on the cylinder prior to the commencement of the venting and purging operation. The vent tube shall be constructed of pipe or tubing materials

approved for use with CNG in accordance with the *International Fire Code*.

The vent tube shall be capable of dispersing the gas not less than 10 feet (3048 mm) above grade level. The vent tube shall not be provided with a rain cap or other feature that would limit or obstruct the gas flow.

At the connection fitting of the vent tube and the CNG cylinder, a *listed* bidirectional detonation flame arrester shall be provided.

❖ The cylinder must be discharged through a vent tube/pipe that is constructed of a material that is compatible with the gas and that has the required strength to withstand the pressure to which it can be exposed. The pressure that the vent can be exposed to must be calculated based on the size of the vent tube, the length of the vent, the friction loss through the vent and the maximum pressure and discharge rate of the cylinder. Vent failure could cause injury resulting from projectile debris and could cause a severe fire/explosion hazard.

[F] 413.9.2.6 Signage. Approved NO SMOKING signs shall be posted within 10 feet (3048 mm) of the cylinder support structure or appurtenance. *Approved* CYLINDER SHALL BE BONDED signs shall be posted on the cylinder support structure or appurtenance.

❖ Approved signs are intended to remind personnel of the potential danger of cylinder venting (see Section 413.9.2.4).

SECTION 414 (IFGC)
SUPPLEMENTAL AND STANDBY GAS SUPPLY

414.1 Use of air or oxygen under pressure. Where air or oxygen under pressure is used in connection with the gas supply, effective means such as a backpressure regulator and relief valve shall be provided to prevent air or oxygen from passing back into the gas *piping*. Where oxygen is used, installation shall be in accordance with NFPA 51.

❖ All sources of oxygen must be isolated from the fuel supply to prevent the possibility of creating a flammable/explosive mixture in the piping system. NFPA 51 applies to oxygen and fuel gas piping systems used for welding, cutting and heat treating metals.

414.2 Interconnections for standby fuels. Where supplementary gas for standby use is connected downstream from a meter or a service regulator where a meter is not provided, a device to prevent backflow shall be installed. A three-way valve installed to admit the standby supply and at the same time shut off the regular supply shall be permitted to be used for this purpose.

❖ Some applications require a continuous supply of fuel gas, and a standby gas supply may be provided that will be used in the event that the normal supply is interrupted. Where a standby fuel gas, such as LP-gas, is used as a fuel for a different gas supply system, a three-way valve or other approved safeguard system must be used to prevent the two different supply systems from being interconnected. The three-way valve is analogous to a single-pole, double-throw switch and provides an opening configuration that allows either fuel supply to be connected to the outlet (load), but does not allow both supplies to be connected simultaneously to the outlet or connected to each other. A manual or automatic valve is used to control which fuel gas line is supplying the fuel to the fuel-burning equipment.

SECTION 415 (IFGS)
PIPING SUPPORT INTERVALS

415.1 Interval of support. *Piping* shall be supported at intervals not exceeding the spacing specified in Table 415.1. Spacing of supports for CSST shall be in accordance with the CSST manufacturer's instructions.

TABLE 415.1
SUPPORT OF PIPING

STEEL PIPE, NOMINAL SIZE OF PIPE (inches)	SPACING OF SUPPORTS (feet)	NOMINAL SIZE OF TUBING (SMOOTH-WALL) (inch O.D.)	SPACING OF SUPPORTS (feet)
$^1/_2$	6	$^1/_2$	4
$^3/_4$ or 1	8	$^5/_8$ or $^3/_4$	6
$1^1/_4$ or larger (horizontal)	10	$^7/_8$ or 1 (horizontal)	8
$1^1/_4$ or larger (vertical)	Every floor level	1 or larger (vertical)	Every floor level

For SI: 1 inch = 25.4 mm, 1 foot = 304.8 mm.

❖ This table is two tables side-by-side. The first two columns are for steel pipe, and the second two columns are for tubing materials (see Section 407). Support for CSST is dictated by the manufacturers' installation instructions.

SECTION 416 (IFGS)
OVERPRESSURE PROTECTION DEVICES

416.1 Where required. Where the serving gas supplier delivers gas at a pressure greater than 2 psi for piping systems serving appliances designed to operate at a gas pressure of 14 inches w.c. or less, overpressure protection devices shall be installed. Piping systems serving equipment designed to operate at inlet pressures greater than 14 inches w.c. shall be equipped with overpressure protection devices as required by the appliance manufacturer's installation instructions.

❖ Overpressure protection devices are used to protect appliances from abnormally high gas pressures that can result from the failure of pressure regulators in the piping system. Overpressure could result in appliance damage and/or hazardous operation (see commentary, Section 410.1)

416.2 Pressure limitation requirements. The requirements for pressure limitation shall be in accordance with Sections 416.2.1 through 416.2.5.

❖ See the referenced sections.

416.2.1 Pressure under 14 inches w.c. Where piping systems serving appliances designed to operate with a gas supply pressure of 14 inches w.c. or less are required to be equipped with overpressure protection by Section 416.1, each overpressure protection device shall be adjusted to limit the gas pressure to each connected appliance to 2 psi or less upon a failure of the line pressure regulator.

❖ Most appliances, other than special purpose and industrial scale appliances, are designed to be operated with a gas supply inlet pressure of 14 inches ($^1/_2$ psi) or less. Section 416.1 requires overpressure protection for such appliances where the gas supplier (utility) delivers gas at a pressure greater than 2 psi. This is because the typical gas appliance gas controls are tested for tolerance of abnormal pressure excursions of up to 2 psi. An unintended pressure increase up to 2 psi maximum is not expected to damage the gas controls for the appliance; however, pressures above 2 psi are capable of causing permanent damage to the gas controls. Where the gas supplier delivers at pressures above 2 psi, overpressure protection is required to prevent damage to the appliance gas controls.

416.2.2 Pressure over 14 inches w.c. Where piping systems serving appliances designed to operate with a gas supply pressure greater than 14 inches w.c. are required to be equipped with overpressure protection by Section 416.1, each overpressure protection device shall be adjusted to limit the gas pressure to each connected appliance as required by the appliance manufacturer's installation instructions.

❖ This section relates to the second sentence of Section 416.1. Overpressure protection would be required where specified by the appliance manufacturer.

416.2.3 Device capability. Each overpressure protection device installed to meet the requirements of this section shall be capable of limiting the pressure to its connected appliance(s) as required by this Section 416.2.1, independently of any other pressure control equipment in the piping system.

❖ The required overpressure protection device must be independently capable of limiting the pressure to the appliance to a maximum of 2 psi in the event that the line pressure regulator fails.

416.2.4 Failure detection. Each gas piping system for which an overpressure protection device is required by Section 416 shall be designed and installed so that a failure of the primary pressure control device(s) is detectable.

❖ Gas supply systems must not be allowed to operate indefinitely in the emergency mode, because in that mode the required overpressure protection device is the only pressure-limiting device capable of protecting the appliances served. Imagine a person continuing to skydive, knowing that they had a defective primary parachute and only an emergency parachute. Failure of the primary pressure-regulating device must be quickly identified so that it can be repaired or replaced and the system can be restored to the intended system design, which consists of a primary pressure control with an overpressure protection device as a backup control.

416.2.5 Relief valve. Where a pressure relief valve is used to meet the requirements of Section 416, it shall have a flow capacity such that the pressure in the protected system is maintained at or below the limits specified in Section 416.2.1 under all of the following conditions:

1. The line pressure regulator for which the relief valve is providing overpressure protection has failed wide open.

2. The gas pressure at the inlet of the line pressure regulator for which the relief valve is providing overpressure protection is not less than the regulator's normal operating inlet pressure.

❖ A pressure relief valve will discharge gas in order to limit the piping system pressure and this device must be sized for worst case conditions. Worst case conditions are: 1) the failed primary pressure-regulating device has failed to fully open, allowing the maximum flow of gas through the regulator; and 2) the gas supply pressure at the inlet of the failed primary pressure-regulating device is not abnormally low, but is instead in the normal pressure range expected for the piping system.

416.3 Devices. Pressure-relieving or pressure-limiting devices shall be one of the following:

1. Pressure relief valve.

2. Monitoring regulator.

3. Series regulator installed upstream from the line regulator and set to continuously limit the pressure on the inlet of the line regulator to the maximum values specified by Section 416.2.1.

4. Automatic shutoff device installed in series with the line pressure regulator and set to shut off when the pressure on the downstream *piping* system reaches the maximum values specified by Section 416.2.1. This device shall be designed so that it will remain closed until manually reset.

The devices specified in this section shall be installed either as an integral part of the service or line pressure regulator or as separate units. Where separate pressure-relieving or pressure-limiting devices are installed, they shall comply with Sections 416.3.1 through 416.3.6.

❖ This section lists the types of devices required by Section 416.2. Item 1 refers to types of relief valves that would discharge gas to the outdoors as the means of limiting the piping system pressure downstream of such devices.

Items 2 and 3 refer to a redundant regulator piped in series with the primary regulator. Series regulators operate continuously in limiting function and monitoring regulators might act only when pressures exceed the intended limits [see Commentary Figure 410.2(2)].

Item 4 refers to combination pressure sensors and shutoff valves that will automatically shut off the gas flow when triggered by the pressure sensor.

The typical utility-owned service regulator has a relief valve built into the body of the regulator, thereby complying with Section 416.2.

416.3.1 Construction and installation. Pressure-relieving and pressure-limiting devices shall be constructed of materials so that the operation of the devices will not be impaired by corrosion of external parts by the atmosphere or of internal parts by the gas. Pressure-relieving and pressure-limiting devices shall be designed and installed so that they can be operated to determine whether the valve is free. The devices shall be designed and installed so that they can be tested to determine the pressure at which they will operate and examined for leakage when in the closed position.

❖ The devices addressed in Section 416 must be compatible with the gas and its contaminants and also must be suitable for the atmosphere to which they are exposed, such as weather exposure. Relief and limiting devices must be capable of being examined and tested for proper operation.

416.3.2 External control piping. External control *piping* shall be designed and installed so that damage to the control *piping* of one device will not render both the regulator and the overpressure protection device inoperative.

❖ Some devices use relatively small diameter control piping or tubing to sense gas pressure at points in the system or to operate pilot valves and regulators that, in turn, operate larger control devices. Such small piping or tubing is more susceptible to damage than larger pipe and damage to control piping/tubing could result in failure of the devices being controlled. Damage could result from falling ice and snow, submersion in water, caving-in of excavations, falling of vegetation and landscaping materials, etc.

416.3.3 Setting. Each pressure-relieving or pressure-limiting device shall be set so that the gas pressure supplied to the connected appliances does not exceed the limits specified in Section 416.2.1.

❖ The maximum allowable working pressure is established for safety reasons. Section 416.2.1 sets the limit at 2 psi and Section 416.2.2 allows the appliance manufacturer to set the pressure limit. All limits must be accomplished by the overpressure protection devices.

416.3.4 Unauthorized operation. Where unauthorized operation of any shutoff valve could render a pressure-relieving valve or pressure-limiting device inoperative, one of the following shall be accomplished:

1. The valve shall be locked in the open position. Authorized personnel shall be instructed in the importance of leaving the shutoff valve open and of being present while the shutoff valve is closed so that it can be locked in the open position before leaving the premises.

2. Duplicate relief valves shall be installed, each having adequate capacity to protect the system, and the isolating valves and three-way valves shall be arranged so that only one relief valve can be rendered inoperative at a time.

❖ Valves are sometimes installed to isolate relief valves and limiting controls from the piping system served to allow maintenance or replacement of such devices.

If such isolation valves are left in the closed position after maintenance or are closed by unauthorized personnel, the relief/limiting devices are useless and protection is lost. Such isolation valves must be either locked in the open position or arranged such that an alternate relief/limiting device is always in the circuit, regardless of the isolation valve position.

416.3.5 Vents. The discharge stacks, vents and outlet parts of all pressure-relieving and pressure-limiting devices shall be located so that gas is safely discharged to the outdoors. Discharge stacks and vents shall be designed to prevent the entry of water, insects and other foreign material that could cause blockage. The discharge stack or vent line shall be not less than the same size as the outlet of the pressure-relieving device.

❖ Discharge piping must terminate in a location where it is unlikely that gas will enter a building opening or be ignited by a source of ignition. Such terminations must be protected from blockage by ice, snow, rodents, insects, debris and other foreign materials. The discharge piping may have to be larger than the size of the outlet on the relief device to compensate for excessive length (see commentary, Section 416.5.6).

416.3.6 Size of fittings, pipe and openings. The fittings, pipe and openings located between the system to be protected and the pressure-relieving device shall be sized to prevent hammering of the valve and to prevent impairment of relief capacity.

❖ Sizing of the piping on the inlet side of relief devices is just as important as sizing of the piping on the discharge side. Undersized piping on the inlet side can reduce the capacity of the relief device and can also cause the valve to rapidly open and close (chatter or hammer), possibly causing damage to the valve and destructive pressure waves in the piping system. This happens when the relief valve opens, thereby reducing the pressure in the undersized inlet piping and the valve reseats (closes) in response to the lowered inlet pressure and then the pressure again climbs in the inlet piping to start the cycle again.

Bibliography

The following resource materials were used in the preparation of the commentary for this chapter of the code.

IBC-2015, *International Building Code*. Washington, DC: International Code Council, 2014.

IFC-2015, *International Fire Code*. Washington, DC: International Code Council, 2014.

Chapter 5:
Chimneys and Vents

General Comments

This chapter intends to regulate the design, construction, installation, maintenance, repair and approval of chimneys, vents and their connections to gas-fired appliances. A properly designed chimney or vent system is needed to conduct to the outdoors the flue gases produced by a gas-fired appliance. In the case of natural-draft appliances (including Category I), the chimney or vent system serving the appliances is depended on to produce a draft at the appliance connection. Draft refers to a phenomenon created by the buoyancy of lighter (less dense) gases in the presence of heavier (more dense) gases. Draft is measured as pressure in inches of water column (kPa) and is a negative pressure, meaning that it is less than the atmospheric pressure at the point of measurement. Draft is produced by the temperature difference between the combustion gases (flue gases) and the ambient atmosphere. Because hotter gases are less dense, they are buoyant and will rise in the chimney, causing negative pressures to develop that are directly related to the height of the chimney or vent and directly related to the temperature difference between the flue gases and the ambient air. Draft produced by a chimney or vent is the motivating force that conveys combustion products from natural-draft appliances to the outside atmosphere through the chimney or vent passageway. It is also the force that draws combustion air into natural-draft atmospheric-burner appliances.

The type of vent or chimney used in a given installation is generally based on the temperature of the flue gases produced by the appliance and the appliance category. The development of higher efficiency appliances continues to produce lower flue gas temperatures, and the diversity of mechanical venting systems continues to expand.

Many vented gas-fired appliances have been further classified into four categories to simplify the process of selecting the appropriate venting system. The four categories of vented appliances are addressed in the definition of "Vented appliance categories." With the advent of new technology and higher efficiency appliances/equipment, venting means are becoming increasingly diversified and complex. Many of today's appliances are designed to be vented by more than one type of venting system.

Purpose

The provisions in this chapter are intended to minimize the hazards associated with the venting of combustion products produced by fuel-gas-fired appliances and equipment. The requirements of this chapter are intended to achieve the complete removal of the products of combustion from vented fuel-gas-fired appliances and equipment. This chapter includes regulations for the proper selection, design, construction and installation of a chimney or vent, along with measures to minimize the related potential fire hazards. A chimney or vent must be designed for the type of appliance or equipment it serves. Chimneys and vents are designed for specific applications depending on the flue gas temperatures and the pressures at which the chimney or vent operates.

The primary hazards associated with chimneys and vents are high temperatures and the chemical nature of combustion byproducts.

SECTION 501 (IFGC)
GENERAL

501.1 Scope. This chapter shall govern the installation, maintenance, repair and approval of factory-built chimneys, chimney liners, vents and connectors and the utilization of masonry chimneys serving gas-fired appliances. The requirements for the installation, maintenance, repair and approval of factory-built chimneys, chimney liners, vents and connectors serving appliances burning fuels other than fuel gas shall be regulated by the *International Mechanical Code*. The construction, repair, maintenance and approval of masonry chimneys shall be regulated by the *International Building Code*.

❖ This section contains requirements for the installation, maintenance, repair and approval of residential, commercial and industrial chimney and venting systems that convey the products of combustion from a gas-fired appliance to the outside atmosphere. Venting systems for fuel-fired appliances other than gas-fired appliances, such as oil and solid-fuel appliances, are covered in *International Mechanical Code®* (IMC®). The construction of masonry chimneys is regulated by Chapter 21 of the *International Building Code®* (IBC®) (see Section 501.3).

501.2 General. Every *appliance* shall discharge the products of combustion to the outdoors, except for appliances exempted by Section 501.8.

❖ Appliances other than those listed in Section 501.8 must be vented to convey the potentially harmful combustion byproducts to the outdoor atmosphere (see Section 501.8).

501.3 Masonry chimneys. Masonry chimneys shall be constructed in accordance with Section 503.5.3 and the *International Building Code*.

❖ A masonry chimney is a field-constructed assembly that can consist of solid masonry units, reinforced concrete, rubble stone, fire-clay liners and mortars. A masonry chimney is permitted to serve low-, medium-, and high-heat appliances. The IBC outlines the general code requirements regarding the construction details for masonry chimneys, including those serving masonry fireplaces.

501.4 Minimum size of chimney or vent. Chimneys and vents shall be sized in accordance with Sections 503 and 504.

❖ Sections 503 and 504 contain the requirements for sizing Category I appliance venting systems. Sizing of venting systems for other categories is dictated by the appliance manufacturer.

501.5 Abandoned inlet openings. Abandoned inlet openings in chimneys and vents shall be closed by an *approved* method.

❖ Abandoned inlet openings result from appliances being disconnected or reconnected at different elevations. Unused openings in chimneys and vents can allow combustion gases to enter the building; can cause loss of draft at other appliances connected to the chimney or vent; can allow conditioned air to escape, resulting in energy losses; and can allow the entry of birds and rodents. For example, consider an appliance that was connected to a chimney and has been replaced with an appliance that vents through an exterior wall. The opening in the chimney left by the disconnection of the appliance must be properly closed to prevent a hazardous condition. Combustion gases from appliances connected to chimneys and vents can escape through openings in the chimney and vent. Those openings can affect the draft of the remaining appliances, causing them to spill combustion gases into the building.

The method used to close an unused opening must not create a protrusion in the flue passageway, which could restrict vent gas flow. It is not uncommon to find multiple openings in chimneys that were used at one time but have since been abandoned and covered or closed in some makeshift fashion, or left open. Inlet openings that are not closed could allow the escape of vent gases and could affect the draft produced by the chimney or vent. Excess dilution air will be admitted through unused openings, which could weaken draft by cooling the vent gases and bypassing the flow through appliance connectors. It is rare to find unused openings in vents because vent openings are easily closed or adapted to different appliance connections; however, the hazard of abandoned openings is the same as for chimneys. Vent and factory-built chimney openings are easily closed by eliminating tees and wyes or by using metal caps. Closure of masonry chimney openings may require the replacement of masonry, especially where the chimney has been structurally weakened by the openings.

501.6 Positive pressure. Where an *appliance* equipped with a mechanical forced draft system creates a positive pressure in the venting system, the venting system shall be designed for positive pressure applications.

❖ Commonly used chimney and vent systems are not designed for positive pressure (above atmospheric). Chimneys and vents are intended to produce a draft and thus operate with negative internal pressures. Specialized vents and chimneys are available that are designed for positive pressure applications (see Commentary Figure 501.6). The misapplication of materials can result in leakage of combustion gases and can cause chimney or vent deterioration from condensate corrosion. Masonry chimneys, Type B vents, Type L vents and most factory-built chimneys are designed for negative pressure applications, meaning that the venting system is producing draft. Negative-pressure venting systems will leak if subjected to positive internal pressures. Leakage can allow combustion gases to enter the building, and condensation can cause rapid deterioration of the chimney or vent.

Figure 501.6
VENT DESIGNED FOR CATEGORY III AND
IV (POSITIVE PRESSURE) APPLICATIONS
(Figure courtesy of Selkirk L.L.C.)

501.7 Connection to fireplace. Connection of appliances to chimney flues serving fireplaces shall be in accordance with Sections 501.7.1 through 501.7.3.

❖ This section regulates installations where fireplace chimneys are used as the venting means for appliances such as room heaters and fireplace-insert-type heaters.

501.7.1 Closure and access. A noncombustible seal shall be provided below the point of connection to prevent entry of room air into the flue. Means shall be provided for *access* to the flue for inspection and cleaning.

❖ Without an air-tight connection between the chimney flue and the appliance chimney connector, chimney draft will draw room air into the flue, thus weakening the draft drawn through the appliance. The chimney connection must be designed for access for inspection and cleaning of the chimney because the chimney flue is no longer open to the fireplace firebox.

501.7.2 Connection to factory-built fireplace flue. An *appliance* shall not be connected to a flue serving a factory-built *fireplace* unless the *appliance* is specifically *listed* for such installation. The connection shall be made in accordance with the *appliance* manufacturer's installation instructions.

❖ Factory-built fireplaces may or may not be tested for any use other than as a traditional fireplace. Installing an appliance or fireplace-insert-type heater could result in abnormally high fireplace or chimney temperatures, component deterioration and hazardous operation. The fireplace manufacturer will provide specific instructions covering what is or is not allowed regarding appliance installations. Typically the manufacturer's instructions will prohibit any modifications and any use of the fireplace other than what it was designed for.

501.7.3 Connection to masonry fireplace flue. A connector shall extend from the *appliance* to the flue serving a masonry *fireplace* such that the flue gases are exhausted directly into the flue. The connector shall be accessible or removable for inspection and cleaning of both the connector and the flue. *Listed* direct connection devices shall be installed in accordance with their listing.

❖ The appliance chimney connector must extend up through the fireplace damper and smoke chamber and terminate in the chimney flue. This helps produce an adequate draft.

501.8 Appliances not required to be vented. The following appliances shall not be required to be vented:

1. Ranges.
2. Built-in domestic cooking units *listed* and marked for optional venting.
3. Hot plates and laundry stoves.
4. Type 1 clothes dryers (Type 1 clothes dryers shall be exhausted in accordance with the requirements of Section 614).
5. A single booster-type automatic instantaneous water heater, where designed and used solely for the sanitiz-ing rinse requirements of a dishwashing machine, provided that the heater is installed in a commercial kitchen having a mechanical exhaust system. Where installed in this manner, the draft hood, if required, shall be in place and unaltered and the draft hood *outlet* shall be not less than 36 inches (914 mm) vertically and 6 inches (152 mm) horizontally from any surface other than the heater.

6. Refrigerators.
7. Counter appliances.
8. Room heaters *listed* for unvented use.
9. Direct-fired makeup air heaters.
10. Other appliances *listed* for unvented use and not provided with flue collars.
11. Specialized appliances of limited input such as laboratory burners and gas lights.

Where the appliances listed in Items 5 through 11 above are installed so that the aggregate input rating exceeds 20 British thermal units (Btu) per hour per cubic foot (207 watts per m³) of volume of the room or space in which such appliances are installed, one or more shall be provided with venting systems or other approved means for conveying the vent gases to the outdoor atmosphere so that the aggregate input rating of the remaining unvented appliances does not exceed 20 Btu per hour per cubic foot (207 watts per m³). Where the room or space in which the *appliance* is installed is directly connected to another room or space by a doorway, archway or other opening of comparable size that cannot be closed, the volume of such adjacent room or space shall be permitted to be included in the calculations.

❖ The specified appliances do not have to be vented unless required by the last paragraph of this section, which addresses Items 5 through 11 and not Items 1 through 4. Section 301.3 requires that all appliances be listed.

Item 4 clarifies that clothes dryers are not vented, but instead, are exhausted. The products of combustion for gas-fired clothes dryers are discharged to the outdoors along with the moisture exhaust; in effect, the dryer is "vented," but not in the sense that other appliances are vented. Clothes dryer exhaust ducts are not referred to as vents because they are exhaust ducts, not a type of vent. Clothes dryer exhaust ducts are erroneously referred to as "vents." Clothes dryers, both gas and electric, must exhaust to the outdoors.

Twenty Btu per hour per cubic foot of room volume (207 W/m³) is the same ratio as 50 cubic feet per 1000 Btu per hour (4.8 m³/W) on which Section 304.5.1 is based. In other words, this section requires that unvented appliances be located in a space that meets the volume requirement for indoor combustion air specified in Section 304.5.1. If the total unvented appliance input rating exceeds the ratio of 20 Btu/h per cubic foot of space volume (207 w/m³), one or more appliances would have to be vented or removed to keep the total at or below the limit. Maintaining a minimum space volume will help control the level of combustion products in the space by dilution with infiltration air. The appliances listed in Items 5 through 11 are allowed to be unvented, only where the total Btu/h input is at or below the limit set for the space volume.

501.9 Chimney entrance. Connectors shall connect to a masonry chimney flue at a point not less than 12 inches (305 mm) above the lowest portion of the interior of the chimney flue.

❖ This requirement is intended to prevent blockage of the connector opening by debris that has collected at the bottom of the flue passage. The 12-inch deep (305 mm) space below the connector functions as a trap (dirt leg) for debris such as dead animals, leaves, flaking tile, mortar and soot. This section is not intended to apply to factory-built chimneys because those chimneys are protected from the entry of debris by the required cap.

501.10 Connections to exhauster. *Appliance* connections to a chimney or vent equipped with a power exhauster shall be made on the inlet side of the exhauster. Joints on the positive pressure side of the exhauster shall be sealed to prevent flue-gas leakage as specified by the manufacturer's installation instructions for the exhauster.

❖ When a mechanical draft device or power exhauster is installed, the resulting "forced draft" creates a positive pressure inside the chimney or vent on the discharge or outlet side of the fan. A hazardous condition will result if an appliance is connected on the outlet or discharge side of a power exhauster. The appliance connector provides an alternate path for the flue gases to travel back into the building. Also, the appliance connected on the discharge side (outlet) of the exhauster will not be supplied with draft, and its combustion products will spill into the building interior.

A hazardous condition can also occur when a vent connector serving an appliance vented by natural draft is connected to the vent of an appliance equipped with integral power-venting means. This section does not prohibit Category I fan-assisted appliances from being common-vented with draft-hood (natural draft) appliances as addressed in Section 504.

Any vent, chimney or connector piping into which an exhauster discharges must be rated and approved for positive pressure and properly installed to prevent leakage of combustion gases. For example, Type B vents cannot be used on the outlet side of an exhauster or a power-vented appliance, regardless of any appliance manufacturer's recommendations. Type B vent material cannot be sealed to prevent leakage when exposed to positive pressure. Positive pressure can force combustion products into the annular space between the pipe walls where condensation could occur, causing deterioration of the piping.

501.11 Masonry chimneys. Masonry chimneys utilized to vent appliances shall be located, constructed and sized as specified in the manufacturer's installation instructions for the appliances being vented and Section 503.

❖ Manufacturers' instructions for today's gas-fired appliances specify very limited conditions under which the appliance is allowed to vent to a masonry chimney. The conditions include chimney size, condition, location and construction. Because many existing masonry chimneys are oversized, unlined, in poor structural condition or partially exposed to the outdoors, they might not be suitable to serve gas-fired appliances. Masonry chimneys serving to vent fan-assisted gas-fired appliances are becoming rare. In many cases, a masonry chimney must be relined with a retrofit metal liner system, or a vent such as Type B vent will have to be installed within the flueway of the chimney. Masonry chimneys have a large thermal mass that has to be heated to above the dew point of the combustion gases to avoid condensation. The flue gas temperatures discharged from many appliances will not be high enough to sufficiently warm the chimney mass, especially where the chimney is oversized and/or exposed to the exterior.

501.12 Residential and low-heat appliances flue lining systems. Flue lining systems for use with residential-type and low-heat appliances shall be limited to the following:

1. Clay flue lining complying with the requirements of ASTM C315 or equivalent. Clay flue lining shall be installed in accordance with the *International Building Code*.

2. *Listed* chimney lining systems complying with UL 1777.

3. Other *approved* materials that will resist, without cracking, softening or corrosion, flue gases and condensate at temperatures up to 1,800°F (982°C).

❖ Flue lining systems for masonry chimneys include clay tile, poured-in-place refractory materials and stainless steel pipe.

501.13 Category I appliance flue lining systems. Flue lining systems for use with Category I appliances shall be limited to the following:

1. Flue lining systems complying with Section 501.12.

2. Chimney lining systems *listed* and *labeled* for use with gas appliances with draft hoods and other Category I gas appliances *listed* and *labeled* for use with Type B vents.

❖ Chimney lining systems for chimneys serving gas-fired appliances include stainless steel and aluminum piping (uninsulated or insulated) designed for both new construction and for relining existing chimneys.

Type B vent is also used as a chimney liner, although it is not considered to be a lining system, and in such cases the chimney is actually a masonry chase containing a vent. Relining of existing chimneys is quite common now because of the incompatibility ᵢ of mid-efficiency appliances and masonry chimneys. Relining systems allow reuse of existing chimneys, thereby saving expense and retaining desirable architectural features of buildings [see Commentary Figures 501.13(1) through 501.13(4)].

501.14 Category II, III and IV appliance venting systems. The design, sizing and installation of vents for Category II, III and IV appliances shall be in accordance with the *appliance* manufacturer's instructions.

❖ Venting systems for these appliances are special vent systems specified by the appliance manufacturer and are not addressed further in the code (see Commentary Figure 501.14).

501.15 Existing chimneys and vents. Where an *appliance* is permanently disconnected from an existing chimney or vent, or where an *appliance* is connected to an existing chimney or vent during the process of a new installation, the chimney or vent shall comply with Sections 501.15.1 through 501.15.4.

❖ Existing chimneys and vents must be reevaluated for continued suitability whenever the conditions of use change. Size, which is covered in Section 501.15.1, is of primary importance. Other considerations are the presence of liner obstructions, combustible deposits in the liner, the structural condition of the liner, the inclusion of a cleanout and clearances to combustibles, which are addressed in Sections 501.15.2 through 501.15.4. It is the intent of this section that whenever a new appliance is connected or an existing appliance is disconnected from a chimney or vent, the chimney or vent is subject to the requirements in Sections 501.15.2 through 501.15.4, which include requirements for inspection, cleaning, possi-

Figure 501.13(2)
MASONRY CHIMNEY RELINING SYSTEM
(Figure courtesy of Simpson Dura-Vent Company, Incorporated)

Figure 501.13(1)
MASONRY CHIMNEY RELINING SYSTEM
(Figure courtesy of Simpson Dura-Vent Company Incorporated)

Figure 501.13(3)
MASONRY CHIMNEY RELINING BY
INSTALLATION OF TYPE B VENT SYSTEM
(Figure courtesy of Simpson Dura-Vent Company, Incorporated)

ble repair, the installation of a cleanout if one is not already there and the establishment of clearances to combustibles in accordance with the requirements for new chimneys or vent installations.

Chimneys, vents and the appliances served are all designed to function together as a system. Any change to an existing chimney or vent system will have an impact on the performance of that system. Something as simple as disconnecting an appliance from a chimney can upset the system balance and cause the venting system to fail to produce a draft for the remaining appliances, and could cause the venting system to produce harmful condensation (see Commentary Figure 501.15.1).

501.15.1 Size. The chimney or vent shall be resized as necessary to control flue gas condensation in the interior of the chimney or vent and to provide the *appliance* or appliances served with the required draft. For Category I appliances, the resizing shall be in accordance with Section 502.

❖ The combined input from multiple appliances, especially older lower-efficiency appliances, can maintain sufficiently high chimney or vent temperatures to provide the necessary draft and to avoid condensation. Changing an existing configuration by disconnecting and eliminating an appliance or by substituting a higher-efficiency appliance can cause a decrease in flue gas temperature, resulting in condensation or poor draft. Also, the elimination of one or more draft-hood-equipped appliances will reduce the amount of dilution air in a venting system, thus increasing the likelihood of condensation. Often, it is necessary to

resize a chimney or vent by replacing it with a smaller-size chimney or vent or by installing a liner system (see Commentary Figure 501.15.1).

A common scenario involves removing chimney-vented furnaces or boilers and leaving a water heater as the only appliance vented to the chimney. In such cases, the chimney would typically be grossly over-sized for the water heater, could fail to produce adequate draft and could be subject to continuous condensation. This scenario has received much attention and has created the phrase "orphaned water heaters."

501.15.2 Flue passageways. The flue gas passageway shall be free of obstructions and combustible deposits and shall be cleaned if previously used for venting a solid or liquid fuel-burning *appliance* or *fireplace*. The flue liner, chimney inner wall or vent inner wall shall be continuous and shall be free of cracks, gaps, perforations or other damage or deterioration that would allow the escape of combustion products, including gases, moisture and creosote.

❖ A chimney or flue must be cleaned to examine the flue liner surface, to eliminate fuel for a possible chimney fire, to reduce frictional resistance to flow and to prevent any residual deposits from falling into lower parts of the venting system. Installation or removal of an appliance without an inspection of the existing flue is prohibited. If the flue served a solid- or liquid-fuel-burning appliance or fireplace, cleaning is required in order to make the required inspection and to remove any combustible deposits. In that case, the addition or removal of an appliance without both

Figure 501.13(4)
MASONRY CHIMNEY RELINING SYSTEM
(Photo courtesy of Simpson Dura-Vent Company, Incorporated)

Figure 501.14
SPECIAL VENT SYSTEM FOR
CATEGORY II, III and IV APPLIANCES
(Photo courtesy of Selkirk L.L.C.)

cleaning and inspection of the existing flue would be a code violation. If inspection shows that the flueway of the vent or chimney is incapable of preventing the leakage of combustion gases, the chimney or vent would have to be repaired or relined.

501.15.3 Cleanout. Masonry chimney flues shall be provided with a cleanout opening having a minimum height of 6 inches (152 mm). The upper edge of the opening shall be located not less than 6 inches (152 mm) below the lowest chimney inlet opening. The cleanout shall be provided with a tight-fitting, noncombustible cover.

❖ This section requires installation of a cleanout in a chimney to facilitate cleaning and inspection. A fireplace inherently provides access to its chimney through the firebox, throat and smoke chamber. The cleanout cover and opening frame must be of an approved material such as cast iron, precast cement or other noncombustible material and must be arranged to remain tightly closed. A loose fitting or unsecured cleanout can allow air to flow into the chimney, affecting draft at the appliance or fireplace, and can result in an energy loss in the building. Under certain conditions, combustion products could escape into the building through an ill-fitting cleanout door or cover. The requirement for placing the cleanout at least 6 inches (152 mm) below the lowest connection to the chimney is intended to minimize the possibility of combustion products exiting the chimney through the cleanout. It is the intent of this section that a cleanout opening be installed, if one is not already provided, for existing masonry chimney flues whenever another appliance is installed or an existing appliance is removed.

501.15.4 Clearances. Chimneys and vents shall have airspace *clearance* to combustibles in accordance with the *International Building Code* and the chimney or vent manufacturer's installation instructions.

Exception: Masonry chimneys without the required airspace clearances shall be permitted to be used if lined or relined with a chimney lining system *listed* for use in chimneys with reduced clearances in accordance with UL 1777. The chimney *clearance* shall be not less than permitted by the terms of the chimney liner listing and the manufacturer's instructions.

❖ This section requires that the existing chimney or vent be inspected for clearances to combustibles for the entire length of the chimney or vent. Clearances to combustibles for masonry chimneys are found in the IBC. The required clearances for a listed factory-built chimney or vent would be in accordance with the existing chimney or vent manufacturer's installation instructions. The clearances to combustibles for vents and factory-built chimneys are typically indicated on the vent or chimney. If they are not, the manufacturer must be contacted and the original installation instructions consulted.

Existing masonry chimneys commonly do not have the clearances required by code, especially if the structure is fairly old. Listed chimney lining systems, which provide insulation value to compensate for the lack of clearance, are available for use in such chimneys. Lining systems listed to UL 1777 include poured-in-place insulating refractory cement and insulated metal liners. The lining system listing will specify what the required clearance must be from the relined chimney to combustibles.

501.15.4.1 Fireblocking. Noncombustible fireblocking shall be provided in accordance with the *International Building Code.*

❖ See Section 2113.20 of the IBC.

Figure 501.15.1
APPLIANCE REPLACEMENTS AFFECTING EXISTING APPLIANCES

SECTION 502 (IFGC)
VENTS

502.1 General. Vents, except as provided in Section 503.7, shall be *listed* and *labeled*. Type B and BW vents shall be tested in accordance with UL 441. Type L vents shall be tested in accordance with UL 641. Vents for Category II and III appliances shall be tested in accordance with UL 1738. Plastic vents for Category IV appliances shall not be required to be *listed* and *labeled* where such vents are as specified by the *appliance* manufacturer and are installed in accordance with the *appliance* manufacturer's instructions.

❖ Vents regulated by this section are factory fabricated and must be listed and labeled. The labeling requirement applies to all components of the system such as the sections of pipe, fittings, terminal caps, supports and spacers.

The provision for unlisted plastic vents is necessary to allow for the venting systems commonly specified in the installation instructions for Category IV condensing appliances. Unlisted plastic pipe (commonly PVC or CPVC plumbing pipe) vents are distinct from listed high-temperature plastic special gas vents that were designed primarily for Category III appliances. Unlisted plastic pipe can be used only when specified by the appliance manufacturer and must be installed as directed in the appliance installation instructions. Unlisted plastic pipe is used only for high efficiency Category IV gas-fired appliances. The pipe functions as an exhaust pipe for combustion gases and as a combustion air intake pipe.

Gas appliances (furnaces and boilers) are categorized so that the proper venting system can be selected. Four categories have been developed based on the pressure produced in the vent passageway and whether or not the flue gas temperatures approach the dew point, making condensation likely (see the definition for "Vented appliance categories").

Category I appliances operate under a negative or neutral vent pressure, are noncondensing appliances, and may generate vent gases with temperatures up to 550° F (288°C). A Category I appliance can use a draft hood or an integral vent blower or draft inducer (fan assisted) [see Commentary Figure 502.1(1)].

Category II appliances also operate under negative or neutral vent pressure, but the vent gas temperature and percent of CO_2 yield a flue loss of less than 17 percent, meaning that condensation of flue gases can occur. Vent systems serving Category II appliances fall under the special vent provisions of Sections 501.14, 503.4 and 503.6.9.3.

Category III appliances operate with a positive vent pressure and can produce condensation, although they are not considered to be "condensing appliances." The positive pressure in the vent system necessitates airtight flue passageways to avoid leakage. Vent systems serving Category III appliances fall under the special vent provisions of Sections 501.14, 503.4, and 503.6.9.3. They are typically constructed of high-temperature plastic or stainless steel

pipe. The type of stainless steel specified today is typically AL29-4C, a special alloy known for its corrosion resistance.

Category IV appliances operate with a positive vent pressure and condense flue gases. Condensing (high-efficiency) appliances require airtight vent passageways and a method of collecting and draining condensation. The vent system serving this type of appliance is usually constructed of plastic, such as PVC or CPVC. Vent systems serving Category IV appliances also fall under the special vent provisions of Sections 501.14, 503.4 and 503.6.9.3 [see Commentary Figure 502.1(2)].

A vent system can be used only in conjunction with gas-burning appliances that are listed and labeled for use with a vent. Appliances equipped with draft hoods and Category I (including fan-assisted) appliances both depend on the vent to produce a draft to convey the combustion products to the outdoors. Solid-fuel-burning appliances must not be connected to a vent system because they produce flue-gas temperatures that are much higher than the temperatures for which vents are designed. The burning of solid fuels also produces creosote. Creosote formation leaves a combustible deposit on the surface of the flue passageway that, if ignited, can produce a fire with temperatures exceeding 2,000°F (1093°C), far exceeding the safe temperature limitations of a vent system.

Type B vents are designed for venting noncondensing gas appliances equipped with a draft hood and fan-assisted appliances that operate with a non-

Figure 502.1(1)
CATEGORY I FAN-ASSISTED FURNACE

positive vent pressure and that are listed and labeled for use with a Type B vent. Type B vents are capable of withstanding continuous firing at flue temperatures not exceeding 400°F (222°C) above room temperature. This section requires that a Type B vent be tested in accordance with UL 441. Type B vents must not be used with incinerators, appliances readily converted to the use of solid or liquid fuels or combination gas/oil burning appliances. Type B vents can serve multiple gas-fired appliance installations [see Commentary Figure 502.1(3)].

Type BW vents must also be tested in accordance with UL 441 and can serve only labeled gas-fired wall furnaces. A Type BW vent is an oval-shaped vent designed and tested for installation in a stud wall cavity. This type of vent is capable of withstanding continuous firing at flue temperatures not exceeding 400°F (222°C) above room temperature. Type BW vents must not be used to vent incinerators, appliances readily converted to the use of solid or liquid fuels or combination gas/oil-burning appliances.

Type L vents must be tested in accordance with UL 641 and are designed to vent oil-burning appliances, which produce a higher flue gas temperature than gas-fired appliances. Type L vents can also be used with appliances listed and labeled for Type B vent systems, meaning that Type L vents can substitute for Type B vents, but Type B vents cannot serve appliances that require Type L vents.

Type B and BW vents are double-walled, air-insulated vent piping systems constructed of galvanized steel outer walls and aluminum inner walls [see Commentary Figure 502.1(4)]. Type L vents are typically double-wall, air-insulated vent piping systems constructed of galvanized and stainless steel.

Type B, BW and L vents are designed for natural draft applications only and must not be used to convey combustion gases under positive pressure. For example, a Type B vent must not be used with Category III or IV gas-fired appliances and must not be used on the discharge side of an exhauster or power-vented appliance. Positive pressures will cause a Type B vent to leak combustion gases into the building, and the leakage can cause condensation to form within the vent system pipe, fittings and joints, resulting in corrosion damage to the vent and eventual vent failure. Fan-assisted Category I appliances are not considered power vented and do not produce positive vent pressures when vented vertically to a properly sized draft-producing vent such as a Type B vent system (see commentary, Section 504). UL 1738 covers factory-built vents for Category II, III and IV appliances. These vents can be constructed of metallic or plastic materials including aluminum alloys, steel (coated, galvanized, stainless or aluminized), PVC and CPVC.

Figure 502.1(3)
TYPE B GAS VENT SERVING
TWO CATEGORY I APPLIANCES
(Photo courtesy of Simpson Dura-Vent Company, Incorporated)

Figure 502.1(2)
CATEGORY IV APPLIANCE (FURNACE)
(Figure courtesy of Carrier Corporation)

Figure 502.1(4)
TYPE B VENT SECTIONS WITH
AIR SPACE BETWEEN PIPE WALLS
(Figure courtesy of Simpson Dura-Vent Company, Incorporated)

502.2 Connectors required. Connectors shall be used to connect appliances to the vertical chimney or vent, except where the chimney or vent is attached directly to the *appliance*. Vent connector size, material, construction and installation shall be in accordance with Section 503.

❖ Connectors designed, constructed and installed in accordance with Section 503 are required except where a chimney or vent serves a single appliance and connects directly to that appliance. In such cases, a connector does not exist. The use of single-wall connectors for fan-assisted appliances is extremely limited by the sizing tables because of the high heat input necessary (minimum capacity column) to offset the heat loss of single-wall pipe.

502.3 Vent application. The application of vents shall be in accordance with Table 503.4.

❖ See Section 503.4.

502.4 Insulation shield. Where vents pass through insulated assemblies, an insulation shield constructed of steel having a minimum thickness of 0.0187 inch (0.4712 mm) (No. 26 gage) shall be installed to provide *clearance* between the vent and the insulation material. The *clearance* shall be not less than the *clearance* to combustibles specified by the vent manufacturer's installation instructions. Where vents pass through attic space, the shield shall terminate not less than 2 inches (51 mm) above the insulation materials and shall be secured in place to prevent displacement. Insulation shields

provided as part of a *listed* vent system shall be installed in accordance with the manufacturer's instructions.

❖ Loose insulation in attic floor assemblies or roof assemblies can fall against vents, creating a fire hazard, and can also cause abnormally high vent temperatures. Even though clearances to combustible construction such as wood framing are maintained, the continued heating of combustible insulation materials that have fallen against a vent could be a source of ignition. This section applies regardless of the combustibility of the insulation. Shields constructed in the field must be sufficient to serve their purpose and must be securely attached to building construction because shields that are merely resting in place could be inadvertently dislodged or removed during maintenance activities.

502.5 Installation. Vent systems shall be sized, installed and terminated in accordance with the vent and *appliance* manufacturer's installation instructions and Section 503.

❖ The standards with which vents and appliances comply require that venting instructions be supplied by the manufacturer of the appliance and the manufacturer of the vent system. These instructions are part of the labeling requirements. Any deviation from them is in violation of the code.

The instructions for Type B vents and for gas appliances approved for use with Type B vents may be specific to a certain appliance or manufacturer's B-vent material. The clearance to combustibles for a vent system is determined by the testing agency and is stated on the component labels and in the manufacturer's instructions. Not all vents have the same required clearances to combustibles. The clearances are determined by the vent's performance during testing in accordance with the applicable standard. Different vent materials and designs impact the vent's ability to control the amount of heat transmitted to surrounding combustibles. It is critical that the vent system be installed in accordance with the clearances listed on the vent's label and in the manufacturer's instructions.

The termination of a natural draft vent must comply with the requirements of the manufacturer's installation instructions and Sections 503.6.3, 503.6.6 and 503.6.7. A vent used in conjunction with a mechanical exhauster must meet the termination requirements established in Sections 503.3.3 and 503.8.

Physical protection of the vent system is required to prevent damage to the vent and to prevent combustibles from coming into contact with or being placed too close to the vents. Such protection is typically provided by enclosing the vent in chases, shafts or cavities in the building construction. Physical protection is not required in the room or space where the vent originates (at the appliance connection) and would not be required in such locations as attics that are not occupied or used for storage. For example, assume that a vent is installed in an existing building and that it extends from the basement, through a first-floor closet, through the attic and through the roof.

The portion of the vent passing through the closet must be protected from damage. The means of physical damage protection should also be designed to maintain separation between the vent and any combustible storage. To prevent the passage of fire and smoke through the annular space around a vent penetration through a floor or ceiling, the vent must be fireblocked with a noncombustible material in accordance with the code. Vent manufacturers provide installation instructions and factory-built components for fireblocking penetrations.

502.6 Support of vents. All portions of vents shall be adequately supported for the design and weight of the materials employed.

❖ Vent manufacturers supply support parts, and manufacturers' instructions contain detailed requirements for support of vent systems. Improper support can cause strain on vent components, appliance connectors and fittings, resulting in vent, appliance or connector damage, as well as loss of required clearance to combustibles and proper slope. Frequently, venting installations suffer from lack of or improperly installed supports, brackets and hangers.

502.7 Protection against physical damage. In concealed locations, where a vent is installed through holes or notches in studs, joists, rafters or similar members less than $1^1/_2$ inches (38 mm) from the nearest edge of the member, the vent shall be protected by shield plates. Protective steel shield plates having a minimum thickness of 0.0575 inch (1.463 mm) (No. 16 gage) shall cover the area of the vent where the member is notched or bored and shall extend not less than 4 inches (102 mm) above sole plates, below top plates and to each side of a stud, joist or rafter.

❖ This section requires protection from nail and screw penetration for vents in the same manner that the code protects gas piping and tubing (see Commentary Figure 502.7). Plastic and metal vent systems typically pass through or are enclosed by construction assemblies, thereby making them highly susceptible to penetration by fasteners. Such penetrations can damage the vent and the damage would not be visible in concealed locations. Possible results from vent penetration include cracking/splitting of pipe, leakage of condensate and/or combustion gases, corrosion failure of metals, structural decay, interference with required clearances and restraint of expansion and contraction movement.

502.7.1 Door swing. Appliance and equipment vent terminals shall be located such that doors cannot swing within 12 inches (305 mm) horizontally of the vent terminal. Door stops or closers shall not be installed to obtain this clearance.

❖ Vent terminals for sidewall-vented appliances (such as direct-vent gas fireplaces and fireplace heaters; direct-vent room heaters; direct-vent water heaters, furnaces and boilers; and also nondirect-vent appliances) are sometimes located where a side-swinging door can impact the vent terminal or swing close to the terminal. This can result in damage to the vent terminal, a fire hazard and interference with the appliance venting and combustion air intake. Also, if the door blocks or deflects the vent discharge such that the combustion products are pulled back into the combustion air intake, the result would be excessive carbon monoxide (CO) production and serious appliance malfunction and sooting. A damaged vent terminal could cause appliance malfunction and carbon monoxide poisoning. It could also cause damage to nearby combustible surfaces and create a fire hazard. Door stops and closer devices cannot be depended on because they are easily defeated or removed [see Commentary Figures 502.7.1(1) and 502.7.1(2)].

Figure 502.7
PLAN VIEW OF STUD WALL

For SI: 1 inch = 25.4 mm.

Figure 502.7.1(1)
DOOR SWING

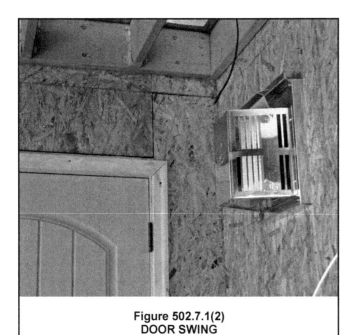

Figure 502.7.1(2)
DOOR SWING

SECTION 503 (IFGS)
VENTING OF APPLIANCES

503.1 General. The venting of appliances shall be in accordance with Sections 503.2 through 503.16.

❖ Section 503 contains general sizing, design and installation requirements for venting systems. Section 504 covers vent system sizing for Category I appliances.

503.2 Venting systems required. Except as permitted in Sections 503.2.1 through 503.2.4 and 501.8, all appliances shall be connected to venting systems.

❖ All appliances must be vented by a venting system except those allowed to be unvented by Section 501.8.
    The appliances addressed in Sections 503.2.1 and 503.2.2 must have a venting system.

503.2.1 Ventilating hoods. Ventilating hoods and exhaust systems shall be permitted to be used to vent appliances installed in commercial applications and to vent industrial appliances, such as where the process itself requires fume disposal.

❖ In commercial and industrial applications only, appliances can to be vented by an exhaust hood instead of a vent or chimney system. This is addressed further in Section 503.3.4. The exhaust system will usually exist for some other purpose such as heat, vapor and/or smoke capture and containment. It is convenient and cost effective to allow the exhaust system to also serve as the venting means for the combustion products produced by the process or appliance. This arrangement is common in manufacturing and food processing. The exhaust system must be interlocked with the appliances to prevent appliance operation when the exhaust system is not operating.

503.2.2 Well-ventilated spaces. Where located in a large and well-ventilated space, industrial appliances shall be permitted to be operated by discharging the flue gases directly into the space.

❖ This section should have extremely limited application. The allowance to operate industrial appliances without combustion product venting is intended only for industrial occupancies such as factories, foundries and manufacturing plants. This section states no limit on Btu/h input, does not specify a minimum space volume and does not define "large" or "well-ventilated"; therefore, one must apply this section with extreme caution, mindful of the intent. The space into which the combustion products are discharged must be large enough, open enough and ventilated enough to prevent the accumulation of harmful contaminants. These criteria might be met in a factory structure with very high ceilings, high infiltration rates and high rates of ventilation air movement. See Section 612, which addresses direct-fired industrial air heaters and specifies a mechanical ventilation rate to control contaminants created by the combustion process.

503.2.3 Direct-vent appliances. *Listed direct-vent appliances* shall be installed in accordance with the manufacturer's instructions and Section 503.8, Item 3.

❖ Direct-vent appliances are vented by a natural or mechanical draft system that is designed as part of the appliance and that usually requires some assembling on the job site. Except for the vent terminal requirements of Section 503.8, Item 3, the code depends on the appliance manufacturer's installation instructions for all aspects of venting direct-vent appliances. Direct-vent appliances include some Category IV condensing furnaces and boilers and power- and gravity-vented direct-vent water heaters, unit heaters, space heaters and packaged terminal units.
    Direct-vent appliances have the distinct advantage of providing their own combustion air supply in addition to their own dedicated venting means [see the definition of "Direct-vent appliance," as well as Commentary Figures 202(6) and 503.2.3(1) through 503.2.3(6)].

503.2.4 Appliances with integral vents. Appliances incorporating integral venting means shall be installed in accordance with the manufacturer's instructions and Section 503.8, Items 1 and 2.

❖ As it does with direct-vent appliances, the code trusts the manufacturer's instructions for the installation of appliances having an integral venting means. Integral-vent appliances such as rooftop heating, ventilation and air-conditioning (HVAC) units, outdoor makeup air heaters, and outdoor heaters for swimming pools have a built-in natural or mechanical venting means. The natural-draft integral venting means is simply a short vertical vent with a factory-supplied vent cap assembly. The mechanical integral-vent appliances use supervised blowers to discharge the combustion gases directly to the atmosphere. The typical means of blower supervision is one or more

**Figure 503.2.3(1)
DIRECT-VENT APPLIANCE
(Figure courtesy of Carrier Corporation)**

**Figure 503.2.3(2)
DIRECT-VENT (SEPARATED
COMBUSTION) UNIT HEATERS
(Figure courtesy of Reznor/Thomas & Betts
Corporation)**

For SI: 1 inch = 25.4 mm.

**Figure 503.2.3(3)
EXAMPLE OF MANUFACTURER'S DETAIL
FOR DIRECT-VENT APPLIANCE CONCENTRIC
VENT-AIR INTAKE TERMINAL
(Figure courtesy of Carrier Corporation)**

**Figure 503.2.3(4)
DIRECT-VENT APPLIANCES
(Photo courtesy of Lochinvar Corporation)**

Figure 503.2.3(5)
GRAVITY DIRECT-VENT WATER HEATER
(Photo courtesy of A.O. Smith Water Products)

Figure 503.2.3(6)
DIRECT-VENT SYSTEM COMPONENTS
(Photo courtesy of Simpson Dura-Vent Company,
Incorporated)

pressure switches or a centrifugal switch mounted to the blower motor shaft. Integral-vent appliances must not be confused with unvented appliances and appliances that need not be vented in accordance with Section 501.8. Integral-vent appliances must be vented to the outside atmosphere and thus are always designed for outdoor use only [see Section 503.8, Items 1 and 2, and Commentary Figures 503.2.4(1) through 503.2.4(3)].

503.2.5 Incinerators. Commercial-industrial-type incinerators shall be vented in accordance with NFPA 82.

❖ The scope of NFPA 82 is commercial/industrial incinerators and excludes incinerators in one- and two-family dwellings. Residential incinerators are not typically installed within dwellings today. The standard covers metal and masonry chimneys for use with incinerators and has provisions for construction, clearances, termination, cleanouts and connectors, among others.

503.3 Design and construction. Venting systems shall be designed and constructed so as to convey all flue and vent gases to the outdoors.

❖ This section is stated in performance language and requires venting systems to convey the products of combustion to the outdoors. The reference to a "positive flow" does not imply that the venting system is under a positive pressure, but rather implies that the vent produces a dependable flow at its design operating pressure.

Figure 503.2.4(1)
HOT WATER BOILER WITH INTEGRAL VENT FOR
OUTDOOR INSTALLATION
(Photo courtesy of A.O. Smith Water Products)

Figure 503.2.4(2)
OUTDOOR DUCT FURNACE WITH INTEGRAL POWER VENTING
(Photo courtesy of Reznor/Thomas & Betts Corporation)

Figure 503.2.4(3)
ROOFTOP HVAC UNIT WITH INTEGRAL POWER VENT
(Photo courtesy of Carrier Corporation)

503.3.1 Appliance draft requirements. A venting system shall satisfy the draft requirements of the *appliance* in accordance with the manufacturer's instructions.

❖ This section is stated in performance language and requires venting systems to produce the draft necessary for the operation of the appliance being vented. An appliance manufacturer will specify the required type of venting means and, to assure the required amount of draft, may specify the size and height of a vent or chimney or may specify a mechanical draft system.

503.3.2 Design and construction. Appliances required to be vented shall be connected to a venting system designed and installed in accordance with the provisions of Sections 503.4 through 503.16.

❖ As defined in Section 202, a venting system includes natural, induced and forced-draft systems, which are addressed in Sections 503.3.3 through 503.15.

503.3.3 Mechanical draft systems. Mechanical draft systems shall comply with the following:

1. Mechanical draft systems shall be *listed* and shall be installed in accordance with the manufacturer's instructions for both the *appliance* and the mechanical draft system.

2. Appliances requiring venting shall be permitted to be vented by means of mechanical draft systems of either forced or induced draft design.

3. Forced draft systems and all portions of induced draft systems under positive pressure during operation shall be designed and installed so as to prevent leakage of flue or vent gases into a building.

4. Vent connectors serving appliances vented by natural draft shall not be connected to any portion of mechanical draft systems operating under positive pressure.

5. Where a mechanical draft system is employed, provisions shall be made to prevent the flow of gas to the main burners when the draft system is not performing so as to satisfy the operating requirements of the *appliance* for safe performance.

6. The exit terminals of mechanical draft systems shall be not less than 7 feet (2134 mm) above finished ground level where located adjacent to public walkways and shall be located as specified in Section 503.8, Items 1 and 2.

❖ The appliances and equipment installations addressed in this section use auxiliary or integral fans and blowers to force the flow of combustion products to the outdoors.

This section applies to externally installed power exhausters, integrally power-exhausted appliances and venting systems equipped with draft inducers. This section does not address fan-assisted Category I appliances and does not address direct-vent appliances. Category III and IV appliances are subject to the requirements of this section if they are not listed as direct-vent appliances. Some Category I appliances are field convertible to Category III. Item 4 also applies to Category I fan-assisted appliances because they are vented by natural draft.

Power exhausters are field-installed pieces of equipment that are independent of, but used in conjunction with, other appliances [see Commentary Figures 503.3.3(1) through 503.3.3(4)].

Power exhausters are typically used where other means of venting are impractical, impossible or uneconomical. Power exhausters are typically designed for use with gas-fired and oil-fired natural draft appliances. Some are also designed for use with fan-assisted appliances and appliances equipped with integral forced-draft (power-vent) fans. Power exhausters produce negative pressures at their inlet connection and positive pressures at their outlet (discharge connection). The most common installation locates the exhauster at the point of termination of the vent or chimney. In such installations, the vent or chimney between the appliance and the exhauster operates under negative pressure [see Commentary Figures 503.3.3(1) through 503.3.3(9)].

Induced draft systems employ separate field-installed units designed to boost draft in a natural (gravity) draft chimney or vent [see Commentary Figures 503.3.3(10) and 503.3.3(11)].

Power-vented (self-venting) appliances are equipped with factory-installed integral blowers that force the combustion products through the special venting systems that are addressed in Sections 501.14 and 503.4. Power-vented appliances include mid- to high-efficiency furnaces, boilers, water heaters and heating units that are designed for through-the-wall or through-the-roof venting [see Commentary Figure 503.3.3(12)]. Appliances for outdoor installation are typically power vented and discharge directly to the atmosphere. They are addressed in Section 503.2.4.

Although not specifically referred to as an "appliance," a power exhauster (mechanical draft system) is considered to be a mechanical appliance appurtenance and, therefore, must bear the label of an approved agency in accordance with Item 1. See the definition of "Appliance." A power exhauster is an essential component of the appliance installation it serves and must be installed in accordance with the manufacturer's instructions for both the power exhauster and the appliance served. Mechanical draft devices that are an integral part of an appliance are covered by the appliance listing.

Vent or chimney systems installed downstream (discharge side) of a power exhauster must be designed and approved for positive-pressure applications. For example, Type B vent cannot be used on the discharge side of an exhauster because such pipe is not designed for positive pressure applications. The application and installation of exhausters (power venters) must comply with the manufacturers' installation instructions for the exhausters and the appliance(s) served by the exhausters (see commentary, Section 501.6).

There are three distinct variations in the use of blowers at the combustion chamber inlet working in conjunction with the fuel burner.

1. Blowers that supply turbulent combustion air to aid fuel-air mixing in a combustion chamber that is under negative pressure. To obtain proper negative over-fire draft (which optimizes combustion) also requires steady negative (below atmospheric) pressure at the flue outlet. This will be produced by a natural draft chimney and may be controlled by a barometric draft regulator. This is tabulated on line "A" in Commentary Figure 503.3.3(13).

2. Blowers that supply sufficient combustion air and pressure to produce flow through the combustion chamber, but the combustion process does not need additional vent or chimney draft. This permits use of gravity or neutral draft venting products such as Type B vent for gas conversion burners. A draft regulator may be used for such equipment to prevent excess draft from affecting combustion efficiency. This is tabulated on line "B" in Commentary Figure 503.3.3(13).

3. Blowers with enough power to overcome internal flue passage pressure losses (i.e., in firetube boilers) that also produce positive pressure at the outlet. This outlet pressure must be added to gravity draft as the motive force for flow in a chimney. If positive outlet pressure exists, the use of a sealed pressure-tight chimney is required. This is tabulated on line "C" in Commentary Figure 503.3.3(13).

These three types of equipment usually have integral blower/burner systems, and all could be considered to be forced combustion systems. Only those described in the third item truly produce forced chimney draft. When power-venting equipment is installed, it becomes an essential part of the appliance it serves. The appliance relies on the power-venting equipment (exhauster) to provide sufficient draft for proper appliance operation and for venting of the combustion products. Power exhausters are electrically interlocked with the appliance or appliances they serve to ensure that the appliances will not operate if there is insufficient draft to vent the products of combustion. If an exhauster fails, the appliances served by the exhauster would discharge the products of combustion into the building unless shut off by controls that monitor draft and draft-hood spillage. An improper or lacking interlock can cause the appliance(s) to malfunction and spill harmful flue gases into the building space. The electrical interlock (interconnection) is typically accomplished with controls provided by the exhauster manufacturer [see Commentary Figure 503.3.3(14)].

The usual sequence of operation is as follows: The call for heat from the appliance operating control starts the exhauster, and pressure controls start the appliance only after adequate draft has been proven to exist. In some cases, temperature sensors are used in addition to pressure controls to sense draft-hood spillage or unusual vent system temperatures and shut off the appliance.

Item 6 addresses the concern for pedestrian safety relative to exposure to combustion gases (see Section 503.8).

Figure 503.3.3(3)
HORIZONTAL POWER EXHAUSTERS
(Photo courtesy of Exhausto, Incorporated)

Figure 503.3.3(1)
VERTICAL DRAFT INDUCER
(Photo courtesy of Tjernlund Products, Incorporated)

Figure 503.3.3(4)
VERTICAL POWER EXHAUSTERS
(Photo courtesy of Exhausto, Incorporated)

Figure 503.3.3(2)
HORIZONTAL POWER EXHAUSTER
(Photo courtesy of Tjernlund Products, Incorporated)

Figure 503.3.3(5)
VERTICAL POWER EXHAUSTERS
(Photo courtesy of Exhausto, Incorporated)

Figure 503.3.3(6)
POWER EXHAUSTER OUTDOOR TERMINAL-TYPE UNIT
(Photo courtesy of Field Controls Company)

Figure 503.3.3(8)
TYPICAL POWER EXHAUSTER
(Photo courtesy of Tjernlund Products, Incorporated)

Figure 503.3.3(7)
EXHAUSTER AND CONTROL PACKAGE
(Photo courtesy of Tjernlund Products, Incorporated)

Figure 503.3.3(10)
DRAFT INDUCER
(Photo courtesy of Tjernlund Products, Incorporated)

Figure 503.3.3(9)
TYPICAL EXHAUSTER APPLICATIONS
(Figure courtesy of Tjernlund Products, Incorporated)

**Figure 503.3.3(11)
DRAFT INDUCER
(Photo courtesy of Field Controls Company)**

**Figure 503.3.3(13)
DRAFT CONDITIONS WITH EQUIPMENT
BURNER-INLET BLOWER
(Figure courtesy of Selkirk, L.L.C.)**

503.3.4 Ventilating hoods and exhaust systems. Ventilating hoods and exhaust systems shall be permitted to be used to vent appliances installed in commercial applications. Where automatically operated appliances, other than commercial cooking appliances, are vented through a ventilating hood or exhaust system equipped with a damper or with a power means of exhaust, provisions shall be made to allow the flow of gas to the main burners only when the damper is open to a position to properly vent the *appliance* and when the power means of exhaust is in operation.

❖ This section addresses commercial occupancies such as restaurant kitchens where the kitchen exhaust system is used to vent gas-fired appliances such as booster water heaters for dishwashers. Commercial gas-fired appliances must be vented (see Section 501.8) either by individual venting means such as Type B vents or exhausters, or by venting the products of combustion into the kitchen exhaust hood system that serves the appliances. This section requires the appliance and the ventilating hood or exhaust system to be interlocked to ensure simultaneous operation of the hood or exhaust system whenever the appliance is firing. See Sections 505.1.1 and 623.6 for a discussion of this type of interlock.

503.3.5 Air ducts and furnace plenums. Venting systems shall not extend into or pass through any fabricated air duct or *furnace plenum.*

❖ If a vent or chimney or a connector passes through or extends into a duct or furnace plenum, it is possible to subject the venting system to negative or positive pressures and/or temperature extremes that could cause the venting system to deteriorate, fail to produce the required draft, produce condensation and/or leak combustion gases. A furnace plenum is a component of the ductwork system constructed as a junc-

**Figure 503.3.3(12)
POWER-VENTED WATER HEATER
(Photo courtesy of A.O. Smith Water Products)**

Figure 503.3.3(14)
SIMPLIFIED SAMPLE INTERLOCK CIRCUIT DIAGRAM FOR APPLIANCE SERVED BY POWER EXHAUSTER
(SHOWN FOR ILLUSTRATIVE PURPOSE ONLY)

tion box for multiple duct connections. The term "furnace plenum" is defined in Section 202 and the term "fabricated air duct" refers to factory-, shop- or site-built supply relief, exhaust and return air ducts and does not refer to nonducted building cavities and plenums, such as spaces above ceilings used to convey air. Section 503.3.6 addresses nonducted interstitial spaces used as plenums.

503.3.6 Above-ceiling air-handling spaces. Where a venting system passes through an above-ceiling air-handling space or other nonducted portion of an air-handling system, the venting system shall conform to one of the following requirements:

1. The venting system shall be a *listed* special gas vent; other venting system serving a Category III or Category IV *appliance;* or other positive pressure vent, with joints sealed in accordance with the *appliance* or vent manufacturer's instructions.

2. The venting system shall be installed such that fittings and joints between sections are not installed in the above-ceiling space.

3. The venting system shall be installed in a conduit or enclosure with sealed joints separating the interior of the conduit or enclosure from the ceiling space.

❖ Vents are allowed to pass through interstitial spaces of a building used to convey environmental air such as above-ceiling return air-handling spaces (plenums), if the penetration/pass-through conforms to one of the three described installations (see Commentary Figure 503.3.6). All three installations are intended to prevent pressure differentials between the air-handling space and the vent interior from causing leakage of combustion products from the vent. See Section 602.2.1 of the IMC regarding combustible vent materials.

503.4 Type of venting system to be used. The type of venting system to be used shall be in accordance with Table 503.4.

❖ The table lists various types of vents and the corresponding types of appliances that can be served by the vents. The vent system must be tested and specifically approved for use with the approved appliance. If the vent system is not a tested and labeled component of the appliance, the material must be approved for use with the appliance and installed in accordance with the manufacturer's instructions.

Gas-burning appliances (listed as Category II, III or IV) require special vent systems that are specific to the type of appliance. Special vent systems are typically associated with mid- to high-efficiency appliances and include vent materials such as PVC, CPVC and special alloys of stainless steel.

A primary consideration for the design of a special vent system that serves a high-efficiency appliance is the selection of a material that is capable of withstanding the corrosive effects of condensate. Flue gas condensate is slightly acidic and corrosive and can deteriorate many vent materials. Sulfur in the fuel and halogens carried in the combustion air can combine with the combustion products to create acids. To prevent the corrosion of a vent and possible escape of flue gas, metallic flue passageways can be protected with appropriate coatings; however, any imperfections, pinholes or discontinuities in the coating can create areas for condensate to collect and accelerate the deterioration of the chimney or vent. The solution to this problem has been to use a material such as PVC, polypropylene (PP), CPVC or high-temperature plastic that is resistant to acidic condensation and compatible with the temperature of the appliance flue gases. Plastic pipe is a suitable material because it is impervious to corrosive condensation and can easily be designed to drain off accumulated condensate.

Figure 503.3.6
VENTS THROUGH AIR PLENUMS

Figure 503.4
DETERMINING VENT HEIGHT

The flue gas temperatures of most condensing-type appliances are not high enough to affect such materials adversely.

Special vent systems must be designed and installed in compliance with the manufacturers' instructions, which will specify installation requirements that are specific to that type of vent and appliance. For example, a maximum developed length, a maximum number of directional fittings, fitting turn radius and specific support requirements will be specified for special vents using plastic pipe. Special vent systems also have special vent termination requirements, and typically incorporate condensate collection and drainage fittings and connections.

Gas- and oil-burning appliances can be used with a vent system if the appliance is so labeled and approved.

The type and size of the vent must be as dictated by the manufacturer's installation instructions for the appliance. The design and installation instructions provided by the vent manufacturer must also be consulted when designing a vent system for any particular application. Commentary Figure 503.4 shows some of the variables that affect vent system sizing. In addition to Section 504, the vent manufacturers' design and application handbooks and installation instructions contain information and tables for the sizing and application of the vent systems.

503.4.1 Plastic piping. Where plastic piping is used to vent an appliance, the appliance shall be listed for use with such venting materials and the appliance manufacturer's installation instructions shall identify the specific plastic piping material.

❖ Plastic pipe has been used as a venting material for Category III and IV appliances for many years and has more recently been used to vent specialized power-vented appliances that are equipped with draft hoods and blowers that introduce enough dilution air to keep the vent gases within the service temperature range of the plastic pipe (see Sections 502.1 and 503.4).

Commonly used plastic venting materials include PVC, PP and CPVC (see Commentary Figure 503.4.1).

In the past, the code required code official approval, but it was difficult to place the approval burden on the code official because he or she was not in a position to determine what venting materials were suitable and safe for venting various appliances. Such decisions should have been left to design engineers, manufacturers and the testing and listing agencies that list the appliances. The previous code text did not actually require that the appliance be listed for use with specific venting system materials, although this was implied by the text and it is required in the appliance standards. The appliance manufacturer's installation instructions must clearly specify exactly what plastic materials are required or allowed for venting an appliance. The installation instructions will be consistent with how the appliance was tested by the listing agency. The product standards for gas appliances contain various testing procedures for plastic venting systems. The appliance manufacturer determines the type of plastic vent that is suitable for venting its product, and the testing and listing agency tests the appliance with that venting system for com-

Figure 503.4.1
DIRECT-VENT APPLIANCE

TABLE 503.4
TYPE OF VENTING SYSTEM TO BE USED

APPLIANCES	TYPE OF VENTING SYSTEM
Listed Category I appliances Listed appliances equipped with draft hood Appliances listed for use with Type B gas vent	Type B gas vent (Section 503.6) Chimney (Section 503.5) Single-wall metal pipe (Section 503.7) Listed chimney lining system for gas venting (Section 503.5.3) Special gas vent listed for these appliances (Section 503.4.2)
Listed vented wall furnaces	Type B-W gas vent (Sections 503.6, 608)
Category II appliances	As specified or furnished by manufacturers of listed appliances (Sections 503.4.1, 503.4.2)
Category III appliances	As specified or furnished by manufacturers of listed appliances (Sections 503.4.1, 503.4.2)
Category IV appliances	As specified or furnished by manufacturers of listed appliances (Sections 503.4.1, 503.4.2)
Incinerators	In accordance with NFPA 82
Appliances that can be converted for use with solid fuel	Chimney (Section 503.5)
Unlisted combination gas and oil-burning appliances	Chimney (Section 503.5)
Listed combination gas and oil-burning appliances	Type L vent (Section 503.6) or chimney (Section 503.5)
Combination gas and solid fuel-burning appliances	Chimney (Section 503.5)
Appliances listed for use with chimneys only	Chimney (Section 503.5)
Unlisted appliances	Chimney (Section 503.5)
Decorative appliances in vented fireplaces	Chimney
Gas-fired toilets	Single-wall metal pipe (Section 626)
Direct-vent appliances	See Section 503.2.3
Appliances with integral vent	See Section 503.2.4

❖ This table lists the currently available types of appliances and the types of venting systems to be used with such appliances. In the first row, the three types of appliances can be matched up with any of the five types of venting systems, if allowed by the appliance manufacturers' installation instructions. The fifth row up from the bottom addresses unlisted appliances that are not allowed by the code except under the alternate approval provisions of Section 105.

pliance with the product standards. There must not be any uncertainty about what type of venting system is required for any appliance so that venting system failures can be avoided. The definition of "Vent" does not include plastic pipe, such as PVC, ABS and CPVC, because such pipes are not currently listed as factory-built venting systems. Section 503.4.1 refers to the plastic pipe as a material used for venting; it is not referred to as a vent conforming to the definition. The PVC, ABS and CPVC pipe manufacturers do not recommend that their pipes be used for appliance venting because such products are not currently listed for such applications. There are polypropylene venting systems on the market that are listed to UL 1738 as appliance venting systems, and they would fall under the definition of "Vent."

503.4.1.1 Plastic vent joints. Plastic pipe and fittings used to vent appliances shall be installed in accordance with the *appliance* manufacturer's instructions. Where a primer is required, it shall be of a contrasting color.

❖ Experience has shown that plastic pipe and fittings used for venting appliances are not always joined together properly and joint failures occur resulting in leakage of combustion products. Piping materials such as PVC and CPVC usually require the use of a primer in the solvent welding process. The color of the primer, typically purple, must contrast with the color of the pipe, fittings and solvent cement so as to be visible for inspection. The use of a clear primer will not be evident on a completed solvent weld joint.

503.4.2 Special gas vent. Special gas vent shall be *listed* and installed in accordance with the special gas vent manufacturer's instructions.

❖ Special gas vents include high-temperature plastic pipe, stainless steel alloy pipe and positive-pressure pipe that are specified by the appliance manufacturer for use with a specific appliance. This section does not necessarily require that all plastic pipe be listed as a venting system component. For example, the installation instructions for Category IV appliances often specify PVC, CPVC and ABS plastic plumbing pipe for venting the appliances, and such pipe is not listed as a component of a venting system (see Section 503.4.1).

503.5 Masonry, metal and factory-built chimneys. Masonry, metal and factory-built chimneys shall comply with Sections 503.5.1 through 503.5.10.

❖ See the commentary to Sections 503.5.1 through 503.5.10.

503.5.1 Factory-built chimneys. Factory-built chimneys shall be installed in accordance with the manufacturer's instructions. Factory-built chimneys used to vent appliances that operate at a positive vent pressure shall be *listed* for such application.

❖ Prefabricated chimney systems must bear the label of an approved agency. A label is required on all components of the chimney system such as the pipe sections, shields, fireblocks, fittings, termination caps

and supports. The label states information such as the type of appliance the chimney was tested for use with, a reference to the manufacturer's installation instructions and the minimum required clearances to combustibles [see Commentary Figures 202(4) and 202(5)].

Most factory-built chimneys are either of the double-walled fiber-insulated design or the triple-walled air-cooled design. Factory-built chimneys are constructed of stainless steel inner liners with stainless steel outer walls. Factory-built chimneys, like vent systems, are composed of components that must be installed as a complete system. Components from different manufacturers are not designed to be mixed and installed together.

The manufacturer's instructions contain sizing criteria and requirements for every aspect of a factory-built chimney installation. The requirements include component assembly, clearances to combustibles, support, terminations, connections, protection from damage and fireblocking.

Any vent or chimney that conveys vent gases under positive pressure must be factory designed for that application. There is no allowable method of taping or sealing a chimney or vent that will convert a nonpositive-pressure chimney or vent into a positive-pressure chimney.

503.5.2 Metal chimneys. Metal chimneys shall be built and installed in accordance with NFPA 211.

❖ This section addresses metal chimneys, which are often referred to as smokestacks. Metal chimneys are used primarily in industrial applications where the vent gases are very high temperature. Discharging high temperature vent gases is a waste of energy and is thus becoming increasingly rare. Note that metal chimneys are different than what the code refers to as factory-built chimneys.

503.5.3 Masonry chimneys. Masonry chimneys shall be built and installed in accordance with NFPA 211 and shall be lined with *approved* clay flue lining, a *listed* chimney lining system or other *approved* material that will resist corrosion, erosion, softening or cracking from vent gases at temperatures up to 1,800°F (982°C).

Exception: Masonry chimney flues serving *listed* gas appliances with draft hoods, Category I appliances and other gas appliances *listed* for use with Type B vents shall be permitted to be lined with a chimney lining system specifically *listed* for use only with such appliances. The liner shall be installed in accordance with the liner manufacturer's instructions. A permanent identifying label shall be attached at the point where the connection is to be made to the liner. The label shall read: "This chimney liner is for appliances that burn gas only. Do not connect to solid or liquid fuel-burning appliances or incinerators."

For installation of gas vents in existing masonry chimneys, see Section 503.6.3.

❖ A chimney liner might be listed for use with solid-, liquid- and gas-fuel-fired appliances. The label required by this section is intended to warn the unknowing installer who sees a masonry chimney and thinks that it can serve a solid- or liquid-fuel-fired appliance. The label would be applicable to chimney liner systems that are designed for gas-fired appliances only (see Section 501.3).

503.5.4 Chimney termination. Chimneys for residential-type or low-heat appliances shall extend not less than 3 feet (914 mm) above the highest point where they pass through a roof of a building and not less than 2 feet (610 mm) higher than any portion of a building within a horizontal distance of 10 feet (3048 mm). Chimneys for medium-heat appliances shall extend not less than 10 feet (3048 mm) higher than any portion of any building within 25 feet (7620 mm). Chimneys shall extend not less than 5 feet (1524 mm) above the highest connected *appliance* draft hood outlet or flue collar. Decorative shrouds shall not be installed at the termination of factory-built chimneys except where such shrouds are *listed* and *labeled* for use with the specific factory-built chimney system and are installed in accordance with the manufacturer's instructions.

❖ Low-heat and residential-type chimneys must extend at least 3 feet (914 mm) above the roof, measured from the highest point of the roof penetration. They must also be at least 2 feet (610 mm) higher than any portion of the roof within a 10-foot (3048 mm) horizontal distance [see Commentary Figures 503.5.4(1) and 503.5.4(2)].

The 2-foot (610 mm) termination requirement is intended to prevent wind and pressure zones from reducing the amount of draft produced by the chimney. Wind and wind-induced eddy currents can react with building structural surfaces, thereby creating air pressure zones that can diminish chimney draft and cause reverse flow (backdraft) in the chimney. Loss of draft will cause the appliance or fireplace served by the chimney to discharge combustion products into the building interior. Locating the chimney outlet well into the undisturbed wind stream and away from the cavity and wake (eddy) zones around the building can counteract the adverse effects and also prevent the reentry of vent gases into the building through openings and fresh air intakes. Terminating a chimney in the eddy current area recirculates the combustion products and allows them to enter the building via infiltration, wall openings and air intakes. A chimney terminal properly located above the eddy current area allows the wind to carry the combustion products away from the building. Factory-built chimneys have termination caps that are part of the labeled chimney system and that must be installed. Like vent systems, the chimney cap is designed to keep out precipitation and animals and minimize the adverse effects of wind.

Decorative shrouds are designed to be aesthetically pleasing and to hide chimney terminations from view. These shrouds have, in some cases, caused fires resulting from overheated combustible construction and the accumulation of debris and animal nesting. Shrouds can also interfere with the functioning of a vent or chimney. The listing of the shroud must state the specific make and model of chimney with which the shroud is intended to be used. Some manufacturers are manufacturing shrouds that are listed for use with a specific make and model of chimney. Most often, shrouds are field-constructed of sheet metal, masonry or stucco on frame.

503.5.5 Size of chimneys. The effective area of a chimney venting system serving *listed* appliances with draft hoods, Category I appliances and other appliances *listed* for use with Type B vents shall be determined in accordance with one of the following methods:

1. The provisions of Section 504.

2. For sizing an individual chimney venting system for a single *appliance* with a draft hood, the effective areas of the vent connector and chimney flue shall be not less than the area of the *appliance* flue collar or draft hood outlet, nor greater than seven times the draft hood outlet area.

3. For sizing a chimney venting system connected to two appliances with draft hoods, the effective area of the chimney flue shall be not less than the area of the larger draft hood outlet plus 50 percent of the area of the smaller draft hood outlet, nor greater than seven times the smallest draft hood outlet area.

4. Chimney venting systems using mechanical draft shall be sized in accordance with *approved* engineering methods.

5. Other *approved* engineering methods.

❖ Chimneys can be sized in accordance with Section 504, by Item 2, 3 or 4 of this section or by an engineering method acceptable to the code official. Some appliances are not categorized; however, appliances with draft hoods and appliances listed for use with Type B vents would fit in Category I if they were categorized. The manufacturer's instructions for some Category I appliances, such as fan-assisted furnaces and boilers, may not allow connecting the appliance to a masonry chimney except under specific limited conditions.

Item 2 applies only to a single appliance that is factory equipped with a draft hood. The "seven times rule" prevents the chimney from being too large, which could result in poor draft and condensation problems.

Like vents, chimneys that are too large in area can be as problematic as chimneys that are too small in area.

Item 3 is limited to only one venting arrangement. It can only be used for a chimney that serves two appliances, both of which must be factory-equipped with draft hoods. The method is based on the areas of the chimneys and draft hood outlets, not on the diameters.

Item 4 recognizes that some manufacturers of mechanical draft equipment (exhausters) have their own sizing methodology specific to their product. Recall that Item 1 of Section 503.3.3 requires listing of such equipment.

For SI: 1 foot = 304.8 mm.

**Figure 503.5.4(1)
LOW-HEAT CHIMNEY TERMINATIONS**

**Figure 503.5.4(2)
DECORATIVE SHROUDS**

503.5.6 Inspection of chimneys. Before replacing an existing *appliance* or connecting a vent connector to a chimney, the chimney passageway shall be examined to ascertain that it is clear and free of obstructions and it shall be cleaned if previously used for venting solid or liquid fuel-burning appliances or fireplaces.

❖ If a chimney is going to serve a new appliance installation or a replacement appliance installation, the chimney must be inspected to determine whether it is still serviceable and free of deposits. Chimneys can become obstructed by debris such as leaves, animal carcasses, loose mortar and pieces of masonry and liner. Combustible deposits can accumulate on chimney walls used to vent liquid- and solid-fuel-fired appliances, and this section mandates that the chimney be cleaned if it served such appliances in the past. Masonry chimneys are especially susceptible to internal deterioration and should be inspected regularly, in addition to when a new or replacement appliance is connected.

503.5.6.1 Chimney lining. Chimneys shall be lined in accordance with NFPA 211.

> **Exception:** Where an existing chimney complies with Sections 503.5.6 through 503.5.6.3 and its sizing is in accordance with Section 503.5.5, its continued use shall be allowed where the *appliance* vented by such chimney is replaced by an *appliance* of similar type, input rating and efficiency.

❖ The liner forms the flue passageway and is the actual conductor of all products of combustion. The chimney liner must be able to withstand exposure to high temperatures and corrosive chemicals. The chimney lining protects the masonry construction of the chimney walls and allows the chimney to be constructed gas tight. This section regulates liners and relining systems. Liners are often used to reline an existing chimney to salvage a masonry chimney or allow connection of higher-efficiency appliances.

Flue lining systems for masonry chimneys include clay tile, poured-in-place refractory materials and stainless steel pipe. Chimney lining systems for chimneys serving gas-fired appliances include stainless steel and aluminum piping (uninsulated or insulated) designed for both new construction and for relining existing chimneys [see Commentary Figures 503.5.6.1(1) and 503.5.6.1(2)]. Type B vent is also used as a chimney liner, although it is not considered a lining system, and in such cases the chimney is actually a masonry chase containing a vent.

Relining of existing chimneys is quite common now because of the incompatibility of mid-efficiency appliances and masonry chimneys. Relining systems allows reuse of existing chimneys, thereby saving expense and retaining desirable architectural features of buildings.

The exception is intended to allow an unlined chimney to serve a new appliance that is installed to replace a previously served appliance if the new appliance does not create any different operating conditions in the chimney; that is, the volume, water vapor content and dilution air content of the new appliance flue gases are the same as those of the old appliance.

The exception assumes that the existing chimney and the existing appliances operated as intended by

Figure 503.5.6.1(1)
CHIMNEY LINING SYSTEM FOR GAS APPLIANCES
(Photo courtesy of Selkirk L.L.C.)

For SI: 1 foot = 304.8 mm.

Figure 503.5.6.1(2)
DIRECT-VENT APPLIANCE FOR
INSTALLATION IN SOLID FUEL FIREPLACE
(Figure courtesy of Quadra-Fire)

this code. If there is evidence that the previous chimney and appliance system was not working as required by the code, the exception is not intended to be applicable. Because this exception would allow an installation that is in conflict with the typical appliance manufacturer's instructions, it is in effect negated because the manufacturer's instructions will prevail in this case (see Sections 305.1 and 503.5.6.3). Before the exception could ever be applied, the chimney would have to be inspected and determined to be clear, clean, free of obstructions, safe for the intended use and properly sized for the appliances served. Also, the replacement appliance(s) would have to be the same type with nearly the same input rating and efficiency to make sure that the same amount of heat and dilution air enters the chimney. This will rarely, if ever, be encountered in the field; therefore, for multiple reasons, the exception will likely never be applicable.

503.5.6.2 Cleanouts. Cleanouts shall be examined to determine if they will remain tightly closed when not in use.

❖ A cleanout door or panel must be capable of securely sealing the opening to prevent air from entering the chimney, which would reduce draft for the appliances served. Under certain conditions, a faulty cleanout could also cause vent gases to leak from the chimney.

503.5.6.3 Unsafe chimneys. Where inspection reveals that an existing chimney is not safe for the intended application, it shall be repaired, rebuilt, lined, relined or replaced with a vent or chimney to conform to NFPA 211 and it shall be suitable for the appliances to be vented.

❖ If a chimney is inspected and found to be unsafe for any reason, it must be made safe or be replaced (see Section 501.3). Some chimneys and appliances are not compatible as dictated by the appliance and/or factory-built chimney manufacturer. Masonry chimneys usually become unsafe as a result of deteriorated mortar joints and liners. Other causes include structural instability, cracks, faulty cleanouts, deposit formation and moisture damage.

503.5.7 Chimneys serving appliances burning other fuels. Chimneys serving *appliances* burning other fuels shall comply with Sections 503.5.7.1 through 503.5.7.4.

❖ See the commentary to Sections 503.5.7.1 through 503.5.7.4.

503.5.7.1 Solid fuel-burning appliances. An *appliance* shall not be connected to a chimney flue serving a separate *appliance* designed to burn solid fuel.

❖ Each solid-fuel-burning appliance or fireplace must be connected to a dedicated independent chimney, or a dedicated independent flue in multiple-flue chimney constructions. Solid-fuel-burning appliances and fireplaces cannot share a common chimney or flueway with any other appliance or fireplace.

Solid-fuel-burning appliances produce creosote deposits on the interior walls of chimney liners. The creosote formation is highly combustible and creates a potential fire hazard. Because of the potential for a chimney fire, other connections to the chimney can allow fire to break out of the chimney into the building.

Also, chimney passageways can become restricted by the creosote formations. If other appliances were vented to such chimneys, combustion products could be discharged into the building and the other appliances might seriously interfere with the draft for the solid-fuel appliance.

This section is also intended to prevent the possibility of creosote leaking out of the chimney through appliance connectors.

Combination (dual fuel) gas- or oil- and solid-fuel-burning appliances are designed to be connected to a single chimney passageway. Such dual-fuel appliances must be listed, labeled and installed in accordance with the manufacturers' instructions.

503.5.7.2 Liquid fuel-burning appliances. Where one chimney flue serves gas appliances and liquid fuel-burning appliances, the appliances shall be connected through separate openings or shall be connected through a single opening where joined by a suitable fitting located as close as practical to the chimney. Where two or more openings are provided into one chimney flue, they shall be at different levels. Where the appliances are automatically controlled, they shall be equipped with safety shutoff devices.

❖ Gas-fired and oil-fired appliances are allowed to share a chimney or flue. The code does not include a sizing method for such arrangements; therefore, the system would have to be engineered or otherwise approved by the code official. Placing the openings at different levels in the chimney flue will minimize interference between the flow of the chimney connectors. The last sentence is consistent with the design of all automatically controlled appliances listed to today's appliance standards as required by this code.

503.5.7.3 Combination gas- and solid fuel-burning appliances. A combination gas- and solid fuel-burning *appliance* shall be permitted to be connected to a single chimney flue where equipped with a manual reset device to shut off gas to the main burner in the event of sustained backdraft or flue gas spillage. The chimney flue shall be sized to properly vent the *appliance*.

❖ A dual fuel appliance, gas and solid fuel, can be served by a single chimney or flue if the appliance is equipped with a safety control that monitors chimney spillage. A solid-fuel fire cannot be turned on and off like other fuel fires; thus, it is possible that the gas burner could be operated while the solid-fuel fire is still burning, and the chimney could be overloaded, causing spillage of combustion gases. The chimney size would have to be engineered or would have to comply with the appliance manufacturer's instructions.

503.5.7.4 Combination gas- and oil fuel-burning appliances. A *listed* combination gas- and oil fuel-burning *appliance* shall be permitted to be connected to a single chimney flue. The chimney flue shall be sized to properly vent the *appliance*.

❖ See the commentary to Section 503.5.6.3.

503.5.8 Support of chimneys. All portions of chimneys shall be supported for the design and weight of the materials employed. Factory-built chimneys shall be supported and spaced in accordance with the manufacturer's installation instructions.

❖ Chimneys, including factory-built, are very heavy compared to vents and need substantial support to carry their weight and prevent displacement. Factory-built chimneys require special support fittings at offsets to prevent the elbows from being damaged by bearing the weight of the chimney sections above the offset. Chimneys, like piping systems, often suffer from lack of adequate support (see Section 502.6).

503.5.9 Cleanouts. Where a chimney that formerly carried flue products from liquid or solid fuel-burning appliances is used with an *appliance* using fuel gas, an accessible cleanout shall be provided. The cleanout shall have a tight-fitting cover and shall be installed so its upper edge is not less than 6 inches (152 mm) below the lower edge of the lowest chimney inlet opening.

❖ If an existing chimney has no cleanout and was used for oil or solid-fuel appliances, a cleanout must be added. The cleanout allows access for a person to inspect and clean the chimney and monitor deposits on the interior walls. The cleanout must be located so that vent gases will not be in contact with the cleanout door/cover.

503.5.10 Space surrounding lining or vent. The remaining space surrounding a chimney liner, gas vent, special gas vent or plastic *piping* installed within a masonry chimney flue shall not be used to vent another *appliance*. The insertion of another liner or vent within the chimney as provided in this code and the liner or vent manufacturer's instructions shall not be prohibited.

The remaining space surrounding a chimney liner, gas vent, special gas vent or plastic *piping* installed within a masonry, metal or factory-built chimney shall not be used to supply *combustion air*. Such space shall not be prohibited from supplying *combustion air* to *direct-vent appliances* designed for installation in a solid fuel-burning *fireplace* and installed in accordance with the manufacturer's instructions.

❖ If a vent or a chimney liner is installed within a chimney flue, there will be space between the vent or liner and the chimney walls. This space must not be used to convey vent gases because of its irregular size and geometry and because the vent gases could damage the vent or liner installed in the chimney. A metallic vent or liner could be corroded by surrounding vent gases, and plastic pipes could be overheated. Liner systems and vent systems are not listed for use within an atmosphere of vent gases. Installing multiple liners or vents within a chimney is not the same

as using the annular space between a vent or liner and the inside chimney walls.

The space between a liner, vent or pipe and the interior chimney walls cannot be used as a duct for conveying combustion air. That space would have poor airflow characteristics, and moving cold combustion air over a liner or vent can cause condensation in the venting system. An exception is made for direct-vent appliances that are listed for such installation in solid-fuel-burning fireplaces.

503.6 Gas vents. Gas vents shall comply with Sections 503.6.1 through 503.6.13 (see Section 202, Definitions).

❖ See the commentary to Sections 503.6 through 503.6.10.

503.6.1 Installation, general. Gas vents shall be installed in accordance with the manufacturer's instructions.

❖ Gas vents, such as Type B vents, come with installation instructions available at the place of purchase or enclosed within the boxed items. The manufacturers also will provide, upon request, very detailed and informative design manuals that discuss sizing and venting fundamentals. These design manuals are invaluable for someone who installs, inspects or designs Type B vent systems (see Sections 502.1 and 502.5).

503.6.2 Type B-W vent capacity. A Type B-W gas vent shall have a *listed* capacity not less than that of the *listed* vented wall furnace to which it is connected.

❖ A Type B-W vent is an oval shaped version of a Type B vent designed to fit within a nominal 2 × 4 inch framed wall and serve a wall furnace (see Section 608).

503.6.3 Gas vents installed within masonry chimneys. Gas vents installed within masonry chimneys shall be installed in accordance with the manufacturer's instructions. Gas vents installed within masonry chimneys shall be identified with a permanent label installed at the point where the vent enters the chimney. The label shall contain the following language: "This gas vent is for appliances that burn gas. Do not connect to solid or liquid fuel-burning appliances or incinerators."

❖ One method of resizing a chimney is to install a Type B vent system within it. When this is done, the chimney is no longer seen as a chimney and becomes nothing more than a chase. This type of installation is practical only for chimneys without offsets. The Type B vent listing conditions and installation instructions will state whether such an installation is allowed and will include installation requirements such as support of the vent, weatherproofing the chimney top and the required condition of the chimney. Special instructions may be applicable for the support of the vent pipe within the chimney. The chimney should be structurally sound to avoid damage to the Type B vent system from falling and/or shifting masonry. There is also a concern for moisture and chimney deposits being corrosive to the Type B vent. The warning label is intended to prevent someone from connecting a solid-fuel or oil-fired appliance to what appears to be a chimney on the outside but is actually a Type B vent in disguise [see Commentary Figure 501.13(3)].

503.6.4 Gas vent terminations. A gas vent shall terminate in accordance with one of the following:

1. Gas vents that are 12 inches (305 mm) or less in size and located not less than 8 feet (2438 mm) from a vertical wall or similar obstruction shall terminate above the roof in accordance with Figure 503.6.4.

2. Gas vents that are over 12 inches (305 mm) in size or are located less than 8 feet (2438 mm) from a vertical wall or similar obstruction shall terminate not less than 2 feet (610 mm) above the highest point where they pass through the roof and not less than 2 feet (610 mm) above any portion of a building within 10 feet (3048 mm) horizontally.

3. As provided for industrial appliances in Section 503.2.2.

4. As provided for direct-vent systems in Section 503.2.3.

5. As provided for appliances with integral vents in Section 503.2.4.

6. As provided for mechanical draft systems in Section 503.3.3.

7. As provided for ventilating hoods and exhaust systems in Section 503.3.4.

❖ This section duplicates typical vent manufacturers' instructions and emphasizes that a vent is a system of components that are all necessary for proper functioning. The cap must be as specified by the vent manufacturer and is a listed component. A vent cap not only keeps out moisture, debris and animals, it also serves to prevent wind interference that could negatively affect the vent's ability to produce the required draft. Vent caps are tested for flow resistance and performance in wind.

This section is consistent with typical vent manufacturers' instructions and applies only to vents equipped with listed caps as required by Section 503.6.6. The relationship between vent terminal height and roof pitch is based on reducing the effects of wind and maintaining a minimum separation between the vent terminal and the roof surface.

See Section 504.2.9 for a discussion of vent terminations that extend above the roof higher than required by the code.

Figure 503.6.4 requires greater vent height above the roof as the roof approaches being a vertical surface. The greater the roof pitch, the greater the effect of wind striking the roof surface and the closer the cap becomes (horizontally) to the roof surface (see Figure 503.6.4).

It is a common misapplication for code users to apply chimney termination height requirements to vents, thereby causing vents to extend above roofs much higher than required in many cases. For example, a 6-inch-diameter (152 mm) Type B vent must extend above a simple gabled roof only 1 foot (305

mm) for roofs having up to 6/12 pitch, if the terminal is not within 8 feet (2438 mm) of a vertical surface such as an upper-story exterior wall. Consideration should be given to vent location in a building in the design/installation phase so that the vent termination will not be difficult or require long extensions of vent pipe exposed to the outdoors. Vent piping exposed to the outdoors encourages condensation in cold weather, could require guy wires or braces depending on height and is considered to be aesthetically unattractive. Another good reason to plan for vent location during the design phase of a building is to allow straight vertical runs of vent and avoid offsets. Good planning can also prevent vents from passing through the roof on the street side of the building. Quite often, vent offsets in attics are installed for the sole purpose of penetrating the roof on the back side of a building.

Vent termination heights are affected by the pitch of the roof and the proximity of walls and other verti-

ROOF SLOPE	H (min) ft
Flat to 6/12	1.0
Over 6/12 to 7/12	1.25
Over 7/12 to 8/12	1.5
Over 8/12 to 9/12	2.0
Over 9/12 to 10/12	2.5
Over 10/12 to 11/12	3.25
Over 11/12 to 12/12	4.0
Over 12/12 to 14/12	5.0
Over 14/12 to 16/12	6.0
Over 16/12 to 18/12	7.0
Over 18/12 to 20/12	7.5
Over 20/12 to 21/12	8.0

For SI: 1 inch = 25.4 mm, 1 foot = 304.8 mm.

FIGURE 503.6.4
TERMINATION LOCATIONS FOR GAS VENTS WITH
LISTED CAPS 12 INCHES OR LESS IN SIZE
AT LEAST 8 FEET FROM A VERTICAL WALL

cal or near vertical structural components. Where vent terminal locations within 8 feet (2438 mm) of a gable end, upper story or other vertical surface cannot be avoided, vent terminations must be regulated in a manner similar to chimneys [i.e., 2 feet (610 mm) minimum height and not less than 2 feet (610 mm) above any part of building within 10 feet (3048 mm)] (see Commentary Figure 503.6.4). Commentary Figure 503.6.4 does not apply to vent sizes larger than 12 inches (305 mm). This is consistent with manufacturers' instructions and relates to the testing of gas vents to the referenced standard.

The intent of Section 504.2.9 is for vents exposed to the outdoors below the roof penetration to be sized by a method other than the tables in Section 504. This means that if a vent is within 8 feet (2438 mm) of a vertical wall or is larger in size than 12 inches (305 mm), it must extend 2 feet (3048 mm) above any part of the building within 10 feet horizontally, and this could expose a substantial run of vent to the outdoors. In such cases, the tables of Section 504 would not apply, and the vent system would have to be engineered.

Figure 503.6.4
TOP LOCATION RULES
(Figure courtesy of Selkirk, L.L.C.)

503.6.4.1 Decorative shrouds. Decorative shrouds shall not be installed at the termination of gas vents except where such shrouds are *listed* for use with the specific gas venting system and are installed in accordance with manufacturer's instructions.

❖ Decorative shrouds have become a popular architectural feature. They are designed to be aesthetically pleasing and to hide chimney and vent terminations. These shrouds have, in some cases, caused fires resulting from overheated combustible construction and the accumulation of debris and animal nesting. Shrouds can also interfere with the functioning of a chimney or vent system. Shrouds are allowed only where they are listed and labeled for use with the specific factory-built chimney system or gas venting system (see Commentary Figure 503.6.4.1).

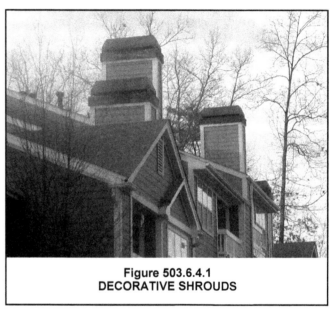

Figure 503.6.4.1
DECORATIVE SHROUDS

503.6.5 Minimum height. A Type B or L gas vent shall terminate not less than 5 feet (1524 mm) in vertical height above the highest connected *appliance* draft hood or flue collar. A Type B-W gas vent shall terminate not less than 12 feet (3658 mm) in vertical height above the bottom of the wall furnace.

❖ The amount of draft produced by a vent is directly related to the height of the vent. A minimum height must be established to produce the minimum draft necessary for the appliance served. This is made evident by looking at the tables in Section 504 that indicate increasing vent capacity as the vent height increases. This is a result of the increase in draft and vent flow velocity. The tables in Section 504 and some vent manufacturers' instructions indicate a minimum height of 6 feet (1829 mm) because this is the lowest entry for height in the sizing tables.

This section is often violated for appliances installed near the roof such as suspended unit heaters and appliances installed under shed roofs.

503.6.6 Roof terminations. Gas vents shall extend through the roof flashing, roof jack or roof thimble and terminate with a *listed* cap or *listed* roof assembly.

❖ This section duplicates the typical vent manufacturer's instructions and emphasizes that a vent is a system of components that are all necessary for proper functioning. The cap must be as specified by the vent manufacturer and is a listed component. A vent cap not only keeps out moisture, debris and animals, it also serves to prevent wind interference that could negatively affect the vent's ability to produce the required draft. Vent caps are tested for flow resistance and performance in wind. The emphasis here is that the vent system must extend through the roof all the way to the cap using components that are listed as part of the listed venting system.

503.6.7 Forced air inlets. Gas vents shall terminate not less than 3 feet (914 mm) above any forced air inlet located within 10 feet (3048 mm).

❖ If the vent termination is 10 feet (3048 mm) or more from the forced air (mechanical intake opening), this section does not apply. The intent is to prevent combustion gases from being drawn into the building through powered air intakes of any type (see Commentary Figure 503.6.7).

503.6.8 Exterior wall penetrations. A gas vent extending through an exterior wall shall not terminate adjacent to the wall or below eaves or parapets, except as provided in Sections 503.2.3 and 503.3.3.

❖ The termination locations prohibited by this section would result in poor draft or back draft because of the effects of wind. Also, termination in such locations could result in a fire hazard, damage to the structure and entry of vent gases into the building. If a gas vent (typically Type B) penetrates an exterior wall, the vent would have to extend vertically above the roof in accordance with Section 503.6.3; however, this exterior extension of vent presents a problem in itself. Gas vents, such as Type B vents, are not designed

for outdoor exposure, except for the short section of vent that passes through and extends above the roof. Sections 504.2.9 and 504.3.16 reinforce this by stating that the vent sizing tables in Section 504 do not apply to vents exposed to the outdoors below the roof line. Vents exposed to the outdoors will suffer from poor draft and condensation production in most climates because of the heat loss through the exposed vent walls. The bottom line is that vents sized in accordance with Section 504 cannot extend up the exterior side of a building as implied by this section and Section 503.6.4. This type of installation could occur only if the vent system was engineered to account for the outdoor exposure. Note that this is also consistent with typical vent manufacturers' design manuals and installation instructions. Even though it is common to see gas vents run up the exterior wall of a building, such installations are not allowed unless they are engineered and allowed by the vent manufacturer, neither of which is likely.

503.6.9 Size of gas vents. Venting systems shall be sized and constructed in accordance with Section 504 or other *approved* engineering methods and the gas vent and *appliance* manufacturer's installation instructions.

❖ Gas vents are to be sized by either Section 504 or an engineering method acceptable to the code official and, in all cases, in accordance with the vent manufacturer's instructions (see Sections 503.6.9.1 through 503.6.9.3).

503.6.9.1 Category I appliances. The sizing of natural draft venting systems serving one or more *listed* appliances equipped with a draft hood or appliances *listed* for use with Type B gas vent, installed in a single story of a building, shall be in accordance with one of the following methods:

1. The provisions of Section 504.

2. For sizing an individual gas vent for a single, draft-hood-equipped *appliance*, the effective area of the vent connector and the gas vent shall be not less than the

For SI: 1 foot = 304.8 mm.

Figure 503.6.7
B-VENT TERMINATIONS WITH RESPECT TO FORCED AIR INTAKES

area of the *appliance* draft hood outlet, nor greater than seven times the draft hood outlet area.

3. For sizing a gas vent connected to two appliances with draft hoods, the effective area of the vent shall be not less than the area of the larger draft hood outlet plus 50 percent of the area of the smaller draft hood outlet, nor greater than seven times the smaller draft hood outlet area.

4. *Approved* engineering practices.

❖ This section reiterates the intent of the preceding section, is specific to Category I appliances, and applies only to appliances installed within the same story. Multiple-story applications are addressed in Section 503.6.10. Category I appliances are vented by natural draft and, depending upon the design and efficiency, are either equipped with a draft hood or a fan-assisted combustion system. Because of the higher thermal efficiencies and lack of dilution air, vent designs for fan-assisted appliances are different from those for draft hood- equipped appliances.

Item 3 allows a very limited application of the old "50-percent rule." Item 2 applies only to a single draft-hood-equipped appliance. Item 3 applies only to a common vent system serving two draft-hood-equipped appliances. Neither Item 2 nor 3 can be applied to the venting of fan-assisted appliances, because the old alternate method (50-percent rule) was not created with fan-assisted appliances in mind and will not work with such appliances. Besides being limited to draft hood appliances, Items 2 and 3 should be further limited to simple venting system configurations having short laterals; few, if any, changes in direction (fittings); short connector lengths with maximum vertical rise and tall vent heights. Vent manufacturers emphatically warn against the use of the alterative sizing rule in Item 3 because this method does not account for many factors now known to be important in vent sizing.

503.6.9.2 Vent offsets. Type B and L vents sized in accordance with Item 2 or 3 of Section 503.6.9.1 shall extend in a generally vertical direction with offsets not exceeding 45 degrees (0.79 rad), except that a vent system having not more than one 60-degree (1.04 rad) *offset* shall be permitted. Any angle greater than 45 degrees (0.79 rad) from the vertical is considered horizontal. The total horizontal distance of a vent plus the horizontal vent connector serving draft-hood-equipped appliances shall be not greater than 75 percent of the vertical height of the vent.

❖ This section addresses vent offsets differently than Section 504 and is in no way related to Section 504. This section would be applied only if Section 504 is not applied; therefore, this section is rarely applied, because Section 504 is necessary for sizing the majority of today's Category I venting systems. The only time that a Type B vent system would not be designed in accordance with Section 504 is if it were an engineered system or if it were designed in accordance with Section 503.6.9.1, Item 2 or 3, applicable only to draft-hood-equipped appliances.

This section regulates vent offsets, but without the tables and design requirements of Section 504, there are only very limited design requirements in Section 503 to accompany the offset provisions. In other words, this section would be only one of many parts necessary to design a vent system. This section is not intended for application in conjunction with the requirements of Section 504.

This section is not applicable to fan-assisted appliances and applies only to Type B and L vents sized in accordance with Item 2 or 3 of Section 503.6.8.1.

503.6.9.3 Category II, III and IV appliances. The sizing of gas vents for Category II, III and IV appliances shall be in accordance with the *appliance* manufacturer's instructions. The sizing of plastic pipe that is specified by the appliance manufacturer as a venting material for Category II, III and IV appliances shall be in accordance with the manufacturer's instructions.

❖ Venting systems for Category II, III and IV appliances are special vent systems specified by the appliance manufacturer and are not addressed further in the code (see Section 503.4).

Because plastic pipes such PVC, ABS and CPVC plumbing pipes are not listed and labeled as appliance vents (see the definition of "Vent"), the code was silent on how to size such pipes. The sizing is covered in the appliance manufacturer's instructions, and the code requires compliance with such instructions. For consistency, however, this section has been modified to address both listed vents and unlisted materials used as vents. For example, PVC pipe that vents a Category IV furnace is not listed as a vent; rather, it is a material that is used as a vent, and the appliance is listed for use with the PVC pipe.

503.6.9.4 Mechanical draft. Chimney venting systems using mechanical draft shall be sized in accordance with *approved* engineering methods.

❖ The manufacturers of exhausters (mechanical draft systems) provide sizing criteria for the vents and chimneys served by such equipment. The code's sizing criteria are based on natural draft chimneys and vents; therefore, engineered sizing methods must be used for other than natural draft (see Section 503.3.3).

503.6.10 Gas vents serving appliances on more than one floor. A common vent shall be permitted in multistory installations to vent Category I appliances located on more than one floor level, provided that the venting system is designed and installed in accordance with *approved* engineering methods. For the purpose of this section, crawl spaces, basements and attics shall be considered as floor levels.

❖ With rare exception, venting systems serve only appliances located within the same story. This section addresses those rare exceptions where a common venting system serves appliances that are installed within different stories or on different floor levels. For example, a single vent serves two appliances on the first floor of a building and also serves two more on the second floor, for a total of four appliances. Such a system requires a unique design approach as dictated in Sections 503.6.10.1 and 503.6.10.2. Note that this section considers crawl spaces, basements and attics as floor levels. For example, if a venting system serves appliances on the main floor of a building and also serves an appliance installed in the attic, that system must comply with this section.

503.6.10.1 Appliance separation. Appliances connected to the common vent shall be located in rooms separated from occupiable space. Each of these rooms shall have provisions for an adequate supply of combustion, ventilation and dilution air that is not supplied from an occupiable space.

❖ This section addresses chimneys or vents that serve multiple appliances located on different floor levels. For example, a common vent may be designed to serve one or more appliances per floor of a multistory building, or a vent that rises up through the furnace rooms of single-story apartments in a multi-story apartment building would be a common vent for the "stacked" apartments. At each level there may be a single appliance or several appliances entering into the vent. Multistory venting involves special design criteria and poses a unique hazard. When a vent is overloaded, blocked or otherwise produces insufficient draft, the flue gases from lower-story appliances could discharge into upper stories through the upper-story appliance connections. The lower-story vent could appear to be functioning normally, while the upper-story occupants would be exposed to a life-threatening hazard.

To avoid this potential hazard, multistory vent systems are designed to isolate the vent system and the appliances from the occupied portions of the building. This isolation is commonly accomplished by locating the common vent and all of the appliances served in mechanical rooms that are accessed from the outdoors only. This is commonly accomplished by placing mechanical rooms on outside walls and making them accessible only from a balcony. The sizing and design of multistory common chimney and vent systems must be in accordance with the vent or chimney manufacturer's installation and design instructions.

This section speaks of "habitable space," but the intent should apply to "occupied/living space" because many spaces such as bathrooms, toilet rooms and halls are not defined as habitable, yet are occupied, even if only for short periods.

This section is not limited to residential occupancies and would apply to all occupancies. The common vent and the appliances served by the common vent have to be located in a space that does not communicate with occupied spaces. In other words, a door cannot be the means of isolating the common vent and appliances from an occupied space because there is no way to control the position of a door or make it gas tight. The most common method of isolation is to locate the common vent and appliances in a space that is accessed only from the outdoors [see Commentary Figures 503.6.10.1(1) and

503.6.10.1(2)]. This location limitation makes it obvious that only appliances such as furnaces, boilers and water heaters are candidates for this type of venting system. (Room heaters and gas fireplace appliances cannot be separated from the occupied space.) Combustion, dilution and ventilation air for the appliances must not be taken from an occupied space because this would allow communication of atmospheres between the appliance space and the occupied space. The method of supplying combustion air must not violate the separation between the appliances and the occupied space.

Figure 503.6.10.1(1)
APPLIANCE AND VENT
SEPARATED FROM LIVING SPACE
(Figure courtesy of Simpson Dura-Vent Company, Incorporated)

503.6.10.2 Sizing. The size of the connectors and common segments of multistory venting systems for appliances *listed* for use with Type B double-wall gas vents shall be in accordance with Table 504.3(1), provided that:

1. The available total height (H) for each segment of a multistory venting system is the vertical distance between the level of the highest draft hood outlet or flue collar on that floor and the centerline of the next highest interconnection tee.

2. The size of the connector for a segment is determined from the *appliance* input rating and available connector rise and shall be not smaller than the draft hood outlet or flue collar size.

3. The size of the common vertical segment, and of the interconnection tee at the base of that segment, shall be based on the total *appliance* input rating entering that segment and its available total height.

❖ Because Table 504.3(1) is referenced, vent connectors within the multistory venting system must be constructed of double-wall gas vent. Manufacturers' instructions may contain additional requirements,

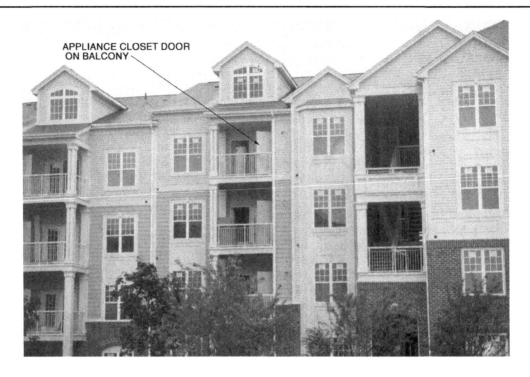

Figure 503.6.10.1(2)
APPLIANCE CLOSETS ACCESSED FROM OUTDOOR BALCONY

such as that the common vent must be entirely vertical with no offsets. The figures in Appendix B clearly depict Items 1, 2, and 3 of this section.

503.6.11 Support of gas vents. Gas vents shall be supported and spaced in accordance with the manufacturer's installation instructions.

❖ Type B vents must be supported to prevent the weight of the system from bearing on appliances and vent fittings. Fittings such as adjustable elbows are not designed to carry the weight of upper or lower sections of vent pipe, nor are appliance draft hoods and flue collars. Support bases, brackets and spacers are available from the vent manufacturer and are used to support the weight of the vent and maintain alignment and clearance to combustibles. On most job sites, the vents will be supported by various field-constructed brackets and straps made of sheet metal scraps, duct slips and drives and plumber's perforated straps. Gas vents are often poorly supported by methods that allow the vent fitting to rotate, that cause excessive loads on fittings and that allow the vent to sag or lose its required clearance. The installations should be carefully examined to determine that such field-constructed supports are adequate. Sheet metal screws must not penetrate the inner liner of any Type B vent component. Type B vents are designed for installation without the use of any additional fasteners that penetrate the vent walls (see Section 503.6.13).

Figure 503.6.13
FASTENER PENETRATIONS

503.6.12 Marking. In those localities where solid and liquid fuels are used extensively, gas vents shall be permanently identified by a label attached to the wall or ceiling at a point where the vent connector enters the gas vent. The determination of where such localities exist shall be made by the code official. The label shall read:

"This gas vent is for appliances that burn gas. Do not connect to solid or liquid fuel-burning appliances or incinerators."

❖ Type B vents installed in localities where oil- and solid-fuel-fired appliances are common must be labeled to warn the installers and occupants that Type B vents cannot be used to vent any appliances other than gas fired. The label is not placed on the vent; rather, it is to be placed on a wall or ceiling where it would be conspicuous to someone attempting to connect to the vent. The code official determines whether their locality is the intended target for this provision. It is assumed that ignorance of gas vents will be more likely in regions where oil and solid fuel have been commonly used for fuel.

503.6.13 Fastener penetrations. Screws, rivets and other fasteners shall not penetrate the inner wall of double-wall gas vents, except at the transition from an *appliance* draft hood outlet, a flue collar or a single-wall metal connector to a double-wall vent.

❖ Gas vents are not designed to be joined with screws or other fasteners, yet installers insist on using screws or rivets to augment the proprietary joint securing methods provided by the vent manufacturer. Manufacturers might not prohibit this practice, but they do not encourage it and definitely do warn against penetrating the inner wall. Screws are also commonly used to attach support (stand-off) brackets and hangers to vent pipe. It is typical to find duct slips and S-drives used as supports for vents with screws used to attach to the vent.

When fasteners penetrate the inner wall, condensation in the vent will corrode the fastener and leak into the airspace between the vent walls causing vent deterioration. Vent manufacturers offer adapter fittings to transition between a vent and an appliance outlet or flue collar, however, installers usually ignore these fittings and connect the vent pipe or vent elbow directly to the appliance. This section permits the inner wall to be penetrated only at the point of transition between single wall pipe and the vent, and between an appliance outlet collar and the vent (see Commentary Figure 503.6.13).

503.7 Single-wall metal pipe. Single-wall metal pipe vents shall comply with Sections 503.7.1 through 503.7.13.

❖ This section addresses single-wall pipe used as a vent, not single-wall pipe used as a vent connector. Although discouraged or prohibited by designers, code officials and appliance and vent manufacturers alike, the code still recognizes the use of unlisted single-wall metal pipe as a vent. All other types of vents are listed systems as required by Section 502.1. As evidenced by Sections 503.7.1 through 503.7.12, single-wall metal pipe is restricted to very limited applications, and extraordinary installation precautions are necessary. Single-wall metal pipe was once common, but it will not be found today except in extremely rare circumstances and in older existing buildings. Appliance manufacturers' instructions will typically prohibit the use of a single-wall metal vent (see Section 503.7.8).

503.7.1 Construction. Single-wall metal pipe shall be constructed of galvanized sheet steel not less than 0.0304 inch (0.7 mm) thick, or other *approved,* noncombustible, corrosion-resistant material.

❖ The galvanized sheet steel thickness specified is equivalent to 22 gage, which is heavier than the pipe commonly used for connectors for most residential solid-fuel-fired appliances.

503.7.2 Cold climate. Uninsulated single-wall metal pipe shall not be used outdoors for venting appliances in regions where the 99-percent winter design temperature is below 32°F (0°C).

❖ This section limits the use of single-wall vents to regions such as the southwestern and southeastern parts of the United States, unless the pipe is an insulated type to compensate for the high heat loss of single-wall pipe. The code contains no guidance for the type of insulation required or what *R*-value is required. At the minimum, insulation applied directly to the pipe would have to be a noncombustible material, the *R*-value would have to exceed that afforded by a double-wall vent, and the insulation would have to be permanently attached and protected from the weather. Insulating single-wall pipe could be considerably more difficult and expensive than using a venting system designed for outdoor use. Bear in mind that air-insulated double-wall vent pipe (Type B) is not designed for use outdoors in cold climates. When applying Section 504, single-wall vents are also prohibited from being exposed to the outdoors below the point of roof penetration.

503.7.3 Termination. Single-wall metal pipe shall terminate at least 5 feet (1524 mm) in vertical height above the highest connected *appliance* draft hood *outlet* or flue collar. Single-wall metal pipe shall extend not less than 2 feet (610 mm) above the highest point where it passes through a roof of a building and not less than 2 feet (610 mm) higher than any portion of a building within a horizontal distance of 10 feet (3048 mm). An *approved* cap or roof assembly shall be attached to the terminus of a single-wall metal pipe.

❖ Single-wall metal vents are not allowed to terminate as allowed by Section 503.6.4 and must comply with the termination requirements for chimneys, except that the minimum height above the roof penetration is 2 feet (610 mm), whereas it is 3 feet (914 mm) for chimneys [see Commentary Figure 503.5.4(1)]. The minimum height and vent cap requirements are consistent with Sections 503.6.4 and 503.6.5.

503.7.4 Limitations of use. Single-wall metal pipe shall be used only for runs directly from the space in which the *appli-*

ance is located through the roof or exterior wall to the outdoor atmosphere.

❖ Single-wall metal pipe is further restricted by this section to only two applications. It can run from the appliance directly to the outdoors through an exterior wall or it can run from the appliance directly to the outdoors through a roof. Single-wall metal pipe cannot pass through any floors, interior walls or partitions and cannot pass through attics or concealed spaces. This means that the appliance served must be in a space with an exterior wall or in the same story as a cathedral or shed-type roof.

503.7.5 Roof penetrations. A pipe passing through a roof shall extend without interruption through the roof flashing, roof jack or roof thimble. Where a single-wall metal pipe passes through a roof constructed of *combustible material*, a noncombustible, nonventilating thimble shall be used at the point of passage. The thimble shall extend not less than 18 inches (457 mm) above and 6 inches (152 mm) below the roof with the annular space open at the bottom and closed only at the top. The thimble shall be sized in accordance with Section 503.7.7.

❖ A thimble is a sheet metal assembly designed to provide clearance between a vent and combustible materials, analogous to a thimble that protects one's fingers from a sewing needle. It is essentially a spacer that provides an annular space around the vent. Annular space is usually ventilated by punched holes in the faces of the thimble assembly. Some thimbles are constructed with insulation in addition to an annular space.

A roof jack is an assembly that passes through the roof and contains a thimble and weather protection all in one. Roof jacks are flashing/thimble combinations and typically mount to a roof curb.

503.7.6 Installation. Single-wall metal pipe shall not originate in any unoccupied attic or concealed space and shall not pass through any attic, inside wall, concealed space or floor. The installation of a single-wall metal pipe through an exterior combustible wall shall comply with Section 503.7.7.

❖ A single-wall metal vent must be exposed for its entire length, except for the section of pipe that is within a thimble or roof jack assembly. Single-wall vents can penetrate only exterior walls and roofs. A roof/ceiling assembly cannot be penetrated if there is an attic or a concealed space between the ceiling and the roof (see Section 503.7.4).

503.7.7 Single-wall penetrations of combustible walls. A single-wall metal pipe shall not pass through a combustible exterior wall unless guarded at the point of passage by a ventilated metal thimble not smaller than the following:

1. For *listed* appliances with draft hoods and appliances *listed* for use with Type B gas vents, the thimble shall be not less than 4 inches (102 mm) larger in diameter than the metal pipe. Where there is a run of not less than 6 feet (1829 mm) of metal pipe in the open between the draft hood outlet and the thimble, the thim-

ble shall be permitted to be not less than 2 inches (51 mm) larger in diameter than the metal pipe.

2. For unlisted appliances having draft hoods, the thimble shall be not less than 6 inches (152 mm) larger in diameter than the metal pipe.

3. For residential and low-heat appliances, the thimble shall be not less than 12 inches (305 mm) larger in diameter than the metal pipe.

Exception: In lieu of thimble protection, all *combustible material* in the wall shall be removed a sufficient distance from the metal pipe to provide the specified *clearance* from such metal pipe to *combustible material*. Any material used to close up such opening shall be noncombustible.

❖ The only single-wall vent system component that is allowed to penetrate an exterior wall is a single-wall vent pipe. Single-wall vent connectors cannot penetrate an exterior wall because if they did, they would be outdoors and exposed to the ambient outdoor temperatures (see Sections 503.7.2 and 503.10.2.2). This section is located in the single-wall vent section because this is the only place that it is applicable.

Items 1 through 3 dictate the size of thimbles used to protect the combustible wall. For example, in accordance with the first sentence of Item 1, a 4-inch (102 mm) metal pipe would require an 8-inch-diameter (203 mm) thimble. Ventilated thimbles have holes in the faces of the thimble to allow air to cool the thimble by convection. Item 2 would rarely apply because unlisted appliances are not allowed by this code unless specifically approved by the code official under Section R104.11. Item 3 would rarely apply because it is addressing appliances other than those addressed in Items 1 and 2. The exception requires that the combustible wall be made noncombustible at the point of penetration. The noncombustible portion of the wall must extend from the metal pipe in all directions a distance equal to the required airspace clearance for the metal pipe.

503.7.8 Clearances. Minimum clearances from single-wall metal pipe to *combustible material* shall be in accordance with Table 503.10.5. The *clearance* from single-wall metal pipe to *combustible material* shall be permitted to be reduced where the *combustible material* is protected as specified for vent connectors in Table 308.2.

❖ Consideration should be given to the fact that single-wall vents will have high surface temperatures compared to insulated vents such as Type B vents. High surface temperatures could be an ignition or burn injury hazard.

503.7.9 Size of single-wall metal pipe. A venting system constructed of single-wall metal pipe shall be sized in accordance with one of the following methods and the appliance manufacturer's instructions:

1. For a draft-hood-equipped *appliance*, in accordance with Section 504.

2. For a venting system for a single *appliance* with a draft hood, the areas of the connector and the pipe each shall

be not less than the area of the *appliance* flue collar or draft hood outlet, whichever is smaller. The vent area shall be not greater than seven times the draft hood outlet area.

3. Other *approved* engineering methods.

❖ As can be seen from Items 1, 2, and 3 and Tables 504.2(5) and 504.3(5), single-wall metal pipe vents are limited to use with draft-hood-equipped appliances, except where the system is engineered. Note that the appliance manufacturer's instructions will generally require connection to a listed vent system.

503.7.10 Pipe geometry. Any shaped single-wall metal pipe shall be permitted to be used, provided that its equivalent effective area is equal to the effective area of the round pipe for which it is substituted, and provided that the minimum internal dimension of the pipe is not less than 2 inches (51 mm).

❖ Single-wall metal pipe does not have to be round; it could be oval, square, or rectangular. Round pipe has the best flow characteristics as evidenced by the fact that a round chimney liner will have a greater capacity than a rectangular liner of equal area.

503.7.11 Termination capacity. The vent cap or a roof assembly shall have a venting capacity of not less than that of the pipe to which it is attached.

❖ Section 503.7.3 requires an approved vent cap or roof termination assembly, and this section requires that the terminus be sized and designed to not reduce the capacity of the vent. An unlisted, untested cap or roof assembly will have an unknown capacity unless it is engineered.

503.7.12 Support of single-wall metal pipe. All portions of single-wall metal pipe shall be supported for the design and weight of the material employed.

❖ Single-wall vents, like all vents and chimneys, must be supported to prevent structural failure and joint separation and to maintain the required clearances.

503.7.13 Marking. Single-wall metal pipe shall comply with the marking provisions of Section 503.6.12.

❖ See the commentary to Section 503.6.12.

503.8 Venting system termination location. The location of venting system terminations shall comply with the following (see Appendix C):

1. A mechanical draft venting system shall terminate not less than 3 feet (914 mm) above any forced-air inlet located within 10 feet (3048 mm).

   Exceptions:

   1. This provision shall not apply to the *combustion air* intake of a direct-vent *appliance*.

   2. This provision shall not apply to the separation of the integral outdoor air inlet and flue gas discharge of *listed* outdoor appliances.

2. A mechanical draft venting system, excluding *direct-vent appliances*, shall terminate not less than 4 feet (1219 mm) below, 4 feet (1219 mm) horizontally from,

or 1 foot (305 mm) above any door, operable window or gravity air inlet into any building. The bottom of the vent terminal shall be located not less than 12 inches (305 mm) above finished ground level.

3. The vent terminal of a direct-vent *appliance* with an input of 10,000 Btu per hour (3 kW) or less shall be located not less than 6 inches (152 mm) from any air opening into a building. Such an *appliance* with an input over 10,000 Btu per hour (3 kW) but not over 50,000 Btu per hour (14.7 kW) shall be installed with a 9-inch (230 mm) vent termination *clearance*, and an *appliance* with an input over 50,000 Btu per hour (14.7 kW) shall have not less than a 12-inch (305 mm) vent termination *clearance*. The bottom of the vent terminal and the air intake shall be located not less than 12 inches (305 mm) above finished ground level.

4. Through-the-wall vents for Category II and IV appliances and noncategorized condensing appliances shall not terminate over public walkways or over an area where condensate or vapor could create a nuisance or hazard or could be detrimental to the operation of regulators, relief valves or other *equipment*. Where local experience indicates that condensate is a problem with Category I and III appliances, this provision shall also apply. Drains for condensate shall be installed in accordance with the appliance and vent manufacturers' instructions.

5. Vent systems for Category IV appliances that terminate through an outside wall of a building and discharge flue gases perpendicular to the adjacent wall shall be located not less than 10 feet (3048 mm) horizontally from an operable opening in an adjacent building. This requirement shall not apply to vent terminals that are 2 feet (607 mm) or more above or 25 feet (7620 mm) or more below operable openings.

❖ This section addresses the terminations of mechanical draft systems and direct-vent appliances. To prevent vent gases from entering the building, Item 1 requires forced air (mechanical) intakes to be at least 10 feet (3048 mm) away from a mechanical draft termination, or the termination must be at least 3 feet (914 mm) above the air intake. The natural buoyancy of vent gases is the justification for allowing a closer distance where the terminal is above the air intake. A combustion air intake for a direct-vent appliance is not what Item 1 intends to address (see Exception 1). Direct-vent appliance exhaust and intake opening locations are dictated by the appliance listing and manufacturer's instructions. Note the distinction between direct-vent and other than direct-vent appliances in the manufacturer's instructions. Item 1 also does not apply to appliances such as rooftop units that have built-in outdoor air intakes and combustion gas exhaust outlets. The code assumes that the design and listing of the outdoor appliance accounts for the location of those openings (see Exception 2).

Item 2 addresses gravity air inlets in the exterior envelope of a building, including doors, operable (openable) windows and intake louvers and grilles.

The appliances addressed by this item use auxiliary or integral fans and blowers to force the flow of combustion products to the outdoors. This item applies to externally installed power exhausters, integrally power-exhausted appliances and venting systems equipped with draft inducers. This item does not address fan-assisted Category I appliances and does not address direct-vent appliances. Category III and IV appliances are subject to the requirements of this item if those appliances are not listed as direct-vent appliances. Some Category I appliances are field convertible to Category III.

Power exhausters are field-installed pieces of equipment that are independent of, but used in conjunction with, other appliances [see Commentary Figure 503.8(1)].

This item is consistent with appliance installation instructions that will contain additional details for the location of the venting system terminals, including but not limited to clearance to grade, inside corners, outside corners, decks, porches, balconies, soffits, roof eaves, utility meters, combustion air intakes, other appliance vents, service regulators, sidewalks and driveways.

There are significant differences between the requirements of Items 2 and 3 as they relate to nondirect-vent and direct-vent appliances, respectively. A Category IV, 90-percent and higher efficient furnace or boiler would be subject to Item 2 if installed without a combustion air intake pipe to the outdoors but would be subject to Item 3, not Item 2, if installed with a combustion air intake pipe to the outdoors. This is because such a furnace can be listed as both a direct-vent and nondirect-vent appliance, and as such can be installed either way. Less restrictive clearances, improved operating efficiency, longer appliance life, immunity from the effects of building exhaust systems and ease of providing combustion air are some of the advantages of direct-vent appliances. For example, according to Item 2, a sidewall vent terminal must be 4 feet (1219 mm) or more to either side of an openable window, or must be 4 feet (1219 mm) or more below the bottom of the window or must be 1 foot (305 mm) or more above the top of the window.

Item 3 is consistent with appliance manufacturers' instructions. Research has shown that the distances associated with the input ranges are necessary to

For SI: 1 inch = 25.4 mm, 1 foot = 304.8 mm.

Figure 503.8(1)
MECHANICAL DRAFT VENTING SYSTEM EXHAUSTER TERMINATIONS

allow the combustion gases to dissipate in the atmosphere, thus avoiding entry into the building through openings. This item is applicable only to direct-vent appliances. For example, a Category IV, 90-percent and higher efficient furnace installed without an outdoor air intake pipe for combustion air would be subject to Item 2, not 3 [see Commentary Figures 503.8(2), 503.8(3) and 503.8(4)].

Item 4 addresses the possible effect of condensation on pedestrians, service equipment, etc. Condensate is corrosive and could create a hazard when it freezes. Condensate ice could be a slip hazard and could obstruct regulator and relief valve vents. Category I and III appliances do not produce as much condensate as appliances in the other two categories,

but they can also cause condensate problems in cold climates. Large masses of ice could build up on or near the vent terminals of Category I and III vents. Such ice masses can fall or slide from roofs and injure someone or cause damage to a building, vehicle or equipment.

Item 5 is specific to sidewall terminations of vents for Category IV appliances, such as furnaces and boilers [see Commentary Figure 503.8(5)].

A common issue arises where buildings, especially homes, are located very close to each other, and sidewall-vented appliances are installed with the vent terminals directed toward the neighboring home. The concern is that combustion gases will enter the adjacent building through openings in the exterior walls

For SI: 1 inch = 25.4 mm.

Figure 503.8(2)
EXAMPLE OF MANUFACTURER'S DETAIL FOR DIRECT-VENT APPLIANCE
(Figure courtesy of Carrier Corporation)

For SI: 1 inch = 25.4 mm.

Figure 503.8(3)
EXAMPLE OF MANUFACTURER'S DETAIL FOR DIRECT-VENT APPLIANCE VENT AND COMBUSTION AIR TERMINALS
(Figure courtesy of Carrier Corporation)

that face the appliance vent terminal. This section applies only to Category IV (condensing) appliances that are sidewall vented with stainless steel or plastic vents.

Computer simulations were conducted as part of a research project, and the results indicated that, in many scenarios, the combustion products would impinge on the neighboring building. Many factors impact the simulated scenarios, including wind speed and direction; the height of the adjacent buildings; and the type of vent terminal (i.e., straight pipe, tee fitting, deflector cap, directional fitting, etc.). The worst-case scenario that this code section addresses is a straight open-ended pipe used as the appliance

vent terminal that is perpendicular to the wall it passes through. This scenario is the most common and the most likely to project combustion gases far enough to be a potential danger to the neighbors. The research project suggested that vent terminals that utilize a tee-fitting outlet, a deflector cap or that are directed at some angle downward are much less likely to interfere with the neighbors because, with these terminals, the combustion gases disperse and lack the velocity to impinge on the adjacent building.

503.9 Condensation drainage. Provisions shall be made to collect and dispose of condensate from venting systems serving Category II and IV appliances and noncategorized condensing appliances in accordance with Section 503.8, Item 4. Where local experience indicates that condensation is a problem, provisions shall be made to drain off and dispose of condensate from venting systems serving Category I and III appliances in accordance with Section 503.8, Item 4.

❖ See Section 503.8, Item 4.

503.10 Vent connectors for Category I appliances. Vent connectors for Category I *appliances* shall comply with Sections 503.10.1 through 503.10.14.

❖ See Sections 503.10.1 through 503.10.14.

503.10.1 Where required. A vent connector shall be used to connect an *appliance* to a gas vent, chimney or single-wall metal pipe, except where the gas vent, chimney or single-wall metal pipe is directly connected to the *appliance*.

❖ Unless the chimney or vent is connected directly to the appliance, a connector, defined in Section 202 as pipe used to connect an approved fuel-burning appliance to a chimney or vent, is necessary. This includes the fittings necessary to make a connection or change in direction. This is usually accomplished with a single-wall metal pipe, but it is also common practice to use listed and labeled chimney and vent pipe or listed factory-built single-wall or double-wall bendable connectors.

Figure 503.8(4)
DIRECT-VENT HOT WATER BOILER
(Photo courtesy of Weil-McLain)

Figure 503.8(5)
VENT TERMINATION

Many factors affect the design and configuration of a connector. The most important of these is the appliance location with respect to the chimney or vent system. This impacts the connector size, length and rise. Another important factor is the number of appliances being vented. The appliance manufacturer's installation instructions may prohibit the use of single-wall connectors. For example, a single-wall vent connector may not be appropriate for fan-assisted appliances because of the possibility of condensation, corrosion and leakage. They are also not permitted in attic or crawl space installations that are subject to cold temperatures. The cold temperatures increase the possibility of condensation, which leads to accelerated material failure. If the vent or chimney extends all the way to a single appliance vent outlet (collar), no connector is needed.

503.10.2 Materials. Vent connectors shall be constructed in accordance with Sections 503.10.2.1 through 503.10.2.5.

❖ See Sections 503.10.2.1 through 503.10.2.4.

503.10.2.1 General. A vent connector shall be made of noncombustible corrosion-resistant material capable of withstanding the vent gas temperature produced by the *appliance* and of sufficient thickness to withstand physical damage.

❖ See Sections 503.10.2.2 through 503.10.2.4.

503.10.2.2 Vent connectors located in unconditioned areas. Where the vent connector used for an *appliance* having a draft hood or a Category I *appliance* is located in or passes through attics, crawl spaces or other unconditioned spaces, that portion of the vent connector shall be *listed* Type B, Type L or *listed* vent material having equivalent insulation properties.

> **Exception:** Single-wall metal pipe located within the exterior walls of the building in areas having a local 99-percent winter design temperature of 5°F (-15°C) or higher shall be permitted to be used in unconditioned spaces other than attics and crawl spaces.

❖ Compared to double-wall pipe, single-wall pipe has a much higher heat loss, and when installed in an unconditioned space, condensation can occur, deteriorating the pipe. A typical scenario is an attic, garage or crawl space appliance installation where a connector would be exposed to cold temperatures, practically ensuring the formation of condensation in the connector. Except as allowed in the exception, this section effectively prohibits the use of single-wall connectors in garages, attics and crawl spaces when the spaces are unheated.

The term "unconditioned" is not defined and the debate over the intent created the text in the exception, which is climate related. Attics and crawl spaces are usually unconditioned spaces; however, the uncertainty has been with residential basements and garages, which are common locations for gas-fired appliances. In many climates, the garage winter temperatures will be cold enough to cause the combus-

tion gases in a connector to reach their dew point, meaning that condensate will form and linger in the connector, causing its rapid destruction. Most residential garages and some basements are considered to be unconditioned insofar as they are not directly or indirectly heated as is the living space.

The typical residential garage is not heated (conditioned) and, in most climates, will certainly have low ambient temperatures. The exception allows single-wall metal pipe in unconditioned spaces (other than attics and crawl spaces) if the climate is as described. The winter design temperatures can be found in the ASHRAE *Handbook of Fundamentals*. To put the climate criterion in perspective, Philadelphia, Pennsylvania; Louisville, Kentucky; Fall River, Massachusetts; and Providence, Rhode Island all have 99-percent winter design temperatures of 5°F (-15°C) or higher. The exception is applicable only to vent connectors that are inside a building (i.e., located within the exterior walls). (see Commentary Figure 503.10.2.2).

**Figure 503.10.2.2
LISTED BENDABLE DOUBLE-WALL CONNECTOR
(Photo courtesy of Selkirk L.L.C.)**

503.10.2.3 Residential-type appliance connectors. Where vent connectors for residential-type appliances are not installed in attics or other unconditioned spaces, connectors for *listed* appliances having draft hoods, appliances having draft hoods and equipped with *listed* conversion burners and Category I appliances shall be one of the following:

1. Type B or L vent material.

2. Galvanized sheet steel not less than 0.018 inch (0.46 mm) thick.

3. Aluminum (1100 or 3003 alloy or equivalent) sheet not less than 0.027 inch (0.69 mm) thick.

4. Stainless steel sheet not less than 0.012 inch (0.31 mm) thick.

5. Smooth interior wall metal pipe having resistance to heat and corrosion equal to or greater than that of Item 2, 3 or 4.

6. A *listed* vent connector.

Vent connectors shall not be covered with insulation.

Exception: *Listed* insulated vent connectors shall be installed in accordance with the manufacturer's instructions.

❖ This section applies to draft-hood-equipped appliances only and lists the allowable materials, including sheet metals, listed connectors and listed vent system pipe. The minimum thickness for single-wall galvanized steel pipe is equivalent to 26 gage. Installers have field-installed insulation on vent connectors in an attempt to control heat loss and the resultant condensation; however, this is prohibited because of the combustibility of most insulation materials and the possibility of excessive connector temperatures. Listed double-wall (air insulated) Type B connectors are available as an option to using Type B vent pipe (see Commentary Figure 503.10.2.2).

503.10.2.4 Low-heat equipment. A vent connector for a nonresidential, low-heat *appliance* shall be a factory-built chimney section or steel pipe having resistance to heat and corrosion equivalent to that for the appropriate galvanized pipe as specified in Table 503.10.2.4. Factory-built chimney sections shall be joined together in accordance with the chimney manufacturer's instructions.

❖ Residential appliances are also low heat by definition, and the provisions of Section 503.10.2.3 apply as well. The table is intended for nonresidential-type appliances, which is apparent when compared to the materials allowed by Section 503.10.2.3.

TABLE 503.10.2.4
MINIMUM THICKNESS FOR GALVANIZED STEEL VENT CONNECTORS FOR LOW-HEAT APPLIANCES

DIAMETER OF CONNECTOR (inches)	MINIMUM THICKNESS (inch)
Less than 6	0.019
6 to less than 10	0.023
10 to 12 inclusive	0.029
14 to 16 inclusive	0.034
Over 16	0.056

For SI: 1 inch = 25.4 mm.

❖ See the commentary to Section 503.10.2.4.

503.10.2.5 Medium-heat appliances. Vent connectors for medium-heat appliances shall be constructed of factory-built medium-heat chimney sections or steel of a thickness not less than that specified in Table 503.10.2.5 and shall comply with the following:

1. A steel vent connector for an *appliance* with a vent gas temperature in excess of 1,000°F (538°C) measured at the entrance to the connector shall be lined with medium-duty fire brick (ASTM C64, Type F), or the equivalent.

2. The lining shall be not less than $2^1/_2$ inches (64 mm) thick for a vent connector having a diameter or greatest cross-sectional dimension of 18 inches (457 mm) or less.

3. The lining shall be not less than $4^1/_2$ inches (114 mm) thick laid on the $4^1/_2$-inch (114 mm) bed for a vent connector having a diameter or greatest cross-sectional dimension greater than 18 inches (457 mm).

4. Factory-built chimney sections, if employed, shall be joined together in accordance with the chimney manufacturer's instructions.

❖ Medium-heat appliances have very high temperature flue gases [1,000°F (538°C)], and because of the waste of enormous amounts of energy are often equipped with secondary heat exchangers to reclaim heat energy and substantially reduce the vent gas temperatures.

TABLE 503.10.2.5
MINIMUM THICKNESS FOR STEEL VENT CONNECTORS FOR MEDIUM-HEAT APPLIANCES

VENT CONNECTOR SIZE		MINIMUM THICKNESS (inch)
Diameter (inches)	Area (square inches)	
Up to 14	Up to 154	0.053
Over 14 to 16	154 to 201	0.067
Over 16 to 18	201 to 254	0.093
Over 18	Larger than 254	0.123

For SI: 1 inch = 25.4 mm, 1 square inch = 645.16 mm².

❖ See the commentary to Section 503.10.2.5.

503.10.3 Size of vent connector. Vent connectors shall be sized in accordance with Sections 503.10.3.1 through 503.10.3.5.

❖ See the commentary to Sections 503.10.3.1 through 503.10.3.5.

503.10.3.1 Single draft hood and fan-assisted. A vent connector for an *appliance* with a single draft hood or for a Category I fan-assisted combustion system *appliance* shall be sized and installed in accordance with Section 504 or other *approved* engineering methods.

❖ The sizing tables in Section 504 will determine connector size with the fundamental requirement that a connector not be smaller than the appliance vent connection, except as allowed by Sections 504.2.2 and 504.3.17. The tables will often require that a connector be larger than the appliance vent outlet (see Section 504).

503.10.3.2 Multiple draft hood. For a single *appliance* having more than one draft hood outlet or flue collar, the manifold shall be constructed according to the instructions of the *appliance* manufacturer. Where there are no instructions, the manifold shall be designed and constructed in accordance with *approved* engineering practices. As an alternate method, the effective area of the manifold shall equal the combined area of the flue collars or draft hood outlets and the vent connectors shall have a minimum 1-foot (305 mm) rise.

❖ Some appliances are designed with more than one venting (flue) outlet. For example, some large furnaces were built with dual heat exchangers and burner sections, and thus have two draft hoods or flue collar outlets. The manufacturer's instructions must be followed. However, in the event that the manufacturer's instructions are not available, this section specifies required sizing criteria. Section 305.1 requires the manufacturer's instructions to be on the job site.

503.10.3.3 Multiple appliances. Where two or more appliances are connected to a common vent or chimney, each vent connector shall be sized in accordance with Section 504 or other *approved* engineering methods.

As an alternative method applicable only when all of the appliances are draft hood equipped, each vent connector shall have an effective area not less than the area of the draft hood outlet of the *appliance* to which it is connected.

❖ The sizing tables in Section 504 will determine connector size with the fundamental requirement that a connector not be smaller than the appliance vent connection, except as allowed by Sections 504.2.2 and 504.3.17. The tables will often require a connector that is larger than the appliance vent outlet (see Section 504). The alternative method paragraph of this section is incompatible with Section 504 and is intended only for use with the alternative sizing method of Section 503.6.8.1.

503.10.3.4 Common connector/manifold. Where two or more appliances are vented through a common vent connector or vent manifold, the common vent connector or vent manifold shall be located at the highest level consistent with available headroom and the required *clearance to combustible materials* and shall be sized in accordance with Section 504 or other *approved* engineering methods.

As an alternate method applicable only where there are two draft hood-equipped appliances, the effective area of the common vent connector or vent manifold and all junction fittings shall be not less than the area of the larger vent connector plus 50 percent of the area of the smaller flue collar outlet.

❖ Vent manufacturers' instructions contain detailed design criteria for manifolds. Connectors serving gas-fired appliances are allowed to join to form a common connector manifold rather than connecting independently to the vertical common vent [see Commentary Figures 503.10.3.4(1) and 503.10.3.4(2)]. Various manufacturers' instructions require a 10-percent to 20-percent reduction in the capacity of a common vent where a manifold is used (see commentary, Section 504.3.4). A manifold is defined as a vent or connector that is a lateral (horizontal) extension of the lower end of a common vent. Because a manifold serves multiple appliances and thus conveys flue gases from two or more appliances, manifolds must

be sized by the common vent tables. This section requires that manifolds be installed as high as possible, respective of ceiling height and clearance-to-combustibles requirements. This will result in the most appliance connector rise, which is always beneficial for venting performance. Connector rise takes advantage of the energy of hot gases discharging directly from the appliance and develops flow velocity and draft in addition to that produced by the common vent [see Commentary Figures 503.10.3.4(1) and 503.10.3.4(2)]. The alternative method in the second paragraph is intended for use only with two draft-hood-equipped appliances (similar to Section 503.6.9.1) and is not compatible with Section 504.

503.10.3.5 Size increase. Where the size of a vent connector is increased to overcome installation limitations and obtain connector capacity equal to the *appliance* input, the size increase shall be made at the *appliance* draft hood outlet.

❖ If a connector must be larger than the appliance vent outlet, the size increase must occur at the appliance connection by use of an increaser fitting. For example, it is common for a water heater with a 3-inch (76 mm) draft-hood outlet to be connected to a 4-inch (102 mm) connector by a 3-inch by 4-inch increaser fitting, as dictated by the vent tables in Section 504.

503.10.4 Two or more appliances connected to a single vent or chimney. Where two or more vent connectors enter a common vent, chimney flue or single-wall metal pipe, the smaller connector shall enter at the highest level consistent with the available headroom or *clearance* to *combustible material*. Vent connectors serving Category I appliances shall not be connected to any portion of a mechanical draft system operating under positive static pressure, such as those serving Category III or IV appliances.

❖ A common vent is somewhat oversized when only a single appliance is operating; therefore, every effort is made to achieve proper vent/chimney operation during all possible operating circumstances. Placing the smallest appliance connection at the highest elevation in the common vent/chimney will allow the greatest connector rise for the smaller appliance connector and will take advantage of any draft priming effect caused by the lower connector. For example, it has been common practice to connect a domestic water heater connector above the furnace or boiler connector in a chimney or vent system. Category I appliances require a vent that produces a draft and therefore cannot be connected to any venting system that produces a positive pressure. For example, a Category I appliance and a Category III appliance cannot share the same vent unless they both connect to the negative pressure (inlet) side of a mechanical draft system in accordance with the appliance and draft system installation instructions (see Section 501.10).

503.10.4.1 Two or more openings. Where two or more openings are provided into one chimney flue or vent, the openings shall be at different levels, or the connectors shall be attached to the vertical portion of the chimney or vent at an angle of 45 degrees (0.79 rad) or less relative to the vertical.

❖ The intent is to prevent turbulence and flow interference that could impede flow from a chimney or vent connector into the chimney or common vent. Openings at the same elevation, such as those directly opposite of each other, could negatively affect the flow from either or both of the connectors (see Commentary Figure 503.10.4.1).

503.10.5 Clearance. Minimum clearances from vent connectors to *combustible material* shall be in accordance with Table 503.10.5.

> **Exception:** The *clearance* between a vent connector and *combustible material* shall be permitted to be reduced where the *combustible material* is protected as specified for vent connectors in Table 308.2.

❖ Vent connectors must have minimum clearances to ignitable materials. The single-wall metal pipe clearance requirements in Table 503.10.5 apply to unlisted single-wall connectors. Connectors that are listed and labeled for this use must be installed with a clearance to combustibles as required by the connector manufacturer's instructions. If the appliance or connector manufacturer's installation instructions

For SI: 1 inch = 25.4 mm.

Figure 503.10.3.4(2)
COMMON VENT OFFSET AND MANIFOLD
(Figure courtesy of Selkirk, L.L.C.)

Figure 503.10.3.4(1)
MANIFOLD SIZING
(Figure courtesy of Selkirk, L.L.C.)

Figure 503.10.4.1
OPPOSING OPENINGS IN CHIMNEY

specify larger clearances than those prescribed by this section, the manufacturer's installation instructions govern. Table 503.10.5 dictates the required airspace clearances between vent connectors and combustible materials and assemblies. The provisions of Section 308 would allow reduction of connector clearances where the prescribed clearance cannot be provided or is impractical. Lack of vent connector clearance is a common code violation. Connectors usually lack the required clearances to wood joists, plastic plumbing piping, building and piping insulation materials and gypsum board. A major disadvantage of single-wall connectors is the large clearance required to combustibles. Listed connectors usually

require far less clearance, in some cases only 1 inch (25 mm).

503.10.6 Joints. Joints between sections of connector *piping* and connections to flue collars and draft hood outlets shall be fastened by one of the following methods:

1. Sheet metal screws.

2. Vent connectors of *listed* vent material assembled and connected to flue collars or draft hood outlets in accordance with the manufacturers' instructions.

3. Other *approved* means.

❖ Single-wall connectors (unlisted) have been traditionally fastened with sheet metal screws and rivets. All connector joints must be fastened, including the joint at the appliance draft hood or flue collar. A displaced connector could result in a life-threatening condition; therefore, connectors must be fastened and supported well. Appliance connectors are often located where they can be impacted or otherwise disturbed by building occupants, making proper fastening even more important. Item 2 speaks of listed connectors such as Type B vent material and factory-built corrugated (bendable) connectors, which use adapter fittings or integral collars that are mechanically fastened to the appliance with screws or rivets (see Commentary Figure 503.10.2.2).

503.10.7 Slope. A vent connector shall be installed without dips or sags and shall slope upward toward the vent or chimney not less than $^1/_4$ inch per foot (21 mm/m).

Exception: Vent connectors attached to a mechanical draft system installed in accordance with the *appliance* and draft system manufacturers' instructions.

❖ The ideal chimney or vent configuration is a totally vertical system, even though it is not always practical. This section requires all portions of a chimney or vent connector to rise vertically a minimum of a $^1/_4$ inch per each foot (21 mm/m) of its horizontal length. The connector slope is intended to induce the flow of flue gases using the natural buoyancy of the hot gases.

TABLE 503.10.5ᵃ
CLEARANCES FOR CONNECTORS

APPLIANCE	MINIMUM DISTANCE FROM COMBUSTIBLE MATERIAL			
	Listed Type B gas vent material	Listed Type L vent material	Single-wall metal pipe	Factory-built chimney sections
Listed appliances with draft hoods and appliances listed for use with Type B gas vents	As listed	As listed	6 inches	As listed
Residential boilers and furnaces with listed gas conversion burner and with draft hood	6 inches	6 inches	9 inches	As listed
Residential appliances listed for use with Type L vents	Not permitted	As listed	9 inches	As listed
Listed gas-fired toilets	Not permitted	As listed	As listed	As listed
Unlisted residential appliances with draft hood	Not permitted	6 inches	9 inches	As listed
Residential and low-heat appliances other than above	Not permitted	9 inches	18 inches	As listed
Medium-heat appliances	Not permitted	Not permitted	36 inches	As listed

For SI: 1 inch = 25.4 mm.

a. These clearances shall apply unless the manufacturer's installation instructions for a listed appliance or connector specify different clearances, in which case the listed clearances shall apply.

Connector slope can promote the priming of a cold venting system and can partially compensate for short connector vertical rise. Low points, dips and sags could also trap condensate and accelerate corrosion of the connector. The exception allows connectors without slope where connected to mechanical draft systems because the connector is on the negative pressure (inlet) side of the exhauster and slope would provide no benefit.

503.10.8 Length of vent connector. The maximum horizontal length of a single-wall connector shall be 75 percent of the height of the chimney or vent except for engineered systems. The maximum horizontal length of a Type B double-wall connector shall be 100 percent of the height of the chimney or vent except for engineered systems.

❖ This section is applied in conjunction with Section 504.3.2. The appliance and chimney or vent must be located to keep the connector length as short as practicable. This section establishes the maximum allowable length for uninsulated chimney and vent connectors, insulated chimney and vent connectors and individual connectors for a chimney or vent system serving multiple appliances. These requirements are based on the heat loss of the connector and the ability of the vent or chimney system to produce a draft.

An insulated connector (double wall) reduces the amount of heat transfer through the connector pipe; thus, the flue gas is maintained at a higher temperature (see commentary, Section 504.3.2). The amount of draft is directly related to the chimney or vent height and the difference in temperature between the flue gases and the ambient air. An uninsulated connector is a run of single-wall pipe. Commentary Figure 503.10.8 shows that the total chimney or vent height is measured from the top of the highest appliance flue outlet connection to the termination point of the chimney or vent. The length limitations of this section are based on the total vertical rise of the chimney or vent and on the developed length of horizontal connectors within the chimney or vent system. The venting tables will often prohibit the use of single-wall connectors with fan-assisted appliances, and most appliance and vent manufacturers recommend against the use of single-wall (uninsulated) connectors.

Connectors are limited in length because of the flow resistance of the connector pipe and because heat loss through the connector is directly related to its length. For a venting system to work, the draft produced by the vertical vent or chimney must be able to overcome the resistance to flow created by the connector. The longer the connector, the longer it takes to prime a cold venting system and develop draft. As stated in Section 504.3.2, connectors should always be as short as the installation conditions will permit.

The sizing methodology and tables in Section 504 for connectors in single-appliance installations might allow connector lengths greater than the general limits given in this section or might require a shorter connector length than would be allowed by these limits, depending on the combination of appliance type and input, connector size and type and vent height.

Because the sizing tables in Section 504.2 are based on specific computer modeling of the installation configuration, the limitations or allowances in the single-appliance tables supersede what would be required in this section. For multiple-appliance systems, the sizing methods and tables do not contain the same limitations on connector length. This section establishes an absolute length limitation for connectors in multiple-appliance systems which still provides liberal installation flexibility without permitting excessively long connectors. This section functions as an absolute cap for the connector length provisions of Sections 504.3.2 and 504.3.3.

The appliance and vent location should be well planned to allow the shortest horizontal run of connector piping (see Sections 504.3.2 and 504.3.3).

The intent of the reference to engineered systems is to allow connector length to be determined by design as part of an engineered venting system.

Figure 503.10.8
VENT HEIGHT
(Figure courtesy of Simpson Dura-Vent Company, Incoroporated)

503.10.9 Support. A vent connector shall be supported for the design and weight of the material employed to maintain clearances and prevent physical damage and separation of joints.

❖ A connector must be supported for the design and weight of the material used. Proper support is necessary to maintain the clearances required by Section

503.10.5, to maintain the required slope and to prevent physical damage and separation of joints. The joints between connectors and appliance draft-hood outlets and flue collars must be secured with screws, rivets or other approved means. The joints between pipe sections also must be secured with screws, rivets or other approved means. Some connectors are available with a proprietary-type fastening method, which does not require screws, rivets or other fasteners if it provides equivalent resistance to disengagement or displacement. This section addresses connectors, not chimneys and vents, and most chimney and vent manufacturers do not require or might even prohibit the use of screws or rivets for securing joints for chimney and vent pipe sections and fittings.

503.10.10 Chimney connection. Where entering a flue in a masonry or metal chimney, the vent connector shall be installed above the extreme bottom to avoid stoppage. Where a thimble or slip joint is used to facilitate removal of the connector, the connector shall be firmly attached to or inserted into the thimble or slip joint to prevent the connector from falling out. Means shall be employed to prevent the connector from entering so far as to restrict the space between its end and the opposite wall of the chimney flue (see Section 501.9).

❖ All of the precautions taken to provide adequate chimney design may be ineffective if the connection between the appliance and the chimney is not properly accomplished. Improper connections can lead to appliance or connector failure and the leakage of vent gases into the building. This section contains requirements for the connection between a chimney and the appliance connector. The connection to a vent is properly accomplished by vent fittings and likewise for factory-built chimneys. Chimney connectors are required to pass through a masonry chimney wall to the inner face of the liner, but not beyond. A connector that extends into a chimney passageway can restrict the flow of vent gases and provide a ledge on which debris can accumulate. The joint between the connector and the chimney must be fastened in a manner that will prevent separation. If the connector enters a masonry chimney, it must be cemented in place with an approved material such as refractory mortar or other heat-resistant cement.

This section also permits the use of thimbles at a masonry chimney opening to provide for easy removal of the connector to facilitate cleaning. When a thimble is installed, it is to be permanently cemented in place with an approved high-temperature cement. A connector must be attached to a thimble in an approved manner to prevent displacement. A collar on the connector or a similar arrangement must be provided to limit the penetration of the connector into the chimney. Otherwise, it is quite possible that a connector will be shoved blindly into a chimney, reaching to the back wall of the chimney, and thus be obstructed.

503.10.11 Inspection. The entire length of a vent connector shall be provided with ready *access* for inspection, cleaning and replacement.

❖ A connector cannot be concealed by any form of building construction, whether or not such construction is removable. Connectors must be inspected periodically for corrosion damage, structural integrity, joint separation and clearances to combustibles (see Section 503.10.13).

503.10.12 Fireplaces. A vent connector shall not be connected to a chimney flue serving a *fireplace* unless the *fireplace* flue opening is permanently sealed.

❖ A fireplace chimney cannot serve any appliance unless the fireplace is retired by permanently sealing off the chimney from the fireplace. If this is not done, the fireplace can affect draft for the connected appliance, and/or vent gases could escape into the building through the fireplace opening.

503.10.13 Passage through ceilings, floors or walls. Single-wall metal pipe connectors shall not pass through any wall, floor or ceiling except as permitted by Section 503.7.4.

❖ This section no longer prohibits double-wall, Type B connectors from passing through walls, floors and ceilings. The prohibition on pass-through is limited to single-wall metal pipe connectors, recognizing that Section 503.7.4 allows single-wall metal pipe vents to extend from the appliance to the outdoors by passing through a roof or exterior wall. A single-wall connector has more than double the heat loss of a listed metal vent or chimney system. This produces high temperatures on the outside surface and cooling of the flue gases on the inside of the connector. The cooling of flue gases produces condensation, which can cause connector deterioration, while the excessive surface temperatures pose a potential fire hazard. For these reasons, a single-wall connector can be located only within the room or space in which the appliance is located. Single-wall connectors are prohibited from passing through any ceiling, floor or wall. If single-wall connectors were allowed to pass through walls, floors or ceilings, they might not be readily observable, and a potential fire hazard or connector deterioration could go undetected. Also, pass-through increase the likelihood that the connector will be excessively long, exposed to low ambient temperatures or subject to contact with combustibles. Connectors that pass through a wall, floor or ceiling would be out of sight and therefore out of mind, meaning that they would not get the attention needed to avoid hazardous conditions (see Commentary Figure 503.10.13).

503.10.14 Medium-heat connectors. Vent connectors for medium-heat appliances shall not pass through walls or partitions constructed of *combustible material*.

❖ When the vent gas temperatures are 1,000°F (538°C) and higher, penetration of combustible assemblies is an unacceptable risk.

Figure 503.10.13
TYPE B VENT CONNECTOR PENETRATIONS

503.11 Vent connectors for Category II, III and IV appliances. Vent connectors for Category II, III and IV appliances shall be as specified for the venting systems in accordance with Section 503.4.

❖ See Sections 503.4, 503.4.1 and 503.4.2.

503.12 Draft hoods and draft controls. The installation of draft hoods and draft controls shall comply with Sections 503.12.1 through 503.12.7.

❖ See Sections 503.12.1 through 503.12.7.

503.12.1 Appliances requiring draft hoods. Vented appliances shall be installed with draft hoods.

> **Exception:** Dual oven-type combination ranges; *direct-vent appliances*; fan-assisted combustion system appliances; appliances requiring chimney draft for operation; single firebox boilers equipped with conversion burners with inputs greater than 400,000 Btu per hour (117 kW); appliances equipped with blast, power or pressure burners that are not *listed* for use with draft hoods; and appliances designed for forced venting.

❖ Draft hoods are integral to or are supplied with natural draft atmospheric-burner gas-fired appliances (see Commentary Figure 503.12.1). With the development of higher appliance efficiencies, fan-assisted combustion, direct-vent appliances and Category III and IV appliances, fewer draft-hood appliances will be used. Draft hoods are still common on conventional tank-type and tankless water heaters and 80-percent efficient boilers. A draft-hood-equipped appliance will allow a continuous flow of air into the venting system, which represents an energy loss except where an automatic vent damper is installed to close off the vent in the off cycle. Without an automatic vent damper, conditioned air is taken from the space in which the appliance is located and residual heat is taken away from the appliance itself, both of which

contribute to energy loss. Not all appliances are categorized, but if they were, draft-hood-equipped appliances would fall under Category I. Draft hoods have multiple functions. They serve as a relief opening in the event of vent or chimney backdraft or blockage, they stabilize/regulate the amount of draft that occurs in the appliance combustion chamber and they allow the introduction of dilution air into the chimney or vent.

Figure 503.12.1
DRAFT-HOOD-EQUIPPED APPLIANCE
(Photo courtesy of A.O. Smith Water Products)

503.12.2 Installation. A draft hood supplied with or forming a part of a *listed* vented *appliance* shall be installed without *alteration*, exactly as furnished and specified by the *appliance* manufacturer.

❖ Draft hoods are often shipped with appliances as a part that must be field installed. Installers have been known to shorten or otherwise modify the draft hoods to compensate for installation constraints and have also installed them in an orientation other than vertical.

503.12.2.1 Draft hood required. If a draft hood is not supplied by the *appliance* manufacturer where one is required, a draft hood shall be installed, shall be of a *listed* or *approved* type and, in the absence of other instructions, shall be of the same size as the *appliance* flue collar. Where a draft hood is required with a conversion burner, it shall be of a *listed* or *approved* type.

❖ In the unlikely event that an appliance manufacturer does not supply the required draft hood, a listed or approved draft hood must be obtained and installed.

503.12.2.2 Special design draft hood. Where it is determined that a draft hood of special design is needed or preferable for a particular installation, the installation shall be in accordance with the recommendations of the *appliance* manufacturer and shall be *approved*.

❖ There could be unusual circumstances, such as where a horizontal-type draft hood is needed, because of low ceilings and/or insufficient connector rise.

503.12.3 Draft control devices. Where a draft control device is part of the *appliance* or is supplied by the *appliance* manufacturer, it shall be installed in accordance with the manufacturer's instructions. In the absence of manufacturer's instructions, the device shall be attached to the flue collar of the *appliance* or as near to the *appliance* as practical.

❖ Draft hoods also serve to control draft; however, this section would be redundant with Section 503.12.2 if it were not addressing draft controls such as barometric dampers. Two types of draft controls are installed in the appliance vent outlet or vent connector: the passive draft hood and the active barometric damper. Barometric dampers consist of a damper plate mounted on an axle that rotates in low friction bearing points such as needle or knife-edge bearings. The device is mounted in a tee fitting arrangement and connects with the appliance vent connector. It admits air in varying amounts into the vent connector to control the amount of draft through the appliance combustion chamber. These devices respond to varying chimney or vent draft caused by changes in indoor and outdoor temperatures, changes in atmospheric (barometric) pressure and changes in wind speed and direction. Barometric dampers are tuned by adjustable balancing weights attached to the damper plate. The damper manufacturer will provide instructions on how and where to install the device [see Commentary Figure 202(2)].

503.12.4 Additional devices. Appliances requiring a controlled chimney draft shall be permitted to be equipped with a *listed* double-acting barometric-draft regulator installed and adjusted in accordance with the manufacturer's instructions.

❖ Like Section 503.12.3, this section addresses barometric dampers, specifically, the double-acting type. These dampers open inward to admit air into the vent system like all other barometric dampers, but can also open outward in the event of backdraft to allow vent gases to be spilled into the room or space in which the appliance is located, just as a draft hood would do. The manufacturer's installation instructions may require or recommend that a sensor (spill switch) be installed that will sense prolonged vent-gas spillage from the damper and shut down the appliance. In such cases, a manufacturer's recommendation should be interpreted as an enforceable requirement.

Some Category I fan-assisted furnaces can be fitted with a factory-supplied conversion part that is, in effect, a retrofit draft hood. This will lower the efficiency of the appliance by converting it to a draft-hood-equipped appliance that will now introduce dilution air (conditioned room air) into the vent. This is done solely to allow fan-assisted furnaces to connect to existing masonry chimneys that would otherwise not be allowed to serve such appliances. These conversion parts use a sensor (spill switch) in the relief/air inlet opening. The manufacturer of the appliance and the draft hood conversion kit will provide specific installation instructions that must be followed for the conversion.

Vent gas spillage must not occur from an incinerator venting system; therefore, a draft regulating device must be of the single-acting type.

503.12.5 Location. Draft hoods and barometric draft regulators shall be installed in the same room or enclosure as the *appliance* in such a manner as to prevent any difference in pressure between the hood or regulator and the *combustion air* supply.

❖ Draft-regulation devices are designed to maintain a constant draft through the appliance combustion chamber and flue passages for the purpose of maintaining proper combustion conditions and achieving the highest possible efficiency. If the draft control sees a different ambient pressure than the appliance, it could allow too much or too little draft through the appliance. The positive, negative or neutral pressures within a building must act equally on both the appliance and the draft control to allow the draft control to be adjusted and perform its intended function.

503.12.6 Positioning. Draft hoods and draft regulators shall be installed in the position for which they were designed with reference to the horizontal and vertical planes and shall be located so that the relief opening is not obstructed by any part of the *appliance* or adjacent construction. The *appliance* and its draft hood shall be located so that the relief opening is accessible for checking vent operation.

❖ Draft hoods and barometric dampers are both affected by their orientation. There are two basic types of draft hoods: vertical draft hoods and horizontal draft hoods. These types of draft hoods are not interchangeable. Barometric dampers can be adjusted to connect to either vertical or horizontal vent connectors; however, they are always positioned so that the closed damper plate is in a vertical plane. The manufacturer's instructions must always be followed. The relief opening on draft controls is also the dilution air inlet, and it must be unobstructed to allow air to enter under normal conditions, allow vent gases to spill in the event of backdraft and allow testing of the vent system for proper draft. For example, for many years, appliance venting/draft has been checked by observing the movement of smoke introduced near the relief/inlet opening.

503.12.7 Clearance. A draft hood shall be located so its relief opening is not less than 6 inches (152 mm) from any surface except that of the *appliance* it serves and the venting system to which the draft hood is connected. Where a greater or lesser *clearance* is indicated on the *appliance* label, the *clearance* shall be not less than that specified on the label. Such clearances shall not be reduced.

❖ The specified clearance is necessary to assure the unimpeded flow of air into the draft hood and flow of vent gas spillage from the hood. The clearance also protects materials from the hot gases that can spill from the draft hood and allows access for testing and observation.

503.13 Manually operated dampers. A manually operated damper shall not be placed in the vent connector for any *appliance*. Fixed baffles shall not be classified as manually operated dampers.

❖ Manual dampers are associated with solid-fuel appliances but not with gas-fired appliances. A manual damper requires manual operation by a human, which cannot be relied on because humans forget. If a damper is left closed or partially closed during appliance operation, a severe hazard could result from vent-gas spillage and/or appliance malfunction (see Section 503.14).

503.14 Automatically operated vent dampers. An automatically operated vent damper shall be of a *listed* type.

❖ Automatic vent dampers are intended for use with gas-fired natural-draft appliances. An automatic vent damper must be installed in strict compliance with the manufacturer's instructions. Because automatic vent-damper failure can result in a hazardous condition, automatic dampers must be listed and labeled. An automatic vent damper is installed on the draft hood outlet of an individual gas-fired appliance (see Commentary Figure 503.14). These dampers must not serve more than one appliance. The manufacturer's instructions require that the damper be installed by a qualified installer in accordance with the terms of the listing and the manufacturer's instructions.

Because the purpose of a vent damper is to close or restrict the flue passageway of an appliance, it is imperative that the device be properly installed to minimize the possibility of failure. A malfunctioning or

improperly installed vent-damper device can cause the appliance to malfunction and could cause the discharge of products of combustion directly into the building interior. Automatic vent dampers are energy-saving devices designed to close off or restrict an appliance flue passageway when the appliance is not operating and is in its "off" cycle. These devices save energy by trapping residual heat in a heat exchanger after the burners shut off and by preventing the escape of conditioned room air up the chimney or vent. Thus, appliance efficiency can be boosted, and building air infiltration can be reduced. Such devices can be field-installed additions to existing equipment and can be a factory-supplied component of an appliance (see Commentary Figure 504.2.1).

A common application for vent dampers today is for mid-efficiency (80 to 83 percent) hot water boilers, which are draft-hood equipped. There are three types of automatic vent dampers: electrical-, mechanical- and thermal-actuating. The electric-motor-operated type is the commonly encountered type today (see Commentary Figures 503.14 and 504.2.1). Electrical- and mechanical-actuating automatic dampers must open automatically prior to main burner ignition and must remain open until completion of the burner cycle. They must also be interlocked electrically with the appliance control circuitry to prevent operation of the appliance in the event of damper failure. Thermal-actuating dampers open in response to flue gas temperature after burner ignition and close after burner shutdown. Thermal-actuated dampers use bimetal components that convert heat energy into mechanical energy. Some thermal-actuated dampers require the installation of a draft-hood spillage sensor that would shut off the appliance if the damper failed to open. The most common type of vent damper uses an electric motor to rotate a damper blade. The device uses switches that prove the damper position and allow the appliance to start only after the damper is opened and verified. The sequence of operation is the same as that described in the commentary to Section 503.3.3.

503.15 Obstructions. Devices that retard the flow of vent gases shall not be installed in a vent connector, chimney or vent. The following shall not be considered as obstructions:

1. Draft regulators and safety controls specifically *listed* for installation in venting systems and installed in accordance with the manufacturer's instructions.

2. *Approved* draft regulators and safety controls that are designed and installed in accordance with *approved* engineering methods.

3. *Listed* heat reclaimers and automatically operated vent dampers installed in accordance with the manufacturer's instructions.

4. *Approved* economizers, heat reclaimers and recuperators installed in venting systems of appliances not required to be equipped with draft hoods, provided that the *appliance* manufacturer's instructions cover the installation of such a device in the venting system and

Figure 503.14
AUTOMATIC VENT DAMPERS

performance in accordance with Sections 503.3 and 503.3.1 is obtained.

5. Vent dampers serving *listed* appliances installed in accordance with Sections 504.2.1 and 504.3.1 or other *approved* engineering methods.

❖ Manual dampers, flow restricting orifices and similar devices are not accounted for in the vent design methods of the code. The tables in Section 504 assume that no such devices are installed in the vent. Item 3 addresses the automatic vent dampers covered in Section 503.14. These dampers open fully before the appliance is allowed to fire and offer negligible flow resistance because the damper plate is parallel to the vent gas flow. Item 4 addresses devices that extract heat from the vent gases to save energy that would otherwise be lost to the outside atmosphere. Appliances without draft hoods, excluding Category I fan-assisted, Category II, III and IV appliances, usually have higher vent gas temperatures than those with draft hoods; therefore, waste

heat can be reclaimed with less chance of diminishing draft (see Section 504.2.1).

503.16 Outside wall penetrations. Where vents, including those for *direct-vent appliances*, penetrate outside walls of buildings, the annular spaces around such penetrations shall be permanently sealed using *approved* materials to prevent entry of combustion products into the building.

❖ Vents that terminate through outside walls are subject to being covered (buried) in snow in many climates and experience has shown that combustion products can enter the building through the annular space between the vent and the wall assembly. In some cases, snow cover can trap combustion gases and cause them to reenter a direct-vent appliance combustion chamber via the combustion air intake. This will generate excessive carbon monoxide levels as a result of oxygen depletion. The snow cover can trap and channel these toxic gases into the building through any unsealed annular space in the wall penetration. The required sealing must be permanent and able to tolerate movement of the vent caused by expansion and contraction of the vent. Plastic pipe used for vents will expand and contract much more than metal pipe, and the annular space seal will be broken if allowances are not made for this movement.

Although not explicitly stated, this section should also be applied to penetrations by combustion air intake pipes that are associated with direct-vent appliances because such intakes are in close proximity to the vent discharge terminal and are affected by the same snow cover.

This section applies to all vents, not just those in regions subject to snowfall. Wind, fences, visual screens, vegetation and other causes could also create the conditions where combustion gases could enter the building.

SECTION 504 (IFGS)
SIZING OF CATEGORY I APPLIANCE
VENTING SYSTEMS

504.1 Definitions. The following definitions apply to the tables in this section.

❖ The definitions in this section are necessary for application of the vent sizing tables and are exclusively related to Section 504. As stated in the main section title, this section applies to Category I appliances. See the definition of "Vented appliance categories." However, it also applies to draft-hood-equipped appliances, which may or may not be categorized at all, even though they could be Category I because they meet the definition of Category I appliances. For example, a draft-hood water heater need not be categorized by its governing standard, but if it were categorized it would be a Category I appliance. A gas-fired appliance with a draft hood depends on gravity flow for supply of combustion air, for flow of combustion products through the heat exchanger and for proper gas venting. The draft hood is designed to maintain proper combustion in the event of draft fluctuations. Its relief opening serves as a flue product exit in the event of a blocked vent or downdraft. The relief opening, however, also allows dilution air to enter the vent during normal appliance/vent operation. When this dilution air is obtained from within the heated space, there is a loss of seasonal efficiency, which can be reduced by installation of a vent damper designed to reduce the flow of air through the vent when the burner is not firing.

To improve annual fuel utilization efficiency (AFUE), many appliance manufacturers have designed fan-assisted combustion using mechanical means (blowers or fans) to obtain either induced or forced flow of combustion air and combustion products. Greater efficiency results from three major effects. First, heat exchange improves because of higher internal flow velocity, which enhances heat transfer by creating turbulent flow and the "scrubbing" of heat exchanger surfaces. Second, induced draft through the appliance combustion chambers and heat exchangers allows the heat exchangers to be constructed with longer passes, which results in greater surface area and thus longer flue gas retention time. This allows extraction of more heat from the combustion gases. Third, there is no longer a draft hood to cause heated air loss up the vent, both when the appliance is operating and when it is not [see Commentary Figures 504.1(1) through 504.1(4)].

A fan-assisted combustion appliance uses pressure-sensing and heat-sensing controls to monitor proper venting. The controls must prove that the appliance flue outlet is at neutral or negative pressure and that adequate flow of flue gases exists. Without a draft hood and dilution air to dilute the combustion products, gases entering the vent will have a higher water vapor content than those from a draft-hood appliance [see Commentary Figure 504.1(3)]. There will likely be a longer period of condensation in the vent (wet time), particularly in the upper and/or colder portions of the venting system. These differences between fan-assisted and draft-hood appliances are reflected in the tables by the lack of minimum capacities for draft hood appliances. The minimums for fan-assisted appliances are based on the heat input needed to control the duration of wet time or condensation during appliance start-up and operation.

The tables do not apply to:

1. Wall furnaces (recessed heaters), which require Type BW vents.

2. Decorative gas appliances, which generally require a specific vent size and are best individually vented.

3. Category II, III or IV gas appliances. If permitted by the appliance manufacturer, some Category III appliances can also be vented in accordance with Category I conditions. Specifically, if the vent for a Category III appliance is sized and configured (connector rise, total height, etc.) for the input, the vent will operate under nonpositive pressure. This allows the use of a Type B gas vent. For a Category III appliance to operate as a Category I, appliance manufacturers' instructions must be followed, and the vent size may need to be larger than the appliance flue collar.

4. Gas-fired appliances listed for use only with chimneys or dual-fuel appliances such as oil/gas, wood/gas, or coal/gas. Dual-fuel appliances require chimneys sized in accordance with the appliance manufacturer's instructions and in accordance with the chimney manufacturer's instructions.

Capacities in the tables assume cold starts such as with an intermittent ignition system (no standing pilot). The type of ignition system does not affect maximum capacities, but it does affect minimums. A standing pilot on an appliance will keep vent gas temperatures slightly above outdoor ambient temperature, whereas a tank of hot water plus water-heater pilot operation will maintain flue temperature ahead of the draft hood at approaching the same temperature as the stored water. This flow of heat into a vent aids in priming as well as reducing wet time. In some cases, the fan-assisted appliance manufacturer and/ or the sizing tables in the code will require connection of a draft-hood-type appliance to the venting system because the relatively inefficient draft-hood appliance contributes heat to the venting system necessary to make it function.

The tables also assume that there are no adverse or building depressurization effects. A strong wind or mechanical ventilation/exhaust might cause a down-draft with draft-hood appliances that will prevent the vent from priming properly.

The input capacity values in the tables are computed for typical natural gas. They can be used for liquefied petroleum gases (LP-gases), such as propane and butane, and mixtures of these with air. With LP-gases, the maximum capacity remains the same, but because these have less hydrogen and produce less water vapor, the possibility of condensation is somewhat lower. To simplify matters, assume that minimum capacities are the same regardless of fuel type.

Oversized or excessively long vents can cause condensation and should be avoided. The tables define limits that minimize wet time in the indoor portions as well as in the upper exposed end of the vent.

The maximum vent lateral lengths for FAN appliances are calculated on the basis of several assumed conditions, including flue gas temperature and composition, and an outdoor ambient temperature of 42°F (6°C), chosen as a representative value. Colder assumed temperatures or greater outdoor exposure of the vent would lead to shorter maximum allowable lengths or to greater possibilities of condensation. To minimize condensation, it is essential to operate closer to maximum than minimum capacity and also to use the smallest allowable vent size.

For SI: 1 cubic foot = 0.0283 m³.

Figure 504.1(2)
FLUE GAS COMPOSITION
(Figure courtesy of Simpson Dura-Vent Company, Incorporated)

Figure 504.1(3)
DRAFT-HOOD APPLIANCE COMPARED TO FAN-ASSISTED APPLIANCE
(Figure courtesy of Simpson Dura-Vent Company, Incorporated)

For SI: 1 cubic foot = 0.0283 m³.

Figure 504.1(1)
COMBUSTION PROCESS FOR NATURAL GAS
(Figure courtesy of Simpson Dura-Vent Company, Incorporated)

Figure 504.1(4)
ILLUSTRATION OF THE BASIC PRINCIPLE OF
FAN-ASSISTED COMBUSTION
(Figure courtesy of Weil-McLain)

Figure 504.1(5)
FAN-ASSISTED HOT WATER BOILER
(Photo courtesy of Weil-McLain)

APPLIANCE CATEGORIZED VENT DIAMETER/ AREA. The minimum vent area/diameter permissible for Category I appliances to maintain a nonpositive vent static pressure when tested in accordance with nationally recognized standards.

❖ The minimum Category I appliance vent area/diameter necessary to maintain a nonpositive vent static pressure when tested in accordance with the applicable standard. The appliance categorized vent diameter/area is determined by the appliance manufacturer and is typically the vent outlet collar installed on the appliance.

FAN-ASSISTED COMBUSTION SYSTEM. An *appliance* equipped with an integral mechanical means to either draw or force products of combustion through the combustion chamber or heat exchanger.

❖ An appliance equipped with an integral mechanical means to either draw or force products of combustion through the combustion chamber or heat exchanger [see Commentary Figures 504.1(4) and (5)].

FAN Min. The minimum input rating of a Category I fan-assisted appliance attached to a vent or connector.

❖ The minimum appliance input rating of a Category I appliance with a fan-assisted combustion system that is capable of being attached to a vent.

FAN Max. The maximum input rating of a Category I fan-assisted appliance attached to a vent or connector.

❖ The maximum appliance input rating of a Category I appliance with a fan-assisted combustion system that is capable of being attached to a vent.

NAT Max. The maximum input rating of a Category I draft-hood-equipped *appliance* attached to a vent or connector.

❖ The maximum appliance input rating of a Category I natural draft appliance that is capable of being attached to a vent.

FAN + FAN. The maximum combined *appliance* input rating of two or more Category I fan-assisted appliances attached to the common vent.

FAN + NAT. The maximum combined *appliance* input rating of one or more Category I fan-assisted appliances and one or more Category I draft-hood-equipped appliances attached to the common vent.

NA. Vent configuration is not allowed due to potential for condensate formation or pressurization of the venting system, or not applicable due to physical or geometric restraints.

❖ Vent configuration is prohibited.

NAT + NAT. The maximum combined *appliance* input rating of two or more Category I draft-hood-equipped appliances attached to the common vent.

504.2 Application of single-appliance vent Tables 504.2(1) through 504.2(6). The application of Tables 504.2(1) through 504.2(6) shall be subject to the requirements of Sections 504.2.1 through 504.2.17.

❖ Note that all of the table titles have been reformatted to make them easier to use and to lessen the possibility for error in choosing the appropriate table. The tables do not apply to Type B-W vents; vents for decorative gas appliances; vents for Category II, III or IV appliances; vents for dual-fuel appliances; vents for appliances not listed for use with Type B vent and appliances listed only for connection to chimneys. Category I appliances can be a fan-assisted design or can be a draft-hood-equipped design, and each design has different vent system design considerations. A noteworthy difference between fan-assisted and draft-hood-equipped designs is that the vents for fan-assisted appliances have both a maximum and a minimum vent capacity. In the past, lower-efficiency draft-hood-type appliances were dominant, and vent sizing was based only on the maximum venting capacity of the vent. In other words, the only concern was to make sure that the vent was big enough. This is no longer the case with the advent of mid-efficiency fan-assisted-type appliances. Today, it is possible for a vent to be either too small or too large. An undersized vent will have insufficient capacity and can allow flue gas spillage or cause positive pressure to occur within the vent. An oversized vent can fail to produce sufficient draft and can be subject to the continuous formation of water vapor condensation on the interior vent surfaces. A typical individual vent is shown in Commentary Figure 504.2(1) for a FAN furnace with an input rating of 150,000 Btu/h (44 kW) and a 5-inch (127 mm) outlet connection.

Procedure for a Type B vent, Table 504.2(1): Go down the height column to 20 feet (6096 mm) and across on the 10-foot (3048 mm) lateral line.

For this furnace, the MAX capacity under the 5-inch (127 mm) size heading is 229,000 Btu/h (67.1 kW) and the MIN capacity is 50,000 Btu/h (14.7 kW). A 5-inch (127 mm) size Type B vent will be correct. See Commentary Figure 504.2(1). In applying the tables, extreme care must be taken to observe the title of the table and to apply the conditions and instructions set forth in Sections 504.2.1 through 504.2.16.

To determine the proper size for an individual vent, apply the table as follows:

1. Determine the total vent height and length of lateral, based on the appliance and vent location and the height to the top of the vent, as indicated in Commentary Figure 504.2(1). If gas appliances, such as a furnace, boiler or water heater, have not been chosen or installed, estimate the height beginning at 6 feet (1829 mm) above the floor. For attic or horizontal furnaces, floor furnaces, room heaters and small boilers, the height location of the draft hood outlet or vent collar should be known.

2. Read down the height column to a height equal to or less than the estimated total height.

3. Select the horizontal row for the appropriate lateral (L) length.

4. Read across to the first column under the type of appliance (FAN or NAT) that shows an appli-

ance input rating equal to or greater than the name plate sea level input rating of the appliance to be vented.

Commentary Figure 504.2(2) illustrates the measurement of height and laterals for draft-hood appliances.

Some vent manufacturers include special requirements regarding certain types of appliances in their installation/design instructions.

504.2.1 Vent obstructions. These venting tables shall not be used where obstructions, as described in Section 503.15, are installed in the venting system. The installation of vents serving *listed* appliances with vent dampers shall be in accordance with the *appliance* manufacturer's instructions or in accordance with the following:

1. The maximum capacity of the vent system shall be determined using the "NAT Max" column.

2. The minimum capacity shall be determined as if the *appliance* were a fan-assisted *appliance*, using the "FAN Min" column to determine the minimum capacity of the vent system. Where the corresponding "FAN Min" is "NA," the vent configuration shall not be permitted and an alternative venting configuration shall be utilized.

❖ An engineered-vent sizing method or appliance and vent/chimney manufacturer's instructions must be used to size vents where devices (obstructions) are installed in a venting system. The tables do not account for the flow resistance caused by such devices, nor do they account for the reduction of flue gas temperatures resulting from heat reclaim exchangers. Section 503.15 lists several items that are not to be viewed as obstructions, including automatic vent dampers. This section recognizes the use of the vent sizing tables with appliances equipped

For SI: 1 foot = 304.8 mm.

Figure 504.2(1)
EXAMPLE OF INDIVIDUAL VENT DIMENSIONS
(Figure courtesy of Selkirk, L.L.C.)

with automatic vent-damper devices. By referring to the appliance manufacturer's installation instructions, this section implies that the appliances are factory equipped with automatic vent dampers. Item 1 recognizes that during the firing cycle, an appliance

Figure 504.2(2)
INDIVIDUAL VENT—LATERAL AND HEIGHT
(Figure courtesy of Selkirk, L.L.C.)

Figure 504.2.1
CATEGORY I NATURAL DRAFT HOT WATER BOILER WITH AUTOMATIC VENT DAMPER
(Photo courtesy of Weil-McLain)

with an automatic vent damper affects a venting system no differently than a draft-hood appliance without one. Item 2 addresses the effect that appliances equipped with automatic vent dampers have on venting systems during the appliance "off" cycle. The appliance acts as if it were an idle fan-assisted appliance because a closed vent damper will allow little or no air to enter the vent system, just as an idle fan-assisted appliance allows no air to enter the vent system. The lack of dilution air entering the venting system is a primary reason that the minimum input columns were created in the vent sizing tables. Boilers and water heaters with draft hoods commonly use automatic vent dampers to boost the efficiency of the appliance by trapping residual heat of the firing cycle, reducing standby losses and reducing the amount of conditioned air that escapes up the vent through the draft hood (see Commentary Figure 504.2.1).

504.2.2 Minimum size. Where the vent size determined from the tables is smaller than the *appliance* draft hood *outlet* or flue collar, the smaller size shall be permitted to be used provided that all of the following requirements are met:

1. The total vent height (H) is not less than 10 feet (3048 mm).

2. Vents for *appliance* draft hood outlets or flue collars 12 inches (305 mm) in diameter or smaller are not reduced more than one table size.

3. Vents for *appliance* draft hood outlets or flue collars larger than 12 inches (305 mm) in diameter are not reduced more than two table sizes.

4. The maximum capacity listed in the tables for a fan-assisted *appliance* is reduced by 10 percent (0.90 × maximum table capacity).

5. The draft hood outlet is greater than 4 inches (102 mm) in diameter. Do not connect a 3-inch-diameter (76 mm) vent to a 4-inch-diameter (102 mm) draft hood outlet. This provision shall not apply to fan-assisted appliances.

❖ If a vent complies with all of items 1 through 5 and is sized in accordance with the appropriate table, it is allowed to be smaller than the appliance connection, primarily because of the increase in capacity of the vent resulting from its height and because vent capacity is based on the Btu/h input rating of the appliance and not the size of the appliance flue collar. This code section suggests that flue collars are sized for worst case vent designs and are thus somewhat oversized for vents with higher venting capacities. Item 4 is necessary to prevent positive pressure from developing in the vent serving fan-assisted appliances. A 6-inch (152 mm) draft hood connected to a 5-inch (127 mm) Type B vent, for example, might look like an installation error, but this section could allow it in some cases, and operating a vent nearer its maximum capacity is always desirable (see Section 504.2.10).

504.2.3 Vent offsets. Single-appliance venting configurations with zero (0) lateral lengths in Tables 504.2(1), 504.2(2) and 504.2(5) shall not have elbows in the venting system. Single-appliance venting configurations with lateral lengths include two 90-degree (1.57 rad) elbows. For each additional elbow up to and including 45 degrees (0.79 rad), the maximum capacity listed in the venting tables shall be reduced by 5 percent. For each additional elbow greater than 45 degrees (0.79 rad) up to and including 90 degrees (1.57 rad), the maximum capacity listed in the venting tables shall be reduced by 10 percent. Where multiple offsets occur in a vent, the total lateral length of all offsets combined shall not exceed that specified in Tables 504.2(1) through 504.2(5).

❖ A vent system with a zero lateral length is an ideal system because it is entirely vertical and thus has the least flow resistance and greatest draft. To create a lateral run of vent or vent connector, one or two elbows (adjustable fittings) are necessary [see Commentary Figure 504.2(2)]. The tables account for two 90-degree (1.57 rad) changes in direction, which will offer considerable resistance to the flow of flue gases. Depending on the appliance flue collar or draft-hood outlet orientation, a single elbow can create a lateral, but the tables assume that a vertical outlet orientation and two elbows are used. Two 45-degree (0.79 rad) fittings joined together, for example, are the equivalent of a single 90-degree (1.57 rad) fitting but would offer slightly less resistance to flow because of the larger turning radius. The capacity reduction penalty for each change in direction in excess of the two "free" 90-degree (1.57 rad) fittings accounts for the increased flow resistance caused by the additional fittings (elbows). For example, a vent using two 90-degree (1.57 rad) fittings and two 45-degree (0.79 rad) fittings would have a 10-percent capacity reduction penalty. A vent using two 90-degree (1.57 rad) fittings and two 30-degree (0.52 rad) fittings would also have a 10-percent capacity penalty. All changes of direction (fittings), regardless of the angle, that are in addition to the two "free" 90-degree fittings are accounted for by a capacity reduction penalty. A fitting adjusted to 45 degrees (0.79 rad) or less will cause a 5-percent penalty and a fitting adjusted to greater than 45 degrees (0.79 rad) and up to 90 degrees (1.57 rad) will cause a 10-percent penalty.

Whenever possible, an individual vent should be located directly over the appliance outlet. If the appliance flue outlet is horizontal, one 90-degree (1.57 rad) elbow should be used with the vent directly over it. A straight vertical vent is more easily supported and has less flow resistance.

There is no need to offset the vertical vent to include a tee and bracket. The use of a tee for cleanout or inspection purposes is quite unnecessary for three reasons. First, using a B-vent cap keeps debris and birds out. Second, clean-burning gas does not produce any deposits needing removal. Third, Type B vent joints are easily opened for inspection of the inside of the piping.

Should an offset be needed, the use of two elbows will provide somewhat greater capacity than an elbow

and a tee. A tee has greater flow resistance than an elbow. The two 90-degree turns that are accounted for by the tables for laterals can occur anywhere within the vent system. This means that if the turns are used where the appliance connects to the system, the rest of the vent must be entirely vertical, or else the penalties of this section must be applied. The two "free" 90-degree (1.57 rad) turns can be composed of any number and angle of fittings that total not more than 180 degrees (3.1 rad). Simply put, there are free turns up to a maximum of 180 degrees (3.1 rad), and after that each additional turn will cause a capacity penalty to apply (see definition of "Offset").

Vent offsets should always be avoided where possible, especially offsets in the upper portion of the vent and in cold spaces such as attics and truss spaces. Offsets can be avoided with simple planning of the vent system and appliance locations. The attic offset is prevalent because the vent and appliance location is often not planned, and the building owners and designers do not want the vent to terminate on the street side of a roof for aesthetic reasons. This type of offset causes the most problems for several reasons. The typical attic will have winter temperatures close to the outdoor temperature; therefore, offsets in attics will cause more vent pipe to be exposed to the cold attic (see commentary, Section 504.2.9). This can increase the production of condensate in the vent, and the fittings used will be points at which the condensate can leak from the inner vent lining. Most condensate leakage from vents can be traced to a fitting used in an offset. The code (Sections 504.2.9 and 504.3.16) assumes that a vent will not be exposed to outdoor temperatures except for the necessary portion that extends above the roof. The temperature in a typical attic today is nearly equivalent to the outdoors.

Studies have shown that offsets in the upper portion of vent systems can produce slight positive pressures because of the flow resistance, and this encourages water vapor to escape from the vent and condense between the vent walls. Offsets in attics usually involve long runs of pipe that increase the surface area of vent pipe exposed to cold temperature. It is quite common to find that attic offsets exceed the horizontal length allowed by Section 504.3.5.

Each offset introduces a horizontal component to a vent and this horizontal component is limited by the "lateral" column of the single appliance tables. The horizontal component of all offsets in a vent must be totaled and the total cannot exceed the lateral length allowance in the single appliance tables.

504.2.4 Zero lateral. Zero (0) lateral (L) shall apply only to a straight vertical vent attached to a top outlet draft hood or flue collar.

❖ The zero lateral rows in the tables refer only to an entirely vertical vent run without offsets, lateral runs and directional fittings. This would be the ideal vent.

Unfortunately, it is seldom seen because of lack of planning and coordination of vent and appliance locations. Architectural features of buildings usually dictate that vents take less than desirable paths to get from the appliances to the outdoors.

504.2.5 High-altitude installations. Sea-level input ratings shall be used when determining maximum capacity for high altitude installation. Actual input (derated for altitude) shall be used for determining minimum capacity for high altitude installation.

❖ Appliances installed at elevations above sea level must have their Btu/h (W) input reduced (derated) because the atmosphere is less dense and oxygen levels decrease as altitude increases. The actual volume of flue gases entering a vent (i.e., primary and secondary combustion air, excess air and combustion byproducts) does not significantly change for different elevations; therefore, the maximum vent capacity is based on the sea-level input rating. The minimum capacity of a vent depends on heat input to the vent to control condensation and maintain the required draft. Therefore, the actual heat input (as derated) is used to determine the minimum capacity of a vent.

504.2.6 Multiple input rate appliances. For appliances with more than one input rate, the minimum vent capacity (FAN Min) determined from the tables shall be less than the lowest *appliance* input rating, and the maximum vent capacity (FAN Max/NAT Max) determined from the tables shall be greater than the highest *appliance* rating input.

❖ To improve efficiency, some appliances are equipped with modulating input or high-low two-stage input. Such appliances typically use two-stage thermostats or timers to change over from low-fire to high-fire operation, thus more closely matching the appliance input to the heating load. The venting system must be capable of operating properly throughout the entire firing range of the appliance. The venting system must be capable of handling the maximum input of the appliance and also must be compatible with the appliance when the appliance is operating at its lowest input. Such appliances will usually operate at the "low-fire" input most of their total operating time. To satisfy the minimum input requirements for fan-assisted appliance vents, the vent design must be based on the lowest Btu/h rating of the appliance because this is what the vent senses as the connected load. The lowest appliance input rate is used for minimum vent capacity, and the highest input rate must be used for maximum vent capacity. This requirement can be easily overlooked if the appliance label is not carefully studied because the appliance will usually be described by its maximum input, and any minimum input will have to obtained from the label.

504.2.7 Liner system sizing and connections. *Listed* corrugated metallic chimney liner systems in masonry chimneys shall be sized by using Table 504.2(1) or 504.2(2) for Type B vents with the maximum capacity reduced by 20 percent (0.80 × maximum capacity) and the minimum capacity as shown in Table 504.2(1) or 504.2(2). Corrugated metallic liner systems installed with bends or offsets shall have their maximum capacity further reduced in accordance with Section 504.2.3. The 20-percent reduction for corrugated metallic chimney liner systems includes an allowance for one long-radius 90-degree (1.57 rad) turn at the bottom of the liner.

Connections between chimney liners and *listed* double-wall connectors shall be made with *listed* adapters designed for such purpose.

❖ A liner system installed within a masonry chimney is sized and designed as if the chimney/liner system combination were a Type B vent. Corrugated metal liners are constructed of semi-rigid piping having a higher flow resistance than smooth-wall rigid pipe. A capacity reduction penalty compensates for the poorer flow characteristics of corrugated pipe. Bends in semirigid liners count as elbows and are regulated by Section 504.2.3. Note that the sweeping 90-degree (1.57 rad) bend in the liner at the point where it exits the bottom of the chimney and connects to the appliance(s) is already accounted for in the 20-percent capacity penalty and no further capacity adjustment is required for such turn/bend. Listed adapter fittings must be used to connect double-wall connectors to the liner to avoid "make shift" field connections that can result in poor fit and leakage.

504.2.8 Vent area and diameter. Where the vertical vent has a larger diameter than the vent connector, the vertical vent diameter shall be used to determine the minimum vent capacity, and the connector diameter shall be used to determine the maximum vent capacity. The flow area of the vertical vent shall not exceed seven times the flow area of the listed appliance categorized vent area, flue collar area or draft hood outlet area unless designed in accordance with *approved* engineering methods.

❖ The maximum capacity of a venting system must be based on the element of the system having the least capacity (i.e., the weakest link in the chain). Likewise, the minimum Btu/h (W) input required for proper vent system operation must be based on the largest element of the system because the largest element will naturally have the highest minimum input requirement. A connector that is smaller than the vent will set the maximum capacity limit for the system because it has the greater flow resistance and smaller cross-sectional area. The minimum Btu/h input is based on the larger vent size because the vent will require more heat to produce adequate draft and avoid condensation. The ratio of appliance vent-connection area to vertical vent area is intended to prevent oversizing of the vent, which can result in poor draft and condensation problems. Because the tables will allow vents to be larger than needed, this section serves as a limit to prevent gross oversizing. For example, in Table 504.2(1) it can be seen that a fan-assisted appliance having an input rating near the maximum end of the minimum/maximum range for any 4-inch (102 mm) venting system configuration can be served by not only a 4-inch (102 mm) vent but also a 5-inch (127 mm) and sometimes a 6-inch (152 mm) or 7-inch (178 mm) vent. Consistent with vent manufacturers' instructions, a venting system should always be designed to use the smallest vent allowed by the tables. Vents that are larger than they need to be usually will not perform as well as accurately sized vents and are more likely to suffer from poor draft and condensate problems.

504.2.9 Chimney and vent locations. Tables 504.2(1), 504.2(2), 504.2(3), 504.2(4) and 504.2(5) shall be used only for chimneys and vents not exposed to the outdoors below the roof line. A Type B vent or listed chimney lining system passing through an unused masonry chimney flue shall not be considered to be exposed to the outdoors. Where vents extend outdoors above the roof more than 5 feet (1524 mm) higher than required by Figure 503.6.4, and where vents terminate in accordance with Section 503.6.4, Item 2, the outdoor portion of the vent shall be enclosed as required by this section for vents not considered to be exposed to the outdoors or such venting system shall be engineered. A Type B vent shall not be considered to be exposed to the outdoors where it passes through an unventilated enclosure or chase insulated to a value of not less than R8.

Table 504.2(3) in combination with Table 504.2(6) shall be used for clay-tile-lined *exterior masonry chimneys*, provided that all of the following are met:

1. Vent connector is a Type B double wall.

2. Vent connector length is limited to 1¹⁄₂ feet for each inch (18 mm per mm) of vent connector diameter.

3. The appliance is draft hood equipped.

4. The input rating is less than the maximum capacity given by Table 504.2(3).

5. For a water heater, the outdoor design temperature is not less than 5°F (-15°C).

6. For a space-heating appliance, the input rating is greater than the minimum capacity given by Table 504.2(6).

❖ Chimneys that have one or more walls exposed to the outdoors and vents exposed to the outdoors can develop condensation and weak draft problems in cold weather. Commentary Figure 504.2.9(1) illustrates why masonry chimneys, including interior chimneys, are more prone to draft and condensation problems. The tables assume that the only portion of the chimney or vent that is exposed to the outdoors is the portion that extends above the roof penetration. Even though an attic is technically not outdoors, the temperature in an attic can be very close to the outdoor temperature. In some climates, the actual attic temperature can be far lower than the assumed temperature used in the computer modeling on which the vent sizing tables are based. Therefore, the portion of a venting system that passes through an attic should be kept to an absolute minimum. If at all possible, the installation should be planned to avoid the need for vent offsets in an attic. Section 504.2.7 states that a

chimney lined with a listed liner or Type B vent is to be sized as if it were a Type B vent. The walls of a chimney will provide some insulating effect for a properly installed Type B vent or lining system installed within the chimney. A chimney lined with a Type B vent is not considered to be a chimney, but is treated as a Type B vent installed within a masonry chase (see Section 504.2.3).

This section intends that the listed tables be applicable only to vents and chimneys that are not exposed to the outdoors below the roof penetration. If the vent or chimney is exposed to the outdoors below the roof penetration, the listed sizing tables do not apply and the venting system would have to be engineered. The intent is that vents and chimneys be enclosed within the building envelope to lessen exposure to low ambient temperatures and wind. It is these conditions that were considered in the development of the vent sizing tables. The sizing tables are based on an assumed outdoor temperature of 42°F (6°C) and the assumption that all portions of a vent or chimney below the roof penetration are exposed to an ambient temperature of 60°F (16°C). This makes it obvious that long runs of vent piping through attics and long extensions above roofs should be avoided in cold climates (see Section 504.3.5).

This section prohibits the use of the vent sizing tables for the installation of a Type B vent run up the outside of an exterior wall of a building. This is consistent with vent manufacturer's instructions. Outdoor vents obviously suffer from poorer draft and increased condensation formation because they are exposed to the elements, low temperatures being the main factor. "Stack effect" in the building interior can cause back drafts (reverse flow) to occur in any vent or chimney, particularly vents exposed to the outdoors. For the same reasons, vents should extend above the roof no more than required by the code because the more vent pipe that extends above the roof, the more vent pipe will be exposed to the cold and wind. In some cases, designers and installers will intentionally raise a vent above a roof more than Section 503.6.4 requires just to increase the vent capacity. In doing so, additional vent pipe is exposed to the outdoors, thus inviting condensation and possibly diminishing draft. Section 503.6.4 could also require a vent to extend well above a lower roof penetration, thus causing the sizing tables of this section to be inapplicable to such a vent. A Type B-vent that is enclosed in an air-tight (unventilated) chase insulated to a value of at least R-8 is protected from the weather and is therefore considered as an indoor vent. R-8 is the same insulation value required by the *International Energy Conservation Code*® (IECC®) for ducts located outside of the building thermal envelope. To be effective, the chase enclosure would have to be built air tight and with an internal volume no larger than necessary to accommodate the vent with the required clearance to combustibles.

Where vents extend outdoors above the roof more than 5 feet higher than required by Figure 503.6.4

and also where vents extend outdoors above the roof as required by Section 503.6.4, Item 2, the outdoor portion of such vents must be enclosed in an insulated and sealed chase/enclosure. This requirement is necessary because such vents are not within the parameters upon which the vent sizing tables were calculated; thus, these vents would be expected to not perform as intended [see Commentary Figures 504.2.9 and 504.2.9(1)]. If the owner, designer, builder or installer wants to avoid the expense and any displeasing aesthetics of these outdoor enclosures, the location of vent terminations will have to be thoughtfully planned, or already installed vents may have to be relocated. The allowance for 5 feet higher than required by Figure 503.6.4 is a necessary tolerance to account for the fact that sections of vent pipe are available only in certain lengths and, in many cases, extending above the roof some distance higher than required by the code will be unavoidable.

The tables for exterior masonry chimneys obviously apply to chimneys that are exposed to the outdoors below the roof line, and the list of conditions intends to compensate for the cold masonry. Conditions such as requiring double-wall connectors and draft-hood-equipped appliances are intended to put as much heat into the chimney as possible because it will be needed to produce draft and minimize condensation in a less-than-ideal chimney. Exterior chimneys require greater heat input to make them work. An alternative to the continued use of an exterior chimney would be relining the chimney or abandoning the chimney and using high-efficiency appliances [see Commentary Figure 504.2.9(2) and example].

Example:

A 40,000 Btu/h (11.7 kW) input draft-hood-equipped water heater is vented by a masonry chimney having one wall exposed to the outdoors (an external chimney has one or more exterior walls) [see Commentary Figure 504.2.9(2)].

In the 20-foot height, 2-foot lateral row of Table 504.2(3) a connector size of 3 inches shows a capacity of 38,000 Btu/h in the "NAT MAX" column. A 4-inch connector shows a capacity of 74,000 Btu/h. Because the water heater input is 40,000 Btu/h, a 4-inch connector is required and because of Item 1 of Section 504.2.9 and the fact that Table 504.2(3) is for double-wall (Type B) connectors, the 4-inch connector must be of Type B vent.

The connector lateral of 2 feet (51 mm) is less than the maximum of 6 feet based on the limit of $1^1/_2$ feet for each inch of diameter specified by Item 2 of Section 504.2.9.

The bottom of Table 504.2(3) indicates that the chimney liner must have an internal cross-sectional area within the range of 19 to 88 square inches. The

Figure 504.2.9
HEIGHT ABOVE ROOF

Figure 504.2.9(1)
MASONRY CHIMNEY COMPARED TO TYPE B VENT
(Figure courtesy of Simpson Dura-Vent Company, Incorporated)

chimney in this example has an internal cross-sectional area of approximately 63 square inches and falls within the range specified by the table for 4-inch connectors. The winter design temperature for the location (Newark, NJ) in this example is above 5°F.

The installation in Commentary Figure 504.2.9(2) complies with Items 1 through 5 of Section 504.2.9 and therefore appears to be code compliant; however, Sections 504.2.8 and 503.5.5 limit the chimney flue liner area to seven times the area of the draft hood outlet. If the draft hood on the water heater in this example has a 3-inch outlet, the chimney would be too large, since seven times the area of a 3-inch draft hood outlet is 49 square inches. If the draft hood has a 4-inch outlet, the installation is code compliant.

504.2.10 Corrugated vent connector size. Corrugated vent connectors shall be not smaller than the listed appliance categorized vent diameter, flue collar diameter or draft hood outlet diameter.

❖ Corrugated vent connectors are becoming more common because they can save installation time and avoid the use of fittings. This section is consistent

For SI: 1 inch = 25.4 mm, 1 foot = 304.8 mm, °C = (°F-32)/1.8,
          1 British thermal unit per hour = 0.2931 W.

Figure 504.2.9(2)
EXTERIOR CHIMNEY SIZING-EXAMPLE

with Section 504.3.17 and applies to both draft-hood-equipped and fan-assisted appliances. Corrugated materials have higher resistance to flow than smooth wall materials; therefore, to ensure the required capacity, full size connectors must be used.

504.2.11 Vent connector size limitation. Vent connectors shall not be increased in size more than two sizes greater than the listed appliance categorized vent diameter, flue collar diameter or draft hood outlet diameter.

❖ The sizing tables of this section will often require that a connector be larger than the appliance vent connection. The appliance manufacturer's instructions may also require that the size of a connector be increased. The increase in size compensates for flow resistance in the connector and prevents positive pressurization. A grossly oversized connector will allow the vent gases to expand and cool too much, thus negatively affecting draft and contributing to the formation of condensation. Connector size is sometimes increased for the sole purpose of gaining more allowable length (Section 504.3.2), and this section will limit that practice.

504.2.12 Component commingling. In a single run of vent or vent connector, different diameters and types of vent and connector components shall be permitted to be used, provided that all such sizes and types are permitted by the tables.

❖ In the unlikely event that an installer wants to mix, for example, Type B vent with single-wall pipe in a single connector (for example, Type B vent from a draft hood, switching to single wall and connecting to Type B vent again), the mixed system must comply with the code based on each type of material as if the run were constructed entirely of that material. There may be some unique situations that would justify commingling of different types and/or sizes of components, but clearances, heat loss, ambient temperatures, joining methods and flow characteristics must be considered.

504.2.13 Draft hood conversion accessories. Draft hood conversion accessories for use with masonry chimneys venting listed Category I fan-assisted appliances shall be listed and installed in accordance with the manufacturer's instructions for such listed accessories.

❖ Some appliance manufacturers offer an accessory kit that allows the appliance to be converted from a fan-assisted Category I appliance to something that performs closely to a draft-hood-equipped Category I appliance. This conversion will result in some loss in appliance efficiency (AFUE) and its sole purpose is to allow the appliance to vent to a masonry chimney, whereas without the conversion, the appliance would be prohibited from venting to a masonry chimney. The net result of the conversion is to put dilution air into the chimney, thus helping to warm it and reduce the possibility of condensation. Such conversion kits come with their own installation instructions and are specific to a particular make and model of appliance.

504.2.14 Table interpolation. Interpolation shall be permitted in calculating capacities for vent dimensions that fall between the table entries.

❖ Because Tables 504.2(1) and 504.2(2) do not contain values for every possible height of a vent or length of a lateral, interpolation could be necessary. The maximum and minimum capacity of vents having heights not listed in the tables can be determined using the following equations.

Interpolated Maximum Vent Capacity = $(A/B \times C) + D$
Interpolated Minimum Vent Capacity = $E - (A/B \times F)$

where:

A = The difference between the design height entry and the next lower height entry in the applicable table.

B = The difference between the closest consecutive height entries in the applicable table.

C = The difference between the maximum capacity column consecutive entries in the applicable table.

D = The lower maximum capacity entry in the applicable table.

E = The higher minimum capacity entry in the applicable table.

F = The difference between the minimum capacity column consecutive entries in the applicable table.

For SI: 1 inch = 25.4 mm, 1 foot = 304.8 mm, 1 Btu/h = 0.2931 W.

Example 1:

Determine the minimum and maximum vent capacity for a single fan-assisted appliance given the following:

Determine the minimum and maximum vent capacity for a single fan-assisted appliance given the following:

Vertical design height of vent	17 feet
Diameter of vent	6 inches
Vent type	Type B
Length of lateral	10 feet

Using Table 504.2(1):

Interpolated Maximum Vent Capacity =
[(17 ft − 15 ft)/(20 ft − 15 ft)
× (351,000 Btu/h − 315,000 Btu/h)]
+ 315,000 Btu/h
= (2/5 × 36,000 Btu/h) + 315,000 Btu/h)]
329,400 Btu/h

Interpolated Minimum Vent Capacity
= 64,000 Btu/h − [(17 ft − 15 ft)/(20 ft − 15 ft)
× (64,000 Btu/h − 62,000 Btu/h)]
= 64,000 Btu/h − (2/5 × 2,000 Btu/h)
= 63,200 Btu/h

Example 2:

Determine the maximum vent capacity for a single fan-assisted furnace given the following:

Input rating of furnace	210,000 Btu/h
Vertical design height of vent	12$\frac{1}{2}$ feet
Diameter of vent	6 inches
Vent type	Type B
Length of lateral	2 feet

Using Table 504.2(1):

Step 1: Find the maximum vent capacity (MAX) at the first height entry greater than 12$\frac{1}{2}$ feet, i.e., 15 feet.

MAX_{15} = 226,000 Btu/h

Step 2: Find the maximum vent capacity (MAX) at the next height entry lower than 12$\frac{1}{2}$ feet, i.e., 10 feet.

MAX_{10} = 194,000 Btu/h

Step 3: Determine the difference between the two maximum vent capacities.

$MAX_{15} - MAX_{10}$ = 226,000 Btu/h – 194,000 Btu/h = 32,000 Btu/h

Step 4: Determine the maximum vent capacity for a 12$\frac{1}{2}$-foot high vent using:

$MAX12\frac{1}{2} = [(A/B) \times (MAX_{15}\ MAX_{10})] + MAX_{10}$

where:

A = The difference between the next higher table vent height and the design vent height.

B = The difference between the consecutive closest table vent heights.

$MAX\ 12\frac{1}{2}$ = [(15 ft –12$\frac{1}{2}$ ft)/(15 ft –10 ft) × 32,000 Btu/h] + 194,000 Btu/h

= [(2$\frac{1}{2}$ ft)/(5 ft) × 32,000 Btu/h] + 194,000 Btu/h

= 16,000 Btu/h + 194,000/h

= 210,000 Btu/h

When a vent height falls between height entries in a table, the code user can choose to interpolate or use the closest table height entry that is lower than the actual height of the vent. These are the only allowable options.

This example also illustrates how the tables may be used to reduce vent size if there is adequate height and capacity to do so (see commentary, Section 504.2.2).

It is important to consider interpolation before assuming that an "in between" height vent installation is not in compliance with the code. For example, a vent with a 25-foot (7620 mm) height could fail to comply with a table when it is viewed as a 20-foot (6096 mm) vent, but that same vent could quite possibly comply with the table when viewed as what it actually is, a 25-foot-high (7620 mm) vent.

This interpolation process can also be used to estimate in-between capacities for intermediate lengths of laterals, as well as in-between minimum capacities. There is no way to estimate in-between capacity between a zero lateral (straight vertical vent) and a 2-foot (610 mm) lateral (which has two elbows). If the vent has just one 90-degree (1.57 rad) turn, the 2-foot (610 mm) lateral capacity applies.

504.2.15 Extrapolation prohibited. Extrapolation beyond the table entries shall not be permitted.

❖ Projecting capacities below the lowest or above the highest boundary entries of the tables is not allowed because the tables are based on the known operational limits of venting systems. The tables cannot predict the operating characteristics of venting systems having heights shorter or taller than the table limits; therefore, any such system must be an engineered system.

504.2.16 Engineering calculations. For vent heights less than 6 feet (1829 mm) and greater than shown in the tables, engineering methods shall be used to calculate vent capacities.

❖ This section parallels the previous section by stating that vents having heights not within the tables must be designed by an engineered method. In other words, extrapolation above or below the table height entries is not allowed (see Section 504.2.14).

504.2.17 Height entries. Where the actual height of a vent falls between entries in the height column of the applicable table in Tables 504.2(1) through 504.2(6), either interpolation shall be used or the lower appliance input rating shown in the table entries shall be used for FAN MAX and NAT MAX column values and the higher appliance input rating shall be used for the FAN MIN column values.

❖ This section makes it very clear what is to be done where the vent being designed is of a height that does not correspond to a height row in the applicable table. For example, it is given that a Type B vent has a height of 17 feet and the venting system is subject to Table 504.2(2). The height of 17 feet falls between the 15-foot row and the 20-foot row in the table, which means that neither the 15- nor 20-foot row provides accurate values for this vent. Often times, the code user will mistakenly choose the 20-foot row because it is close to 17, but this is incorrect because a 17-foot-high vent will have less capacity than a 20-foot-high vent, as can be seen by comparing the FAN MAX and NAT MAX entries for the 15- and 20-foot rows and also because a 17-foot-high vent requires a greater minimum input than a 20-foot-high vent, as can be seen by comparing the FAN MIN entries. Clearly, the 20-foot row is inappropriate. Because the 20-foot row is not a choice, either interpolation must be used to mathematically construct entries for a 17-foot row, or the 15-foot row must be used. When the 15-foot row is used, the FAN MAX and NAT MAX values will be lower than the corresponding values in the 20-foot row and the FAN MIN value will be higher than the corresponding value in the 20-foot row. This approach is a safe alternative to interpolation because the 17-foot vent will perform better than a 15-foot vent. In other words, the 17-foot vent has a higher maximum appliance input capacity and a lower minimum appliance input requirement than a 15-foot vent. Using the 15-foot row corresponds to the instructions of this section to use the lower appliance input rating given in each of the MAX columns and to use the higher appliance input rating for the FAN MIN column.

504.3 Application of multiple appliance vent Tables 504.3(1) through 504.3(7). The application of Tables 504.3(1) through 504.3(7) shall be subject to the requirements of Sections 504.3.1 through 504.3.28.

❖ The application of Tables 504.3(1) through 504.3(4) shall be subject to the requirements of Sections 504.3.1 through 504.3.23. This part of Section 504 regulates vents and chimneys that serve more than one appliance. These venting systems can be described as combined vents and include multiple connectors. A "combined vent" is a vent for two or more appliances at one level served by a common vent. "Least total height" is the vertical distance from the highest appliance outlet (draft hood or flue collar) to the lowest discharge opening of the vent top. This height dimension is illustrated in Commentary Figure 504.3(1) for a typical (FAN + NAT) system. The same height measurement applies to FAN + FAN and NAT + NAT systems. Least total height is used for vent sizing of all connected appliances on one level. "Connector rise" for any appliance is the vertical distance from its outlet connection to the level at which it joins the common vent, as shown in Commentary Figure 504.3(1).

A "connector" for purposes of designing a combined vent is that part of the vent piping between the appliance outlet and its junction with or interconnection to the rest of the system. A chimney is referred to as a venting means, although a chimney and a vent are distinct. An improperly sized common vent or chimney could fail to vent the combustion products of the appliances served. Many factors influence the sizing of a vent or chimney, including total height, offsets, lateral lengths, the type of appliance, the appliance energy input and the appliance connector configurations. The "common vent" is that portion of the system serving two or more connected appliances. If connectors are joined before reaching the vertical vent, the run between the last entering connector and the vertical portion is also treated as part of the common vent. In Commentary Figure 504.3(1), the vertical common vent is cross-hatched beginning at the interconnection tee.

For each connector, the correct size must be found from the applicable tables based on its appliance input, rise and least total vent height. For draft-hood (NAT) appliances, the outlet size may be too small if there is not enough rise; therefore, connector design involves choosing the correct size and verifying that the use of a connector the same size as the outlet is within input rating limits.

For the common vent, the capacity table shows maximum combined ratings only. The size of the common vent is thus based on least total height and the combination of attached appliances: FAN + FAN, FAN + NAT or NAT + NAT. To find each connector

size, use Table 504.3(1) for Type B vent connectors or Table 504.3(2) for single-wall metal connectors and proceed as follows:

1. Determine "least total height" for the system.

2. Determine connector rise for each appliance.

3. Enter the applicable vent connector table at the least total height. Continue across on the line for appliance connector rise to a MAX input rating equal to or greater than that of the appliance. For a FAN appliance, this input rating should also be greater than shown in the MIN column. Read the connector size at the top of the column. If Table 504.3(2) for single wall shows NA, use a double-wall connector in accordance with Table 504.3(1). In some cases, the table will dictate a connector size that is larger than the appliance vent outlet (flue collar).

To find the common vent size, proceed as follows:

1. Add all appliance Btu input ratings to get the total Btu input.

2. If one or both connectors are single-wall metal, use Table 504.3(2) for the common vent.

3. Enter the common vent table at the same least total height used for connectors.

4. Continue across and stop at the first applicable column of combined appliance input rating equal to or greater than the total. If Table 504.3(2) for the common vent shows NA, the entire system must be double-wall Type B vent.

5. Read the size of the common vent at the top of the applicable column (NAT + NAT, FAN + NAT or FAN + FAN).

   (a) Regardless of table results for size, the common vent must be at least as large as the largest connector (see Section 504.3.8).

   (b) In accordance with vent manufacturer's instructions, for NAT + NAT appliance combinations, if both connectors are the same size, the common vent must be at least one size larger.

Example 1:

Commentary Figure 504.3(1) shows a two-appliance FAN + NAT system, combining a draft-hood water heater (NAT appliance) with a fan-assisted combustion Category I (FAN) furnace. The system is designed in the following steps using Table 504.3(1) for double-wall connectors.

1. For the water heater, enter the vent connector table at a least total height of 15 feet (4572 mm) and a connector rise of 1 foot (305 mm). Read across to the MAX Btu/h rating for a NAT appliance vent higher than 35,000 Btu/h (10.3 kW). The table shows that a 4-inch (102 mm) connector is needed. This size must be used begin-

ning at the draft hood, regardless of draft hood size, which might be 3 inches (76 mm) [see Commentary Figure 504.3(1)].

2. For the 105,000 Btu/h (30.8 kW) furnace, enter the vent connector table at the same least total height [15 feet (4572 mm)] at a connector rise of 3 feet (914 mm). Read across to 163,000 under the FAN MAX column. The MIN is 51,000. Therefore, 5 inches (127 mm) is the correct connector size [see Commentary Figure 504.3(1)].

3. For the common vent, the sum of the two ratings is 140,000 Btu/h (41.0 kW). Enter the common vent table at 15 foot (4572 mm) least total height.

For a FAN + NAT combination, the maximum input of 5-inch (127 mm) vent is 164,000 (48.1 kW), so 5 inches (127 mm) is the proper size, as shown in Commentary Figure 504.3(1). Both draft-hood and fan-assisted appliances may be common vented in any combination, as indicated by the headings in the common vent table. Appliance types include the following:

- Central heating furnaces,
- Central heating boilers (hot water and steam),
- Water heaters,
- Unit heaters,
- Duct furnaces,
- Room heaters*, and
- Floor furnaces*.

*If these have draft hoods, a design input increase of 40 percent is recommended by some vent manufacturers in order to use the tables (see commentary, Section 504.2).

The common vent tables do not apply to the following:

1. Gas cooking appliances, which should be vented into an exhaust hood or vented individually in accordance with the appliance manufacturer's instructions.

2. Forced draft and commercial or industrial hot water or steam boilers without draft hoods. For this equipment, see the boiler and vent/chimney manufacturers' instructions.

3. Categories II, III and IV gas-burning equipment, for which the equipment manufacturer's venting instructions must be used.

4. Gas-fired incinerators.

The manufacturers of chimney and vent systems provide installation and design information as part of their installation instructions. The common chimney or vent must be designed to properly vent the products of combustion when any one, any combination or all of the connected appliances operate. [see Com-

mentary Figure 504.3(2)]. Solid-fuel-burning appliances must not connect to common chimneys (see Section 504.5.6.1).

Combined vent systems for two or more gas appliances of either type (FAN or NAT) must be designed to prevent draft-hood spillage for natural draft (NAT) appliances and to avoid positive pressure for fan-assisted (FAN) appliances. The connector and common vent tables have been computed by examining the most critical situations for any operating combination. A common vent must function properly when all, one or any combination of the connected appliances are firing. It is particularly demanding of a vent system to operate properly when only a single appliance having the lowest input rating is firing, thus creating the greatest potential for poor draft and condensation. Oversizing of venting systems should be avoided.

The connector tables are based on the most critical condition for a particular appliance when operating by itself, whereas the common vent tables show sizes that assure adequate capacity and draft, whether one or all appliances are operating simultaneously.

All parts of a combined vent must be checked for capacity. For connectors, the size must be determined from the tables, particularly for low-height vents or where headroom restricts available connector rise.

The connector in a combined-vent system is defined as the piping from a draft hood or flue collar to the junction of the common vent or to a junction in a vent manifold. Proper connector design is vital to obtaining adequate capacity. The connector must produce its share of the total draft for its NAT or FAN appliance and must deliver enough heat to the common vent so that it can contribute the balance of draft needed.

For application of the code's installation provisions, the Type B gas vent connector is a "gas vent". It is essential, however, for system design purposes to use the word "connector" so that its rise and configuration may be explained and tabulated. The connector tables for combined vents show MIN and MAX capacities only for FAN appliances because no minimums are needed for NAT appliances with Type B gas-vent connectors. Important factors in connector design include the following:

- Connector material,
- Connector length,
- Connector rise,
- Number,
- Appliance location as it affects the piping arrangement,
- Number of attached appliances or different connector sizes, and
- Connection to an offset or manifold rather than directly to the vertical common vent.

Example 2:

This example is a common scenario [see Commentary Figure 504.3(2)].

Given:

- Single-family house with Type B vent serving two appliances;
- Category I fan-assisted furnace, 80 percent efficient, 75,000 Btu/h input (4-inch vent collar);
- Furnace connector single-wall pipe with 3-foot rise;
- Draft-hood-equipped water heater, 40,000 Btu/h (3-inch draft hood outlet);
- Water heater connector single-wall with 2-foot rise;
- Vent height: 25 feet; and
- A 45-degree offset occurs in the common vent in the attic.

1. Choose the applicable table. Table 504.3(2) addresses two or more Category I appliances with single-wall connectors.

2. Using the top portion of the table, size the appliance connectors. Because the table has no row for the 25-foot height, the 20-foot row must be used, or interpolation must be used to calculate the capacities at 25 feet. To simplify the procedure, first proceed using the 20-foot row until interpolation is necessary.

   (a) Looking in the "NAT" columns, see that a 4-inch connector is required for the

water heater because the water heater input of 40,000 Btu/h exceeds the maximum capacity (36,000) of a 3-inch connector with a 2-foot rise.

   (b) Looking in the "FAN" columns, see that the furnace input of 75,000 Btu/h is less than the minimum capacity of 4-inch and larger connectors with a 3-foot rise. This means that single-wall pipe cannot be used for the furnace connector. The furnace connector must be double-wall (Type B or equivalent) and must be sized using Table 504.3(1). In Table 504.3(1), see that a 4-inch double-wall connector is suitable for the furnace because the 75,000 Btu/h input fits within the range of 35,000 to 110,000 Btu/h for a 3-foot rise.

   The furnace connector has been determined to be 4-inch double-wall pipe, and the water heater connector has been determined to be 4-inch single-wall pipe.

3. Using the bottom portion of Table 504.3(2), the common vent is sized. (Because both single- and double-wall connectors are used within the same system, Section 504.3.18 requires use of the single-wall table.) The water heater and furnace inputs total 115,000 Btu/h. Under the 4-inch vent heading, FAN + NAT column, see that 4-inch

Figure 504.3(1)
COMBINED VENT DIMENSIONS
(Figure courtesy of Selkirk, L.L.C.)

For SI: 1 inch = 25.4 mm, 1 foot = 304.8 mm, 1 Btu/h = 0.2931 W.

Type B vent has a capacity of 118,000 Btu/h, which exceeds the appliance total of 115,000. However, the system contains an offset, which will reduce this capacity by 20 percent in accordance with Section 504.3.5 (0.80 × 118,000) = 94,400 Btu/h.

The actual capacity of a 4-inch common vent is 94,400 Btu/h, which is less than the 115,000 appliance total; therefore, a 5-inch common vent is required.

Verify 5-inch capacity: (0.80 × 177) = 141,600 Btu/h.

From the 30-foot height rows, it is apparent that interpolation for the actual height of 25 feet would not have changed the outcome of this system design.

Appendix B contains additional sizing examples.

For SI: 1 foot = 304.8 mm,
1 British thermal unit per hour = 0.2831 W.

Figure 504.3(2)
FAN AND NAT VENT SIZING EXAMPLE

504.3.1 Vent obstructions. These venting tables shall not be used where obstructions, as described in Section 503.15, are installed in the venting system. The installation of vents serving listed appliances with vent dampers shall be in accordance with the appliance manufacturer's instructions or in accordance with the following:

1. The maximum capacity of the vent connector shall be determined using the NAT Max column.

2. The maximum capacity of the vertical vent or chimney shall be determined using the FAN+NAT column when the second appliance is a fan-assisted appliance, or the NAT+NAT column when the second appliance is equipped with a draft hood.

3. The minimum capacity shall be determined as if the appliance were a fan-assisted appliance.

    3.1. The minimum capacity of the vent connector shall be determined using the FAN Min column.

    3.2. The FAN+FAN column shall be used where the second appliance is a fan-assisted appliance, and the FAN+NAT column shall be used where the second appliance is equipped with a draft hood, to determine whether the vertical vent or chimney configuration is not permitted (NA). Where the vent configuration is NA, the vent configuration shall not be permitted and an alternative venting configuration shall be utilized.

❖ This section is a bit more complex than Section 504.2.1 because the tables for multiple appliances are two-part tables. The difference lies in Items 2 and 3.2 dealing with the common vent portion of the sizing tables. Item 2 determines the maximum capacity of the common vent by treating the vent-damper-equipped appliance as a draft-hood appliance coupled with either another draft-hood appliance or a fan-assisted appliance. Item 3.2 is a test wherein the vent-damper- equipped appliance is treated as a fan-assisted appliance. If the table says NA, the vent system must be redesigned, regardless of the fact that it may have been allowed by the table under Item 2 (see commentary, Section 504.2.1).

504.3.2 Connector length limit. The vent connector shall be routed to the vent utilizing the shortest possible route. Except as provided in Section 504.3.3, the maximum vent connector horizontal length shall be $1^1/_2$ feet for each inch (18 mm per mm) of connector diameter as shown in Table 504.3.2.

❖ Connectors are necessary for chimney and vent systems that serve multiple appliances. To reduce friction loss and flue-gas heat loss and to conserve available draft, connectors must be kept to a minimum length. Thoughtful placement of appliances and chimneys or vents can reduce connector lengths and achieve optimum venting performance. Unlike Tables 504.2(1) and 504.2(2), Tables 504.3(1) through 504.3(4) do not provide a range of permissible lateral lengths. Therefore, Section 503.10.8 places an overall maximum limit on connector length. Section 503.10.8 provides a necessary limiting function because Tables 504.3(1) through 504.3(4) would appear to allow (unintentionally) a connector to be considerably longer than the height of the vent.

Increasing the length of the connector increases the required minimum capacity of a vent and decreases its maximum capacity. Some vent manufacturers appear to measure only the horizontal (lateral) run of a connector in their design manuals. The more conservative approach would be to consider the friction loss and heat loss for the entire run of a connector throughout its developed length (see commentary, Section 503.10.8). The vent sizing tables were developed with the assumption that vent connectors are no longer than 18 inches per inch of connector diameter. This section establishes the basic rule, which is modified by Sections 504.3.3 and 503.10.8. Simply put, connector length is limited by this section to 18 inches per inch of diameter without capacity penalties or is limited in length by Section 503.10.8 if the capacity penalties of Section 504.3.3 are applied and the minimum capacity is met for fan-assisted appliances. Section 504.3.3 allows longer connector lengths, but this section and the others mentioned still require connectors to take the shortest route to the vent.

TABLE 504.3.2
MAXIMUM VENT CONNECTOR LENGTH

CONNECTOR DIAMETER (inches)	CONNECTOR MAXIMUM HORIZONTAL LENGTH (feet)
3	$4^1/_2$
4	6
5	$7^1/_2$
6	9
7	$10^1/_2$
8	12
9	$13^1/_2$
10	15
12	18
14	21
16	24
18	27
20	30
22	33
24	36

For SI: 1 inch = 25.4 mm, 1 foot = 304.8 mm.

❖ See the commentary to Section 504.3.2.

504.3.3 Connectors with longer lengths. Connectors with longer horizontal lengths than those listed in Section 504.3.2 are permitted under the following conditions:

1. The maximum capacity (FAN Max or NAT Max) of the vent connector shall be reduced 10 percent for each additional multiple of the length allowed by Section 504.3.2. For example, the maximum length listed in Table 504.3.2 for a 4-inch (102 mm) connector is 6 feet (1829 mm). With a connector length greater than 6 feet (1829 mm) but not exceeding 12 feet (3658 mm), the maximum capacity must be reduced by 10 percent (0.90 × maximum vent connector capacity). With a connector length greater than 12 feet (3658 mm) but not exceeding 18 feet (5486 mm), the maximum capacity must be reduced by 20 percent (0.80 × maximum vent capacity).

2. For a connector serving a fan-assisted appliance, the minimum capacity (FAN Min) of the connector shall be determined by referring to the corresponding single-appliance table. For Type B double-wall connectors, Table 504.2(1) shall be used. For single-wall connectors, Table 504.2(2) shall be used. The height (H) and lateral (L) shall be measured according to the procedures for a single-appliance vent, as if the other appliances were not present.

❖ Section 503.10.8 still applies and serves to limit the application of this section and Section 504.3.2. For fan-assisted appliances, both Items 1 and 2 apply. Item 1 allows the maximum length permitted by Section 504.3.2 to be increased (doubled, tripled, etc.) if the vent connector capacity is reduced by 10 percent when the maximum connector length is doubled, by 20 percent when tripled, by 30 percent when quadrupled, etc. Regardless of the length permitted, connectors should always be as short as practicable.

Item 2 applies to fan-assisted appliances and allows connector lengths to be as long as the lateral entries in Table 504.2(1) or 504.2(2) if the limits of Section 503.10.8 are not exceeded and the appliance complies with the minimum input requirement of the respective table.

When horizontal lengths in excess of those stated in Section 504.3.2 are necessary, the minimum capacity of the system is determined by referring to the corresponding single-appliance table. In this case, for each appliance the entire vent connector and common vent from appliance to vent termination is treated as a single-appliance vent (of the same size as the common vent), as if the others were not present. Any appliance failing to meet the Min input may be prone to creating excessive condensation and/or insufficient draft within the vent system if the appliance is operated by itself. In that case, options may include relocation of appliances, selective sequential or simultaneous operation of appliances or separate vent installations. This section is intended for use only where a longer connector cannot be avoided as required by Section 503.10.8 and the connector takes the shortest possible route to the vent. The longer the connector, the greater the flow resistance and the greater the heat loss, both of which negatively affect vent system operation. The capacity penalties account for the increased flow resistance and Item 2 accounts for the increased heat loss. Item 2 addresses the worst case condition when a single appliance is operating and treats a connector serving a fan-assisted appliance as if the appliance were being independently vented, hence, the reference to the single appliance tables. Recall that the single appliance tables limit lateral length, which can be, in effect, a connector length limitation.

In many cases, the sizing tables will prevent someone from taking advantage of the allowances of this section for connectors. For example, a single-wall connector serving a fan-assisted appliance may not be able to meet the minimum input requirement (FAN

Min) if the capacity penalty of this section forces increasing the connector size. Note that Section 503.10.8 will not allow a connector length to exceed the vent or chimney height in any case.

504.3.4 Vent connector manifold. Where the vent connectors are combined prior to entering the vertical portion of the common vent to form a common vent manifold, the size of the common vent manifold and the common vent shall be determined by applying a 10-percent reduction (0.90 × maximum common vent capacity) to the common vent capacity part of the common vent tables. The length of the common vent connector manifold (L_m) shall not exceed $1^1/_2$ feet for each inch (18 mm per mm) of common vent connector manifold diameter (D).

❖ This section inflicts a 10-percent capacity penalty on the common vent as a result of the lateral (offset) effect of a manifold. A connector manifold is a horizontal extension of the lower end of a common vent [see Commentary Figures 503.10.3.4(1) and 503.10.3.4(2)]. The capacity reduction accounts for the additional turn in the common vent. The manifold portion of a vent is sized based on the common vent portion of the tables. The length of a connector manifold is measured from the vertical common vent to the most upstream interconnection fitting.

504.3.5 Common vertical vent offset. Where the common vertical vent is *offset*, the maximum capacity of the common vent shall be reduced in accordance with Section 504.3.6. The horizontal length of the common vent *offset* (L_o) shall not exceed $1^1/_2$ feet for each inch (18 mm per mm) of common vent diameter (D). Where multiple offsets occur in a common vent, the total horizontal length of all offsets combined shall not exceed $1^1/_2$ feet for each inch (18 mm per mm) of common vent diameter (D).

❖ Offsets, as opposed to a single change in direction, require two fittings (elbows) [see Commentary Figures 202(14) and 503.10.3.4(2)]. An offset offers resistance to flow as does any change of direction in a vent chimney or connector, and the capacity reduction penalty compensates for this flow resistance. Common vent locations should be well planned in a building to avoid the need for an offset. Offsets also introduce the need for additional supports to prevent stress on fittings (elbows). The term "Offset (Vent)" is defined in Section 202 (see commentary, Sections 504.2.3 and 504 2.9).

Offsets should be avoided because of the flow resistance, capacity reduction, additional support requirements and possibility of condensation leakage. The length limitation of this section is absolute with no allowance for greater lengths. This provision is commonly violated in attic space offsets with some violations resulting in offsets that are as long or longer than the vent height. The length of an offset is measured in a horizontal plane, not along the developed length of the vent pipe. For example, a 45-degree offset with 7 feet (2134 mm) of pipe between the fittings will have an offset length of 5 feet (1524 mm) for application of this section. [For a 45-degree

(0.79 rad) offset, actual pipe developed length is approximately 1.4 times the horizontal distance between the parallel vertical sections of pipe] (see Commentary Figure 504.3.5).

The total length of all horizontal offsets combined cannot exceed the maximum limit of $1^1/_2$ feet per inch of vent diameter. In other words, horizontal offsets are accumulative.

504.3.6 Elbows in vents. For each elbow up to and including 45 degrees (0.79 rad) in the common vent, the maximum common vent capacity listed in the venting tables shall be reduced by 5 percent. For each elbow greater than 45 degrees (0.79 rad) up to and including 90 degrees (1.57 rad), the maximum common vent capacity listed in the venting tables shall be reduced by 10 percent.

❖ This section applies to the common vent portion of a vent system (see commentary, Section 504.2.3). For example, a common vent may have two 30-degree (0.52 rad) changes of direction at its base and might also have an offset constructed with 45-degree (0.79 rad) fittings in an attic space, thus having a total capacity reduction of 20 percent. If that same vent system had two 45-degree (0.79 rad) changes of direction at the base and an offset constructed with 60-degree (1.05 rad) fittings in the attic, the total capacity penalty would be 30 percent. In Commentary Figure 504.3.6(1), an offset is created in the common vent by the use of two fittings (elbows) adjusted to 45 degree (0.79 rad) angles. The common vent capacity would be reduced by 10 percent total (5 percent for each fitting/elbow). If the fittings were adjusted to angles greater than 45 degrees (0.79 rad), the total capacity reduction would be 10 percent per fitting for a total of 20 percent (see Commentary Figures 504.3.6(2) and 504.3.6(3)].

504.3.7 Elbows in connectors. The vent connector capacities listed in the common vent sizing tables include allowance for two 90-degree (1.57 rad) elbows. For each additional elbow up to and including 45 degrees (0.79 rad), the maximum vent connector capacity listed in the venting tables shall be reduced by 5 percent. For each elbow greater than 45 degrees (0.79 rad) up to and including 90 degrees (1.57 rad), the maximum vent connector capacity listed in the venting tables shall be reduced by 10 percent.

❖ In Commentary Figure 504.3.6(1), the connector for the fan-assisted appliance is offset with two 60-degree (1.05 rad) fittings (elbows). Because the tables include an allowance for two "free" 90-degree (1.57 rad) fittings, no capacity reduction would be required for this connector. The draft-hood-equipped appliance has a connector with a total of four 90-degree (1.57 rad) fittings, including the tee in the common vent This would require a total capacity reduction of 20 percent. Note that Section 503.10.8 intends to prohibit connectors with excessive length and fittings/elbows. Excess fittings and length are often the result of poor planning, improper appliance location and obstructions, such as ducts, structural members and other appliances. Capacity reduction

For SI: 1 inch = 25.4 mm, 1 foot = 304.8 mm.

Figure 504.3.5
COMMON VENT OFFSET IN ATTIC FOR THE PURPOSE OF AVOIDING A STREET-SIDE ROOF PENETRATION

penalties can force an increase in connector size, and in the case of fan-assisted appliances, increasing the connector size might raise the minimum required input rating above the appliance rating, thus forcing redesign of the system.

504.3.8 Common vent minimum size. The cross-sectional area of the common vent shall be equal to or greater than the cross-sectional area of the largest connector.

❖ This requirement is consistent with vent manufacturers' instructions and is necessary to prevent overloading of the common vent. It is apparent that a vent's maximum capacity could be exceeded where multiple connectors are involved, especially if one of the connectors is larger than the common vent itself. Vent manufacturers may also require that the common vent be increased one size where more than one appliance connector and the common vent are the same size.

504.3.9 Common vent fittings. At the point where tee or wye fittings connect to a common vent, the opening size of the fitting shall be equal to the size of the common vent. Such fittings shall not be prohibited from having reduced-size openings at the point of connection of appliance vent connectors.

❖ This section specifically allows use of tee and wye fittings with reduced-size openings to join appliance vent connectors to the common vent. The fittings must be the same size as the common vent at the point where they join, but are allowed to have smaller branch openings and smaller openings on the opposite end of the run of the fitting. A common question regarding interconnection fittings is whether or not the code intends to allow single-wall interconnection fittings such as tees and wyes. These fittings have commonly been used with single-wall connectors to join two appliances to a common Type B vent 5 × 4 × 3-inch wyes, 6 × 6 × 4-inch tees, etc. Vent manufacturers' opinion on this issue is that because the interconnection fitting is part of the common vent, it should be constructed as required for the common vent, meaning Type B vent tees and wyes. Certainly, double-wall vent connectors are not allowed to discharge into a single-wall fitting. The code was previously silent with regard to single-wall connectors discharging into single-wall fittings; however, Section 504.3.9.1 has settled this issue. Double-wall tees and wyes are preferable because they have far less heat loss and less required clearance to combustibles [see Commentary Figures 504.3.9(1) and 504.3.9(2) and Section 504.3.9.1].

For SI: 1 degree = 0.0175 rad.

Figure 504.3.6(1)
OFFSET IN COMMON VENT AND CONNECTORS

Figure 504.3.6(2)
TYPE B VENT ADJUSTABLE
45-DEGREE ELBOW FITTING
(Figure courtesy of Simpson Dura-Vent Company,
Incorporated)

Figure 504.3.6(3)
TYPE B VENT ADJUSTABLE
90-DEGREE ELBOW FITTING
(Figure courtesy of Simpson Dura-Vent Company,
Incorporated)

Figure 504.3.9(1)
TYPE B VENT REDUCING TEE
(Photo courtesy of Simpson Dura-Vent Company,
Incorporated)

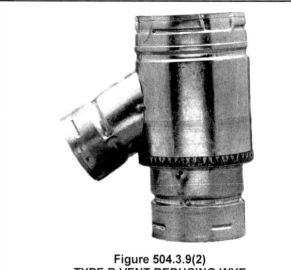

Figure 504.3.9(2)
TYPE B VENT REDUCING WYE
(Photo courtesy of Simpson Dura-Vent Company,
Incorporated)

504.3.9.1 Tee and wye fittings. Tee and wye fittings connected to a common gas vent shall be considered as part of the common gas vent and shall be constructed of materials consistent with that of the common gas vent.

❖ Because the tee and wye fittings are part of the common vent, they must be of the same construction as the common vent. This is consistent with the vent manufacturer's recommendations [see Commentary Figures 504.3.9.1, 504.3.9(1) and 504.3.9(2)]. Tee and wye fittings used with Type B common vents are not allowed to be single wall fittings.

504.3.10 High-altitude installations. Sea-level input ratings shall be used when determining maximum capacity for high-altitude installation. Actual input (derated for altitude) shall be used for determining minimum capacity for high-altitude installation.

❖ See the commentary to Section 504.2.5.

504.3.11 Connector rise measurement. Connector rise (R) for each appliance connector shall be measured from the draft hood outlet or flue collar to the centerline where the vent gas streams come together.

❖ It is desirable to achieve the greatest connector rise that conditions will permit [see Commentary Figures 504.3.11 and 504.3(2)].

504.3.12 Vent height measurement. For multiple appliances all located on one floor, available total height (H) shall be measured from the highest draft hood outlet or flue collar up to the level of the outlet of the common vent.

❖ The vent design is based on the "worst case" (most conservative) height so that a vent is not given credit for being taller than it actually is [see Commentary Figure 504.3(2)].

504.3.13 Multistory height measurement. For multistory installations, available total height (H) for each segment of the system shall be the vertical distance between the highest

draft hood outlet or flue collar entering that segment and the centerline of the next higher interconnection tee.

❖ See Appendix B and Section 503.6.10.

504.3.14 Multistory lowest portion sizing. The size of the lowest connector and of the vertical vent leading to the lowest interconnection of a multistory system shall be in accordance with Table 504.2(1) or 504.2(2) for available total height (H) up to the lowest interconnection.

❖ See Appendix B and Section 503.6.10. The lowest appliance vent is sized as an individual vent.

504.3.15 Multistory common vents. Where used in multistory systems, vertical common vents shall be Type B double wall and shall be installed with a *listed* vent cap.

❖ The design and sizing criteria for multistory venting systems are based on the known performance of listed Type B vents with the listed cap (See Sections 502.1 and 503.6.6).

504.3.16 Multistory common vent offsets. *Offsets* in multistory common vent systems shall be limited to a single *offset* in each system, and systems with an *offset* shall comply with all of the following:

1. The *offset* angle shall not exceed 45 degrees (0.79 rad) from vertical.

2. The horizontal length of the *offset* shall not exceed $1^1/_2$ feet for each inch (18 mm per mm) of common vent diameter of the segment in which the *offset* is located.

3. For the segment of the common vertical vent containing the *offset,* the common vent capacity listed in the common venting tables shall be reduced by 20 percent (0.80 × maximum common vent capacity).

4. A multistory common vent shall not be reduced in size above the *offset.*

❖ This requirement is consistent with vent manufacturers' instructions and is intended to eliminate flow restrictions and pressure differentials in the venting

Figure 504.3.9.1
SINGLE-WALL TEES AND WYES PROHIBITED IN TYPE B COMMON VENT SYSTEMS

system. Because appliances connect to the multistory common vent at different floor levels, flow disturbances must be avoided in the entire extent of the common vent to prevent having an effect on lower-floor appliances. The criteria for a single offset are based on research conducted by the Gas Research Institute, which used a specialized computer program to investigate the performance of multistory venting systems with offsets. The research indicated that multistory Type B vents with a single offset can perform as well as a nonoffset vent performs if the adjustments and limitations in this section are applied. Those limitations involve a reduction of the vent capacity found in the tables and a limitation on the horizontal length of the offset based on vent size (see commentary, Section 504.3.5).

A multistory common vertical vent is allowed to have a single offset if all of the following requirements are met:

1. The offset angle does not exceed 45 degrees (0.79 rad).

2. The horizontal length of the offset does not exceed 1$^1/_2$ feet for each inch (18 mm per mm) of common vent diameter of the segment in which the offset is located.

3. For the segment of the common vertical vent containing the offset, the common vent capacity listed in the common venting tables is reduced by 20 percent (0.80 × maximum common vent capacity).

4. A multistory common vent must not be reduced in size above the offset.

504.3.17 Vertical vent maximum size. Where two or more appliances are connected to a vertical vent or chimney, the flow area of the largest section of vertical vent or chimney shall not exceed seven times the smallest listed appliance categorized vent areas, flue collar area or draft hood outlet area unless designed in accordance with *approved* engineering methods.

❖ Vent manufacturer's instructions address ratios of areas of appliance connectors connecting to a com-

18 FT
LEAST TOTAL
HEIGHT

1 FT
CONNECTOR
RISE

2 FT
CONNECTOR
RISE

WATER HEATER
45,000 Btu/h

FAN-ASSISTED
FURNACE
100,000 Btu/h

For SI: 1 Btu/h = 0.2931 W, 1 foot = 304.8 mm.

**Figure 504.3.11
CONNECTOR RISE
(Figure courtesy of Simpson Dura-Vent Company,
Incorporated)**

mon vent. This section relates the area ratio limitation to the smallest appliance connection and the area of the vertical common vent. The concern is that the smaller or smallest appliance may not be able to provide enough heat to the common vent to develop sufficient draft and prevent a condensation problem. It can be seen that the common vent tables would allow gross oversizing if it were not for the limit of this section. For example, if the table permits a 5-inch common vent, it would also permit a 6-, 7-, 8-, 9-, and 10-inch common vent. This section serves to limit the size of the common vent, whereas the table does not. The larger the common vent, the more difficult it becomes to maintain draft and avoid condensation, especially when the smallest appliance is firing by itself.

504.3.18 Multiple input rate appliances. For appliances with more than one input rate, the minimum vent connector capacity (FAN Min) determined from the tables shall be less than the lowest appliance input rating, and the maximum vent connector capacity (FAN Max or NAT Max) determined from the tables shall be greater than the highest appliance input rating.

❖ See the commentary to Section 504.2.6.

504.3.19 Liner system sizing and connections. Listed, corrugated metallic chimney liner systems in masonry chimneys shall be sized by using Table 504.3(1) or 504.3(2) for Type B vents, with the maximum capacity reduced by 20 percent (0.80 × maximum capacity) and the minimum capacity as

shown in Table 504.3(1) or 504.3(2). Corrugated metallic liner systems installed with bends or offsets shall have their maximum capacity further reduced in accordance with Sections 504.3.5 and 504.3.6. The 20-percent reduction for corrugated metallic chimney liner systems includes an allowance for one long-radius 90-degree (1.57 rad) turn at the bottom of the liner. Where double-wall connectors are required, tee and wye fittings used to connect to the common vent chimney liner shall be listed double-wall fittings. Connections between chimney liners and listed double-wall fittings shall be made with listed adapter fittings designed for such purpose.

❖ See the commentary to Section 504.2.7.

504.3.20 Chimney and vent location. Tables 504.3(1), 504.3(2), 504.3(3), 504.3(4) and 504.3(5) shall be used only for chimneys and vents not exposed to the outdoors below the roof line. A Type B vent or *listed* chimney lining system passing through an unused masonry chimney flue shall not be considered to be exposed to the outdoors. Where vents extend outdoors above the roof more than 5 feet (1524 mm) higher than required by Figure 503.6.4 and where vents terminate in accordance with Section 503.6.4, Item 2, the outdoor portion of the vent shall be enclosed as required by this section for vents not considered to be exposed to the outdoors or such venting system shall be engineered. A Type B vent shall not be considered to be exposed to the outdoors where it passes through an unventilated enclosure or chase insulated to a value of not less than R8.

Tables 504.3(6a), 504.3(6b), 504.3(7a) and 504.3(7b) shall be used for clay-tile-lined *exterior masonry chimneys,* provided that all of the following conditions are met:

1. Vent connectors are Type B double wall.

2. Not less than one appliance is draft hood equipped.

3. The combined appliance input rating is less than the maximum capacity given by Table 504.3(6a) for NAT+NAT or Table 504.3(7a) for FAN+NAT.

4. The input rating of each space-heating appliance is greater than the minimum input rating given by Table 504.3(6b) for NAT+NAT or Table 504.3(7b) for FAN+NAT.

5. The vent connector sizing is in accordance with Table 504.3(3).

❖ See the commentary to Section 504.2.9, Commentary Figure 504.3.20 and the example.

Example:

A 40,000 Btu/h input draft-hood-equipped water heater and a 100,000 Btu/h input Category I boiler are common vented by a masonry chimney having one wall exposed to the outdoors (see Commentary Figure 504.3.20).

Table 504.3(7a) indicates that the maximum combined appliance input rating for 20-foot (6096 mm) tall, 63-square inch (46 cm²) internal area, exterior masonry chimney is 668,000 Btu/h (196 kW). The appliances in this example have a combined input of 140,000 Btu/h (41 kW), which is less than the maximum.

Table 504.3(7a) indicates that a 20-foot (6096 mm) tall chimney with a 63-square-inch (46 cm²) liner in a location with a 5 to 16°F (45° to -9°C) winter design temperature needs a minimum space-heating appliance input of 408,000 Btu/h. In this example, the space-heating appliance input is far less than the minimum required, therefore, the installation does not comply with code and an alternative venting system must be used, such as relining the chimney. (Note that the appliance manufacturer might prohibit the use of an exterior chimney to vent their appliance.)

504.3.21 Connector maximum and minimum size. Vent connectors shall not be increased in size more than two sizes greater than the listed appliance categorized vent diameter, flue collar diameter or draft hood outlet diameter. Vent connectors for draft hood-equipped appliances shall not be smaller than the draft hood outlet diameter. Where a vent connector size(s) determined from the tables for a fan-assisted appliance(s) is smaller than the flue collar diameter, the use of the smaller size(s) shall be permitted provided that the installation complies with all of the following conditions:

1. Vent connectors for fan-assisted appliance flue collars 12 inches (305 mm) in diameter or smaller are not reduced by more than one table size [e.g., 12 inches to 10 inches (305 mm to 254 mm) is a one-size reduction] and those larger than 12 inches (305 mm) in diameter are not reduced more than two table sizes [e.g., 24 inches to 20 inches (610 mm to 508 mm) is a two-size reduction].

2. The fan-assisted appliance(s) is common vented with a draft-hood-equipped appliance(s).

3. The vent connector has a smooth interior wall.

❖ Using a connector that is larger than the appliance outlet connection may sometimes be necessary to comply with the requirements of the sizing tables for multiple-appliance systems. However, having a connector significantly larger than the appliance outlet connection could cause reduced flow velocity, excessive loss of heat and could result in the allowance of excessively long connectors.

In accordance with Section 504.2.2, a venting system serving a single appliance can, under certain conditions, be smaller than the appliance outlet connection. However, this is prohibited by this section for draft hood appliances in common venting systems serving more than one appliance (see commentary,

For SI: 1 foot = 304.8 mm, °C = (°F-32)/1.8,
1 British thermal unit per hour = 0.2931 W.

Figure 504.3.20
EXTERIOR CHIMNEY SIZING—EXAMPLE

Section 504.2.11). The reason for this prohibition is that a common vent or chimney, being large enough for multiple appliances, may not be able to produce adequate draft or avoid condensation where a single appliance is operating with a reduced-size vent connector.

Recent research involving computerized modeling has shown that connectors for fan-assisted appliances in multiple-appliance systems can be reduced under conditions similar to those for single-appliance installations. Therefore, this section makes an exception for systems with fan-assisted appliances commonly vented with appliances equipped with a draft hood (see commentary, Section 504.2.2). The allowance for a connector to be smaller than an appliance flue collar is applicable only to fan-assisted appliances. A reduction in size is not allowed for corrugated metal connectors.

504.3.22 Component commingling. All combinations of pipe sizes, single-wall and double-wall metal pipe shall be allowed within any connector run(s) or within the common vent, provided that all of the appropriate tables permit all of the desired sizes and types of pipe, as if they were used for the entire length of the subject connector or vent. Where single-wall and Type B double-wall metal pipes are used for vent connectors within the same venting system, the common vent must be sized using Table 504.3(2) or 504.3(4), as appropriate.

❖ Because this section parallels Section 504.2.12, see the commentary for that section. This section addresses the installation that uses both single-wall connectors and double-wall connectors in the same venting system. The penalty for mixing is that the common vent must be sized from the more restrictive single-wall connector tables. Double-wall pipe is not designed to be interchangeable with single-wall pipe, and all such transitions would have to made with fittings designed for the transition.

504.3.23 Draft hood conversion accessories. Draft hood conversion accessories for use with masonry chimneys venting listed Category I fan-assisted appliances shall be listed and installed in accordance with the manufacturer's instructions for such listed accessories.

❖ See the commentary to Section 504.2.13.

504.3.24 Multiple sizes permitted. Where a table permits more than one diameter of pipe to be used for a connector or vent, all the permitted sizes shall be permitted to be used.

❖ This section broadly states that any size pipe permitted by the sizing tables may be used. However, do not overlook Sections 504.3.8, 504.3.17 and 504.3.21, which contain provisions that can override this section. Vent manufacturers stress that the smallest size pipe allowed by the code should be used rather than any larger sizes that may be permitted. Operating a vent or chimney at nearer to its maximum capacity will produce stronger draft and result in better protection against the formation of condensation.

504.3.25 Table interpolation. Interpolation shall be permitted in calculating capacities for vent dimensions that fall between table entries.

❖ See Section 504.2.14.

504.3.26 Extrapolation prohibited. Extrapolation beyond the table entries shall not be permitted.

❖ See Section 504.2.15.

504.3.27 Engineering calculations. For vent heights less than 6 feet (1829 mm) and greater than shown in the tables, engineering methods shall be used to calculate vent capacities.

❖ See Section 504.2.16.

504.3.28 Height entries. Where the actual height of a vent falls between entries in the height column of the applicable table in Tables 504.3(1) through 504.3(7b), either interpolation shall be used or the lower appliance input rating shown in the table shall be used for FAN MAX and NAT MAX column values and the higher appliance input rating shall be used for the FAN MIN column values.

❖ See the commentary to Section 504.2.17.

SECTION 505 (IFGC)
DIRECT-VENT, INTEGRAL VENT, MECHANICAL VENT AND VENTILATION/EXHAUST HOOD VENTING

505.1 General. The installation of direct-vent and integral vent appliances shall be in accordance with Section 503. Mechanical venting systems and exhaust hood venting systems shall be designed and installed in accordance with Section 503.

❖ See Sections 503.2.1, 503.2.3, 503.2.4, 503.3.3 and 503.8.

505.1.1 Commercial cooking appliances vented by exhaust hoods. Where commercial cooking appliances are vented by means of the Type I or II kitchen exhaust hood system that serves such appliances, the exhaust system shall be fan powered and the appliances shall be interlocked with the exhaust hood system to prevent appliance operation when the exhaust hood system is not operating. The method of interlock between the exhaust hood system and the appliances equipped with standing pilot burner ignition systems shall not cause such pilots to be extinguished. Where a solenoid valve is installed in the gas *piping* as part of an interlock system, gas *piping* shall not be installed to bypass such valve. Dampers shall not be installed in the exhaust system.

Exception: An interlock between the cooking appliance(s) and the exhaust hood system shall not be required where heat sensors or other *approved* methods automatically activate the exhaust hood system when cooking operations occur.

❖ This section is an appliance-specific relative of Section 503.3.4. Commercial cooking appliances must be vented (see Section 501.2) but are rarely vented by means of direct connection to a vent. Instead, such appliances are typically vented by the Type I or II exhaust system that is required by the IMC. A gravity (nonpowered) exhaust system is not allowed because its ability to vent combustion products is not consistently dependable and there is no practical method of interlocking the appliances with a gravity exhaust system [see Commentary Figure 505.1.1(1)].

The appliance control system must be interlocked with the kitchen exhaust hood system to permit appliance operation only when the power means of exhaust is in operation. This requirement is similar to that of Section 503.3.3, Item 5 and Section 507.2.1.1 of the IMC, which requires Type I hoods to automatically activate the exhaust fan when cooking operations occur. The appliances that depend on the exhaust system must be shut down in the event of exhaust system failure or halted operation.

This interlock must not allow appliance operation until the exhaust hood system has been started. Such an interlock is easily accomplished with electrically heated appliances and gas-fired appliances having electronic ignition; however, it is more difficult in the case of gas-fired appliances having traditional standing pilot ignition systems. If the interlock method uses an electrically actuated (solenoid) main fuel valve, that valve will close when the exhaust fan system is shut down, thus extinguishing any standing pilots and necessitating the relighting of pilots each day or each time the kitchen is "off line."

Besides the obvious inconvenience, there is also a safety concern with having employees and owners routinely relighting pilots that may be difficult to access. This is why the code prohibits a method of interlock that would cause standing pilot burners to be extinguished. Although some attempts have been made to circumvent this problem, a practical and safe

solution to the standing pilot problem is not known at this time.

This section specifically prohibits the installation of bypass piping/tubing around a solenoid-operated gas valve [see Commentary Figure 505.1.1(2)]. Bypass piping/tubing around the solenoid valve has been used to keep the standing pilots burning when the exhaust system is shut down and the solenoid valve is thereby closed. This approach may solve the pilot outage problem, but it also creates potential hazards when the solenoid valve is closed, all related to low gas pressure and operator confusion. The potential hazards include: delayed ignition and flashbacks, ignition within burner venturi passages and at the orifices, poor burner aspiration and incomplete combustion, erratic appliance operation and the potential for explosions where appliances are left on and unattended.

Note that this section does not require that the interlock be accomplished by means of a gas solenoid valve. For example, for appliances that have electrical power connections (nonstanding-pilot appliances), contactors/relays could be used to disable the appliances when the exhaust system is not operating. The exception captures the basic intent of this section, which is simply to prevent cooking without the exhaust system being on. Remember that the appliances depend on the exhaust hood to vent combustion products in addition to venting cooking byproducts. Some exhaust hood technology allows the hoods to start automatically when heat, infrared or light interference sensors detect that cooking is occurring. The code official could accept still other arrangements that accomplish the intent (see IMC commentary, Section 507.2.1.1).

Figure 505.1.1(1)
EXHAUST HOOD INTERLOCK WITH COOKING APPLIANCES (SOLENOID)

Figure 505.1.1(2)
PROHIBITED SOLENOID VALVE BYPASS

Note that if a solenoid gas valve arrangement is used to comply with this section, it should be powered through a manual reset-type relay so that in the event of a power failure, the valve will not reopen automatically when power is restored and the appliances are unattended.

SECTION 506 (IFGC)
FACTORY-BUILT CHIMNEYS

506.1 Building heating appliances. Factory-built chimneys for building heating appliances producing flue gases having a temperature not greater than 1,000°F (538°C), measured at the entrance to the chimney, shall be listed and *labeled* in accordance with UL 103 and shall be installed and terminated in accordance with the manufacturer's instructions.

❖ Factory-built chimneys can be used with gas-fired appliances if allowed by the chimney manufacturer. The sizing tables in Section 504 do not apply to factory-built chimneys; therefore, sizing must be engineered or as specified by the chimney and appliance manufacturers' instructions. Factory-built chimneys will be too large for many residential gas appliances [see Commentary Figures 506.1(1) and 506.1(2)].

506.2 Support. Where factory-built chimneys are supported by structural members, such as joists and rafters, such members shall be designed to support the additional load.

❖ Factory-built chimneys are supported by the building structure and can impose considerable weight on structural components. Joists, rafters and other structural members must be designed to support the additional loading. Structural evaluation may be necessary, especially where framing for a chimney requires cutting and heading of joists and rafters.

Figure 506.1(1)
FACTORY-BUILT CHIMNEY
(Photo courtesy of Selkirk, L.L.C.)

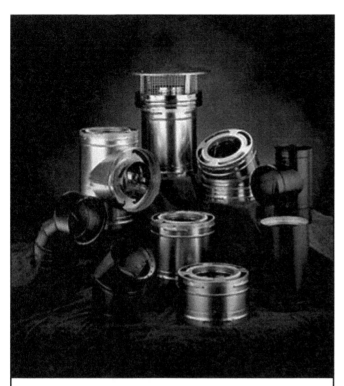

**Figure 506.1(2)
AIR-COOLED FACTORY-BUILT CHIMNEY SYSTEM
(Photo courtesy of Simpson Dura-Vent Company,
Incorporated)**

506.3 Medium-heat appliances. Factory-built chimneys for medium-heat appliances producing flue gases having a temperature above 1,000°F (538°C), measured at the entrance to the chimney, shall be listed and *labeled* in accordance with UL 959 and shall be installed and terminated in accordance with the manufacturer's instructions.

❖ Factory-built chimneys serving medium-heat appliances in which the maximum continuous flue gas temperature does not exceed 2,000°F (1093°C) must be tested in accordance with UL 959. Such chimneys are associated with factory and industrial equipment and incinerators. Chimneys serving medium-heat appliances may require greater termination heights and clearances to combustibles.

Bibliography

The following resource materials were used in the preparation of the commentary for this chapter of the code.

Code Changes Resource Collection. Washington, DC: International Code Council, 2014.

IBC-2015, *International Building Code*. Washington, DC: International Code Council, 2014.

IECC-2015, *International Energy Conservation Code*. Washington, DC: International Code Council, 2014.

IMC-2015, *International Mechanical Code*. Washington, DC: International Code Council, 2014.

IPC-2015, *International Plumbing Code*. Washington, DC: International Code Council, 2014.

TABLE 504.2(1)
TYPE B DOUBLE-WALL GAS VENT

Number of Appliances	Single
Appliance Type	Category I
Appliance Vent Connection	Connected directly to vent

VENT DIAMETER—(D) inches

APPLIANCE INPUT RATING IN THOUSANDS OF BTU/H

HEIGHT (H) (feet)	LATERAL (L) (feet)	3 FAN Min	3 FAN Max	3 NAT Max	4 FAN Min	4 FAN Max	4 NAT Max	5 FAN Min	5 FAN Max	5 NAT Max	6 FAN Min	6 FAN Max	6 NAT Max	7 FAN Min	7 FAN Max	7 NAT Max	8 FAN Min	8 FAN Max	8 NAT Max	9 FAN Min	9 FAN Max	9 NAT Max
6	0	0	78	46	0	152	86	0	251	141	0	375	205	0	524	285	0	698	370	0	897	470
6	2	13	51	36	18	97	67	27	157	105	32	232	157	44	321	217	53	425	285	63	543	370
6	4	21	49	34	30	94	64	39	153	103	50	227	153	66	316	211	79	419	279	93	536	362
6	6	25	46	32	36	91	61	47	149	100	59	223	149	78	310	205	93	413	273	110	530	354
8	0	0	84	50	0	165	94	0	276	155	0	415	235	0	583	320	0	780	415	0	1,006	537
8	2	12	57	40	16	109	75	25	178	120	28	263	180	42	365	247	50	483	322	60	619	418
8	5	23	53	38	32	103	71	42	171	115	53	255	173	70	356	237	83	473	313	99	607	407
8	8	28	49	35	39	98	66	51	164	109	64	247	165	84	347	227	99	463	303	117	596	396
10	0	0	88	53	0	175	100	0	295	166	0	447	255	0	631	345	0	847	450	0	1,096	585
10	2	12	61	42	17	118	81	23	194	129	26	289	195	40	402	273	48	533	355	57	684	457
10	5	23	57	40	32	113	77	41	187	124	52	280	188	68	392	263	81	522	346	95	671	446
10	10	30	51	36	41	104	70	54	176	115	67	267	175	88	376	245	104	504	330	122	651	427
15	0	0	94	58	0	191	112	0	327	187	0	502	285	0	716	390	0	970	525	0	1,263	682
15	2	11	69	48	15	136	93	20	226	150	22	339	225	38	475	316	45	633	414	53	815	544
15	5	22	65	45	30	130	87	39	219	142	49	330	217	64	463	300	76	620	403	90	800	529
15	10	29	59	41	40	121	82	51	206	135	64	315	208	84	445	288	99	600	386	116	777	507
15	15	35	53	37	48	112	76	61	195	128	76	301	198	98	429	275	115	580	373	134	755	491
20	0	0	97	61	0	202	119	0	349	202	0	540	307	0	776	430	0	1,057	575	0	1,384	752
20	2	10	75	51	14	149	100	18	250	166	20	377	249	33	531	346	41	711	470	50	917	612
20	5	21	71	48	29	143	96	38	242	160	47	367	241	62	519	337	73	697	460	86	902	599
20	10	28	64	44	38	133	89	50	229	150	62	351	228	81	499	321	95	675	443	112	877	576
20	15	34	58	40	46	124	84	59	217	142	73	337	217	94	481	308	111	654	427	129	853	557
20	20	48	52	35	55	116	78	69	206	134	84	322	206	107	464	295	125	634	410	145	830	537

(continued)

TABLE 504.2(1)—continued
TYPE B DOUBLE-WALL GAS VENT

Number of Appliances	Single
Appliance Type	Category I
Appliance Vent Connection	Connected directly to vent

VENT DIAMETER—(D) inches

APPLIANCE INPUT RATING IN THOUSANDS OF BTU/H

HEIGHT (H) (feet)	LATERAL (L) (feet)	3 FAN Min	3 FAN Max	3 NAT Max	4 FAN Min	4 FAN Max	4 NAT Max	5 FAN Min	5 FAN Max	5 NAT Max	6 FAN Min	6 FAN Max	6 NAT Max	7 FAN Min	7 FAN Max	7 NAT Max	8 FAN Min	8 FAN Max	8 NAT Max	9 FAN Min	9 FAN Max	9 NAT Max
30	0	0	100	64	0	213	128	0	374	220	0	587	336	0	853	475	0	1,173	650	0	1,548	855
	2	9	81	56	13	166	112	14	283	185	18	432	280	27	613	394	33	826	535	42	1,072	700
	5	21	77	54	28	160	108	36	275	176	45	421	273	58	600	385	69	811	524	82	1,055	688
	10	27	70	50	37	150	102	48	262	171	59	405	261	77	580	371	91	788	507	107	1,028	668
	15	33	64	NA	44	141	96	57	249	163	70	389	249	90	560	357	105	765	490	124	1,002	648
	20	56	58	NA	53	132	90	66	237	154	80	374	237	102	542	343	119	743	473	139	977	628
	30	NA	NA	NA	73	113	NA	88	214	NA	104	346	219	131	507	321	149	702	444	171	929	594
50	0	0	101	67	0	216	134	0	397	232	0	633	363	0	932	518	0	1,297	708	0	1,730	952
	2	8	86	61	11	183	122	14	320	206	15	497	314	22	715	445	26	975	615	33	1,276	813
	5	20	82	NA	27	177	119	35	312	200	43	487	308	55	702	438	65	960	605	77	1,259	798
	10	26	76	NA	35	168	114	45	299	190	56	471	298	73	681	426	86	935	589	101	1,230	773
	15	59	70	NA	42	158	NA	54	287	180	66	455	288	85	662	413	100	911	572	117	1,203	747
	20	NA	NA	NA	50	149	NA	63	275	169	76	440	278	97	642	401	113	888	556	131	1,176	722
	30	NA	NA	NA	69	131	NA	84	250	NA	99	410	259	123	605	376	141	844	522	161	1,125	670
100	0	NA	NA	NA	0	218	NA	0	407	NA	0	665	400	0	997	560	0	1,411	770	0	1,908	1,040
	2	NA	NA	NA	10	194	NA	12	354	NA	13	566	375	18	831	510	21	1,155	700	25	1,536	935
	5	NA	NA	NA	26	189	NA	33	347	NA	40	557	369	52	820	504	60	1,141	692	71	1,519	926
	10	NA	NA	NA	33	182	NA	43	335	NA	53	542	361	68	801	493	80	1,118	679	94	1,492	910
	15	NA	NA	NA	40	174	NA	50	321	NA	62	528	353	80	782	482	93	1,095	666	109	1,465	895
	20	NA	NA	NA	47	166	NA	59	311	NA	71	513	344	90	763	471	105	1,073	653	122	1,438	880
	30	NA	NA	NA	NA	NA	NA	78	290	NA	92	483	NA	115	726	449	131	1,029	627	149	1,387	849
	50	NA	NA	NA	NA	NA	NA	NA	NA	NA	147	428	NA	180	651	405	197	944	575	217	1,288	787

(continued)

TABLE 504.2(1)—continued
TYPE B DOUBLE-WALL GAS VENT

Number of Appliances	Single
Appliance Type	Category I
Appliance Vent Connection	Connected directly to vent

APPLIANCE INPUT RATING IN THOUSANDS OF BTU/H — VENT DIAMETER—(D) inches

HEIGHT (H) (feet)	LATERAL (L) (feet)	10 FAN Min	10 FAN Max	10 NAT Max	12 FAN Min	12 FAN Max	12 NAT Max	14 FAN Min	14 FAN Max	14 NAT Max	16 FAN Min	16 FAN Max	16 NAT Max	18 FAN Min	18 FAN Max	18 NAT Max	20 FAN Min	20 FAN Max	20 NAT Max	22 FAN Min	22 FAN Max	22 NAT Max	24 FAN Min	24 FAN Max	24 NAT Max
6	0	0	1,121	570	0	1,645	850	0	2,267	1,170	0	2,983	1,530	0	3,802	1,960	0	4,721	2,430	0	5,737	2,950	0	6,853	3,520
	2	75	675	455	103	982	650	138	1,346	890	178	1,769	1,170	225	2,250	1,480	296	2,782	1,850	360	3,377	2,220	426	4,030	2,670
	4	110	668	445	147	975	640	191	1,338	880	242	1,761	1,160	300	2,242	1,475	390	2,774	1,835	469	3,370	2,215	555	4,023	2,660
	6	128	661	435	171	967	630	219	1,330	870	276	1,753	1,150	341	2,235	1,470	437	2,767	1,820	523	3,363	2,210	618	4,017	2,650
8	0	0	1,261	660	0	1,858	970	0	2,571	1,320	0	3,399	1,740	0	4,333	2,220	0	5,387	2,750	0	6,555	3,360	0	7,838	4,010
	2	71	770	515	98	1,124	745	130	1,543	1,020	168	2,030	1,340	212	2,584	1,700	278	3,196	2,110	336	3,882	2,560	401	4,634	3,050
	5	115	758	503	154	1,110	733	199	1,528	1,010	251	2,013	1,330	311	2,563	1,685	398	3,180	2,090	476	3,863	2,545	562	4,612	3,040
	8	137	746	490	180	1,097	720	231	1,514	1,000	289	2,000	1,320	354	2,552	1,670	450	3,163	2,070	537	3,850	2,530	630	4,602	3,030
10	0	0	1,377	720	0	2,036	1,060	0	2,825	1,450	0	3,742	1,925	0	4,782	2,450	0	5,955	3,050	0	7,254	3,710	0	8,682	4,450
	2	68	852	560	93	1,244	850	124	1,713	1,130	161	2,256	1,480	202	2,868	1,890	264	3,556	2,340	319	4,322	2,840	378	5,153	3,390
	5	112	839	547	149	1,229	829	192	1,696	1,105	243	2,238	1,461	300	2,849	1,871	382	3,536	2,318	458	4,301	2,818	540	5,132	3,371
	10	142	817	525	187	1,204	795	238	1,669	1,080	298	2,209	1,430	364	2,818	1,840	459	3,504	2,280	546	4,268	2,780	641	5,099	3,340
15	0	0	1,596	840	0	2,380	1,240	0	3,323	1,720	0	4,423	2,270	0	5,678	2,900	0	7,099	3,620	0	8,665	4,410	0	10,393	5,300
	2	63	1,019	675	86	1,495	985	114	2,062	1,350	147	2,719	1,770	186	3,467	2,260	239	4,304	2,800	290	5,232	3,410	346	6,251	4,080
	5	105	1,003	660	140	1,476	967	182	2,041	1,327	229	2,696	1,748	283	3,442	2,235	355	4,278	2,777	426	5,204	3,385	501	6,222	4,057
	10	135	977	635	177	1,446	936	227	2,009	1,289	283	2,659	1,712	346	3,402	2,193	432	4,234	2,739	510	5,159	3,343	599	6,175	4,019
	15	155	953	610	202	1,418	905	257	1,976	1,250	318	2,623	1,675	385	3,363	2,150	479	4,192	2,700	564	5,115	3,300	665	6,129	3,980
20	0	0	1,756	930	0	2,637	1,350	0	3,701	1,900	0	4,948	2,520	0	6,376	3,250	0	7,988	4,060	0	9,785	4,980	0	11,753	6,000
	2	59	1,150	755	81	1,694	1,100	107	2,343	1,520	139	3,097	2,000	175	3,955	2,570	220	4,916	3,200	269	5,983	3,910	321	7,154	4,700
	5	101	1,133	738	135	1,674	1,079	174	2,320	1,498	219	3,071	1,978	270	3,926	2,544	337	4,885	3,174	403	5,950	3,880	475	7,119	4,662
	10	130	1,105	710	172	1,641	1,045	220	2,282	1,460	273	3,029	1,940	334	3,880	2,500	413	4,835	3,130	489	5,896	3,830	573	7,063	4,600
	15	150	1,078	688	195	1,609	1,018	248	2,245	1,425	306	2,988	1,910	372	3,835	2,465	459	4,786	3,090	541	5,844	3,795	631	7,007	4,575
	20	167	1,052	665	217	1,578	990	273	2,210	1,390	335	2,948	1,880	404	3,791	2,430	495	4,737	3,050	585	5,792	3,760	689	6,953	4,550

(continued)

TABLE 504.2(1)—continued
TYPE B DOUBLE-WALL GAS VENT

Number of Appliances	Single
Appliance Type	Category I
Appliance Vent Connection	Connected directly to vent

VENT DIAMETER—(D) inches

APPLIANCE INPUT RATING IN THOUSANDS OF BTU/H

HEIGHT (H) (feet)	LATERAL (L) (feet)	10 FAN Min	10 FAN Max	10 NAT Max	12 FAN Min	12 FAN Max	12 NAT Max	14 FAN Min	14 FAN Max	14 NAT Max	16 FAN Min	16 FAN Max	16 NAT Max	18 FAN Min	18 FAN Max	18 NAT Max	20 FAN Min	20 FAN Max	20 NAT Max	22 FAN Min	22 FAN Max	22 NAT Max	24 FAN Min	24 FAN Max	24 NAT Max
30	0	0	1,977	1,060	0	3,004	1,550	0	4,252	2,170	0	5,725	2,920	0	7,420	3,770	0	9,341	4,750	0	11,483	5,850	0	13,848	7,060
	2	54	1,351	865	74	2,004	1,310	98	2,786	1,800	127	3,696	2,380	159	4,734	3,050	199	5,900	3,810	241	7,194	4,650	285	8,617	5,600
	5	96	1,332	851	127	1,981	1,289	164	2,759	1,775	206	3,666	2,350	252	4,701	3,020	312	5,863	3,783	373	7,155	4,622	439	8,574	5,552
	10	125	1,301	829	164	1,944	1,254	209	2,716	1,733	259	3,617	2,300	316	4,647	2,970	386	5,803	3,739	456	7,090	4,574	535	8,505	5,471
	15	143	1,272	807	187	1,908	1,220	237	2,674	1,692	292	3,570	2,250	354	4,594	2,920	431	5,744	3,695	507	7,026	4,527	590	8,437	5,391
	20	160	1,243	784	207	1,873	1,185	260	2,633	1,650	319	3,523	2,200	384	4,542	2,870	467	5,686	3,650	548	6,964	4,480	639	8,370	5,310
	30	195	1,189	745	246	1,807	1,130	305	2,555	1,585	369	3,433	2,130	440	4,442	2,785	540	5,574	3,565	635	6,842	4,375	739	8,239	5,225
50	0	0	2,231	1,195	0	3,441	1,825	0	4,934	2,550	0	6,711	3,440	0	8,774	4,460	0	11,129	5,635	0	13,767	6,940	0	16,694	8,430
	2	41	1,620	1,010	66	2,431	1,513	86	3,409	2,125	113	4,554	2,840	141	5,864	3,670	171	7,339	4,630	209	8,980	5,695	251	10,788	6,860
	5	90	1,600	996	118	2,406	1,495	151	3,380	2,102	191	4,520	2,813	234	5,826	3,639	283	7,295	4,597	336	8,933	5,654	394	10,737	6,818
	10	118	1,567	972	154	2,366	1,466	196	3,332	2,064	243	4,464	2,767	295	5,763	3,585	355	7,224	4,542	419	8,855	5,585	491	10,652	6,749
	15	136	1,536	948	177	2,327	1,437	222	3,285	2,026	274	4,409	2,721	330	5,701	3,534	396	7,155	4,511	465	8,779	5,546	542	10,570	6,710
	20	151	1,505	924	195	2,288	1,408	244	3,239	1,987	300	4,356	2,675	361	5,641	3,481	433	7,086	4,479	506	8,704	5,506	586	10,488	6,670
	30	183	1,446	876	232	2,214	1,349	287	3,150	1,910	347	4,253	2,631	412	5,523	3,431	494	6,953	4,421	577	8,557	5,444	672	10,328	6,603
100	0	0	2,491	1,310	0	3,925	2,050	0	5,729	2,950	0	7,914	4,050	0	10,485	5,300	0	13,454	6,700	0	16,817	8,600	0	20,578	10,300
	2	30	1,975	1,170	44	3,027	1,820	72	4,313	2,550	95	5,834	3,500	120	7,591	4,600	138	9,577	5,800	169	11,803	7,200	204	14,264	8,800
	5	82	1,955	1,159	107	3,002	1,803	136	4,282	2,531	172	5,797	3,475	208	7,548	4,566	245	9,528	5,769	293	11,748	7,162	341	14,204	8,756
	10	108	1,923	1,142	142	2,961	1,775	180	4,231	2,500	223	5,737	3,434	268	7,478	4,509	318	9,447	5,717	374	11,658	7,100	436	14,105	8,683
	15	126	1,892	1,124	163	2,920	1,747	206	4,182	2,469	252	5,678	3,392	304	7,409	4,451	358	9,367	5,665	418	11,569	7,037	487	14,007	8,610
	20	141	1,861	1,107	181	2,880	1,719	226	4,133	2,438	277	5,619	3,351	330	7,341	4,394	387	9,289	5,613	452	11,482	6,975	523	13,910	8,537
	30	170	1,802	1,071	215	2,803	1,663	265	4,037	2,375	319	5,505	3,267	378	7,209	4,279	446	9,136	5,509	514	11,310	6,850	592	13,720	8,391
	50	241	1,688	1,000	292	2,657	1,550	350	3,856	2,250	415	5,289	3,100	486	6,956	4,050	572	8,841	5,300	659	10,979	6,600	752	13,354	8,100

For SI: 1 inch = 25.4 mm, 1 foot = 304.8 mm, 1 British thermal unit per hour = 0.2931 W.

TABLE 504.2(2)
TYPE B DOUBLE-WALL GAS VENT

Number of Appliances	Single
Appliance Type	Category I
Appliance Vent Connection	Single-wall metal connector

VENT DIAMETER—(D) inches

APPLIANCE INPUT RATING IN THOUSANDS OF BTU/H

HEIGHT (H) (feet)	LATERAL (L) (feet)	3 FAN Min	3 FAN Max	3 NAT Max	4 FAN Min	4 FAN Max	4 NAT Max	5 FAN Min	5 FAN Max	5 NAT Max	6 FAN Min	6 FAN Max	6 NAT Max	7 FAN Min	7 FAN Max	7 NAT Max	8 FAN Min	8 FAN Max	8 NAT Max	9 FAN Min	9 FAN Max	9 NAT Max	10 FAN Min	10 FAN Max	10 NAT Max	12 FAN Min	12 FAN Max	12 NAT Max
6	0	38	77	45	59	151	85	85	249	140	126	373	204	165	522	284	211	695	369	267	894	469	371	1,118	569	537	1,639	849
6	2	39	51	36	60	96	66	85	156	104	123	231	156	159	320	213	201	423	284	251	541	368	347	673	453	498	979	648
6	4	NA	NA	33	74	92	63	102	152	102	146	225	152	187	313	208	237	416	277	295	533	360	409	664	443	584	971	638
6	6	NA	NA	31	83	89	60	114	147	99	163	220	148	207	307	203	263	409	271	327	526	352	449	656	433	638	962	627
8	0	37	83	50	58	164	93	83	273	154	123	412	234	161	580	319	206	777	414	258	1,002	536	360	1,257	658	521	1,852	967
8	2	39	56	39	59	108	75	83	176	119	121	261	179	155	363	246	197	482	321	246	617	417	339	768	513	486	1,120	743
8	5	NA	NA	37	77	102	69	107	168	114	151	252	171	193	352	235	245	470	311	305	604	404	418	754	500	598	1,104	730
8	8	NA	NA	33	90	95	64	122	161	107	175	243	163	223	342	225	280	458	300	344	591	392	470	740	486	665	1,089	715
10	0	37	87	53	57	174	99	82	293	165	120	444	254	158	628	344	202	844	449	253	1,093	584	351	1,373	718	507	2,031	1,057
10	2	39	61	41	59	117	80	82	193	128	119	287	194	153	400	272	193	531	354	242	681	456	332	849	559	475	1,242	848
10	5	52	56	39	76	111	76	105	185	122	148	277	186	190	388	261	241	518	344	299	667	443	409	834	544	584	1,224	825
10	10	NA	NA	34	97	100	68	132	171	112	188	261	171	237	369	241	296	497	325	363	643	423	492	808	520	688	1,194	788
15	0	36	93	57	56	190	111	80	325	186	116	499	283	153	713	388	195	966	523	244	1,259	681	336	1,591	838	488	2,374	1,237
15	2	38	69	47	57	136	93	80	225	149	115	337	224	148	473	314	187	631	413	232	812	543	319	1,015	673	457	1,491	983
15	5	51	63	44	75	128	86	102	216	140	144	326	217	182	459	298	231	616	400	287	795	526	392	997	657	562	1,469	963
15	10	NA	NA	39	95	116	79	128	201	131	182	308	203	228	438	284	284	592	381	349	768	501	470	966	628	664	1,433	928
15	15	NA	NA	NA	NA	NA	72	158	186	124	220	290	192	272	418	269	334	568	367	404	742	484	540	937	601	750	1,399	894
20	0	35	96	60	54	200	118	78	346	201	114	537	306	149	772	428	190	1,053	573	238	1,379	750	326	1,751	927	473	2,631	1,346
20	2	37	74	50	56	148	99	78	248	165	113	375	248	144	528	344	182	708	468	227	914	611	309	1,146	754	443	1,689	1,098
20	5	50	68	47	73	140	94	100	239	158	141	363	239	178	514	334	224	692	457	279	896	596	381	1,126	734	547	1,665	1,074
20	10	NA	NA	41	93	129	86	125	223	146	177	344	224	222	491	316	277	666	437	339	866	570	457	1,092	702	646	1,626	1,037
20	15	NA	NA	NA	NA	NA	80	155	208	136	216	325	210	264	469	301	325	640	419	393	838	549	526	1,060	677	730	1,587	1,005
20	20	NA	NA	NA	NA	NA	NA	186	192	126	254	306	196	309	448	285	374	616	400	448	810	526	592	1,028	651	808	1,550	973

(continued)

TABLE 504.2(2)—continued
TYPE B DOUBLE-WALL GAS VENT

Number of Appliances	Single
Appliance Type	Category I
Appliance Vent Connection	Single-wall metal connector

APPLIANCE INPUT RATING IN THOUSANDS OF BTU/H

HEIGHT (H) (feet)	LATERAL (L) (feet)	3 FAN Min	3 FAN Max	3 NAT Max	4 FAN Min	4 FAN Max	4 NAT Max	5 FAN Min	5 FAN Max	5 NAT Max	6 FAN Min	6 FAN Max	6 NAT Max	7 FAN Min	7 FAN Max	7 NAT Max	8 FAN Min	8 FAN Max	8 NAT Max	9 FAN Min	9 FAN Max	9 NAT Max	10 FAN Min	10 FAN Max	10 NAT Max	12 FAN Min	12 FAN Max	12 NAT Max
30	0	34	99	63	53	211	127	76	372	219	110	584	334	144	849	472	184	1,168	647	229	1,542	852	312	1,971	1,056	454	2,996	1,545
	2	37	80	56	55	164	111	76	281	183	109	429	279	139	610	392	175	823	533	219	1,069	698	296	1,346	863	424	1,999	1,308
	5	49	74	52	72	157	106	98	271	173	136	417	271	171	595	382	215	806	521	269	1,049	684	366	1,324	846	524	1,971	1,283
	10	NA	NA	NA	91	144	98	122	255	168	171	397	257	213	570	367	265	777	501	327	1,017	662	440	1,287	821	620	1,927	1,234
	15	NA	NA	NA	115	131	NA	151	239	157	208	377	242	255	547	349	312	750	481	379	985	638	507	1,251	794	702	1,884	1,205
	20	NA	NA	NA	NA	NA	NA	181	223	NA	246	357	228	298	524	333	360	723	461	433	955	615	570	1,216	768	780	1,841	1,166
	30	NA	NA	NA	NA	NA	NA	NA	NA	NA	NA	NA	NA	389	477	NA	461	670	426	541	895	574	704	1,147	720	937	1,759	1,101
50	0	33	99	66	51	213	133	73	394	230	105	629	361	138	928	515	176	1,292	704	220	1,724	948	295	2,223	1,189	428	3,432	1,818
	2	36	84	61	53	181	121	73	318	205	104	495	312	133	712	443	168	971	613	209	1,273	811	280	1,615	1,007	401	2,426	1,509
	5	48	80	NA	70	174	117	94	308	198	131	482	305	164	696	435	204	953	602	257	1,252	795	347	1,591	991	496	2,396	1,490
	10	NA	NA	NA	89	160	NA	118	292	186	162	461	292	203	671	420	253	923	583	313	1,217	765	418	1,551	963	589	2,347	1,455
	15	NA	NA	NA	112	148	NA	145	275	174	199	441	280	244	646	405	299	894	562	363	1,183	736	481	1,512	934	668	2,299	1,421
	20	NA	NA	NA	NA	NA	NA	176	257	NA	236	420	267	285	622	389	345	866	543	415	1,150	708	544	1,473	906	741	2,251	1,387
	30	NA	NA	NA	NA	NA	NA	NA	NA	NA	315	376	NA	373	573	NA	442	809	502	521	1,086	649	674	1,399	848	892	2,159	1,318
100	0	NA	NA	NA	49	214	NA	69	403	NA	100	659	395	131	991	555	166	1,404	765	207	1,900	1,033	273	2,479	1,300	395	3,912	2,042
	2	NA	NA	NA	51	192	NA	70	351	NA	98	563	373	125	828	508	158	1,152	698	196	1,532	933	259	1,970	1,168	371	3,021	1,817
	5	NA	NA	NA	67	186	NA	90	342	NA	125	551	366	156	813	501	194	1,134	688	240	1,511	921	322	1,945	1,153	460	2,990	1,796
	10	NA	NA	NA	85	175	NA	113	324	NA	153	532	354	191	789	486	238	1,104	672	293	1,477	902	389	1,905	1,133	547	2,938	1,763
	15	NA	NA	NA	132	162	NA	138	310	NA	188	511	343	230	764	473	281	1,075	656	342	1,443	884	447	1,865	1,110	618	2,888	1,730
	20	NA	NA	NA	NA	NA	NA	168	295	NA	224	487	NA	270	739	458	325	1,046	639	391	1,410	864	507	1,825	1,087	690	2,838	1,696
	30	NA	NA	NA	NA	NA	NA	231	264	NA	301	448	NA	355	685	NA	418	988	NA	491	1,343	824	631	1,747	1,041	834	2,739	1,627
	50	NA	NA	NA	NA	NA	NA	NA	NA	NA	NA	NA	NA	540	584	NA	617	866	NA	711	1,205	NA	895	1,591	NA	1,138	2,547	1,489

For SI: 1 inch = 25.4 mm, 1 foot = 304.8 mm, 1 British thermal unit per hour = 0.2931 W.

TABLE 504.2(3)
MASONRY CHIMNEY

Number of Appliances	Single
Appliance Type	Category I
Appliance Vent Connection	Type B double-wall connector

TYPE B DOUBLE-WALL CONNECTOR DIAMETER—(D) Inches to be used with chimney areas within the size limits at bottom

APPLIANCE INPUT RATING IN THOUSANDS OF BTU/H

HEIGHT (H) (feet)	LATERAL (L) (feet)	3 FAN Min	3 FAN Max	3 NAT Max	4 FAN Min	4 FAN Max	4 NAT Max	5 FAN Min	5 FAN Max	5 NAT Max	6 FAN Min	6 FAN Max	6 NAT Max	7 FAN Min	7 FAN Max	7 NAT Max	8 FAN Min	8 FAN Max	8 NAT Max	9 FAN Min	9 FAN Max	9 NAT Max	10 FAN Min	10 FAN Max	10 NAT Max	12 FAN Min	12 FAN Max	12 NAT Max
6	2	NA	NA	28	NA	NA	52	NA	NA	86	NA	NA	130	NA	NA	180	NA	NA	247	NA	NA	320	NA	NA	401	NA	NA	581
6	5	NA	NA	25	NA	NA	49	NA	NA	82	NA	NA	117	NA	NA	165	NA	NA	231	NA	NA	298	NA	NA	376	NA	NA	561
8	2	NA	NA	29	NA	NA	55	NA	NA	93	NA	NA	145	NA	NA	198	NA	NA	266	84	590	350	100	728	446	139	1,024	651
8	5	NA	NA	26	NA	NA	52	NA	NA	88	NA	NA	134	NA	NA	183	NA	NA	247	NA	NA	328	149	711	423	201	1,007	640
8	8	NA	NA	24	NA	NA	48	NA	NA	83	NA	NA	127	NA	NA	175	NA	NA	239	NA	NA	318	173	695	410	231	990	623
10	2	NA	NA	31	NA	NA	61	NA	NA	103	NA	NA	162	NA	NA	221	68	519	298	82	655	388	98	810	491	136	1,144	724
10	5	NA	NA	28	NA	NA	57	NA	NA	96	NA	NA	148	NA	NA	204	NA	NA	277	124	638	365	146	791	466	196	1,124	712
10	10	NA	NA	25	NA	NA	50	NA	NA	87	NA	NA	139	NA	NA	191	NA	NA	263	155	610	347	182	762	444	240	1,093	668
15	2	NA	NA	35	NA	NA	67	NA	NA	114	NA	NA	179	53	475	250	64	613	336	77	779	441	92	968	562	127	1,376	841
15	5	NA	NA	35	NA	NA	62	NA	NA	107	NA	NA	164	NA	NA	231	99	594	313	118	759	416	139	946	533	186	1,352	828
15	10	NA	NA	28	NA	NA	55	NA	NA	97	NA	NA	153	NA	NA	216	126	565	296	148	727	394	173	912	567	229	1,315	777
15	15	NA	NA	NA	NA	NA	48	NA	NA	89	NA	NA	141	NA	NA	201	NA	NA	281	171	698	375	198	880	485	259	1,280	742
20	2	NA	NA	38	NA	NA	74	NA	NA	124	NA	NA	201	51	522	274	61	678	375	73	867	491	87	1,083	627	121	1,548	953
20	5	NA	NA	36	NA	NA	68	NA	NA	116	NA	NA	184	80	503	254	95	658	350	113	845	463	133	1,059	597	179	1,523	933
20	10	NA	NA	NA	NA	NA	60	NA	NA	107	NA	NA	172	NA	NA	237	122	627	332	143	811	440	167	1,022	566	221	1,482	879
20	15	NA	NA	NA	NA	NA	NA	NA	NA	97	NA	NA	159	NA	NA	220	NA	NA	314	165	780	418	191	987	541	251	1,443	840
20	20	NA	NA	NA	NA	NA	NA	NA	NA	83	NA	NA	148	NA	NA	206	NA	NA	296	186	750	397	214	955	513	277	1,406	807

(continued)

TABLE 504.2(3)—continued
MASONRY CHIMNEY

Number of Appliances	Single
Appliance Type	Category I
Appliance Vent Connection	Type B double-wall connector

TYPE B DOUBLE-WALL CONNECTOR DIAMETER —(D) Inches to be used with chimney areas within the size limits at bottom

APPLIANCE INPUT RATING IN THOUSANDS OF BTU/H

HEIGHT (H) (feet)	LATERAL (L) (feet)	3 FAN Min	3 FAN Max	3 NAT Max	4 FAN Min	4 FAN Max	4 NAT Max	5 FAN Min	5 FAN Max	5 NAT Max	6 FAN Min	6 FAN Max	6 NAT Max	7 FAN Min	7 FAN Max	7 NAT Max	8 FAN Min	8 FAN Max	8 NAT Max	9 FAN Min	9 FAN Max	9 NAT Max	10 FAN Min	10 FAN Max	10 NAT Max	12 FAN Min	12 FAN Max	12 NAT Max
30	2	NA	NA	41	NA	NA	82	NA	NA	137	NA	NA	216	47	581	303	57	762	421	68	985	558	81	1,240	717	111	1,793	1,112
	5	NA	NA	NA	NA	NA	76	NA	NA	128	NA	NA	198	75	561	281	90	741	393	106	962	526	125	1,216	683	169	1,766	1,094
	10	NA	NA	NA	NA	NA	67	NA	NA	115	NA	NA	184	NA	NA	263	115	709	373	135	927	500	158	1,176	648	210	1,721	1,025
	15	NA	NA	NA	NA	NA	NA	NA	NA	107	NA	NA	171	NA	NA	243	NA	NA	353	156	893	476	181	1,139	621	239	1,679	981
	20	NA	NA	NA	NA	NA	NA	NA	NA	91	NA	NA	159	NA	NA	227	NA	NA	332	176	860	450	203	1,103	592	264	1,638	940
	30	NA	NA	NA	NA	NA	NA	NA	NA	NA	NA	NA	NA	NA	NA	188	NA	NA	288	NA	NA	416	249	1,035	555	318	1,560	877
50	2	NA	NA	NA	NA	NA	92	NA	NA	161	NA	NA	251	NA	NA	351	51	840	477	61	1,106	633	72	1,413	812	99	2,080	1,243
	5	NA	NA	NA	NA	NA	NA	NA	NA	151	NA	NA	230	NA	NA	323	83	819	445	98	1,083	596	116	1,387	774	155	2,052	1,225
	10	NA	NA	NA	NA	NA	NA	NA	NA	138	NA	NA	215	NA	NA	304	NA	NA	424	126	1,047	567	147	1,347	733	195	2,006	1,147
	15	NA	NA	NA	NA	NA	NA	NA	NA	127	NA	NA	199	NA	NA	282	NA	NA	400	146	1,010	539	170	1,307	702	222	1,961	1,099
	20	NA	NA	NA	NA	NA	NA	NA	NA	NA	NA	NA	185	NA	NA	264	NA	NA	376	165	977	511	190	1,269	669	246	1,916	1,050
	30	NA	NA	NA	NA	NA	NA	NA	NA	NA	NA	NA	NA	NA	NA	NA	NA	NA	327	NA	NA	468	233	1,196	623	295	1,832	984
Minimum Internal Area of Chimney (square inches)		12			19			28			38			50			63			78			95			132		
Maximum Internal Area of Chimney (square inches)		Seven times the listed appliance categorized vent area, flue collar area or draft hood outlet area.																										

For SI: 1 inch = 25.4 mm, 1 square inch = 645.16 mm², 1 foot = 304.8 mm, 1 British thermal unit per hour = 0.2931 W.

TABLE 504.2(4)
MASONRY CHIMNEY

Number of Appliances	Single
Appliance Type	Category I
Appliance Vent Connection	Single-wall metal connector

SINGLE-WALL METAL CONNECTOR DIAMETER—(D) Inches to be used with chimney areas within the size limits at bottom

APPLIANCE INPUT RATING IN THOUSANDS OF BTU/H

HEIGHT (H) (feet)	LATERAL (L) (feet)	3 FAN Min	3 FAN Max	3 NAT Max	4 FAN Min	4 FAN Max	4 NAT Max	5 FAN Min	5 FAN Max	5 NAT Max	6 FAN Min	6 FAN Max	6 NAT Max	7 FAN Min	7 FAN Max	7 NAT Max	8 FAN Min	8 FAN Max	8 NAT Max	9 FAN Min	9 FAN Max	9 NAT Max	10 FAN Min	10 FAN Max	10 NAT Max	12 FAN Min	12 FAN Max	12 NAT Max
6	2	NA	NA	28	NA	NA	52	NA	NA	86	NA	NA	130	NA	NA	180	NA	NA	247	NA	NA	319	NA	NA	400	NA	NA	580
6	5	NA	NA	25	NA	NA	48	NA	NA	81	NA	NA	116	NA	NA	164	NA	NA	230	NA	NA	297	NA	NA	375	NA	NA	560
8	2	NA	NA	29	NA	NA	55	NA	NA	93	NA	NA	145	NA	NA	197	NA	NA	265	NA	NA	349	382	725	445	549	1,021	650
8	5	NA	NA	26	NA	NA	51	NA	NA	87	NA	NA	133	NA	NA	182	NA	NA	246	NA	NA	327	NA	NA	422	673	1,003	638
8	8	NA	NA	23	NA	NA	47	NA	NA	82	NA	NA	126	NA	NA	174	NA	NA	237	NA	NA	317	NA	NA	408	747	985	621
10	2	NA	NA	31	NA	NA	61	NA	NA	102	NA	NA	161	NA	NA	220	216	518	297	271	654	387	373	808	490	536	1,142	722
10	5	NA	NA	28	NA	NA	56	NA	NA	95	NA	NA	147	NA	NA	203	NA	NA	276	334	635	364	459	789	465	657	1,121	710
10	10	NA	NA	24	NA	NA	49	NA	NA	86	NA	NA	137	NA	NA	189	NA	NA	261	NA	NA	345	547	758	441	771	1,088	665
15	2	NA	NA	35	NA	NA	67	NA	NA	113	NA	NA	178	166	473	249	211	611	335	264	776	440	362	965	560	520	1,373	840
15	5	NA	NA	32	NA	NA	61	NA	NA	106	NA	NA	163	NA	NA	230	261	591	312	325	775	414	444	942	531	637	1,348	825
15	10	NA	NA	27	NA	NA	54	NA	NA	96	NA	NA	151	NA	NA	214	NA	NA	294	392	722	392	531	907	504	749	1,309	774
15	15	NA	NA	NA	NA	NA	46	NA	NA	87	NA	NA	138	NA	NA	198	NA	NA	278	452	692	372	606	873	481	841	1,272	738
20	2	NA	NA	38	NA	NA	73	NA	NA	123	NA	NA	200	163	520	273	206	675	374	258	864	490	252	1,079	625	508	1,544	950
20	5	NA	NA	35	NA	NA	67	NA	NA	115	NA	NA	183	80	NA	252	255	655	348	317	842	461	433	1,055	594	623	1,518	930
20	10	NA	NA	NA	NA	NA	59	NA	NA	105	NA	NA	170	NA	NA	235	312	622	330	382	806	437	517	1,016	562	733	1,475	875
20	15	NA	NA	NA	NA	NA	NA	NA	NA	95	NA	NA	156	NA	NA	217	NA	NA	311	442	773	414	591	979	539	823	1,434	835
20	20	NA	NA	NA	NA	NA	NA	NA	NA	80	NA	NA	144	NA	NA	202	NA	NA	292	NA	NA	392	663	944	510	911	1,394	800

(continued)

TABLE 504.2(4)—continued
MASONRY CHIMNEY

Number of Appliances	Single
Appliance Type	Category I
Appliance Vent Connection	Single-wall metal connector

SINGLE-WALL METAL CONNECTOR DIAMETER—(D) Inches to be used with chimney areas within the size limits at bottom

APPLIANCE INPUT RATING IN THOUSANDS OF BTU/H

HEIGHT (H) (feet)	LATERAL (L) (feet)	3 FAN Min	3 FAN Max	3 NAT Max	4 FAN Min	4 FAN Max	4 NAT Max	5 FAN Min	5 FAN Max	5 NAT Max	6 FAN Min	6 FAN Max	6 NAT Max	7 FAN Min	7 FAN Max	7 NAT Max	8 FAN Min	8 FAN Max	8 NAT Max	9 FAN Min	9 FAN Max	9 NAT Max	10 FAN Min	10 FAN Max	10 NAT Max	12 FAN Min	12 FAN Max	12 NAT Max
30	2	NA	NA	41	NA	NA	81	NA	NA	136	NA	NA	215	158	578	302	200	759	420	249	982	556	340	1,237	715	489	1,789	1,110
30	5	NA	NA	NA	NA	NA	75	NA	NA	127	NA	NA	196	NA	NA	279	245	737	391	306	958	524	417	1,210	680	600	1,760	1,090
30	10	NA	NA	NA	NA	NA	66	NA	NA	113	NA	NA	182	NA	NA	260	300	703	370	370	920	496	500	1,168	644	708	1,713	1,020
30	15	NA	NA	NA	NA	NA	NA	NA	NA	105	NA	NA	168	NA	NA	240	NA	NA	349	428	884	471	572	1,128	615	798	1,668	975
30	20	NA	NA	NA	NA	NA	NA	NA	NA	88	NA	NA	155	NA	NA	223	NA	NA	327	NA	NA	445	643	1,089	585	883	1,624	932
30	30	NA	NA	NA	NA	NA	NA	NA	NA	NA	NA	NA	NA	NA	NA	182	NA	NA	281	NA	NA	408	NA	NA	544	1,055	1,539	865
50	2	NA	NA	NA	NA	NA	91	NA	NA	160	NA	NA	250	NA	NA	350	191	837	475	238	1,103	631	323	1,408	810	463	2,076	1,240
50	5	NA	NA	NA	NA	NA	NA	NA	NA	149	NA	NA	228	NA	NA	321	NA	NA	442	293	1,078	593	398	1,381	770	571	2,044	1,220
50	10	NA	NA	NA	NA	NA	NA	NA	NA	136	NA	NA	212	NA	NA	301	NA	NA	420	355	1,038	562	447	1,337	728	674	1,994	1,140
50	15	NA	NA	NA	NA	NA	NA	NA	NA	124	NA	NA	195	NA	NA	278	NA	NA	395	NA	NA	533	546	1,294	695	761	1,945	1,090
50	20	NA	NA	NA	NA	NA	NA	NA	NA	NA	NA	NA	180	NA	NA	258	NA	NA	370	NA	NA	504	616	1,251	660	844	1,898	1,040
50	30	NA	NA	NA	NA	NA	48	NA	NA	NA	NA	NA	NA	NA	NA	NA	NA	NA	318	NA	NA	458	NA	NA	610	1,009	1,805	970
Minimum Internal Area of Chimney (square inches)		12			19			28			38			50			63			78			95			132		
Maximum Internal Area of Chimney (square inches)		Seven times the listed appliance categorized vent area, flue collar area or draft hood outlet area.																										

For SI: 1 inch = 25.4 mm, 1 square inch = 645.16 mm², 1 foot = 304.8 mm, 1 British thermal unit per hour = 0.2931 W.

TABLE 504.2(5)
SINGLE-WALL METAL PIPE OR TYPE B ASBESTOS CEMENT VENT

Number of Appliances	Single
Appliance Type	Draft hood equipped
Appliance Vent Connection	Connected directly to pipe or vent

HEIGHT (H) (feet)	LATERAL (L) (feet)	VENT DIAMETER—(D) inches							
		3	4	5	6	7	8	10	12
		MAXIMUM APPLIANCE INPUT RATING IN THOUSANDS OF BTU/H							
6	0	39	70	116	170	232	312	500	750
	2	31	55	94	141	194	260	415	620
	5	28	51	88	128	177	242	390	600
8	0	42	76	126	185	252	340	542	815
	2	32	61	102	154	210	284	451	680
	5	29	56	95	141	194	264	430	648
	10	24	49	86	131	180	250	406	625
10	0	45	84	138	202	279	372	606	912
	2	35	67	111	168	233	311	505	760
	5	32	61	104	153	215	289	480	724
	10	27	54	94	143	200	274	455	700
	15	NA	46	84	130	186	258	432	666
15	0	49	91	151	223	312	420	684	1,040
	2	39	72	122	186	260	350	570	865
	5	35	67	110	170	240	325	540	825
	10	30	58	103	158	223	308	514	795
	15	NA	50	93	144	207	291	488	760
	20	NA	NA	82	132	195	273	466	726
20	0	53	101	163	252	342	470	770	1,190
	2	42	80	136	210	286	392	641	990
	5	38	74	123	192	264	364	610	945
	10	32	65	115	178	246	345	571	910
	15	NA	55	104	163	228	326	550	870
	20	NA	NA	91	149	214	306	525	832
30	0	56	108	183	276	384	529	878	1,370
	2	44	84	148	230	320	441	730	1,140
	5	NA	78	137	210	296	410	694	1,080
	10	NA	68	125	196	274	388	656	1,050
	15	NA	NA	113	177	258	366	625	1,000
	20	NA	NA	99	163	240	344	596	960
	30	NA	NA	NA	NA	192	295	540	890
50	0	NA	120	210	310	443	590	980	1,550
	2	NA	95	171	260	370	492	820	1,290
	5	NA	NA	159	234	342	474	780	1,230
	10	NA	NA	146	221	318	456	730	1,190
	15	NA	NA	NA	200	292	407	705	1,130
	20	NA	NA	NA	185	276	384	670	1,080
	30	NA	NA	NA	NA	222	330	605	1,010

For SI: 1 inch = 25.4 mm, 1 foot = 304.8 mm, 1 British thermal unit per hour = 0.2931 W.

TABLE 504.2(6)
EXTERIOR MASONRY CHIMNEY

Number of Appliances	Single
Appliance Type	NAT
Appliance Vent Connection	Type B double-wall connector

MINIMUM ALLOWABLE INPUT RATING OF SPACE-HEATING APPLIANCE IN THOUSANDS OF BTU PER HOUR									
VENT HEIGHT (feet)	Internal area of chimney (square inches)								
	12	19		28	38	50	63	78	113
37°F or Greater	Local 99% Winter Design Temperature: 37°F or Greater								
6	0	0		0	0	0	0	0	0
8	0	0		0	0	0	0	0	0
10	0	0		0	0	0	0	0	0
15	NA	0		0	0	0	0	0	0
20	NA	NA		123	190	249	184	0	0
30	NA	NA		NA	NA	NA	393	334	0
50	NA	NA		NA	NA	NA	NA	NA	579
27 to 36°F	Local 99% Winter Design Temperature: 27 to 36°F								
6	0	0		68	116	156	180	212	266
8	0	0		82	127	167	187	214	263
10	0	51		97	141	183	201	225	265
15	NA	NA		NA	NA	233	253	274	305
20	NA	NA		NA	NA	NA	307	330	362
30	NA	NA		NA	NA	NA	419	445	485
50	NA	NA		NA	NA	NA	NA	NA	763
17 to 26°F	Local 99% Winter Design Temperature: 17 to 26°F								
6	NA	NA		NA	NA	NA	215	259	349
8	NA	NA		NA	NA	197	226	264	352
10	NA	NA		NA	NA	214	245	278	358
15	NA	NA		NA	NA	NA	296	331	398
20	NA	NA		NA	NA	NA	352	387	457
30	NA	NA		NA	NA	NA	NA	507	581
50	NA	NA		NA	NA	NA	NA	NA	NA
5 to 16°F	Local 99% Winter Design Temperature: 5 to 16°F								
6	NA	NA		NA	NA	NA	NA	NA	416
8	NA	NA		NA	NA	NA	NA	312	423
10	NA	NA		NA	NA	NA	289	331	430
15	NA	NA		NA	NA	NA	NA	393	485
20	NA	NA		NA	NA	NA	NA	450	547
30	NA	NA		NA	NA	NA	NA	NA	682
50	NA	NA		NA	NA	NA	NA	NA	972
-10 to 4°F	Local 99% Winter Design Temperature: -10 to 4°F								
6	NA	NA		NA	NA	NA	NA	NA	484
8	NA	NA		NA	NA	NA	NA	NA	494
10	NA	NA		NA	NA	NA	NA	NA	513
15	NA	NA		NA	NA	NA	NA	NA	586
20	NA	NA		NA	NA	NA	NA	NA	650
30	NA	NA		NA	NA	NA	NA	NA	805
50	NA	NA		NA	NA	NA	NA	NA	1,003
-11°F or Lower	Local 99% Winter Design Temperature: -11°F or Lower								
	Not recommended for any vent configurations								

For SI: °C = (°F - 32)/1.8, 1 inch = 25.4 mm, 1 foot = 304.8 mm, 1 British thermal unit per hour = 0.2931 W.

Note: See Figure B-19 in Appendix B for a map showing local 99-percent winter design temperatures in the United States.

TABLE 504.3(1)
TYPE B DOUBLE-WALL VENT

Number of Appliances	Two or more
Appliance Type	Category I
Appliance Vent Connection	Type B double-wall connector

VENT CONNECTOR CAPACITY

VENT HEIGHT (H) (feet)	CONNECTOR RISE (R) (feet)	3 FAN Min	3 FAN Max	3 NAT Max	4 FAN Min	4 FAN Max	4 NAT Max	5 FAN Min	5 FAN Max	5 NAT Max	6 FAN Min	6 FAN Max	6 NAT Max	7 FAN Min	7 FAN Max	7 NAT Max	8 FAN Min	8 FAN Max	8 NAT Max	9 FAN Min	9 FAN Max	9 NAT Max	10 FAN Min	10 FAN Max	10 NAT Max
6	1	22	37	26	35	66	46	46	106	72	58	164	104	77	225	142	92	296	185	109	376	237	128	466	289
	2	23	41	31	37	75	55	48	121	86	60	183	124	79	253	168	95	333	220	112	424	282	131	526	345
	3	24	44	35	38	81	62	49	132	96	62	199	139	82	275	189	97	363	248	114	463	317	134	575	386
8	1	22	40	27	35	72	48	49	114	76	64	176	109	84	243	148	100	320	194	118	408	248	138	507	303
	2	23	44	32	36	80	57	51	128	90	66	195	129	86	269	175	103	356	230	121	454	294	141	564	358
	3	24	47	36	37	87	64	53	139	101	67	210	145	88	290	198	105	384	258	123	492	330	143	612	402
10	1	22	43	28	34	78	50	49	123	78	65	189	113	89	257	154	106	341	200	125	436	257	146	542	314
	2	23	47	33	36	86	59	51	136	93	67	206	134	91	282	182	109	374	238	128	479	305	149	596	372
	3	24	50	37	37	92	67	52	146	104	69	220	150	94	303	205	111	402	268	131	515	342	152	642	417
15	1	21	50	30	33	89	53	47	142	83	64	220	120	88	298	163	110	389	214	134	493	273	162	609	333
	2	22	53	35	35	96	63	49	153	99	66	235	142	91	320	193	112	419	253	137	532	323	165	658	394
	3	24	55	40	36	102	71	51	163	111	68	248	160	93	339	218	115	445	286	140	565	365	167	700	444
20	1	21	54	31	33	99	56	46	157	87	62	246	125	86	334	171	107	436	224	131	552	285	158	681	347
	2	22	57	37	34	105	66	48	167	104	64	259	149	89	354	202	110	463	265	134	587	339	161	725	414
	3	23	60	42	35	110	74	50	176	116	66	271	168	91	371	228	113	486	300	137	618	383	164	764	466
30	1	20	62	33	31	113	59	45	181	93	60	288	134	83	391	182	103	512	238	125	649	305	151	802	372
	2	21	64	39	33	118	70	47	190	110	62	299	158	85	408	215	105	535	282	129	679	360	155	840	439
	3	22	66	44	34	123	79	48	198	124	64	309	178	88	423	242	108	555	317	132	706	405	158	874	494
50	1	19	71	36	30	133	64	43	216	101	57	349	145	78	477	197	97	627	257	120	797	330	144	984	403
	2	21	73	43	32	137	76	45	223	119	59	358	172	81	490	234	100	645	306	123	820	392	148	1,014	478
	3	22	75	48	33	141	86	46	229	134	61	366	194	83	502	263	103	661	343	126	842	441	151	1,043	538
100	1	18	82	37	28	158	66	40	262	104	53	442	150	73	611	204	91	810	266	112	1,038	341	135	1,285	417
	2	19	83	44	30	161	79	42	267	123	55	447	178	75	619	242	94	822	316	115	1,054	405	139	1,306	494
	3	20	84	50	31	163	89	44	272	138	57	452	109	78	627	272	97	834	355	118	1,069	455	142	1,327	555

COMMON VENT CAPACITY

VENT HEIGHT (H) (feet)	4 FAN+FAN	4 FAN+NAT	4 NAT+NAT	5 FAN+FAN	5 FAN+NAT	5 NAT+NAT	6 FAN+FAN	6 FAN+NAT	6 NAT+NAT	7 FAN+FAN	7 FAN+NAT	7 NAT+NAT	8 FAN+FAN	8 FAN+NAT	8 NAT+NAT	9 FAN+FAN	9 FAN+NAT	9 NAT+NAT	10 FAN+FAN	10 FAN+NAT	10 NAT+NAT
6	92	81	65	140	116	103	204	161	147	309	248	200	404	314	260	547	434	335	672	520	410
8	101	90	73	155	129	114	224	178	163	339	275	223	444	348	290	602	480	378	740	577	465
10	110	97	79	169	141	124	243	194	178	367	299	242	477	377	315	649	522	405	800	627	495
15	125	112	91	195	164	144	283	228	206	427	352	280	556	444	365	753	612	465	924	733	565
20	136	123	102	215	183	160	314	255	229	475	394	310	621	499	405	842	688	523	1,035	826	640
30	152	138	118	244	210	185	361	297	266	547	459	360	720	585	470	979	808	605	1,209	975	740
50	167	153	134	279	244	214	421	353	310	641	547	423	854	706	550	1,164	977	705	1,451	1,188	860
100	175	163	NA	311	277	NA	489	421	NA	751	658	479	1,025	873	625	1,408	1,215	800	1,784	1,502	975

(continued)

 2015 INTERNATIONAL FUEL GAS CODE® COMMENTARY

TABLE 504.3(1)—continued
TYPE B DOUBLE-WALL VENT

Number of Appliances	Two or more
Appliance Type	Category I
Appliance Vent Connection	Type B double-wall connector

VENT CONNECTOR CAPACITY

VENT HEIGHT (H) (feet)	CONNECTOR RISE (R) (feet)	12 FAN Min	12 FAN Max	12 NAT Max	14 FAN Min	14 FAN Max	14 NAT Max	16 FAN Min	16 FAN Max	16 NAT Max	18 FAN Min	18 FAN Max	18 NAT Max	20 FAN Min	20 FAN Max	20 NAT Max	22 FAN Min	22 FAN Max	22 NAT Max	24 FAN Min	24 FAN Max	24 NAT Max
6	2	174	764	496	223	1,046	653	281	1,371	853	346	1,772	1,080	NA	NA	NA	NA	NA	NA	NA	NA	NA
	4	180	897	616	230	1,231	827	287	1,617	1,081	352	2,069	1,370	NA	NA	NA	NA	NA	NA	NA	NA	NA
	6	NA	NA	NA	NA	NA	NA	NA	NA	NA	NA	NA	NA	NA	NA	NA	NA	NA	NA	NA	NA	NA
8	2	186	822	516	238	1,126	696	298	1,478	910	365	1,920	1,150	NA	NA	NA	NA	NA	NA	NA	NA	NA
	4	192	952	644	244	1,307	884	305	1,719	1,150	372	2,211	1,460	471	2,737	1,800	560	3,319	2,180	662	3,957	2,590
	6	198	1,050	772	252	1,445	1,072	313	1,902	1,390	380	2,434	1,770	478	3,018	2,180	568	3,665	2,640	669	4,373	3,130
10	2	196	870	536	249	1,195	730	311	1,570	955	379	2,049	1,205	NA	NA	NA	NA	NA	NA	NA	NA	NA
	4	201	997	664	256	1,371	924	318	1,804	1,205	387	2,332	1,535	486	2,887	1,890	581	3,502	2,280	686	4,175	2,710
	6	207	1,095	792	263	1,509	1,118	325	1,989	1,455	395	2,556	1,865	494	3,169	2,290	589	3,849	2,760	694	4,593	3,270
15	2	214	967	568	272	1,334	790	336	1,760	1,030	408	2,317	1,305	NA	NA	NA	NA	NA	NA	NA	NA	NA
	4	221	1,085	712	279	1,499	1,006	344	1,978	1,320	416	2,579	1,665	523	3,197	2,060	624	3,881	2,490	734	4,631	2,960
	6	228	1,181	856	286	1,632	1,222	351	2,157	1,610	424	2,796	2,025	533	3,470	2,510	634	4,216	3,030	743	5,035	3,600
20	2	223	1,051	596	291	1,443	840	357	1,911	1,095	430	2,533	1,385	NA	NA	NA	NA	NA	NA	NA	NA	NA
	4	230	1,162	748	298	1,597	1,064	365	2,116	1,395	438	2,778	1,765	554	3,447	2,180	661	4,190	2,630	772	5,005	3,130
	6	237	1,253	900	307	1,726	1,288	373	2,287	1,695	450	2,984	2,145	567	3,708	2,650	671	4,511	3,190	785	5,392	3,790
30	2	216	1,217	632	286	1,664	910	367	2,183	1,190	461	2,891	1,540	NA	NA	NA	NA	NA	NA	NA	NA	NA
	4	223	1,316	792	294	1,802	1,160	376	2,366	1,510	474	3,110	1,920	619	3,840	2,365	728	4,861	2,860	847	5,606	3,410
	6	231	1,400	952	303	1,920	1,410	384	2,524	1,830	485	3,299	2,340	632	4,080	2,875	741	4,976	3,480	860	5,961	4,150
50	2	206	1,479	689	273	2,023	1,007	350	2,659	1,315	435	3,548	1,665	NA	NA	NA	NA	NA	NA	NA	NA	NA
	4	213	1,561	860	281	2,139	1,291	359	2,814	1,685	447	3,730	2,135	580	4,601	2,633	709	5,569	3,185	851	6,633	3,790
	6	221	1,631	1,031	290	2,242	1,575	369	2,951	2,055	461	3,893	2,605	594	4,808	3,208	724	5,826	3,885	867	6,943	4,620
100	2	192	1,923	712	254	2,644	1,050	326	3,490	1,370	402	4,707	1,740	NA	NA	NA	NA	NA	NA	NA	NA	NA
	4	200	1,984	888	263	2,731	1,346	336	3,606	1,760	414	4,842	2,220	523	5,982	2,750	639	7,254	3,330	769	8,650	3,950
	6	208	2,035	1,064	272	2,811	1,642	346	3,714	2,150	426	4,968	2,700	539	6,143	3,350	654	7,453	4,070	786	8,892	4,810

COMMON VENT CAPACITY

VENT HEIGHT (H) (feet)	12 FAN+FAN	12 FAN+NAT	12 NAT+NAT	14 FAN+FAN	14 FAN+NAT	14 NAT+NAT	16 FAN+FAN	16 FAN+NAT	16 NAT+NAT	18 FAN+FAN	18 FAN+NAT	18 NAT+NAT	20 FAN+FAN	20 FAN+NAT	20 NAT+NAT	22 FAN+FAN	22 FAN+NAT	22 NAT+NAT	24 FAN+FAN	24 FAN+NAT	24 NAT+NAT
6	900	696	588	1,284	990	815	1,735	1,336	1,065	2,253	1,732	1,345	2,838	2,180	1,660	3,488	2,677	1,970	4,206	3,226	2,390
8	994	773	652	1,423	1,103	912	1,927	1,491	1,190	2,507	1,936	1,510	3,162	2,439	1,860	3,890	2,998	2,200	4,695	3,616	2,680
10	1,076	841	712	1,542	1,200	995	2,093	1,625	1,300	2,727	2,113	1645	3,444	2,665	2,030	4,241	3,278	2,400	5,123	3,957	2,920
15	1,247	986	825	1,794	1,410	1,158	2,440	1,910	1,510	3,184	2,484	1,910	4,026	3,133	2,360	4,971	3,862	2,790	6,016	4,670	3,400
20	1,405	1,116	916	2,006	1,588	1,290	2,722	2,147	1,690	3,561	2,798	2,140	4,548	3,552	2,640	5,573	4,352	3,120	6,749	5,261	3,800
30	1,658	1,327	1,025	2,373	1,892	1,525	3,220	2,558	1,990	4,197	3,326	2,520	5,303	4,193	3,110	6,539	5,157	3,680	7,940	6,247	4,480
50	2,024	1,640	1,280	2,911	2,347	1,863	3,964	3,183	2,430	5,184	4,149	3,075	6,567	5,240	3,800	8,116	6,458	4,500	9,837	7,813	5,475
100	2,569	2,131	1,670	3,732	3,076	2,450	5,125	4,202	3,200	6,749	5,509	4,050	8,597	6,986	5,000	10,681	8,648	5,920	13,004	10,499	7,200

For SI: 1 inch = 25.4 mm, 1 foot = 304.8 mm, 1 British thermal unit per hour = 0.2931 W.

TABLE 504.3(2)
TYPE B DOUBLE-WALL VENT

Number of Appliances	Two or more
Appliance Type	Category I
Appliance Vent Connection	Single-wall metal connector

VENT CONNECTOR CAPACITY

VENT HEIGHT (H) (feet)	CONNECTOR RISE (R) (feet)	SINGLE-WALL METAL VENT CONNECTOR DIAMETER—(D) inches																						
		3			4			5			6			7			8			9			10		
		APPLIANCE INPUT RATING LIMITS IN THOUSANDS OF BTU/H																						
		FAN		NAT	FAN		NAT	FAN		NAT	FAN		NAT	FAN		NAT	FAN		NAT	FAN		NAT	FAN		NAT
		Min	Max	Max	Min	Max	Max	Min	Max	Max	Min	Max	Max	Min	Max	Max	Min	Max	Max	Min	Max	Max	Min	Max	Max
6	1	NA	NA	26	NA	NA	46	NA	NA	71	NA	NA	102	207	223	140	262	293	183	325	373	234	447	463	286
	2	NA	NA	31	NA	NA	55	NA	NA	85	168	182	123	215	251	167	271	331	219	334	422	281	458	524	344
	3	NA	NA	34	NA	NA	62	121	131	95	175	198	138	222	273	188	279	361	247	344	462	316	468	574	385
8	1	NA	NA	27	NA	NA	48	NA	NA	75	NA	NA	106	226	240	145	285	316	191	352	403	244	481	502	299
	2	NA	NA	32	NA	NA	57	125	126	89	184	193	127	234	266	173	293	353	228	360	450	292	492	560	355
	3	NA	NA	35	NA	NA	64	130	138	100	191	208	144	241	287	197	302	381	256	370	489	328	501	609	400
10	1	NA	NA	28	NA	NA	50	119	121	77	182	186	110	240	253	150	302	335	196	372	429	252	506	534	308
	2	NA	NA	33	84	85	59	124	134	91	189	203	132	248	278	183	311	369	235	381	473	302	517	589	368
	3	NA	NA	36	89	91	67	129	144	102	197	217	148	257	299	203	320	398	265	391	511	339	528	637	413
15	1	NA	NA	29	79	87	52	116	138	81	177	214	116	238	291	158	312	380	208	397	482	266	556	596	324
	2	NA	NA	34	83	94	62	121	150	97	185	230	138	246	314	189	321	411	248	407	522	317	568	646	387
	3	NA	NA	39	87	100	70	127	160	109	193	243	157	255	333	215	331	438	281	418	557	360	579	690	437
20	1	49	56	30	78	97	54	115	152	84	175	238	120	233	325	165	306	425	217	390	538	276	546	664	336
	2	52	59	36	82	103	64	120	163	101	182	252	144	243	346	197	317	453	259	400	574	331	558	709	403
	3	55	62	40	87	107	72	125	172	113	190	264	164	252	363	223	326	476	294	412	607	375	570	750	457
30	1	47	60	31	77	110	57	112	175	89	169	278	129	226	380	175	296	497	230	378	630	294	528	779	358
	2	51	62	37	81	115	67	117	185	106	177	290	152	236	397	208	307	521	274	389	662	349	541	819	425
	3	54	64	42	85	119	76	122	193	120	185	300	172	244	412	235	316	542	309	400	690	394	555	855	482
50	1	46	69	34	75	128	60	109	207	96	162	336	137	217	460	188	284	604	245	364	768	314	507	951	384
	2	49	71	40	79	132	72	114	215	113	170	345	164	226	473	223	294	623	293	376	793	375	520	983	458
	3	52	72	45	83	136	82	119	221	123	178	353	186	235	486	252	304	640	331	387	816	423	535	1,013	518
100	1	45	79	34	71	150	61	104	249	98	153	424	140	205	585	192	269	774	249	345	993	321	476	1,236	393
	2	48	80	41	75	153	73	110	255	115	160	428	167	212	593	228	279	788	299	358	1,011	383	490	1,259	469
	3	51	81	46	79	157	85	114	260	129	168	433	190	222	603	256	289	801	339	368	1,027	431	506	1,280	527

COMMON VENT CAPACITY

VENT HEIGHT (H) (feet)	TYPE B DOUBLE-WALL COMMON VENT DIAMETER—(D) inches																				
	4			5			6			7			8			9			10		
	COMBINED APPLIANCE INPUT RATING IN THOUSANDS OF BTU/H																				
	FAN +FAN	FAN +NAT	NAT +NAT	FAN +FAN	FAN +NAT	NAT +NAT	FAN +FAN	FAN +NAT	NAT +NAT	FAN +FAN	FAN +NAT	NAT +NAT	FAN +FAN	FAN +NAT	NAT +NAT	FAN +FAN	FAN +NAT	NAT +NAT	FAN +FAN	FAN +NAT	NAT +NAT
6	NA	78	64	NA	113	99	200	158	144	304	244	196	398	310	257	541	429	332	665	515	407
8	NA	87	71	NA	126	111	218	173	159	331	269	218	436	342	285	592	473	373	730	569	460
10	NA	94	76	163	137	120	237	189	174	357	292	236	467	369	309	638	512	398	787	617	487
15	121	108	88	189	159	140	275	221	200	416	343	274	544	434	357	738	599	456	905	718	553
20	131	118	98	208	177	156	305	247	223	463	383	302	606	487	395	824	673	512	1,013	808	626
30	145	132	113	236	202	180	350	286	257	533	446	349	703	570	459	958	790	593	1,183	952	723
50	159	145	128	268	233	208	406	337	296	622	529	410	833	686	535	1,139	954	689	1,418	1,157	838
100	166	153	NA	297	263	NA	469	398	NA	726	633	464	999	846	606	1,378	1,185	780	1,741	1,459	948

For SI: 1 inch = 25.4 mm, 1 foot = 304.8 mm, 1 British thermal unit per hour = 0.2931 W.

TABLE 504.3(3)
MASONRY CHIMNEY

Number of Appliances	Two or more
Appliance Type	Category I
Appliance Vent Connection	Type B double-wall connector

VENT CONNECTOR CAPACITY

VENT HEIGHT (H) (feet)	CONNEC-TOR RISE (R) (feet)	3 FAN Min	3 FAN Max	3 NAT Max	4 FAN Min	4 FAN Max	4 NAT Max	5 FAN Min	5 FAN Max	5 NAT Max	6 FAN Min	6 FAN Max	6 NAT Max	7 FAN Min	7 FAN Max	7 NAT Max	8 FAN Min	8 FAN Max	8 NAT Max	9 FAN Min	9 FAN Max	9 NAT Max	10 FAN Min	10 FAN Max	10 NAT Max
6	1	24	33	21	39	62	40	52	106	67	65	194	101	87	274	141	104	370	201	124	479	253	145	599	319
	2	26	43	28	41	79	52	53	133	85	67	230	124	89	324	173	107	436	232	127	562	300	148	694	378
	3	27	49	34	42	92	61	55	155	97	69	262	143	91	369	203	109	491	270	129	633	349	151	795	439
8	1	24	39	22	39	72	41	55	117	69	71	213	105	94	304	148	113	414	210	134	539	267	156	682	335
	2	26	47	29	40	87	53	57	140	86	73	246	127	97	350	179	116	473	240	137	615	311	160	776	394
	3	27	52	34	42	97	62	59	159	98	75	269	145	99	383	206	119	517	276	139	672	358	163	848	452
10	1	24	42	22	38	80	42	55	130	71	74	232	108	101	324	153	120	444	216	142	582	277	165	739	348
	2	26	50	29	40	93	54	57	153	87	76	261	129	103	366	184	123	498	247	145	652	321	168	825	407
	3	27	55	35	41	105	63	58	170	100	78	284	148	106	397	209	126	540	281	147	705	366	171	893	463
15	1	24	48	23	38	93	44	54	154	74	72	277	114	100	384	164	125	511	229	153	658	297	184	824	375
	2	25	55	31	39	105	55	56	174	89	74	299	134	103	419	192	128	558	260	156	718	339	187	900	432
	3	26	59	35	41	115	64	57	189	102	76	319	153	105	448	215	131	597	292	159	760	382	190	960	486
20	1	24	52	24	37	102	46	53	172	77	71	313	119	98	437	173	123	584	239	150	752	312	180	943	397
	2	25	58	31	39	114	56	55	190	91	73	335	138	101	467	199	126	625	270	153	805	354	184	1,011	452
	3	26	63	35	40	123	65	57	204	104	75	353	157	104	493	222	129	661	301	156	851	396	187	1,067	505
30	1	24	54	25	37	111	48	52	192	82	69	357	127	96	504	187	119	680	255	145	883	337	175	1,115	432
	2	25	60	32	38	122	58	54	208	95	72	376	145	99	531	209	122	715	287	149	928	378	179	1,171	484
	3	26	64	36	40	131	66	56	221	107	74	392	163	101	554	233	125	746	317	152	968	418	182	1,220	535
50	1	23	51	25	36	116	51	51	209	89	67	405	143	92	582	213	115	798	294	140	1,049	392	168	1,334	506
	2	24	59	32	37	127	61	53	225	102	70	421	161	95	604	235	118	827	326	143	1,085	433	172	1,379	558
	3	26	64	36	39	135	69	55	237	115	72	435	80	98	624	260	121	854	357	147	1,118	474	176	1,421	611
100	1	23	46	24	35	108	50	49	208	92	65	428	155	88	640	237	109	907	334	134	1,222	454	161	1,589	596
	2	24	53	31	37	120	60	51	224	105	67	444	174	92	660	260	113	933	368	138	1,253	497	165	1,626	651
	3	25	59	35	38	130	68	53	237	118	69	458	193	94	679	285	116	956	399	141	1,282	540	169	1,661	705

COMMON VENT CAPACITY

VENT HEIGHT (H) (feet)	12 FAN+FAN	12 FAN+NAT	12 NAT+NAT	19 FAN+FAN	19 FAN+NAT	19 NAT+NAT	28 FAN+FAN	28 FAN+NAT	28 NAT+NAT	38 FAN+FAN	38 FAN+NAT	38 NAT+NAT	50 FAN+FAN	50 FAN+NAT	50 NAT+NAT	63 FAN+FAN	63 FAN+NAT	63 NAT+NAT	78 FAN+FAN	78 FAN+NAT	78 NAT+NAT	113 FAN+FAN	113 FAN+NAT	113 NAT+NAT
6	NA	74	25	NA	119	46	NA	178	71	NA	257	103	NA	351	143	NA	458	188	NA	582	246	1,041	853	NA
8	NA	80	28	NA	130	53	NA	193	82	NA	279	119	NA	384	163	NA	501	218	724	636	278	1,144	937	408
10	NA	84	31	NA	138	56	NA	207	90	NA	299	131	NA	409	177	606	538	236	776	686	302	1,226	1,010	454
15	NA	NA	36	NA	152	67	NA	233	106	NA	334	152	523	467	212	682	611	283	874	781	365	1,374	1,156	546
20	NA	NA	41	NA	NA	75	NA	250	122	NA	368	172	565	508	243	742	668	325	955	858	419	1,513	1,286	648
30	NA	NA	NA	NA	NA	NA	NA	270	137	NA	404	198	615	564	278	816	747	381	1,062	969	496	1,702	1,473	749
50	NA	NA	NA	NA	NA	NA	NA	NA	NA	NA	NA	NA	NA	620	328	879	831	461	1,165	1,089	606	1,905	1,692	922
100	NA	NA	NA	NA	NA	NA	NA	NA	NA	NA	NA	NA	NA	348	NA	NA	499	NA	NA	NA	669	2,053	1,921	1,058

For SI:1 inch = 25.4 mm, 1 square inch = 645.16 mm², 1 foot = 304.8 mm, 1 British thermal unit per hour = 0.2931 W.

TABLE 504.3(4)
MASONRY CHIMNEY

Number of Appliances	Two or more
Appliance Type	Category I
Appliance Vent Connection	Single-wall metal connector

VENT CONNECTOR CAPACITY

VENT HEIGHT (H) (feet)	CONNECTOR RISE (R) (feet)	3 FAN Min	3 FAN Max	3 NAT Max	4 FAN Min	4 FAN Max	4 NAT Max	5 FAN Min	5 FAN Max	5 NAT Max	6 FAN Min	6 FAN Max	6 NAT Max	7 FAN Min	7 FAN Max	7 NAT Max	8 FAN Min	8 FAN Max	8 NAT Max	9 FAN Min	9 FAN Max	9 NAT Max	10 FAN Min	10 FAN Max	10 NAT Max
6	1	NA	NA	21	NA	NA	39	NA	NA	66	179	191	100	231	271	140	292	366	200	362	474	252	499	594	316
	2	NA	NA	28	NA	NA	52	NA	NA	84	186	227	123	239	321	172	301	432	231	373	557	299	509	696	376
	3	NA	NA	34	NA	NA	61	134	153	97	193	258	142	247	365	202	309	491	269	381	634	348	519	793	437
8	1	NA	NA	21	NA	NA	40	NA	NA	68	195	208	103	250	298	146	313	407	207	387	530	263	529	672	331
	2	NA	NA	28	NA	NA	52	137	139	85	202	240	125	258	343	177	323	465	238	397	607	309	540	766	391
	3	NA	NA	34	NA	NA	62	143	156	98	210	264	145	266	376	205	332	509	274	407	663	356	551	838	450
10	1	NA	NA	22	NA	NA	41	130	151	70	202	225	106	267	316	151	333	434	213	410	571	273	558	727	343
	2	NA	NA	29	NA	NA	53	136	150	86	210	255	128	276	358	181	343	489	244	420	640	317	569	813	403
	3	NA	NA	34	97	102	62	143	166	99	217	277	147	284	389	207	352	530	279	430	694	363	580	880	459
15	1	NA	NA	23	NA	NA	43	129	151	73	199	271	112	268	376	161	349	502	225	445	646	291	623	808	366
	2	NA	NA	30	92	103	54	135	170	88	207	295	132	277	411	189	359	548	256	456	706	334	634	884	424
	3	NA	NA	34	96	112	63	141	185	101	215	315	151	286	439	213	368	586	289	466	755	378	646	945	479
20	1	NA	NA	23	87	99	45	128	167	76	197	303	117	265	425	169	345	569	235	439	734	306	614	921	347
	2	NA	NA	30	91	111	55	134	185	90	205	325	136	274	455	195	355	610	266	450	787	348	627	986	443
	3	NA	NA	35	96	119	64	140	199	103	213	343	154	282	481	219	365	644	298	461	831	391	639	1,042	496
30	1	NA	NA	24	86	108	47	126	187	80	193	347	124	259	492	183	338	665	250	430	864	330	600	1,089	421
	2	NA	NA	31	91	119	57	132	203	93	201	366	142	269	518	205	348	699	282	442	908	372	613	1,145	473
	3	NA	NA	35	95	127	65	138	216	105	209	381	160	277	540	229	358	729	312	452	946	412	626	1,193	524
50	1	NA	NA	24	85	113	50	124	204	87	188	392	139	252	567	208	328	778	287	417	1,022	383	582	1,302	492
	2	NA	NA	31	89	123	60	130	218	100	196	408	158	262	588	230	339	806	320	429	1,058	425	596	1,346	545
	3	NA	NA	35	94	131	68	136	231	112	205	422	176	271	607	255	349	831	351	440	1,090	466	610	1,386	597
100	1	NA	NA	23	84	104	49	122	200	89	182	410	151	243	617	232	315	875	328	402	1,181	444	560	1,537	580
	2	NA	NA	30	88	115	59	127	215	102	190	425	169	253	636	254	326	899	361	415	1,210	488	575	1,570	634
	3	NA	NA	34	93	124	67	133	228	115	199	438	188	262	654	279	337	921	392	427	1,238	529	589	1,604	687

COMMON VENT CAPACITY

VENT HEIGHT (H) (feet)	12 FAN+FAN	12 FAN+NAT	12 NAT+NAT	19 FAN+FAN	19 FAN+NAT	19 NAT+NAT	28 FAN+FAN	28 FAN+NAT	28 NAT+NAT	38 FAN+FAN	38 FAN+NAT	38 NAT+NAT	50 FAN+FAN	50 FAN+NAT	50 NAT+NAT	63 FAN+FAN	63 FAN+NAT	63 NAT+NAT	78 FAN+FAN	78 FAN+NAT	78 NAT+NAT	113 FAN+FAN	113 FAN+NAT	113 NAT+NAT
6	NA	NA	25	NA	118	45	NA	176	71	NA	255	102	NA	348	142	NA	455	187	NA	579	245	NA	846	NA
8	NA	NA	28	NA	128	52	NA	190	81	NA	276	118	NA	380	162	NA	497	217	NA	633	277	1,136	928	405
10	NA	NA	31	NA	136	56	NA	205	89	NA	295	129	NA	405	175	NA	532	234	171	680	300	1,216	1,000	450
15	NA	NA	36	NA	NA	66	NA	230	105	NA	335	150	NA	400	210	677	602	280	866	772	360	1,359	1,139	540
20	NA	NA	NA	NA	NA	74	NA	247	120	NA	362	170	NA	503	240	765	661	321	947	849	415	1,495	1,264	640
30	NA	NA	NA	NA	NA	NA	NA	NA	135	NA	398	195	NA	558	275	808	739	377	1,052	957	490	1,682	1,447	740
50	NA	NA	NA	NA	NA	NA	NA	NA	NA	NA	NA	NA	NA	612	325	NA	821	456	1,152	1,076	600	1,879	1,672	910
100	NA	NA	NA	NA	NA	NA	NA	NA	NA	NA	NA	NA	NA	NA	NA	NA	NA	494	NA	NA	663	2,006	1,885	1,046

For SI: 1 inch = 25.4 mm, 1 square inch = 645.16 mm^2, 1 foot = 304.8 mm, 1 British thermal unit per hour = 0.2931 W.

TABLE 504.3(5)
SINGLE-WALL METAL PIPE OR TYPE ASBESTOS CEMENT VENT

Number of Appliances	Two or more
Appliance Type	Draft hood-equipped
Appliance Vent Connection	Direct to pipe or vent

VENT CONNECTOR CAPACITY

TOTAL VENT HEIGHT (H) (feet)	CONNECTOR RISE (R) (feet)	VENT CONNECTOR DIAMETER—(D) inches					
		3	4	5	6	7	8
		MAXIMUM APPLIANCE INPUT RATING IN THOUSANDS OF BTU/H					
6-8	1	21	40	68	102	146	205
	2	28	53	86	124	178	235
	3	34	61	98	147	204	275
15	1	23	44	77	117	179	240
	2	30	56	92	134	194	265
	3	35	64	102	155	216	298
30 and up	1	25	49	84	129	190	270
	2	31	58	97	145	211	295
	3	36	68	107	164	232	321

COMMON VENT CAPACITY

TOTAL VENT HEIGHT (H) (feet)	COMMON VENT DIAMETER—(D) inches						
	4	5	6	7	8	10	12
	COMBINED APPLIANCE INPUT RATING IN THOUSANDS OF BTU/H						
6	48	78	111	155	205	320	NA
8	55	89	128	175	234	365	505
10	59	95	136	190	250	395	560
15	71	115	168	228	305	480	690
20	80	129	186	260	340	550	790
30	NA	147	215	300	400	650	940
50	NA	NA	NA	360	490	810	1,190

For SI: 1 inch = 25.4 mm, 1 foot = 304.8 mm, 1 British thermal unit per hour = 0.2931 W.

Number of Appliances	Two or more
Appliance Type	NAT + NAT
Appliance Vent Connection	Type B double-wall connector

TABLE 504.3(6a)
EXTERIOR MASONRY CHIMNEY

**Combined Appliance Maximum
Input Rating in Thousands of Btu per Hour**

VENT HEIGHT (feet)	INTERNAL AREA OF CHIMNEY (square inches)							
	12	19	28	38	50	63	78	113
6	25	46	71	103	143	188	246	NA
8	28	53	82	119	163	218	278	408
10	31	56	90	131	177	236	302	454
15	NA	67	106	152	212	283	365	546
20	NA	NA	NA	NA	NA	325	419	648
30	NA	NA	NA	NA	NA	NA	496	749
50	NA	NA	NA	NA	NA	NA	NA	922
100	NA	NA	NA	NA	NA	NA	NA	NA

Number of Appliances	Two or more
Appliance Type	NAT + NAT
Appliance Vent Connection	Type B double-wall connector

TABLE 504.3(6b)
EXTERIOR MASONRY CHIMNEY

**Minimum Allowable Input Rating of
Space-heating Appliance in Thousands of Btu per Hour**

VENT HEIGHT (feet)	INTERNAL AREA OF CHIMNEY (square inches)							
	12	19	28	38	50	63	78	113
37°F or Greater	Local 99% Winter Design Temperature: 37°F or Greater							
6	0	0	0	0	0	0	0	NA
8	0	0	0	0	0	0	0	0
10	0	0	0	0	0	0	0	0
15	NA	0	0	0	0	0	0	0
20	NA	NA	NA	NA	NA	184	0	0
30	NA	NA	NA	NA	NA	393	334	0
50	NA	NA	NA	NA	NA	NA	NA	579
100	NA	NA	NA	NA	NA	NA	NA	NA
27 to 36°F	Local 99% Winter Design Temperature: 27 to 36°F							
6	0	0	68	NA	NA	180	212	NA
8	0	0	82	NA	NA	187	214	263
10	0	51	NA	NA	NA	201	225	265
15	NA	NA	NA	NA	NA	253	274	305
20	NA	NA	NA	NA	NA	307	330	362
30	NA	NA	NA	NA	NA	NA	445	485
50	NA	NA	NA	NA	NA	NA	NA	763
100	NA	NA	NA	NA	NA	NA	NA	NA

(continued)

TABLE 504.3(6b)
EXTERIOR MASONRY CHIMNEY—continued

**Minimum Allowable Input Rating of
Space-heating Appliance in Thousands of Btu per Hour**

VENT HEIGHT (feet)	INTERNAL AREA OF CHIMNEY (square inches)							
	12	19	28	38	50	63	78	113
17 to 26°F	Local 99% Winter Design Temperature: 17 to 26°F							
6	NA	NA	NA	NA	NA	NA	NA	NA
8	NA	NA	NA	NA	NA	NA	264	352
10	NA	NA	NA	NA	NA	NA	278	358
15	NA	NA	NA	NA	NA	NA	331	398
20	NA	NA	NA	NA	NA	NA	387	457
30	NA	NA	NA	NA	NA	NA	NA	581
50	NA	NA	NA	NA	NA	NA	NA	862
100	NA	NA	NA	NA	NA	NA	NA	NA
5 to 16°F	Local 99% Winter Design Temperature: 5 to 16°F							
6	NA	NA	NA	NA	NA	NA	NA	NA
8	NA	NA	NA	NA	NA	NA	NA	NA
10	NA	NA	NA	NA	NA	NA	NA	430
15	NA	NA	NA	NA	NA	NA	NA	485
20	NA	NA	NA	NA	NA	NA	NA	547
30	NA	NA	NA	NA	NA	NA	NA	682
50	NA	NA	NA	NA	NA	NA	NA	NA
100	NA	NA	NA	NA	NA	NA	NA	NA
4°F or Lower	Local 99% Winter Design Temperature: 4°F or Lower							
	Not recommended for any vent configurations							

For SI: °C = (°F - 32)/1.8, 1 inch = 25.4 mm, 1 square inch = 645.16 mm², 1 foot = 304.8 mm, 1 British thermal unit per hour = 0.2931 W.

Note: See Figure B-19 in Appendix B for a map showing local 99-percent winter design temperatures in the United States.

TABLE 504.3(7a)
EXTERIOR MASONRY CHIMNEY

Number of Appliances	Two or more
Appliance Type	FAN + NAT
Appliance Vent Connection	Type B double-wall connector

Combined Appliance Maximum Input Rating in Thousands of Btu per Hour

VENT HEIGHT (feet)	INTERNAL AREA OF CHIMNEY (square inches)							
	12	19	28	38	50	63	78	113
6	74	119	178	257	351	458	582	853
8	80	130	193	279	384	501	636	937
10	84	138	207	299	409	538	686	1,010
15	NA	152	233	334	467	611	781	1,156
20	NA	NA	250	368	508	668	858	1,286
30	NA	NA	NA	404	564	747	969	1,473
50	NA	NA	NA	NA	NA	831	1,089	1,692
100	NA	NA	NA	NA	NA	NA	NA	1,921

Number of Appliances	Two or more
Appliance Type	FAN + NAT
Appliance Vent Connection	Type B double-wall connector

TABLE 504.3(7b)
EXTERIOR MASONRY CHIMNEY

Minimum Allowable Input Rating of Space-heating Appliance in Thousands of Btu per Hour

VENT HEIGHT (feet)	INTERNAL AREA OF CHIMNEY (square inches)							
	12	19	28	38	50	63	78	113
37°F or Greater	Local 99% Winter Design Temperature: 37°F or Greater							
6	0	0	0	0	0	0	0	0
8	0	0	0	0	0	0	0	0
10	0	0	0	0	0	0	0	0
15	NA	0	0	0	0	0	0	0
20	NA	NA	123	190	249	184	0	0
30	NA	NA	NA	334	398	393	334	0
50	NA	NA	NA	NA	NA	714	707	579
100	NA	NA	NA	NA	NA	NA	NA	1,600
27 to 36°F	Local 99% Winter Design Temperature: 27 to 36°F							
6	0	0	68	116	156	180	212	266
8	0	0	82	127	167	187	214	263
10	0	51	97	141	183	201	225	265
15	NA	111	142	183	233	253	274	305
20	NA	NA	187	230	284	307	330	362
30	NA	NA	NA	330	319	419	445	485
50	NA	NA	NA	NA	NA	672	705	763
100	NA	NA	NA	NA	NA	NA	NA	1,554

TABLE 504.3(7b)
EXTERIOR MASONRY CHIMNEY—continued

Minimum Allowable Input Rating of Space-heating Appliance in Thousands of Btu per Hour

VENT HEIGHT (feet)	INTERNAL AREA OF CHIMNEY (square inches)							
	12	19	28	38	50	63	78	113
17 to 26°F	Local 99% Winter Design Temperature: 17 to 26°F							
6	0	55	99	141	182	215	259	349
8	52	74	111	154	197	226	264	352
10	NA	90	125	169	214	245	278	358
15	NA	NA	167	212	263	296	331	398
20	NA	NA	212	258	316	352	387	457
30	NA	NA	NA	362	429	470	507	581
50	NA	NA	NA	NA	NA	723	766	862
100	NA	NA	NA	NA	NA	NA	NA	1,669
5 to 16°F	Local 99% Winter Design Temperature: 5 to 16°F							
6	NA	78	121	166	214	252	301	416
8	NA	94	135	182	230	269	312	423
10	NA	111	149	198	250	289	331	430
15	NA	NA	193	247	305	346	393	485
20	NA	NA	NA	293	360	408	450	547
30	NA	NA	NA	377	450	531	580	682
50	NA	NA	NA	NA	NA	797	853	972
100	NA	NA	NA	NA	NA	NA	NA	1,833
-10 to 4°F	Local 99% Winter Design Temperature: -10 to 4°F							
6	NA	NA	145	196	249	296	349	484
8	NA	NA	159	213	269	320	371	494
10	NA	NA	175	231	292	339	397	513
15	NA	NA	NA	283	351	404	457	586
20	NA	NA	NA	333	408	468	528	650
30	NA	NA	NA	NA	NA	603	667	805
50	NA	NA	NA	NA	NA	NA	955	1,003
100	NA	NA	NA	NA	NA	NA	NA	NA
-11°F or Lower	Local 99% Winter Design Temperature: -11°F or Lower							
	Not recommended for any vent configurations							

For SI: °C = (°F - 32)/1.8, 1 inch = 25.4 mm, 1 square inch = 645.16 mm², 1 foot = 304.8 mm, 1 British thermal unit per hour = 0.2931 W.

Note: See Figure B-19 in Appendix B for a map showing local 99-percent winter design temperatures in the United States.

(continued)

Chapter 6:
Specific Appliances

General Comments

Chapter 6 regulates the design and installation of specifically named gas-fired appliances such as furnaces, boilers, water heaters, heaters, cooking and lighting appliances and clothes dryers. This chapter lists all of the appliances that the code regulates within its scope. The chapters of the code, with the exceptions of Chapter 3 and this chapter, are dedicated to single subjects. This chapter, however, is a collection of requirements for various appliances and equipment. The only commonality in this chapter is that all subjects use fuel gas to perform some task or function. Because none of the subjects in this chapter have the volume of text to warrant an individual chapter, they have been combined into a single chapter.

Purpose

This chapter contains requirements for specific appliances and their installation. In addition to being listed and labeled, many appliances have special installation or location requirements. Unlike Chapter 3, titled General Regulations, this chapter contains and organizes the specific regulations applicable to certain appliances.

SECTION 601 (IFGC)
GENERAL

601.1 Scope. This chapter shall govern the approval, design, installation, construction, maintenance, *alteration* and repair of the appliances and *equipment* specifically identified herein.

❖ This chapter regulates all aspects of appliances and equipment to the extent found in each of Sections 602 through 636.

SECTION 602 (IFGC)
DECORATIVE APPLIANCES
FOR INSTALLATION IN FIREPLACES

602.1 General. Decorative appliances for installation in *approved* solid fuel-burning fireplaces shall be tested in accordance with ANSI Z21.60 and shall be installed in accordance with the manufacturer's instructions. Manually lighted natural gas decorative appliances shall be tested in accordance with ANSI Z21.84.

❖ These appliances include gas log sets that are designed to simulate wood fires (see Commentary Figure 602.1). Sections 301 through 310 address the requirements for testing, labeling and installing mechanical equipment and appliances. The gas-burning appliance must be tested to the standard or standards appropriate for the equipment. The testing agency is responsible for determining the standard to be used to test the equipment. In the case of decorative gas log sets, ANSI Z21.60 or Z21.84 is the applicable test standard.

Labeling is the code official's assurance that the subject product is a representative duplication of the product that the testing agency tested in the laboratory. The label indicates that an independent agency has conducted inspections at the plant to verify that all units conform to the specifications that the quality control manual sets forth for fabricating the gas appliances. Information that must be contained on the label is described in Section 301.5 and includes the manufacturer's identification, the third-party inspection agency's identification, the model number, the serial number, the input ratings and the type of fuel the appliance is designed to burn.

The code requires that the appliance be installed in accordance with the manufacturer's instructions. This requirement is also linked to the laboratory testing of the appliance because the laboratory used the same instructions to install the prototype appliance being tested. When the appliance has been tested and evaluated for code compliance and judged to meet the performance and construction requirements of the applicable standard, the installation instructions

Figure 602.1
VENTED GAS LOG SET
(Photo by D.F. Noyes Studio, courtesy of DESA International)

become an integral part of the labeling requirements and must be strictly adhered to.

The intent of this section is to regulate gas-burning appliances that are accessory to, and designed for installation in, vented solid-fuel-burning fireplaces. Gas-fired decorative log sets and log lighters are examples of accessory appliances that are designed for installation in solid-fuel-burning fireplaces. Gas log sets provide some radiant heat; however, their primary function is to create an aesthetically pleasing simulation of a wood log fire. This section addresses vented appliances and does not address "unvented gas log sets" (room heaters).

Decorative gas-burning appliances, such as some gas log sets, are designed to simulate wood fires by intentionally causing incomplete combustion as necessary to yield yellow or yellow-tipped flames. This incomplete combustion results in an increase in the amount of carbon monoxide produced. The highly toxic carbon monoxide is odorless and colorless, and the accompanying products of combustion produced by a gas-burning appliance do not have a strong odor or an odor that is readily recognized by an untrained person.

If a decorative gas appliance were operated with the fireplace damper closed, the carbon monoxide levels could be dangerously high before the building occupants became aware of the hazard. To prevent harm and the possible asphyxiation of the occupants, it is imperative that the fireplace damper be open whenever the appliance is burning. The manufacturer's installation instructions will specify the minimum free area of damper opening required to vent the appliance combustion products. The damper area is proportional to the appliance's input rating. The fireplace damper plate must be removed or permanently fixed in a position that provides the opening area required by the appliance manufacturer's installation instructions.

602.2 Flame safeguard device. Decorative appliances for installation in *approved* solid fuel-burning fireplaces, with the exception of those tested in accordance with ANSI Z21.84, shall utilize a direct ignition device, an ignitor or a pilot flame to ignite the fuel at the main burner, and shall be equipped with a flame safeguard device. The flame safeguard device shall automatically shut off the fuel supply to a main burner or group of burners when the means of ignition of such burners becomes inoperative.

❖ To eliminate the hazards associated with manually igniting a decorative gas appliance, the code requires that the main burner be ignited by a supervised means of direct ignition, such as a standing pilot. The exception to this requirement is for manually lighted appliances listed to ANSI Z21.84. The use of automatic ignition controls prevents the operator of the appliance from being exposed to the potential hazard of trying to ignite the main burner with a match or other manual lighting device. The flame safeguard device consists of a control valve assembly with an integral means of pilot or igniter supervision and is

designed to prevent or shut off the flow of gas to the appliance main burner in the event that the source of ignition is extinguished or otherwise fails.

The typical flame safeguard device used with gas log set appliances is a combination manual control valve, pressure regulator, pilot feed and magnetic pilot safety mechanism with a thermocouple generator. If the pilot flame is extinguished, the drop in thermocouple output voltage will cause the control valve to "lock out" in the closed position, thereby preventing the flow of gas to the main burner and the pilot burner. It is not the intent of this section to require flame safeguard devices for manually operated log lighter appliances used to kindle wood fires.

A flame safeguard, also known as a safety shutoff device, functions to automatically shut off the fuel supply to the main burner or burners of an appliance when the source of ignition becomes inoperative. Gas-burning equipment may operate using standing pilot ignition, which is a small flame kept constantly burning for the sole purpose of igniting the main burner when the thermostat calls for heat or a manual gas valve is opened by the operator. If the pilot is not lit, for whatever reason, the flame safeguard will automatically activate to prevent the flow of fuel to the main burner or burners of the appliance. Appliances that use automatic sparking devices or other means for ignition are subject to the same requirements for providing flame safeguards, and the method of operation of the flame safeguard will differ depending on the type of ignition system in use.

Automatically operated gas-burning equipment and appliances are equipped with some means of monitoring the means of ignition or the main burner flame because fuel introduced into a combustion chamber and not ignited and burned could cause a dangerous condition.

The most commonly used type of flame safeguard device consists of a thermocouple and an electromechanical device that function together to supervise a standing pilot flame. In the event of pilot failure, the thermocouple and flame safeguard device will function to prevent the flow of fuel to the main burners, and if the safeguard control is of the 100-percent shutoff type, it will also shut off the flow of fuel to the pilot burner.

Modern appliances seldom have standing pilots and instead use electric-spark-ignited intermittent pilots or direct ignition devices such as hot surface igniters and glow coils. The methods of detecting the presence of a pilot flame include thermocouples, millivolt (power pile) generators, liquid-filled capillary tubes, bimetal mechanisms and flame rectification circuits. Where only the main burner is monitored to verify ignition and continued combustion, the methods of detection include flame rectification, infrared detectors, ultraviolet detectors and bimetal stack sensors. The majority of today's appliances rely on electronic flame rectification or radiant energy detectors (infrared or ultraviolet) as the means of proving ignition of a pilot burner or the main burner(s). The flame rectification method of sensing the presence of flames is very common today and involves two electrodes: a flame rod and the burner itself. An ac voltage is applied between the flame rod and the burner. If a flame is present and impinges on both of these electrodes, a few microamps of current will flow through the ionized gases in the flame, but almost entirely in one direction because of the relative size of the surface areas of the two electrodes. This unidirectional flow rectifies the ac current to half-wave dc current, which is then recognized by electronic circuitry, thereby proving the presence of a flame. Manually lighted natural gas decorative appliances tested and listed to ANSI Z21.84 do not have to be equipped with a flame safeguard device.

Flame safeguard devices are commonly integral with a combination gas control. These controls typically incorporate multiple solenoid valves, a manual valve, an appliance gas-pressure regulator, a pressure operated valve and flame safeguard devices all in one device body. It is also common for flame safeguard devices to be incorporated into an electronic integrated circuit control module that is separate from the combination gas control. For example, spark ignition and hot surface ignition systems incorporate the means of ignition and ignition supervision in an electronic control module that may also contain components and circuitry for control of combustion fans, furnace blowers, combustion chamber purging and monitoring of multiple system pressure and temperature sensors located throughout an appliance. Commentary Figure 602.2(1) illustrates a type of combination gas control designed for use with an outboard intermittent ignition control module. Commentary Figure 602.2(2) illustrates the internal workings that are typical of combination gas controls. These controls have redundant valve mechanisms as an additional safety feature.

Figure 602.2(1)
COMBINATION GAS CONTROL
(Photo courtesy of Emerson Climate Technologies, White Rodgers)

Figure 602.2(2)
COMBINATION GAS CONTROL FOR APPLIANCES
(Shown for illustrative purposes only)
(Figure courtesy of Emerson Climate Technologies, White Rodgers)

602.3 Prohibited installations. Decorative appliances for installation in fireplaces shall not be installed where prohibited by Section 303.3.

❖ Section 303.3 lists, with exceptions, various locations where gas appliances cannot be installed, such as bedrooms and bathrooms. Those prohibitions also apply to decorative appliances.

SECTION 603 (IFGC)
LOG LIGHTERS

603.1 General. Log lighters shall be tested in accordance with CSA 8 and installed in accordance with the manufacturer's instructions.

❖ Log lighters are simple manually operated burners used to start wood fires. Log lighters are functional rather than decorative appliances. The heat produced by the log lighter flames raises the temperature of the wood fuel to its ignition temperature. Log lighters are designed to be turned off manually after the wood fire is capable of sustaining combustion. The code, rather than providing prescriptive installation requirements, requires that log lighters be installed in accordance with the manufacturer's installation instructions. Log lighters must comply with the standard CSA 8. Note that log lighters are often packaged with a T-handle shutoff valve. The package will indicate that the device is listed, which refers to the shutoff valve. The code requires that the log lighter itself be listed. The lighter must be designed for the type of gas, i.e., natural or propane.

SECTION 604 (IFGC)
VENTED GAS FIREPLACES
(DECORATIVE APPLIANCES)

604.1 General. Vented gas fireplaces shall be tested in accordance with ANSI Z21.50, shall be installed in accordance with the manufacturer's instructions and shall be designed and equipped as specified in Section 602.2.

❖ Unlike the appliances addressed in Section 602, these appliances are self-contained and do not rely on a fireplace to contain or vent them. Such appliances are referred to as gas fireplaces because they are designed to simulate a solid-fuel-burning fireplace. The standard, ANSI Z21.50, that regulates these appliances has recently been retitled as "Vented Gas Fireplaces." Therefore, these appliances will be referred to as gas fireplaces even though they do not fall under the definition of "fireplace" [see Commentary Figures 604.1(1), 604.1(2) and 604.1(3)]. These appliances are designed for various methods of venting, including direct-venting through the wall or roof and conventional venting with Type B vent or factory-supplied vent material. Section 303.3 controls the type of appliance that is allowed in bathrooms, toilet rooms and bed-

rooms. Direct-vent appliances have the advantage of a closed combustion chamber that does not communicate with the room in which they are installed, as well as the advantage of an outdoor combustion air supply.

604.2 Access. Panels, grilles and *access* doors that are required to be removed for normal servicing operations shall not be attached to the building.

❖ Access to the appliance for maintenance and repair may be through panels, grilles or doors, but such panels must be removable as intended by the manufacturer of the appliance and not a permanent part of the structure (see the definition of "Access" in Section 202).

Figure 604.1(1)
VENTED DECORATIVE APPLIANCE
(Photo by D.F. Noyes Studio, courtesy of DESA International)

Figure 604.1(2)
DIRECT-VENT DECORATIVE APPLIANCE
(Photo by D.F. Noyes Studio, courtesy of DESA International)

Figure 604.1(3)
DIRECT-VENT GAS FIREPLACE

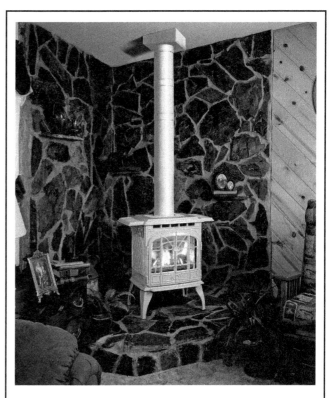

Figure 605.1
VENTED GAS FIREPLACE HEATER
(Photo courtesy of Quadra-Fire)

SECTION 605 (IFGC)
VENTED GAS FIREPLACE HEATERS

605.1 General. Vented gas fireplace heaters shall be installed in accordance with the manufacturer's instructions, shall be tested in accordance with ANSI Z21.88 and shall be designed and equipped as specified in Section 602.2.

❖ These appliances are similar to those addressed in Section 604, the main difference being that vented gas fireplace heaters are designed with more emphasis on space heating while maintaining the decorative features. These heaters must comply with minimum thermal efficiency requirements (see Commentary Figure 605.1).

SECTION 606 (IFGC)
INCINERATORS AND CREMATORIES

606.1 General. Incinerators and crematories shall be installed in accordance with the manufacturer's instructions.

❖ Incinerators used today are installed in hospitals, universities and other relatively large institutions that must dispose of medical wastes, toxic and contaminated substances, pathological waste and human and animal tissues. A crematory is a furnace that is used exclusively to reduce human and animal cadavers to ashes.

Modern incinerator installations take advantage of the heat produced by incineration by passing the flue gases through boilers or other heat exchangers for space heating, cooling or processing operations. The reclamation of heat from incinerators can substantially reduce the energy costs of the facility served by the incinerator.

This section requires factory-built incinerators and crematories to conform to the requirements of Chapter 3 for equipment approval and installation. The incinerator or crematory must be tested to the standard that governs its use, and evidence must be provided indicating that the manufacturer has arranged for a third-party inspection of the manufacturer's facility to determine that the units being produced are the same as those that were tested. As required by Section 301.4.1, the testing agency is responsible for determining the appropriate test protocol, subject ultimately to approval by the code official. When tested in the laboratory, the equipment is installed in accordance with the manufacturer's instructions. For the equipment to perform as designed and tested, the equipment must be installed in the same manner as the tested installation. Therefore, the manufacturer's installation instructions become an integral part of the approval process.

SECTION 607 (IFGC)
COMMERCIAL-INDUSTRIAL INCINERATORS

607.1 Incinerators, commercial-industrial. Commercial-industrial-type incinerators shall be constructed and installed in accordance with NFPA 82.

❖ Commercial-industrial incinerators are large versions of the incinerators described in Section 606. These units burn large amounts of refuse that reduce the load on land fills and, in many instances, capture the heat of combustion for use in another process such as commercial steam heating, generation of electricity or heat for a manufacturing process. The reclamation of heat from incinerators can substantially reduce the energy costs of the facility served by the incinerator.

Commercial-industrial incinerators used today are installed in hospitals, universities and other relatively large institutions that must dispose of medical wastes, toxic and contaminated substances, pathological waste and human and animal tissues. To achieve a more complete incineration and thereby reduce the amount of pollutants released into the atmosphere, multiple-chamber incinerators are used rather than single-chamber incinerators. There are two basic designs of multiple-chamber incinerators: the in-line type and the retort type. The in-line type is characterized by the flow of combustion gases always traveling straight through the incinerator, with 90-degree (1.57 rad) turns only in the vertical direction. The retort type is characterized by a design that causes the combustion gases to flow through 90-degree (1.57 rad) turns in both lateral and vertical directions.

SECTION 608 (IFGC)
VENTED WALL FURNACES

608.1 General. Vented wall furnaces shall be tested in accordance with ANSI Z21.86/CSA 2.32 and shall be installed in accordance with the manufacturer's instructions.

❖ Wall furnaces are a type of room heater usually designed to be installed within a 2- by 4-inch (51 mm by 102 mm) stud cavity in frame construction. They are typically used in cottages, room additions and homes in mild climates. Some units are designed to serve a single room, and others are designed as through-the-wall units to serve adjacent rooms. Wall furnaces are ductless; however, some units are listed for use with a surface-mounted supply outlet extension. Wall furnaces can be either gravity or forced air type.

608.2 Venting. Vented wall furnaces shall be vented in accordance with Section 503.

❖ Because wall furnaces are designed to fit within a wall cavity, a special type of oval-shaped vent is required to fit within the same wall. Type BW vent is specially designed for wall furnace applications (see commentary, Section 503).

608.3 Location. Vented wall furnaces shall be located so as not to cause a fire hazard to walls, floors, combustible furnishings or doors. Vented wall furnaces installed between bathrooms and adjoining rooms shall not circulate air from bathrooms to other parts of the building.

❖ Wall furnaces, like all room heaters, can present a fire hazard if improperly located. The heat discharged or directly radiated from these units can ignite nearby wall or floor surfaces, furniture, trim items, window treatments and doors (see commentary, Section 608.4). A through-the-wall unit serving both a bathroom and an adjacent room must not be capable of recirculating air between those spaces.

608.4 Door swing. Vented wall furnaces shall be located so that a door cannot swing within 12 inches (305 mm) of an air inlet or air outlet of such furnace measured at right angles to the opening. Doorstops or door closers shall not be installed to obtain this *clearance*.

❖ A combustible door that swings close to a wall furnace could be a fire hazard if at some position the door would be within 12 inches (305 mm) of an air inlet or outlet. Because door closers and door stops are easily defeated, they must not be depended upon to secure the required clearance. If the door swing cannot comply with this section, the door would have to be removed or rehung to change its swing direction.

608.5 Ducts prohibited. Ducts shall not be attached to wall furnaces. Casing extension boots shall not be installed unless *listed* as part of the *appliance*.

❖ Most wall furnaces are not designed to force air through ducts, especially those gravity types that do not use fans. Attachment of ducts would add resistance to the flow of air through the furnace, thereby causing an abnormally high temperature rise across the heat exchanger and abnormally high temperatures on furnace surfaces. Some wall furnaces are listed and designed for use with supply duct extensions intended for wall mounting and intended to improve heat distribution. Such duct extensions are factory-built and supplied only by the furnace manufacturer.

608.6 Access. Vented wall furnaces shall be provided with *access* for cleaning of heating surfaces, removal of burners, replacement of sections, motors, controls, filters and other working parts, and for adjustments and lubrication of parts requiring such attention. Panels, grilles and *access* doors that are required to be removed for normal servicing operations shall not be attached to the building construction.

❖ All access panels and doors must be removable as intended by the manufacturer of the furnace.

SECTION 609 (IFGC)
FLOOR FURNACES

609.1 General. Floor furnaces shall be tested in accordance with ANSI Z21.86/CSA 2.32 and shall be installed in accordance with the manufacturer's instructions.

❖ Floor furnaces are vented appliances that are installed in an opening in the floor. These units heat the room by gravity convection and direct radiation and usually serve as the sole source of space heating. Such furnaces are common in cottages, small homes, seasonally occupied structures and rural homes. Because the floor grille can become hot, extreme care must be exercised to prevent occupants, especially children, from contacting the grille by walking or falling on it. Also, care must be taken to avoid a fire hazard caused by placement of materials or furnishings on or near the furnace floor grille (see Commentary Figure 609.1).

Figure 609.1
FLOOR FURNACE
(Photo courtesy of Empire Comfort Systems Incorporated)

609.2 Placement. The following provisions apply to floor furnaces:

1. Floors. Floor furnaces shall not be installed in the floor of any doorway, stairway landing, aisle or passageway of any enclosure, public or private, or in an exitway from any such room or space.

2. Walls and corners. The register of a floor furnace with a horizontal warm-air outlet shall not be placed closer than 6 inches (152 mm) to the nearest wall. A distance of not less than 18 inches (457 mm) from two adjoining sides of the floor furnace register to walls shall be provided to eliminate the necessity of occupants walking over the warm-air discharge. The remaining sides shall

be permitted to be placed not closer than 6 inches (152 mm) to a wall. Wall-register models shall not be placed closer than 6 inches (152 mm) to a corner.

3. Draperies. The furnace shall be placed so that a door, drapery or similar object cannot be nearer than 12 inches (305 mm) to any portion of the register of the furnace.

4. Floor construction. Floor furnaces shall not be installed in concrete floor construction built on grade.

5. Thermostat. The controlling thermostat for a floor furnace shall be located within the same room or space as the floor furnace or shall be located in an adjacent room or space that is permanently open to the room or space containing the floor furnace.

❖ 1. Floor furnaces must not be located where they would interfere with or impede egress. In an emergency egress situation, a floor furnace could be a tripping hazard and could collapse under the live load of many occupants.

2. The floor furnace register must be at least 6 inches (152 mm) from walls to avoid creating a fire hazard by raising the wall temperature to a combustible level. The same reasoning applies to the 6-inch (152 mm) distance from wall registers to a corner. The 18-inch (457 mm) clearance from adjoining sides of a register to walls provides space for the occupants to walk around the grille, which can reach temperatures high enough to cause burns.

3. As stated in Section 608.3, the furnace location must not create a fire hazard by being too close to wall surfaces, trim items, furnishings and window treatments.

4. A floor furnace installed in a slab-on-grade would have to be in a pit, would be subject to flooding and corrosion, and would be inaccessible for service and inspection.

5. If the controlling thermostat does not sense the air temperature in the room in which the furnace is installed, dangerous overheating could result. A thermostat isolated from the source of heat it controls would not respond to the condition in the space served by the furnace.

609.3 Bracing. The floor around the furnace shall be braced and headed with a support framework designed in accordance with the *International Building Code*.

❖ The framing around the floor opening must be capable of supporting the floor system, the anticipated floor loads and the weight of the furnace. The structural requirements of the *International Building Code®* (IBC®) must be complied with.

609.4 Clearance. The lowest portion of the floor furnace shall have not less than a 6-inch (152 mm) *clearance* from the grade level; except where the lower 6-inch (152 mm) portion of the floor furnace is sealed by the manufacturer to prevent entrance of water, the minimum *clearance* shall be not less than 2 inches (51 mm). Where such clearances cannot be provided, the ground below and to the sides shall be excavated to form a pit under the furnace so that the required *clearance* is provided beneath the lowest portion of the furnace. A 12-inch (305 mm) minimum *clearance* shall be provided on all sides except the control side, which shall have an 18-inch (457 mm) minimum *clearance*.

❖ This section specifies clearances between the ground and the furnace, which apply to crawl space installations. The clearances allow access for service and inspection and help prevent corrosion of the furnace assembly.

609.5 First floor installation. Where the basement story level below the floor in which a floor furnace is installed is utilized as habitable space, such floor furnaces shall be enclosed as specified in Section 609.6 and shall project into a nonhabitable space.

❖ Where a floor furnace is installed above a habitable basement space, the furnace must project into an uninhabitable space and be separated from that space by noncombustible construction (see commentary, Section 609.6).

609.6 Upper floor installations. Floor furnaces installed in upper stories of buildings shall project below into nonhabitable space and shall be separated from the nonhabitable space by an enclosure constructed of *noncombustible materials*. The floor furnace shall be provided with *access, clearance* to all sides and bottom of not less than 6 inches (152 mm) and *combustion air* in accordance with Section 304.

❖ Where the floor furnace is installed in an upper floor of a building, the furnace must project into a nonhabitable space, similar to the installation in Section 609.5, and must be separated from that space by noncombustible construction. Access and clearance must be provided for maintenance and servicing. An adequate source of combustion air must also supply the furnace.

SECTION 610 (IFGC)
DUCT FURNACES

610.1 General. Duct furnaces shall be tested in accordance with ANSI Z83.8 or UL 795 and shall be installed in accordance with the manufacturer's instructions.

❖ Duct furnaces are designed to be installed "in-line" in an air duct served by an external blower. Such units are basically a heat exchanger and burner assembly in a cabinet and rely upon an independent air handler [see Commentary Figures 610.1(1) through 610.1(4)].

Figure 610.1(1)
INTEGRAL POWER-VENTED
OUTDOOR DUCT FURNACE
(Photo courtesy of Reznor/Thomas & Betts Corporation)

Figure 610.1(2)
INTEGRAL GRAVITY-VENTED
OUTDOOR DUCT FURNACE
(Photo courtesy of Reznor/Thomas & Betts Corporation)

Figure 610.1(3)
OUTDOOR DUCT FURNACES WITH AIR HANDLER
(Figure courtesy of Reznor/Thomas & Betts Corporation)

Figure 610.1(4)
DUCT FURNACE
(Photo courtesy of Reznor/Thomas & Betts Corporation)

610.2 Access panels. Ducts connected to duct furnaces shall have removable *access* panels on both the upstream and downstream sides of the furnace.

❖ Access panels in the ducts on both sides of the furnace provide full access to the furnace for inspection, repair and regular maintenance of the heat exchanger.

610.3 Location of draft hood and controls. The controls, *combustion air* inlets and draft hoods for duct furnaces shall be located outside of the ducts. The draft hood shall be located in the same enclosure from which *combustion air* is taken.

❖ For access to the controls without the necessity of opening the duct access panels on a regular basis, the furnace controls must be located outside the ducts. The combustion air inlets and draft hood cannot operate properly inside the ducts when exposed to the pressurized airflow from the circulating fan.

610.4 Circulating air. Where a duct furnace is installed so that supply ducts convey air to areas outside the space containing the furnace, the return air shall be conveyed by a duct(s) sealed to the furnace casing and terminating outside the space containing the furnace.

The duct furnace shall be installed on the positive pressure side of the circulating air blower.

❖ Where the furnace supplies air to areas outside the space containing the furnace, return air must also be supplied from the space outside the furnace area. This will prevent the creation of negative pressure in the space containing the furnace.

SECTION 611 (IFGC)
NONRECIRCULATING DIRECT-FIRED
INDUSTRIAL AIR HEATERS

611.1 General. *Nonrecirculating direct-fired industrial air heaters* shall be *listed* to ANSI Z83.4/CSA 3.7 and shall be installed in accordance with the manufacturer's instructions.

❖ Direct-fired air-heating equipment is unique in that there is no heat exchanger to separate the burners from the air being heated. As the name implies, the heat of combustion and all byproducts are directly introduced into the airstream. Because there is no heat exchanger, chimney or vent, direct-fired heaters are close to 100-percent energy efficient (see Commentary Figure 611.1). The term "nonrecirculating" refers to the fact that the same air is not passed through the combustion zone more than once. In other words, these units heat 100-percent outdoor air and are what used to be referred to as make-up air heaters. The air that passes across the burner must be 100-percent outdoor air.

611.2 Installation. *Nonrecirculating direct-fired industrial air heaters* shall not be used to supply any area containing sleeping quarters. *Nonrecirculating direct-fired industrial air heaters* shall be installed only in industrial or commercial occupancies. *Nonrecirculating direct-fired industrial air heaters* shall be permitted to provide ventilation air.

❖ Direct-fired make-up air heaters are used to preheat outdoor air that is supplied as make-up air for large exhaust systems such as those in a commercial kitchen or manufacturing facility. Direct-fired make-up air heaters must not be used to supply any area con-

Figure 611.1
DIRECT-FIRED MAKE-UP AIR HEATER
(Figure courtesy of Reznor/Thomas & Betts Corporation)

taining sleeping quarters. The intent of this section is to prohibit the use of such equipment to serve sleeping quarters because direct-fired heaters are unvented and discharge all combustion byproducts into the air being heated. There is a health concern because of the possible buildup of carbon monoxide, nitrogen oxides and other combustion gases.

Direct-fired heaters are well suited for large heating loads where thermal efficiency is the primary concern. The heating of outdoor areas, make-up air to be exhausted, and buildings with high infiltration rates is considered energy wasteful; therefore, the highest possible combustion and thermal efficiency is sought for such applications.

611.3 Clearance from *combustible materials*. *Nonrecirculating direct-fired industrial air heaters* shall be installed with a *clearance* from *combustible materials* of not less than that shown on the rating plate and in the manufacturer's instructions.

❖ Like all listed appliances, the manufacturer's instructions and nameplate (label) will specify the minimum required clearances to combustibles.

611.4 Supply air. All air handled by a *nonrecirculating direct-fired industrial air heater*, including *combustion air*, shall be ducted directly from the outdoors.

❖ See the commentary to Section 611.1.

611.5 Outdoor air louvers. If outdoor air louvers of either the manual or automatic type are used, such devices shall be proven to be in the open position prior to allowing the main burners to operate.

❖ Nonrecirculating direct-fired industrial (make-up) air heaters convey 100 percent outdoor air; therefore, it is imperative that any air intake dampers be proven open prior to operation of the heater. As with all interlocks required by the code, the intent is to require a proof-of-operation sequence as opposed to a simple

parallel operation arrangement. In other words, the dampers must be electrically proven to be open before heater operation begins.

611.6 Atmospheric vents and gas reliefs or bleeds. *Nonrecirculating direct-fired industrial air heaters* with valve train components equipped with atmospheric vents or gas reliefs or bleeds shall have their atmospheric vent lines or gas reliefs or bleeds lead to the outdoors. Means shall be employed on these lines to prevent water from entering and to prevent blockage by insects and foreign matter. An atmospheric vent line shall not be required to be provided on a valve train component equipped with a *listed* vent limiter.

❖ Large gas-fired equipment is commonly equipped with a complicated gas control arrangement (fuel/valve train components) that can include high- and low-gas pressure switches, gas relief valves, diaphragm valve gas bleed lines, "tell-tale" valve leakage indicator ports and pressure regulators. Control requirements are dictated by equipment standards and historically have also been dictated by insurance underwriters such as Factory Mutual (FM) and Industrial Risk Insurance (IRI).

Reliefs, vents and bleeds can convey raw fuel gas under normal or abnormal conditions and therefore must be piped to the outdoors to avoid a fire or explosion hazard in the building. This piping/tubing is usually joined to an appropriately sized manifold that is vented to the outdoors by one or more pipes.

Atmospheric vent piping is commonly terminated with a screened and turned down outlet or specially designed vent terminal fitting. Blockage by debris, water, ice or insects could cause dangerous control malfunctions of the equipment served by the vent. Where allowed by the listing of a device, such as a pressure regulator; for example, a flow-restrictor device, can be substituted for the vent to the outdoors. A vent-limiter device allows the device served to function as intended, and in the event of a diaphragm failure, the volume of gas escaping through the vent would be controlled by the vent-limiter device.

611.7 Relief opening. The design of the installation shall include provisions to permit *nonrecirculating direct-fired industrial air heaters* to operate at rated capacity without overpressurizing the space served by the heaters by taking into account the structure's designed infiltration rate, providing properly designed relief openings or an interlocked power exhaust system, or a combination of these methods. The structure's designed infiltration rate and the size of relief openings shall be determined by *approved* engineering methods. Relief openings shall be permitted to be louvers or counterbalanced gravity dampers. Motorized dampers or closable louvers shall be permitted to be used, provided they are verified to be in their full open position prior to main burner operation.

❖ Most direct-fired heater installations convey 100-percent outdoor air or a mixture of outdoor and recirculated air. Without a means of exhausting the outdoor air being introduced into a building, the building would pressurize, and the airflow through the heater would be retarded. This must not be allowed to happen because the heater depends on a certain flow-through rate for proper burner operation. Exhaust systems, relief air openings, calculated exfiltration rates of the building, or a combination of these must be used to allow the heater to operate at its required airflow rate (see commentary, Section 611.5 for discussion of interlock requirements).

611.8 Access. *Nonrecirculating direct-fired industrial air heaters* shall be provided with *access* for removal of burners; replacement of motors, controls, filters and other working parts; and for adjustment and lubrication of parts requiring maintenance.

❖ See the commentary to Section 306.

611.9 Purging. Inlet ducting, where used, shall be purged by not less than four air changes prior to an ignition attempt.

❖ Purging will require that burner ignition be delayed for the time necessary to move a volume of outdoor air through the duct equal to four times the volume of the entire inlet duct. Purging will ensure that no flammable fuel gas/air mixtures are present in the combustion zone.

SECTION 612 (IFGC)
RECIRCULATING DIRECT-FIRED
INDUSTRIAL AIR HEATERS

612.1 General. *Recirculating direct-fired industrial air heaters* shall be *listed* to ANSI Z83.18 and shall be installed in accordance with the manufacturer's instructions.

❖ Recirculating direct-fired industrial air heaters are usually used to condition very large volume buildings used as factories, mills, warehouses, etc. They most often consist of ductless blower and burner assemblies that turn over the air in the building through a vertical air discharge configuration (see commentary, Section 611.1). Commentary Figures 612.1(1) and 612.1(2) illustrate the basic difference between direct-fired and indirect-fired appliances. Direct-fired appliances do not use a heat exchanger to separate the combustion (flue) gases from the air being conditioned.

Figure 612.1(1)
BASIC PRINCIPLE OF DIRECT-FIRED SPACE HEATING APPLIANCE
(Figure courtesy of Reznor/Thomas & Betts Corporation)

Figure 612.1(2)
BASIC PRINCIPLE OF INDIRECT-FIRED SPACE HEATING APPLIANCE
(Figure courtesy of Reznor/Thomas & Betts Corporation)

612.2 Location. *Recirculating direct-fired industrial air heaters* shall be installed only in industrial and commercial occupancies. *Recirculating direct-fired air heaters* shall not serve any area containing sleeping quarters. *Recirculating direct-fired industrial air heaters* shall not be installed in hazardous locations or in buildings that contain flammable solids, liquids or gases, explosive materials or substances that can become toxic when exposed to flame or heat.

❖ Industrial air heaters are large space heaters, commonly called air turnover units, used to temper the air within factory, industrial and storage facilities. The intent of this section is to prohibit the use of such equipment to serve sleeping quarters because direct-fired heaters are unvented and discharge all combustion byproducts into the air being heated. There is a health concern because of the possible buildup of carbon monoxide, nitrogen oxides and other combustion gases.

Air recirculated across the burner of a direct-fired heater could be an ignition source for flammable or explosive atmospheres or could create toxins by chemically altering substances that pass through the combustion zone.

Heaters used in hazardous areas and areas having substances present that are made toxic by exposure to flames must not recirculate air and therefore must condition only 100-percent outdoor air or transfer air from nonhazardous areas and areas with suitable atmospheres. For example, a direct-fired heater could serve as a 100-percent outdoor make-up air unit for a paint spray booth exhaust system but could not recirculate any air from within the booth. Passing flammable-vapor-laden air across the burner would pose a serious hazard.

Because of their high thermal efficiency, direct-fired heaters are well suited for the heating of outdoor areas, makeup air heating and the heating of buildings with high ventilation rates.

612.3 Installation. Direct-fired industrial air heaters shall be permitted to be installed in accordance with their listing and the manufacturer's instructions. Direct-fired industrial air heaters shall be installed only in industrial or commercial occupancies. Direct-fired industrial air heaters shall be permitted to provide fresh air ventilation.

❖ The code does not include prescriptive installation requirements for industrial air heaters, but instead relies on the manufacturer's installation instructions and the listing requirements. This will ensure that the equipment is installed just as it was tested by the listing agency. The heaters are allowed only in industrial or commercial occupancies because of the products of combustion that are introduced into the occupied space (see commentary, Section 612.2). The code allows outdoor air heated by the direct-fired heaters to supply some of the ventilation air required by Section 403 of the *International Mechanical Code®* (IMC®) (see commentary, Section 612.5).

612.4 Clearance from *combustible materials*. Direct-fired industrial air heaters shall be installed with a *clearance* from *combustible material* of not less than that shown on the label and in the manufacturer's instructions.

❖ The code relies on the listing and the manufacturer's installation instructions for the required clearance to combustibles, just as it does for other installation requirements in Section 612.3.

612.5 Air supply. Air to direct-fired industrial air heaters shall be taken from the building, ducted directly from outdoors, or a combination of both. Direct-fired industrial air heaters shall incorporate a means to supply outside ventilation air to the space at a rate of not less than 4 cubic feet per minute per 1,000 Btu per hour (0.38 m³ per min per kW) of rated input of the heater. If a separate means is used to supply ventilation air, an interlock shall be provided so as to lock out the main burner operation until the mechanical means is verified. Where outside air dampers or closing louvers are used, they shall be verified to be in the open position prior to main burner operation.

❖ The required ventilation air controls the concentration of combustion byproducts in the indoor air. The prescribed ventilation rate is intended to dilute the combustion gases introduced by the direct-fired equipment. If means other than the heater's air handler itself is used to supply the ventilation air required by this section, it must be interlocked with the heating equipment to ensure that the heater cannot operate if the ventilation air supply is inoperative.

612.6 Atmospheric vents, gas reliefs or bleeds. Direct-fired industrial air heaters with valve train components equipped with atmospheric vents, gas reliefs or bleeds shall have their atmospheric vent lines and gas reliefs or bleeds lead to the outdoors.

Means shall be employed on these lines to prevent water from entering and to prevent blockage by insects and foreign matter. An atmospheric vent line shall not be required to be provided on a valve train component equipped with a *listed* vent limiter.

❖ See the commentary to Section 611.6.

612.7 Relief opening. The design of the installation shall include adequate provision to permit direct-fired industrial air heaters to operate at rated capacity by taking into account the structure's designed infiltration rate, providing properly designed relief openings or an interlocked power exhaust system, or a combination of these methods. The structure's designed infiltration rate and the size of relief openings shall be determined by *approved* engineering methods. Relief openings shall be permitted to be louvers or counterbalanced gravity dampers. Motorized dampers or closable louvers shall be permitted to be used, provided they are verified to be in their full open position prior to main burner operation.

❖ Most direct-fired heater installations convey 100-percent outdoor air or a mixture of outdoor and recirculated air. Without a means of exhausting the outdoor air being introduced into a building, the building would pressurize, and the airflow through the heater would be retarded. This must not be allowed to happen because the heater depends on a certain flow-through rate for proper burner operation. Exhaust systems, relief air openings, calculated exfiltration rates of the building or a combination of these must

be used to allow the heater to operate at its required airflow rate (see commentary, Section 611.5 for a discussion of interlock requirements).

SECTION 613 (IFGC)
CLOTHES DRYERS

613.1 General. Clothes dryers shall be tested in accordance with ANSI Z21.5.1 or ANSI Z21.5.2 and shall be installed in accordance with the manufacturer's instructions.

❖ This section addresses clothes dryer appliances, Types 1 and 2 (see the definition of "Clothes dryer"). Dryers are tested to the applicable safety standard for the appliance, and the manufacturer's installation instructions convey the information needed to duplicate the installation configuration that was tested and found to meet the requirements of the safety standard. The manufacturer's installation instructions are evaluated by the agency responsible for testing, listing and labeling the appliance, and will therefore prescribe an installation that is consistent with the appliance installation that was tested. Clothes dryers must be exhausted in compliance with Section 614 and the manufacturer's installation instructions (see Commentary Figure 613.1).

SECTION 614 (IFGC)
CLOTHES DRYER EXHAUST

[M] 614.1 Installation. Clothes dryers shall be exhausted in accordance with the manufacturer's instructions. Dryer exhaust systems shall be independent of all other systems, and shall convey the moisture and any products of combustion to the outside of the building.

❖ Clothes dryer exhaust systems must convey the moisture and any products of combustion directly to the exterior of the building (outdoors). The code does not use the term "dryer vent." Gas-fired clothes dryers are not "vented" as are other fuel-fired appliances, but rather are exhausted.

Clothes dryer exhaust ducts must be installed to comply with the dryer manufacturer's instructions and the requirements of this section. Dryers are designed and built to meet industry safety standards.

The clothes dryer manufacturer's installation instructions control the type of exhaust duct material allowed and the method of installation. For example, typical dryer installation instructions will require

---

Figure 613.1
TYPE 1 CLOTHES DRYER LABEL
(Figure courtesy of Maytag Corporation)

metallic duct materials and will impose more stringent length limitations for flexible metallic ducts than for rigid ducts because of the poorer flow characteristics of flexible duct materials.

Because clothes dryer exhaust contains high concentrations of combustible lint, debris and water vapor, dryer exhaust systems must be independent of all other systems. This requirement prevents the fire hazards associated with such an exhaust system from extending into or affecting other systems or other areas in the building. Additionally, this section intends to prevent products of combustion from entering the building through other systems (see Section 614.6.1).

Dryer exhaust ducts must be independent of other dryer exhaust ducts unless connected to an engineered exhaust system specifically designed to serve multiple dryers. Type I domestic dryers are not designed for connection to a common exhaust duct serving multiple dryers. For example, connecting multiple Type I dryers to a common duct riser would pressurize the riser and cause exhaust to back up into any dryer that was not operating. Also, the low flow velocities and duct temperature losses could result in water vapor condensation and the buildup of lint and debris at low points (see Section 614.8).

Clothes dryer exhaust systems cannot terminate in or discharge to any enclosed space such as an attic or crawl space, regardless of whether the space is ventilated through openings to the outdoors. The high levels of moisture in the exhaust air can cause condensation to form on exposed surfaces or in insulation materials. Water vapor condensation can cause structural damage and deterioration of building materials and contribute to the growth of mold and fungus. Clothes dryer exhausts that discharge to enclosed spaces will also cause an accumulation of combustible lint and debris, creating a significant fire hazard. An improperly installed clothes dryer exhaust system not only reduces dryer efficiency and increases running time but can also cause a significant increase in exhaust temperature, in turn causing the dryer to cycle on its high limit control, which is an unsafe operating condition.

[M] 614.2 Duct penetrations. Ducts that exhaust clothes dryers shall not penetrate or be located within any fireblocking, draftstopping or any wall, floor/ceiling or other assembly required by the *International Building Code* to be fire-resistance rated, unless such duct is constructed of galvanized steel or aluminum of the thickness specified in Table 603.4 of the *International Mechanical Code* and the fire-resistance rating is maintained in accordance with the *International Building Code*. Fire dampers shall not be installed in clothes dryer exhaust duct systems.

❖ Rigid clothes dryer exhaust ducts are permitted to penetrate assemblies that are not fire-resistance rated and building elements not used as fireblocking or draftstopping. In all other cases, ducts must be constructed of galvanized steel or aluminum of the thickness specified in the IMC and the penetration

must be protected to maintain the fire-resistance rating and integrity of the assembly or element being penetrated. Because of the strength and rigidity differences between steel and aluminum, aluminum ducts generally must be of a heavier (thicker) gage than steel ducts for a given application.

The metal thickness requirements of the IMC necessitate rigid pipe and rule out the use of flexible duct where it must penetrate fireblocking, draftstopping or a fire-resistance-rated assembly. Where penetrating fireblocking or draftstopping, the exhaust duct must be constructed of galvanized steel or aluminum, and the annular space around the duct must be fireblocked in accordance with the IBC.

Where penetrating a fire-resistance-rated assembly, the penetration must be protected. The requirements of this section, combined with Section 614.6.5, make a compelling case for always placing clothes dryers against outside walls to avoid long duct runs and penetrations of other than exterior walls.

[M] 614.3 Cleaning access. Each vertical duct riser for dryers *listed* to ANSI Z21.5.2 shall be provided with a cleanout or other means for cleaning the interior of the duct.

❖ This section applies to Type 2 dryers. Lint and debris carried in the dryer exhaust will adhere to duct walls and settle in the lowest point of any vertical riser in the system; therefore, an accessible means for removing accumulations in the system must be provided to prevent duct blockages and to eliminate the fire hazard of any combustible accumulations. Because of the difficulty of transporting suspended solids vertically against gravity, a vertical section of exhaust duct will eventually require maintenance. The exhaust duct connection to an individual dryer outlet is typically considered as a cleanout because the code requires a "means for cleaning" without specifying what that means can or cannot be. Where exhaust ducts can be accessed and/or readily disassembled, the intent of this section has been met. This section does not require that a tee and cap or similar arrangement be installed where ducts can otherwise be accessed for cleaning.

[M] 614.4 Exhaust installation. Exhaust ducts for clothes dryers shall terminate on the outside of the building and shall be equipped with a backdraft damper. Screens shall not be installed at the duct termination. Ducts shall not be connected or installed with sheet metal screws or other fasteners that will obstruct the flow. Clothes dryer exhaust ducts shall not be connected to a vent connector, vent or chimney. Clothes dryer exhaust ducts shall not extend into or through ducts or plenums.

❖ Exhaust ducts must connect directly to terminals that pass through the building envelope to the outdoor atmosphere. Attics and crawl spaces are not considered to be outdoors, and exhaust ducts cannot terminate in those spaces (see commentary, Section 614.1). Backdraft dampers must be installed in dryer exhaust ducts to avoid outdoor air infiltration during periods when the dryer is not operating and to prevent the entry of animals. These dampers should be designed and installed to provide an adequate seal when in a closed position to minimize air leakage (infiltration). A backdraft damper is usually of the gravity type that is opened by the energy of the exhaust discharge. Some dryer manufacturers prohibit the use of magnetic backdraft dampers because of the extra resistance that the exhaust flow must overcome.

Exhaust terminal opening size is also governed by the dryer manufacturer's instructions. Full-opening terminals present less resistance to flow and might be mandated by the dryer manufacturer. A "full opening" is considered to be an opening having no dimension less than the diameter of the exhaust duct. Dryer exhaust flow must not be restricted by screens or fastening devices such as sheet metal screws. See the allowance in Section 614.8.2 for fasteners. Any type of screen, including so-called bird screens, could become completely blocked with fibers in a very short time. Consider that the filter screen integral with the appliance becomes restricted with lint in each cycle of operation. These restrictions and projections will promote the accumulation of combustible lint and debris in the exhaust duct, thereby creating a potential fire hazard and causing flow resistance. Duct tape should not be relied upon as the sole means of joining dryer exhaust ducts because adhesives can deteriorate with age and when exposed to high temperatures, causing joints to separate. However, sealing joints is still desirable to limit leakage of lint, fibers, moisture vapor and combustion products into the occupied space. Ducts can also separate during duct cleaning operations when brushes are used.

[M] 614.5 Dryer exhaust duct power ventilators. Domestic dryer exhaust duct power ventilators shall be listed and labeled to UL 705 for use in dryer exhaust duct systems. The dryer exhaust duct power ventilator shall be installed in accordance with the manufacturer's instructions.

❖ The previous code editions did not recognize dryer exhaust duct power ventilators (DEDPVs) as an option for clothes dryer installations. DEDPVs are loosely referred to as "dryer booster fans" in the marketplace. Prior to the 2015 code, the designer's choices for exhaust duct length were to: 1) limit the duct length to 35 feet; 2) follow the length limits in the clothes dryer manufacturer's instructions; or, if neither of those choices work, 3) relocate the dryer. A fourth option was to get the code official to approve the installation of a DEDPV under the alternative approval provision in Section 105. Exhaust ducts that exceed the developed length allowed by the code are a potential fire hazard, create maintenance problems, increase drying times and cause the dryer to be inefficient and waste energy. Dryer exhaust systems are commonly installed improperly with excessive lengths, too many elbows and the wrong duct materials. Because of the high incidence of reported dryer fires, the code strictly regulates the installation. DEDPVs are listed to a revised version of UL 705 that now

contains tests and construction requirements that are specific to these devices. DEDPVs have been around for years, but until recently, were not listed to a national consensus standard that was specific to these devices. The UL 705 standard contains requirements for the construction, testing and installation of DEDPVs and requires them to be equipped with features such as interlocks, limit controls, monitoring controls and enunciator devices to make certain that the dryers or dryer operators are aware of the operating status of the DEDPVs [see Commentary Figures 614.5(1) and 614.5(2)].

Figure 614.5(1)
DEDPV

[M] 614.6 Makeup air. Installations exhausting more than 200 cfm (0.09 m^3/s) shall be provided with makeup air. Where a closet is designed for the installation of a clothes dryer, an opening having an area of not less than 100 square inches (645 mm^2) for makeup air shall be provided in the closet enclosure, or makeup air shall be provided by other *approved* means.

❖ Makeup air must be supplied to compensate for the air exhausted by the dryer exhaust system. A typical domestic clothes dryer will exhaust less than 200 cubic feet per minute (0.09 m^3/s). For closet installations, the closet door or the closet enclosure must have an opening with a minimum area of 100 square inches. Where louvers or grilles are used, the solid portion of the louver or grille should be evaluated in accordance with Section 304.10. Makeup air is necessary to prevent the room or space housing the dryer(s) from developing a negative pressure with respect to adjacent spaces or the outdoors, which could result in the improper and dangerous operation of the dryer and other fuel-burning appliances. The required amount of makeup air should be approximately equal to the amount exhausted and is normally supplied by infiltration of air from outdoors or through openings to the outdoors.

The makeup air not only supplies the air that is to be exhausted from the dryer, but also supplies combustion air for gas-fired appliances.

For SI: 1 inch = 25.4 mm.

Figure 614.5(2)
DEDPV

[M] 614.7 Protection required. Protective shield plates shall be placed where nails or screws from finish or other work are likely to penetrate the clothes dryer exhaust duct. Shield plates shall be placed on the finished face of all framing members where there is less than $1^1/_4$ inches (32 mm) between the duct and the finished face of the framing member. Protective shield plates shall be constructed of steel, shall have a minimum thickness of 0.062 inch (1.6 mm) and shall extend not less than 2 inches (51 mm) above sole plates and below top plates.

❖ Protection from nail and screw penetration has been required for wiring, vent systems, plumbing piping, fuel piping, hydronic piping and refrigerant piping for some time in the code, and now it is also required for dryer exhaust ducts. Depending upon the depth of penetration, a nail or screw could create a serious obstruction problem and potential fire hazard.

[M] 614.8 Domestic clothes dryer exhaust ducts. Exhaust ducts for domestic clothes dryers shall conform to the requirements of Sections 614.8.1 through 614.8.6.

❖ Section 305.1 states that gas-fired equipment and appliances must be installed in accordance with the manufacturer's instructions for the listed and labeled equipment. An installation complying with the manufacturer's instructions is required, except where the code requirements are more stringent. The code specifically addresses dryer exhaust ducts, a requirement that is also addressed in the dryer manufacturer's installation instructions. The code requirements for the items addressed in these sections, such as maximum total developed length of dryer exhaust ducts, equivalent lengths of directional fittings and minimum nominal size to name a few, may parallel or exceed the applicable requirements in the manufacturer's installation instructions. The manufacturer's installation instructions could also contain requirements that exceed those in the code.

In all cases, the more restrictive requirements would apply. Therefore, the clothes dryer exhaust duct size must be determined by the clothes dryer

manufacturer but must not be less than 4 inches (102 mm) in diameter.

Dryer exhaust ducts cannot be larger than 4 inches in diameter unless specifically allowed by the manufacturer. Note that enlarging a duct will cause a reduction in flow velocity, and dryer exhaust ducts must maintain sufficient flow velocity to transport lint and fibers through the duct to the discharge terminal.

The duct length, the number and degree of directional fittings, the smoothness of the duct interior wall and the type of exhaust outlet terminal all contribute to the overall friction loss of a clothes dryer exhaust system. When the friction loss is high enough to restrict the required exhaust flow, the duct system must be redesigned to allow the required flow rate. An exhaust duct may have to be shortened or rerouted to compensate for the flow resistance.

Excessive friction losses will also result in reduced flow velocities, which means that the exhaust ducts would be much more likely to collect debris. An improperly designed exhaust system will result in poor dryer performance and poor energy efficiency and can cause the appliance to cycle on its high temperature limit control, which can be hazardous.

Joints in the dryer exhaust system must be reasonably air tight, must have a smooth interior finish and must run in the direction of airflow. For example, the male end of each section of duct must point away from the dryer. This permits the dryer exhaust system to function as intended so that joints and connections do not serve as collection points for lint and debris.

Proper support is required and necessary for maintaining alignment of the dryer exhaust duct system and to prevent excessive stress on ducts and duct joints. A sagging duct will increase the internal resistance to airflow, reduce the efficiency of the system and cause the accumulation of lint and debris.

Section 614.6.4 specifically addresses transition duct connectors. Within the context of this section, a transition duct is a flexible connector used as a transition between the dryer outlet and the connection point to the exhaust duct system. Transition duct connectors must be listed and labeled as transition ducts for clothes dryer application. Transition ducts are currently listed to comply with UL 2158A and are not considered to be flexible air ducts or flexible air connectors subject to the material requirements of Chapter 16.

Transition ducts are flexible ducts constructed of a metalized (foil) fabric supported on a spiral wire frame. They are more fire resistant than the typical plastic spiral duct. Transition duct connectors are limited to 8 feet (1829 mm) in length and must be installed in compliance with their listing and the manufacturer's instructions. These duct connectors must not be concealed by any portion of the structure's permanent finish materials such as drywall, plaster, paneling, built-in furniture or cabinets or any other similar permanently affixed building component; they must remain entirely within the room in which the appliance is installed. Transition duct connectors cannot be joined together to extend beyond the 8-foot (1829 mm) maximum length limit. Transition ducts are to be cut to length as needed to avoid excess duct and unnecessary bends.

Transition duct connectors are necessary for domestic dryers because of appliance movement, vibration and outlet location. In many cases, connecting a domestic clothes dryer directly to rigid duct would be difficult.

[M] 614.8.1 Material and size. Exhaust ducts shall have a smooth interior finish and shall be constructed of metal a minimum 0.016 inch (0.4 mm) thick. The exhaust duct size shall be 4 inches (102 mm) nominal in diameter.

❖ This section requires that the diameter of the exhaust duct be 4 inches (102 mm), no larger and no smaller. A 4-inch (102 mm) duct is the basis for the design of the appliance exhaust system. Ducts that are too small would restrict air movement and would cause the dryer to consume more energy because of the longer drying time. They could also cause appliance overheating. Ducts that are too large will cause a decrease in the flow velocity, causing lint and debris to drop out of the exhaust air stream. The exhaust duct material must be smooth-wall metal with the specified minimum wall thickness. This translates to rigid metal duct of not less than 28 gage. Plastic ducts, corrugated ducts and flexible ducts are all prohibited with the exception of transition ducts in accordance with Section 614.6.4.

[M] 614.8.2 Duct installation. Exhaust ducts shall be supported at 4-foot (1219 mm) intervals and secured in place. The insert end of the duct shall extend into the adjoining duct or fitting in the direction of airflow. Ducts shall not be joined with screws or similar fasteners that protrude more than 1/8 inch (3.2 mm) into the inside of the duct.

❖ Clothes dryer ducts must be well supported to prevent deformation and joint separation. The male end of the duct and fittings must point in the direction of airflow to lessen flow resistance, joint leakage and

collection of lint. This is common practice for all ducts, regardless of application. The last sentence of this section presents somewhat of a dilemma because it says that duct fasteners cannot protrude into the inside of the ducts. All ducts have been required by tradition, codes or installation standards to be mechanically fastened at all joints, however, this section departs from that practice by specifically prohibiting the only means of joining such ducts other than spot welding (i.e., screws and rivets). Joint sealing is not equivalent to or a substitute for mechanical fastening, yet this section has, from a practical standpoint, forced the installation of dryer exhaust ducts that rely solely on joint sealants to hold them together. Dryer exhaust ducts must be exceptionally well supported and sealed so that the joints will not separate. This is especially important where the ducts are to be concealed within the building construction and where they will be mechanically cleaned by duct cleaning apparatus.

Section 614.4 states that fasteners used to join fittings and sections of dryer exhaust ducts together must not obstruct the airflow. This could be interpreted as a prohibition of screws and rivets, or it could mean that such fasteners must not penetrate too far into the duct. The revision to Section 614.8.2 now makes it clear how Section 614.4 is to be interpreted. A fastener protrusion of $^{1}/_{8}$ inch or less will collect some lint, but it is believed to be insignificant. Actually, smooth duct walls collect lint also. The trade-off for allowing tiny amounts of lint to collect is improved duct construction. If dryer exhaust ducts are not allowed to be mechanically fastened, the only method to prevent separation of joints is duct tape. Duct tape should never be depended upon as the sole means of securing duct systems because it is a sealing means, not a fastening means. Now ducts can be properly and securely fastened and then sealed with tapes or mastics. Section M1502 of the *International Residential Code*® (IRC®) requires dryer exhaust ducts to be mechanically fastened and allows the same $^{1}/_{8}$-inch maximum penetration (see the commentary to Section 614.8.2 and Commentary Figure 614.8.2).

[M] 614.8.3 Transition ducts. Transition ducts used to connect the dryer to the exhaust duct system shall be a single length that is *listed* and *labeled* in accordance with UL 2158A. Transition ducts shall be not more than 8 feet (2438 mm) in length, and shall not be concealed within construction.

❖ Within the context of this section, a transition duct is a flexible connector used as a transition between the dryer outlet and the connection point to the exhaust duct system. Transition duct connectors must be listed and labeled as transition ducts for clothes dryer application. Transition ducts are currently listed to comply with UL 2158A. Such transition ducts are not considered as flexible air ducts or flexible air connectors.

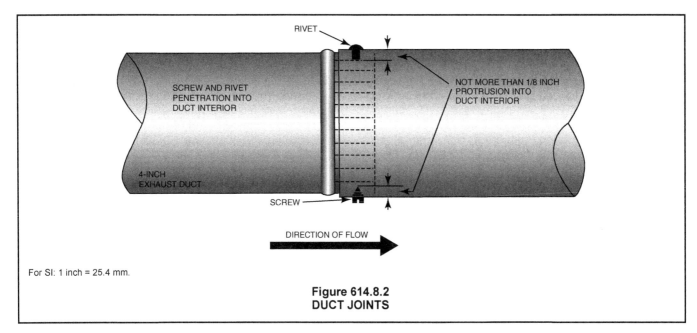

RIVET

SCREW AND RIVET
PENETRATION INTO
DUCT INTERIOR

NOT MORE THAN 1/8 INCH
PROTRUSION INTO
DUCT INTERIOR

4-INCH
EXHAUST DUCT

SCREW

DIRECTION OF FLOW

For SI: 1 inch = 25.4 mm.

Figure 614.8.2
DUCT JOINTS

Transition ducts are flexible ducts constructed of a metalized (foil) laminated fabric supported on a spiral wire frame. They are more fire resistant than the typical plastic spiral duct. Transition duct connectors are necessary for domestic dryers because of appliance movement, vibration and outlet location. In many cases, connecting a domestic clothes dryer directly to rigid duct would be difficult.

Transition duct connectors are limited to 8 feet (2438 mm) in length and must be installed in accordance with their listing and the manufacturer's instructions. These duct connectors must not be concealed by any portion of the structure's permanent finish materials, such as drywall, plaster, paneling, built-in furniture or cabinets or any other similar permanently affixed building component; they must remain entirely within the room in which the appliance is installed. Transition duct connectors cannot be joined to extend beyond the 8-foot (2438 mm) maximum length limit. Transition ducts are to be cut to length as needed to avoid excess duct and unnecessary bends. Note that in the application of Section 614.6.5.1, the length of the transition duct is not counted in the overall duct system length, despite the fact that the transition duct does contribute airflow resistance to the exhaust duct system.

[M] 614.8.4 Duct length. The maximum allowable exhaust duct length shall be determined by one of the methods specified in Sections 614.8.4.1 through 614.8.4.3.

❖ The duct length, the number and angle of fittings and the smoothness of the duct interior all contribute to the total friction loss of the airflow. When the friction loss is excessive, a reduction of the air velocity occurs, allowing lint and debris to accumulate in the duct and thus creating a fire hazard. Impeded airflow will also increase drying time, decrease appliance efficiency and raise system temperatures. The manufacturer's instructions and the code limitations con-

cerning duct length, the number of elbows installed, and the type of termination fittings must be adhered to help prevent a fire hazard and allow proper appliance operation.

Note that this section introduces two distinct choices in its two subsections.

[M] 614.8.4.1 Specified length. The maximum length of the exhaust duct shall be 35 feet (10 668 mm) from the connection to the transition duct from the dryer to the outlet terminal. Where fittings are utilized, the maximum length of the exhaust duct shall be reduced in accordance with Table 614.8.4.1.

❖ The maximum exhaust duct length of 25 feet (7620 mm) was the traditional installation requirement for domestic clothes dryers. The 25-foot (7620 mm) limit was based on the worst case scenario where the dryer is rated for a maximum duct length of 25 feet (7620 mm). The intent was to ensure that the least capable dryer available would be compatible with a 25-foot (7620 mm) duct system. If the exhaust duct system is designed for worst case, it should work properly in all cases. Modern appliance design justifies the increase from 25 to 35 feet. Note that some appliances may not allow a duct length of 35 feet. Because the maximum length is based on equivalent length, all fittings must be accounted for in accordance with Table 614.8.4.1. The length is measured from the point where the transition duct (if used) connects to the rigid exhaust duct system to the terminal outdoors.

[M] 614.8.4.2 Manufacturer's instructions. The maximum length of the exhaust duct shall be determined by the dryer manufacturer's installation instructions. The code official shall be provided with a copy of the installation instructions for the make and model of the dryer. Where the exhaust duct is to be concealed, the installation instructions shall be provided to the code official prior to the concealment inspection. In the absence of fitting equivalent length calculations from

[M] TABLE 614.8.4.1
DRYER EXHAUST DUCT FITTING EQUIVALENT LENGTH

[M] TABLE 614.8.4.1
DRYER EXHAUST DUCT FITTING EQUIVALENT LENGTH

DRYER EXHAUST DUCT FITTING TYPE	EQUIVALENT LENGTH
4 inch radius mitered 45-degree elbow	2 feet, 6 inches
4 inch radius mitered 90-degree elbow	5 feet
6 inch radius smooth 45-degree elbow	1 foot
6 inch radius smooth 90-degree elbow	1 foot, 9 inches
8 inch radius smooth 45-degree elbow	1 foot
8 inch radius smooth 90-degree elbow	1 foot, 7 inches
10 inch radius smooth 45-degree elbow	9 inches
10 inch radius smooth 90-degree elbow	1 foot, 6 inches

For SI: 1 inch = 25.4 mm, 1 foot = 304.8 mm, 1 degree = 0.01745 rad.

the clothes dryer manufacturer, Table 614.8.4.1 shall be utilized.

❖ This section allows the 35-foot (7620 mm) limit of the previous section to be exceeded where longer exhaust duct lengths are allowed by the appliance manufacturer's instructions. Today's appliances often, but not always, permit longer distances, therefore this section allows the installer to take advantage of those longer distances where specified by the manufacturer. The make and model of the dryer must be provided to the code official, along with the respective installation instructions, to permit the code official to inspect the duct installation based on the manufacturer's instructions. Because the installation is specific to and dependent upon a certain appliance make and model, the code official must follow through to make sure that that appliance did get installed. Remember that the prime objective is to prevent a dangerous mismatch between a dryer and its exhaust system.

Where not otherwise specified by the dryer manufacturer, the flow resistance contributed by fittings installed in the duct system must be in accordance with Table 614.8.4.1. Each type of fitting and its angle is equated to a certain length of straight duct and this "equivalent length" is added to the actual straight duct length in the system to obtain the total length. The table recognizes newer fitting designs that have smooth bore interiors and much larger turning radii than traditionally used fittings. Such designs greatly reduce friction loss as can be seen by their much smaller equivalent lengths. If an installer is unable to stay within the allowable length on a particular job, he or she may find that substituting better fittings will correct the problem (see Section 614.6.6).

[M] 614.8.4.3 Dryer exhaust duct power ventilator length. The maximum length of the exhaust duct shall be determined by the dryer exhaust duct power ventilator manufacturer's installation instructions.

❖ See the commentary to Section 614.5.

[M] 614.8.5 Length identification. Where the exhaust duct equivalent length exceeds 35 feet (10 668 mm), the equivalent length of the exhaust duct shall be identified on a perma-

nent label or tag. The label or tag shall be located within 6 feet (1829 mm) of the exhaust duct connection.

❖ A point to consider is the situation that will exist when the original dryer is replaced or the original occupant moves out and the new occupant installs a different dryer. In either case, the replacement dryer might not be compatible with the existing duct length. If the duct system equivalent length is not identified in a conspicuous manner, the new resident will be unaware of the potentially hazardous situation that has been created by the dryer and exhaust duct mismatch. It is assumed that the installer will be aware of the equivalent length of exhaust ducts that are entirely exposed and observable (see Commentary Figure 614.8.5).

[M] 614.8.6 Exhaust duct required. Where space for a clothes dryer is provided, an exhaust duct system shall be installed.

Where the clothes dryer is not installed at the time of occupancy, the exhaust duct shall be capped at the location of the future dryer.

Exception: Where a *listed* condensing clothes dryer is installed prior to occupancy of the structure.

❖ Where a space is provided for the future installation of a clothes dryer, an exhaust duct system must be installed so that such system can be inspected and approved. Otherwise, it is likely that the dryer and exhaust duct system will be installed later without the code official's knowledge and it will not be inspected for proper installation. Note that Section 614.8.4.2 does not appear to be an option when an exhaust duct system is "roughed-in" for future use because the make, model and exhaust capability of the dryer will be unknown.

[M] 614.9 Commercial clothes dryers. The installation of dryer exhaust ducts serving Type 2 clothes dryers shall comply with the *appliance* manufacturer's instructions. Exhaust fan motors installed in exhaust systems shall be located outside of the airstream. In multiple installations, the fan shall operate continuously or be interlocked to operate when any individual unit is operating. Ducts shall have a minimum *clearance* of 6 inches (152 mm) to *combustible materials*.

❖ The code classifies clothes dryers into two categories: Type 1 and Type 2. Type 1 clothes dryers are

Sample of typical placard to be placed behind all future dryer installation locations
Approx. size: 8" x 6" (heavy plastic-like laminate material)
Type: .014" poly carbonate label with area for length display to be punched out
with typical hole punch and exhibit a paint-resistant laminate coating
Permanent pressure-sensitive 3M adhesive and 4 holes 3/32"
Provided and installed by HVAC or sheet metal tradesman at trim-out

Figure 614.8.5
EXAMPLE OF A DRYER EXHAUST LENGTH LABEL

primarily used in the family living environment and do not have the capacity to generate the amount of heat necessary to dry large volumes of clothing. The scope of this section does not regulate Type 1 clothes dryers.

Type 2 clothes drying appliances and exhaust ducts are designed for the higher volumes, temperatures and frequency of use found in public access and commercial laundries. Type 2 dryers must be installed in accordance with the manufacturer's instructions. These dryers are tested to the appropriate safety standard for the appliance, and the manufacturer's installation instructions convey the information needed to duplicate the installation configuration that was tested and found to meet the requirements of the safety standard. The manufacturer's installation instructions are evaluated by the agency responsible for testing the appliance and, therefore, the instructions will prescribe an installation that is consistent with Section 304.1, which addresses the installation of all mechanical equipment. Common exhaust systems for multiple dryers must be designed and engineered in accordance with the equipment and the appliance manufacturer's installation instructions [see Commentary Figure 614.8.6].

The potential fire hazards associated with Type 2 clothes dryers are similar to those of Type 1 clothes dryers (combustible lint, debris and high tempera-

tures). However, the complex exhaust systems, intense heat and larger volume of combustible lint and combustion byproducts associated with commercial dryers call for additional safety requirements.

Commercial and multiple-dryer installations can rely on a single exhaust fan to prevent back pressure, overcome duct friction and induce the proper flow within the common exhaust system. Commentary Figure 614.8.6 shows a four-unit installation with the exhaust fan located at the roof level. This section requires that exhaust fans for Type 2 clothes dryer installations be operated continuously or be interlocked to operate when any individual unit is operating.

It is possible to exhaust multiple Type 2 dryers through a common duct system, such as a common riser or common manifold serving groups of adjacent units. This configuration will usually require an exhaust fan to prevent back pressure and to overcome static pressure in the duct system. In such installations, the common duct system will be connected to the inlet side of an exhaust fan and the fan will be located at the duct system termination to the outdoors.

The motor of the exhaust fan must be positioned outside of the exhaust airstream so that lint, debris and condensation do not accumulate in or on the motor. This requirement is generally satisfied by belt-driven fans with externally mounted motors. Location in the exhaust stream could result in deterioration and failure of the motor, as well as a fire hazard associated with the accumulation of combustible fibers in contact with a motor that may be hot enough to cause ignition. Failure of the exhaust fan motor could result in an unsafe operating condition for the dryers served by the fan.

Where multiple dryer units vent to a common venting system using a common fan, the common exhaust fan must operate when any one or more of the individual dryer unit(s) are operating. This can be achieved in two ways.

The first, continuous exhaust fan operation, is not desirable and is rarely done because of increased energy usage and the unreliability of the design. Continuous operation without interlocks (fan supervision) is not reliable because the dryer units would not respond to or be "aware" of an exhaust fan failure. Any mechanical failure of the exhaust fan or opening of an overcurrent protection device could create an unsafe condition. The second and preferred method is to install exhaust fan supervision interlocked with each dryer unit so that the dryers would be stopped or prevented from starting in the event of exhaust fan failure. The supervision/interlock design should first prove exhaust fan operation by sensing duct pressures before enabling dryer operation. The supervision/interlock logic is identical to that discussed in the commentary to Section 804.3.2 of the IMC and Sections 503.3.3 and 503.3.4 of this code.

Transition ducts are listed factory-built connectors designed to connect an appliance discharge outlet to the exhaust duct system.

[M] 614.10 Common exhaust systems for clothes dryers located in multistory structures. Where a common multistory duct system is designed and installed to convey exhaust from multiple clothes dryers, the construction of such system shall be in accordance with all of the following:

1. The shaft in which the duct is installed shall be constructed and fire-resistance rated as required by the *International Building Code*.

2. Dampers shall be prohibited in the exhaust duct. Penetrations of the shaft and ductwork shall be protected in accordance with Section 607.5.5, Exception 2, of the *International Mechanical Code*.

3. Rigid metal ductwork shall be installed within the shaft to convey the exhaust. The ductwork shall be constructed of sheet steel having a minimum thickness of 0.0187 inch (0.471 mm) (No. 26 gage) and in accordance with SMACNA *Duct Construction Standards*.

4. The ductwork within the shaft shall be designed and installed without offsets.

5. The exhaust fan motor design shall be in accordance with Section 503.2 of the *International Mechanical Code*.

6. The exhaust fan motor shall be located outside of the airstream.

7. The exhaust fan shall run continuously, and shall be connected to a standby power source.

8. The exhaust fan operation shall be monitored in an *approved* location and shall initiate an audible or visual signal when the fan is not in operation.

Figure 614.8.6
MULTIPLE DRYER EXHAUST SYSTEM

9. Makeup air shall be provided for the exhaust system.

10. A cleanout opening shall be located at the base of the shaft to provide *access* to the duct to allow for cleaning and inspection. The finished opening shall be not less than 12 inches by 12 inches (305 mm by 305 mm).

11. Screens shall not be installed at the termination.

❖ This type of exhaust shaft for clothes dryers was used in buildings, but the code was previously silent on this design. This section specifies the design and construction of multistory exhaust shafts for dryers and addresses most of the concerns that code officials have expressed. In this system, an exhaust fan suitable for such duty must be installed at the top of shaft to draw air upward through the ductwork within the shaft. This system is to serve cothes dryers only and must be independent of all other types of exhaust.

Item 2 intends to protect each dryer duct penetration of the exhaust shaft by means of 22-inch subducts and by requiring continuous upward airflow in the shaft. Both Exceptions 1 and 2 of Section 607.5.5 of the IMC employ subducts as an alternative to fire and smoke dampers. Dampers would not be reliable when exposed to moisture and debris in the exhaust stream.

Item 8 requires exhaust fan supervision to alert personnel in the event of fan failure or power failure.

Item 9 intends to require a source of makeup air to prevent the exhaust fan on top of the shaft from producing negative pressures in the shaft such that air will be drawn through the connected clothes dryers when they are idle. This will likely require some type of modulating gravity damper or mechanical makeup air fan to supply air into the bottom of the shaft at a rate that will acheive an approximate flow balance in the shaft. The rate of makeup air would be inversely proportional to the rate of dryer exhaust flow into the shaft duct.

The type of fan chosen for the terminus of the shaft must be suitable for operation with moisture and lint-laden exhaust. Some fans will load up with debris (clothes fibers) and their performance will degrade.

SECTION 615 (IFGC)
SAUNA HEATERS

615.1 General. Sauna heaters shall be installed in accordance with the manufacturer's instructions.

❖ Sauna heaters are used in steam baths and similar rooms to generate heat and steam. They must be installed in compliance with the manufacturer's instructions to ensure they are installed as designed and tested. The code relies on these installation instructions rather than stating prescriptive requirements that might contradict the manufacturer.

615.2 Location and protection. Sauna heaters shall be located so as to minimize the possibility of accidental contact by a person in the room.

❖ Sauna heaters produce steam by passing water over a heated surface. The choice of locations must consider the fact that a heat-producing appliance will be exposed within a small, limited-visibility room with unclothed occupants.

615.2.1 Guards. Sauna heaters shall be protected from accidental contact by an *approved* guard or barrier of material having a low coefficient of thermal conductivity. The guard shall not substantially affect the transfer of heat from the heater to the room.

❖ Guards must be installed to protect the occupants from being burned. The guards must be constructed of a material that is a poor conductor of heat (e.g., wood) so that the guard itself will not present a burn hazard. The design of the guard must protect the occupants but impede the flow of heat into the room from the heater.

615.3 Access. Panels, grilles and *access* doors that are required to be removed for normal servicing operations shall not be attached to the building.

❖ Access panels, covers and doors must not be made unusable by trim, woodwork or room enclosures that impede or interfere with access.

615.4 Combustion and dilution air intakes. Sauna heaters of other than the direct-vent type shall be installed with the draft hood and *combustion air* intake located outside the sauna room. Where the *combustion air* inlet and the draft hood are in a dressing room adjacent to the sauna room, there shall be provisions to prevent physically blocking the *combustion air* inlet and the draft hood inlet, and to prevent physical contact with the draft hood and vent assembly, or warning notices shall be posted to avoid such contact. Any warning notice shall be easily readable, shall contrast with its background and the wording shall be in letters not less than $^1/_4$ inch (6.4 mm) high.

❖ Combustion air and dilution air must not be taken from the sauna room because of the excessive water vapor in the air and the fact that the sauna room would probably be incapable of providing the required volume of combustion air. If a draft hood or combustion air inlet is obstructed, the appliance could malfunction, which could threaten the occupants. For example, a combustion-air-starved heater would produce high levels of carbon monoxide that could enter the sauna room in the event of venting failure. In general, the code expresses concern for the use of fuel-fired appliances in small closed rooms, especially where the occupants are sleeping or would have impaired senses or a diminished ability to recognize danger.

615.5 Combustion and ventilation air. *Combustion air* shall not be taken from inside the sauna room. Combustion and ventilation air for a sauna heater not of the direct-vent type shall be provided to the area in which the *combustion air* inlet and draft hood are located in accordance with Section 304.

❖ This section would require the heater to be either a direct-vent type or a separated-combustion type in which the combustion chamber does not communicate with the sauna room (see commentary, Section 615.4).

615.6 Heat and time controls. Sauna heaters shall be equipped with a thermostat which will limit room temperature to 194°F (90°C). If the thermostat is not an integral part of the sauna heater, the heat-sensing element shall be located within 6 inches (152 mm) of the ceiling. If the heat-sensing element is a capillary tube and bulb, the assembly shall be attached to the wall or other support, and shall be protected against physical damage.

❖ A thermostat is required to limit the temperature in the sauna for fire safety. The control must sense the warmest air near the ceiling and must be protected from physical damage.

615.6.1 Timers. A timer, if provided to control main burner operation, shall have a maximum operating time of 1 hour. The control for the timer shall be located outside the sauna room.

❖ Timers are not required by the code but provide an extra level of protection when installed. To protect both the occupants and the building, timers must limit the heater operating time to 1 hour. Resetting the timer would require the occupant to exit the sauna, thus lessening the chances of overexposure.

615.7 Sauna room. A ventilation opening into the sauna room shall be provided. The opening shall be not less than 4 inches by 8 inches (102 mm by 203 mm) located near the top of the door into the sauna room.

❖ A ventilation opening is required to allow the escape of steam and heat and to supply ventilation for the occupants. The opening is required near the ceiling to take advantage of the natural tendency of heat and steam to rise, facilitating the exhaust.

615.7.1 Warning notice. The following permanent notice, constructed of *approved* material, shall be mechanically attached to the sauna room on the outside:

WARNING: DO NOT EXCEED 30 MINUTES IN SAUNA. EXCESSIVE EXPOSURE CAN BE HARMFUL TO HEALTH. ANY PERSON WITH POOR HEALTH SHOULD CONSULT A PHYSICIAN BEFORE USING SAUNA.

The words shall contrast with the background and the wording shall be in letters not less than $^1/_4$ inch (6.4 mm) high.

Exception: This section shall not apply to one- and two-family dwellings.

❖ A warning sign must be posted to alert users of the potential hazard associated with prolonged exposure to high temperature and humidity levels. The sign is not necessary in a one- and two-family dwelling where the user of the sauna is more familiar with the equipment.

SECTION 616 (IFGC)
ENGINE AND GAS
TURBINE-POWERED EQUIPMENT

616.1 Powered equipment. Permanently installed *equipment* powered by internal combustion engines and turbines shall be installed in accordance with the manufacturer's instructions and NFPA 37. Stationary engine generator assemblies shall meet the requirements of UL 2200.

❖ This section addresses gas-fired internal combustion engines and turbines. Engine- and turbine-driven electrical generators for private use are becoming more popular as are engine-driven cooling appliances and heat pumps. This equipment is used to power fire pumps, generators, water pumps, refrigeration machines and other stationary equipment.

NFPA 37 addresses the fire safety of this equipment, including requirements for enclosures, controls, fuel supplies, exhaust systems, cooling systems and combustion air.

616.2 Gas supply connection. *Equipment* powered by internal combustion engines and turbines shall not be rigidly connected to the gas supply *piping*.

❖ Rigid pipe connections to equipment that vibrates, shakes or otherwise moves on the equipment mountings would be subjected to stress, vibration and noise transmission. To prevent piping damage, loosening of fittings and noise transmission, flexible connectors, suitable for the environment, must be used to connect the distribution piping system to the equipment having engines or turbines. The typical appliance connector could fail from metal fatigue if used with engine-driven equipment. Special hose-type connectors may be required (see commentary, Section 411.1).

SECTION 617 (IFGC)
POOL AND SPA HEATERS

617.1 General. Pool and spa heaters shall be tested in accordance with ANSI Z21.56 and shall be installed in accordance with the manufacturer's instructions.

❖ Pool and spa heaters are specialized water heaters very similar in design to hot water supply boilers and are used with swimming pools, recreational or therapeutic spas and hot tubs. These heaters are usually of the water-tube type and are designed for either indoor or outdoor installation.

SECTION 618 (IFGC)
FORCED-AIR WARM-AIR FURNACES

618.1 General. Forced-air warm-air furnaces shall be tested in accordance with ANSI Z21.47 or UL 795 and shall be installed in accordance with the manufacturer's instructions.

❖ Forced-air warm-air furnaces are considered to be central heating units and consist of burners or heating elements, heat exchangers, blowers and associated controls. Forced-air furnaces are made in many different configurations, including upflow, counterglow (down flow), horizontal flow, multi-position flow and indoor and outdoor units. Commentary Figure 502.1(2) is an example of a multiposition flow (i.e., up, down and horizontal flow) forced-air warm-air furnace.

618.2 Forced-air furnaces. The minimum unobstructed total area of outdoor and return air ducts or openings to a forced-air warm-air furnace shall be not less than 2 square inches for each 1,000 Btu/h ($4402 \text{ mm}^2/\text{W}$) output rating capacity of the furnace and not less than that specified in the furnace manufacturer's installation instructions. The minimum unobstructed total area of supply ducts from a forced-air warm-air furnace shall be not less than 2 square inches for each 1,000 Btu/h ($4402 \text{ mm}^2/\text{W}$) output rating capacity of the furnace and not less than that specified in the furnace manufacturer's installation instructions.

Exception: The total area of supply air ducts and outdoor and return air ducts shall not be required to be larger than the minimum size required by the furnace manufacturer's installation instructions.

❖ The aggregate area of all ducts or openings that convey supply air from the furnace or return air back to the furnace must be adequate to allow the required airflow through the furnace. A furnace that is "starved" for return air or is restricted by an inadequate supply air duct size will produce an abnormal temperature rise across the heat exchanger, which is both a fire hazard and detrimental to the furnace. The furnace output rating is not usually indicated on the label and would be determined as approximately the input rating in Btu/h (W) times the efficiency rating of the furnace. For example, 100,000 Btu/h (29 310 W) input times 0.80 (80-percent efficiency) is 80,000 Btu/h (23 448 W) output. Return or supply air openings required by this section must not be less than that specified by the furnace manufacturer's installation instructions.

The exception intends to clarify that if the furnace installation instructions specify a lesser return or supply area than this section, that lesser area is permitted. The code-specified minimum area applies where the furnace manufacturer does not specify a minimum area.

618.3 Dampers. Volume dampers shall not be placed in the air inlet to a furnace in a manner that will reduce the required air to the furnace.

❖ Dampers are usually avoided in return air ducts and openings because of the risk of starving the furnace for return air. If dampers are installed, the total unrestricted return air duct or opening area must be as required by the code with all of the dampers in the fully closed position.

618.4 Prohibited sources. Outdoor or return air for forced-air heating and cooling systems shall not be taken from the following locations:

1. Closer than 10 feet (3048 mm) from an *appliance* vent outlet, a vent opening from a plumbing drainage system or the discharge outlet of an exhaust fan, unless the outlet is 3 feet (914 mm) above the outside air inlet.

2. Where there is the presence of objectionable odors, fumes or flammable vapors; or where located less than 10 feet (3048 mm) above the surface of any abutting public way or driveway; or where located at grade level by a sidewalk, street, alley or driveway.

3. A hazardous or insanitary location or a refrigeration machinery room as defined in the *International Mechanical Code*.

4. A room or space, the volume of which is less than 25 percent of the entire volume served by such system. Where connected by a permanent opening having an area sized in accordance with Section 618.2, adjoining rooms or spaces shall be considered as a single room or space for the purpose of determining the volume of such rooms or spaces.

   Exception: The minimum volume requirement shall not apply where the amount of return air taken from a room or space is less than or equal to the amount of supply air delivered to such room or space.

5. A room or space containing an *appliance* where such a room or space serves as the sole source of return air.

   Exception: This shall not apply where:

   1. The *appliance* is a direct-vent *appliance* or an *appliance* not requiring a vent in accordance with Section 501.8.

   2. The room or space complies with the following requirements:

      2.1. The return air shall be taken from a room or space having a volume exceeding 1 cubic foot for each 10 Btu/h (9.6L/W) of combined input rating of all fuel-burning appliances therein.

      2.2. The volume of supply air discharged back into the same space shall be approximately equal to the volume of return air taken from the space.

      2.3. Return-air inlets shall not be located within 10 feet (3048 mm) of a draft hood in the same room or space or the combustion chamber of any atmospheric burner *appliance* in the same room or space.

   3. Rooms or spaces containing solid fuel-burning appliances, provided that return-air inlets are

located not less than 10 feet (3048 mm) from the firebox of such appliances.

6. A closet, bathroom, toilet room, kitchen, garage, boiler room, furnace room or unconditioned attic.

Exceptions:

1. Where return air intakes are located not less than 10 feet (3048 mm) from cooking appliances and serve only the kitchen area, taking return air from a kitchen area shall not be prohibited.

2. Dedicated forced air systems serving only a garage shall not be prohibited from obtaining return air from the garage.

7. A crawl space by means of direct connection to the return side of a forced-air system. Transfer openings in the crawl space enclosure shall not be prohibited.

❖ This section prohibits outdoor air and return air from being taken from locations that are potential sources of contamination, odor, flammable vapors or toxic substances and also from locations that would negatively affect the operation of the furnace itself or other fuel-burning appliances.

Items 1, 2 and 3 are contaminant related.

Part of the intent of Items 4 and 5 is to prevent the system from being starved for return-air by the placement of the main or only return-air intake in an area not meeting the volume requirements. Also there is concern for the possibility of combustion gases leaking from fuel-fired appliances because of negative pressure in the space produced by an air handler drawing return air from such space. It is not the intent of this section to prohibit the common practice of installing return air intakes in bedrooms and similarly sized rooms that typically have a volume that is far less than 25 percent of the total volume of the space served by the furnace. The return-air system must be able to convey the required airflow to the furnace regardless of the position of any doors to any rooms in the building served by the furnace. Item 6 is contaminant related and space pressurization related. Exception 1 recognizes that taking return air from a kitchen is of no concern if the air-handling system serves only the kitchen and the intake does not interfere with cooking exhaust systems or entrain cooking effluent from cooking appliances. Exception 2 recognizes that it is illogical to prohibit return air from being taken from a garage where the air handler is serving only the garage.

Item 7 is contaminant related and recognizes that some crawl spaces are designed to be within the thermal envelope of a building.

It is the intent of this section to avoid arrangements that cause an air pressure imbalance, which can cause fuel-fired appliances to spill combustion products into the occupied space. Pressure imbalances can be avoided by making sure that the amount of supply air discharge to a room or space is approximately equal to the amount of return air taken from the room or space and by locating return-air inlets where they will not cause depressurization of an enclosed space.

618.5 Screen. Required outdoor air inlets for residential portions of a building shall be covered with a screen having $^1/_4$-inch (6.4 mm) openings. Required outdoor air inlets serving a nonresidential portion of a building shall be covered with screen having openings larger than $^1/_4$ inch (6.4 mm) and not larger than 1 inch (25 mm).

❖ The inlet openings for outdoor air must be covered with screen or mesh material with a mesh opening size as specified. The screen openings must be small enough to keep out insects, rodents, birds, etc., while being large enough to prevent blockage of airflow by lint, debris and plant fibers.

618.6 Return-air limitation. Return air from one *dwelling unit* shall not be discharged into another *dwelling unit*.

❖ This section prohibits a forced-air heating/cooling system from serving more than one dwelling unit. Any arrangement in which dwelling units share all or part of an air distribution system would allow a communication of atmospheres between the units. This type of communication would spread odors, smoke, allergens, contaminants and disease-causing organisms from one dwelling unit to another and therefore must be avoided.

618.7 (IFGS) Furnace plenums and air ducts. Where a furnace is installed so that supply ducts carry air circulated by the furnace to areas outside of the space containing the furnace, the return air shall be handled by a duct(s) sealed to the furnace casing and terminating outside of the space containing the furnace.

❖ This section is somewhat redundant with Section 618.4 and has the same intent. That is, to prevent a furnace room, closet, alcove, etc., from developing a negative pressure with respect to the outside of the enclosure (see commentary, Section 618.4). It is a typical design to have a furnace or air handler in a closet with a water heater and the closet door is louvered or a grille is placed in the closet wall. If none of the appliances in the closet are fuel fired, this design is harmless, but, a real hazard is created if any are fuel fired.

SECTION 619 (IFGC)
CONVERSION BURNERS

619.1 Conversion burners. The installation of conversion burners shall conform to ANSI Z21.8.

❖ The referenced standard is an installation standard that, in addition to the manufacturers' instructions, would govern the installation of conversion burners. Conversion burners are an assembly of components including burners, gas controls, blowers, safety devices and supporting means. These units are designed to convert an existing appliance from another fuel to gas, commonly from fuel oil or coal.

Conversion of an existing appliance to a different fuel can involve much more than installation of a conversion burner. It can also include the addition of safety controls and limits, combustion air and secondary air supplies, fuel gas piping installation, chimney or vent alterations and other modifications to the existing appliance and its control system.

SECTION 620 (IFGC)
UNIT HEATERS

620.1 General. Unit heaters shall be tested in accordance with ANSI Z83.8 and shall be installed in accordance with the manufacturer's instructions.

❖ Unit heaters are ductless warm-air space heaters that are self-contained and usually suspended from a ceiling or roof structure. Garages, workshops, warehouses, factories, gymnasiums, mercantile spaces and similar large, open spaces are the most common locations for unit heaters [see Commentary Figures 620.1(1) through 620.1(5)].

620.2 Support. Suspended-type unit heaters shall be supported by elements that are designed and constructed to accommodate the weight and dynamic loads. Hangers and brackets shall be of *noncombustible material*.

❖ As with all suspended fuel-fired appliances, a support failure can result in a fire, explosion or injury to building occupants. The supports themselves must be properly designed. Equally important are the structural members to which the supports are attached, such as rafters, beams, joists and purlins. Brackets, pipes, rods, angle iron, structural members and fasteners must be designed for the dead and dynamic loads of the suspended appliance (see Commentary Figure 620.2).

620.3 Ductwork. Ducts shall not be connected to a unit heater unless the heater is *listed* for such installation.

❖ Unit heaters are usually not designed to move air through ductwork. Unless specifically listed for the application, the fans or blowers on unit heaters are designed only for moving air across the heat exchanger without the added friction of ductwork. The addition of ductwork could create a hazard by restricting airflow through the heater.

620.4 Clearance. Suspended-type unit heaters shall be installed with clearances to *combustible materials* of not less than 18 inches (457 mm) at the sides, 12 inches (305 mm) at the bottom and 6 inches (152 mm) above the top where the unit heater has an internal draft hood or 1 inch (25 mm) above the top of the sloping side of the vertical draft hood.

Floor-mounted-type unit heaters shall be installed with clearances to *combustible materials* at the back and one side only of not less than 6 inches (152 mm). Where the flue gases are vented horizontally, the 6-inch (152 mm) *clearance* shall be measured from the draft hood or vent instead of the rear wall of the unit heater. Floor-mounted-type unit heaters shall not be installed on combustible floors unless *listed* for such installation.

Clearances for servicing all unit heaters shall be in accordance with the manufacturer's installation instructions.

Exception: Unit heaters *listed* for reduced *clearance* shall be permitted to be installed with such clearances in accordance with their listing and the manufacturer's instructions.

❖ The required clearances to combustible materials for suspended and floor-mounted unit heaters are specified in this section and must be adhered to unless the unit is listed for less clearance. The clearance reduction methods detailed in Section 308 may be applied where applicable. See the commentary to Section 308 for a complete discussion concerning clearance requirements and clearance reduction methods.

Figure 620.1(1)
DIRECT-VENT UNIT HEATER
(Photo courtesy of Modine Manufacturing Company)

Figure 620.1(2)
POWER-VENTED UNIT HEATER
(Photo courtesy of Modine Manufacturing Company)

Figure 620.1(3)
UNIT HEATER VENTING (TYPICAL)
(Figure courtesy of Reznor/Thomas & Betts Corporation)

SERIES 100
MODEL F
PHOTO I.D. 1997

Figure 620.1(4)
DRAFT-HOOD-EQUIPPED UNIT HEATER
(Photo courtesy of Reznor/Thomas & Betts Corporation)

620.5 (IFGS) Installation in commercial garages and aircraft hangars. Unit heaters installed in garages for more than three motor vehicles or in aircraft hangars shall be installed in accordance with Sections 305.9, 305.10 and 305.11.

❖ See the commentary to Sections 305.9 through 309.11.

SECTION 621 (IFGC)
UNVENTED ROOM HEATERS

621.1 General. Unvented room heaters shall be tested in accordance with ANSI Z21.11.2 and shall be installed in accordance with the conditions of the listing and the manufacturer's instructions. Unvented room heaters utilizing fuels other than fuel gas shall be regulated by the *International Mechanical Code.*

❖ Unvented room heaters are limited-size gas-fired space heaters that discharge the combustion byproducts into the space being heated. Like all appliances regulated by the code, unvented gas-fired room heaters must be listed and labeled, and their installation must comply with the manufacturer's installation instructions [see Commentary Figures 621.1(1) and 621.1(2)].

621.2 Prohibited use. One or more unvented room heaters shall not be used as the sole source of comfort heating in a *dwelling unit.*

❖ Unvented room heaters are designed to supplement a central heating system to allow zone heating of particular rooms and spaces. Unvented room heaters

Figure 620.1(5)
FUNDAMENTAL COMPONENTS OF UNIT HEATER
(Figure courtesy of Reznor/Thomas & Betts Corporation)

are not intended for continuous use as would occur if they were the only source of heat in a building. One or more unvented room heaters used as the sole source of heat would not provide adequate heat distribution in most building arrangements and, depending on the heating load, could require continuous operation of the appliance. Unvented gas-log heaters are listed as room heaters and their installation in factory-built fireplaces is addressed in UL 127 [see Commentary Figure 621.1(2)]. Fireplaces built to the current edition of UL 127 will have a label and installation instructions that will either allow or disallow the installation of an unvented gas log in the fireplace firebox (see Commentary Figure 621.2).

Figure 620.2
UNIT HEATER SUPPORT
(Photo courtesy of Reznor/Thomas & Betts Corporation)

621.3 Input rating. Unvented room heaters shall not have an input rating in excess of 40,000 Btu/h (11.7 kW).

❖ The input rating limitation is consistent with the industry standard for such appliances and allows unvented room heaters to be categorized in the supplemental room heater classification.

621.4 Prohibited locations. Unvented room heaters shall not be installed within occupancies in Groups A, E and I. The location of unvented room heaters shall comply with Section 303.3.

❖ In accordance with this section and Section 303.3, unvented gas-fired heaters are prohibited in assembly, educational and institutional occupancies and in sleeping rooms, bathrooms and toilet rooms in all occupancies (see commentary, Section 303.3 for exceptions). These heaters are considered to be an unacceptable risk in such occupancies because of occupant density, occupant age, occupant physical condition and awareness and small room volumes.

621.5 Room or space volume. The aggregate input rating of all unvented appliances installed in a room or space shall not exceed 20 Btu/h per cubic foot (207 W/m^3) of volume of such room or space. Where the room or space in which the appliances are installed is directly connected to another room or space by a doorway, archway or other opening of comparable size that cannot be closed, the volume of such adjacent room or space shall be permitted to be included in the calculations.

❖ The Btu/h (W) input to room volume ratio limits the accumulation of combustion byproducts in the building interior. Combustion byproducts include carbon monoxide, nitrogen oxides and water vapor. The required room volume would allow dilution of the combustion byproducts by infiltration (see commentary, Section 501.8).

Figure 621.1(1)
UNVENTED ROOM HEATER
(Photo courtesy of Empire Comfort Systems Incorporated)

Figure 621.2
UNVENTED ROOM HEATER (DECORATIVE)
(Photo by D.F. Noyes Studio, courtesy of DESA International)

Figure 621.1(2)
FIREBOX FOR UNVENTED ROOM HEATER
(Photo by D.F. Noyes Studio, courtesy of DESA International)

621.6 Oxygen-depletion safety system. Unvented room heaters shall be equipped with an oxygen-depletion-sensitive safety shutoff system. The system shall shut off the gas supply to the main and pilot burners when the oxygen in the surrounding atmosphere is depleted to the percent concentration specified by the manufacturer, but not lower than 18 percent. The system shall not incorporate field adjustment means

capable of changing the set point at which the system acts to shut off the gas supply to the room heater.

❖ Because unvented heaters are not vented to the outdoors, the appliance standard requires them to incorporate this extra safety feature. This safety system consists of a special pilot burner device that is incorporated with the appliance's flame safeguard device (see commentary, Section 602.2). The oxygen-depletion sensor is basically a pilot burner that is extremely sensitive to the oxygen content in the combustion air (see definition, "Oxygen depletion safety shutoff system"). If the oxygen content (approximately 20-percent normal) drops to a predetermined level, the pilot flame will become "lazy," shifting from a stable horizontal flame to a less stable, more vertical flame, which is incapable of sufficiently heating the thermocouple or thermopile generator, resulting in main gas control valve shutdown and lockout. The predetermined oxygen level that activates burner shutdown is above the level at which incomplete combustion would start to occur.

The monitoring of oxygen levels in this manner is an indicator of approaching insufficient combustion air and is designed to prevent the formation of excess carbon monoxide. If the oxygen level reaches a low enough percentage of air, the level of oxygen will have an inverse relationship with the level of carbon monoxide. This means that as burners are increasingly starved for oxygen, the amount of carbon monoxide produced by the flames is increased. This relationship is valid for appliances in good working order. An improperly maintained, maladjusted or defective burner could produce abnormal (elevated) amounts of carbon monoxide regardless of the oxygen level in the combustion air, and in such cases, the oxygen-depletion sensor device will react only to the oxygen level because it cannot directly detect burner malfunction or abnormal carbon monoxide production.

621.7 Unvented decorative room heaters. An unvented decorative room heater shall not be installed in a factory-built *fireplace* unless the *fireplace* system has been specifically tested, *listed* and *labeled* for such use in accordance with UL 127.

❖ Because the fireplace chimney damper can be closed while an unvented decorative (log) heater (room heater) is operating, the firebox might reach surface temperatures higher than would occur with the damper open. The lack of chimney draft with the resultant dilution air introduction into the firebox would allow higher temperatures to occur in the fireplace components. To prevent surface temperatures from exceeding that allowed by UL 127, that standard now requires factory-built fireplace manufacturers to either test their units for use with unvented log heaters and meet the standard criteria or provide instructions and labels that prohibit such use.

If a factory-built fireplace has not been tested for use with an unvented appliance, the installation instructions and labels must state that the use is prohibited. Fireplaces manufactured before the UL standard added coverage for unvented decorative (log) heaters could not have been tested for that use, and the manufacturers did not have the opportunity to include instructions and labels covering unvented log heaters for use with their units. The installation instructions for fireplace units that were built prior to the unvented heater coverage being added to the UL 127 standard stated that the fireplace was permitted for use only with solid fuel and decorative gas logs listed to ANSI Z21.60. Therefore, regardless of when the fireplace unit was manufactured, unvented decorative heaters can be installed only in factory-built fireplaces that specifically state that they are tested for use with unvented decorative heaters. The fireplace manufacturers will currently indicate whether or not their units are intended for use with unvented decorative heaters. If listed for the application, an unvented decorative (log) heater can be used as a vented decorative appliance by opening the fireplace chimney damper.

621.7.1 Ventless firebox enclosures. Ventless firebox enclosures used with unvented decorative room heaters shall be *listed* as complying with ANSI Z21.91.

❖ Ventless firebox enclosures are designed and tested to accommodate the appliances addressed in Section 621.7. Such enclosures have no chimney or vent connection [see Commentary Figure 621.1(2)].

SECTION 622 (IFGC)
VENTED ROOM HEATERS

622.1 General. Vented room heaters shall be tested in accordance with ANSI Z21.86/CSA 2.32, shall be designed and equipped as specified in Section 602.2 and shall be installed in accordance with the manufacturer's instructions.

❖ This section addresses vented gas-fired space/room heaters, including direct-vent and vent- or chimney-connected appliances, which are limited-size room heaters designed to connect to a vent or chimney. As required by the appliance standard, these appliances are equipped with safety controls that will prevent gas flow to the burners in the event of ignition system failure [see Commentary Figures 622.1(1) and 622.1(2)].

Figure 622.1(1)
VENTED GAS FIREPLACE HEATER
(Photo courtesy of Quadra-Fire)

Figure 622.1(2)
VENTED ROOM HEATER
(Photo courtesy of Empire Comfort Systems Incorporated)

SECTION 623 (IFGC)
COOKING APPLIANCES

623.1 Cooking appliances. Cooking appliances that are designed for permanent installation, including ranges, ovens, stoves, broilers, grills, fryers, griddles, hot plates and barbecues, shall be tested in accordance with ANSI Z21.1, ANSI Z21.58 or ANSI Z83.11 and shall be installed in accordance with the manufacturer's instructions.

❖ This section addresses cooking appliances in all occupancies that are designed for permanent installation, including, but not limited to, ranges, ovens, stoves, broilers, grills, fryers, griddles and barbecues. These appliances must be installed in accordance with the listing and the manufacturer's instructions [see Commentary Figures 623.1(1) and 623.1(2)].

The code intends to regulate the design, construction and installation of cooking appliances that are designed for permanent installation-heated countertop appliances. Appliances that are not readily moveable to another location because of a gas-fuel supply connection would be considered permanently installed, even if they were on casters. Line equipment under a Type I hood, for example, is usually on casters and connected with quick-disconnect-type fuel supply lines to allow movement for routine cleaning. This kind of equipment would be considered permanently installed.

623.2 Prohibited location. Cooking appliances designed, tested, *listed* and *labeled* for use in commercial occupancies shall not be installed within dwelling units or within any area where domestic cooking operations occur.

Exception: Appliances that are also listed as domestic cooking appliances.

❖ Commercial cooking appliances are tested and labeled to different standards than those listed for domestic use. Commercial cooking appliances generally are not insulated to the same level, have higher surface operating temperatures and require a much greater clearance to combustible material. The safety measures inherent to household cooking appliances, such as child-safe push-to-turn knobs and insulated oven doors, are not usually found in commercial cooking appliances. Commercial cooking appliances also have a greater ventilation air requirement for safe operation than household-type cooking appliances. For these reasons, the installation of commercial-type cooking appliances in dwellings is prohibited. Note that many cooking appliances are dual listed as both commercial and domestic (household) cooking appliances. The code does not prohibit such dual-listed appliances in dwelling units because such appliances have the safety features that are necessary for the domestic environment [see Commentary Figures 623.2(1) and 623.2(2)].

623.3 Domestic appliances. Cooking appliances installed within dwelling units and within areas where domestic cooking operations occur shall be *listed* and *labeled* as household-type appliances for domestic use.

❖ Cooking appliances used in dwelling units or in areas where domestic cooking operations occur require a greater degree of user protection and must be listed and labeled as household-type appliances for domestic use (see commentary, Section 623.2). To satisfy residential consumer demand for commercial appliances in the home, some manufacturers are producing listed household-type appliances that have the appearance of commercial application cooking appliances [see Commentary Figures 623.3(1) and 623.3(2)]. Note that appliances could exist that are dual listed to both commercial and domestic standards and as such, would be suitable for either occupancy.

Figure 623.1(1)
COMMERCIAL OVEN
(Photo courtesy of Blodgett Ovens)

Figure 623.1(2)
COMMERCIAL DECK OVEN
(Photo courtesy of Blodgett Ovens)

Figure 623.2(1)
DOMESTIC RANGE

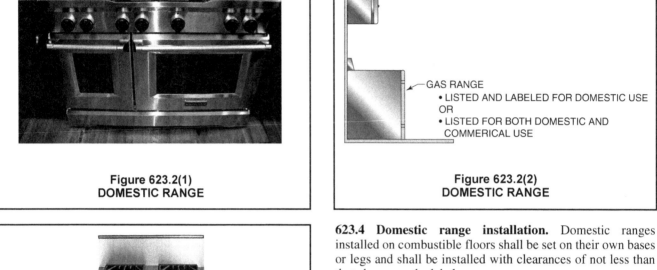

GAS RANGE
• LISTED AND LABELED FOR DOMESTIC USE OR
• LISTED FOR BOTH DOMESTIC AND COMMERICAL USE

Figure 623.2(2)
DOMESTIC RANGE

Figure 623.3(1)
DOMESTIC (HOUSEHOLD) RANGE
(Photo courtesy of Maytag Corporation)

623.4 Domestic range installation. Domestic ranges installed on combustible floors shall be set on their own bases or legs and shall be installed with clearances of not less than that shown on the label.

❖ This section requires the installation of a domestic range using the legs or base provided by the manufacturer where the range is resting on a combustible surface. These supports were part of the test set-up when the range was tested and must be included in the installation to ensure compliance with the listing. Clearances behind and to the sides of the range must also comply with the listing.

623.5 Open-top broiler unit hoods. A ventilating hood shall be provided above a domestic open-top broiler unit, unless otherwise *listed* for forced down draft ventilation.

❖ Open top broiler units generate smoke and grease-laden vapors as part of the cooking process. A ventilating hood is required to evacuate the effluent from the kitchen. Broilers using forced downdraft ventila-

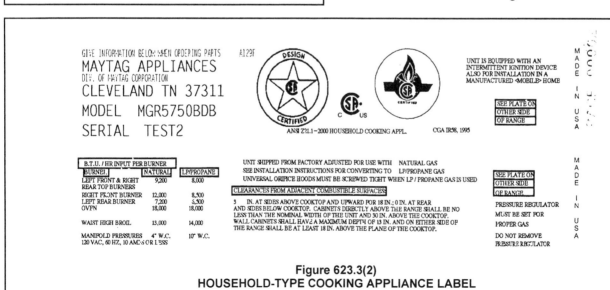

Figure 623.3(2)
HOUSEHOLD-TYPE COOKING APPLIANCE LABEL
(Figure courtesy of Maytag Corporation)

tion are exempt from the hood requirement because they exhaust the effluent down through exhaust ducts installed in or under the floor.

623.5.1 Clearances. A minimum *clearance* of 24 inches (610 mm) shall be maintained between the cooking top and *combustible material* above the hood. The hood shall be at least as wide as the open-top broiler unit and be centered over the unit.

❖ The ventilation hood must be large enough and positioned in such a way that cooking effluent will be captured. The clearance to combustibles above the broiler surface must be maintained to avoid a potential fire hazard.

623.6 Commercial cooking appliance venting. Commercial cooking appliances, other than those exempted by Section 501.8, shall be vented by connecting the *appliance* to a vent or chimney in accordance with this code and the *appliance* manufacturer's instructions or the *appliance* shall be vented in accordance with Section 505.1.1.

❖ Unless exempted by Section 501.8, commercial cooking appliances must be vented by a chimney or venting system or by the ventilation exhaust hood (Type I or II) required by the IMC (see commentary, Section 505.1.1).

Because the hood will be installed in almost all cases, it is taken advantage of and used as the means of venting the appliances. In this case, the Type I or II hood required by the IMC serves a dual

purpose as an exhaust system for cooking effluent and as a combustion product venting means.

623.7 (IFGS) Vertical clearance above cooking top. Household cooking appliances shall have a vertical *clearance* above the cooking top of not less than 30 inches (760 mm) to *combustible material* and metal cabinets. A minimum *clearance* of 24 inches (610 mm) is permitted where one of the following is installed:

1. The underside of the *combustible material* or metal cabinet above the cooking top is protected with not less than $^1/_4$-inch (6.4 mm) insulating millboard covered with sheet metal not less than 0.0122 inch (0.3 mm) thick.

2. A metal ventilating hood constructed of sheet metal not less than 0.0122 inch (0.3 mm) thick is installed above the cooking top with a *clearance* of not less than $^1/_4$ inch (6.4 mm) between the hood and the underside of the *combustible material* or metal cabinet. The hood shall have a width not less than the width of the *appliance* and shall be centered over the *appliance*.

3. A *listed* cooking *appliance* or microwave oven is installed over a *listed* cooking *appliance* and in compliance with the terms of the manufacturer's installation instructions for the upper appliance.

❖ This section addresses the fire hazard created by having hot surfaces, open flames and, possibly, a cooking fire located under combustible cabinets, soffits, etc., and also under metal cabinets with combustible contents (see Commentary Figure 623.7).

For SI: 1 inch = 25.4 mm.

Figure 623.7
CLEARANCE ABOVE COOKING TOP

SECTION 624 (IFGC)
WATER HEATERS

624.1 General. Water heaters shall be tested in accordance with ANSI Z21.10.1 and ANSI Z21.10.3 and shall be installed in accordance with the manufacturer's instructions. Water heaters utilizing fuels other than fuel gas shall be regulated by the *International Mechanical Code*.

❖ Water heaters are recognized by the codes as both a plumbing and a mechanical appliance. Chapter 5 of the *International Plumbing Code*® (IPC®) and Chapter 10 of the IMC contain more detailed information concerning the testing and installation of water heaters than is given in the code. This section identifies those aspects of water heaters specifically related to fuel gas. The IMC addresses those aspects related to other fuel or power sources.

The standards listed are specific to gas-fired water heaters. ANSI Z21.10.1 is for water heaters with an input rating less than or equal to 75,000 Btu per hour (22 kW). ANSI Z21.10.3 is for water heaters with an input rating greater than 75,000 Btu per hour (22 kW) and for circulating and instantaneous water heaters.

624.1.1 Installation requirements. The requirements for water heaters relative to sizing, relief valves, drain pans and scald protection shall be in accordance with the *International Plumbing Code*.

❖ This section refers the user to the IPC for generic requirements related to water heaters not related to the fuel source. These include connections to the potable water system, safety devices such as relief valves, drain pans for protection of the structure and sizing of the water heaters.

624.2 Water heaters utilized for space heating. Water heaters utilized both to supply potable hot water and provide hot water for space-heating applications shall be *listed* and *labeled* for such applications by the manufacturer and shall be installed in accordance with the manufacturer's instructions and the *International Plumbing Code*.

❖ Water heaters serving the dual purpose of supplying potable hot water and serving as a heat source for a space-heating system must be listed and labeled for that dual application. This section does not address water heaters used solely for space-heating applications, but rather addresses water heaters that serve a secondary purpose of space heating. The label will indicate whether the water heater is suitable for space heating [see Commentary Figures 624.2(1) and 624.2(2)].

The IPC contains additional requirements for the proper installation of these appliances.

SECTION 625 (IFGC)
REFRIGERATORS

625.1 General. Refrigerators shall be tested in accordance with ANSI Z21.19 and shall be installed in accordance with the manufacturer's instructions.

Figure 624.2(1)
DIRECT-VENT WATER HEATER DESIGNED FOR BOTH POTABLE WATER HEATING AND SPACE HEATING APPLICATIONS

Refrigerators shall be provided with adequate clearances for ventilation at the top and back, and shall be installed in accordance with the manufacturer's instructions. If such instructions are not available, not less than 2 inches (51 mm) shall be provided between the back of the refrigerator and the wall and not less than 12 inches (305 mm) above the top.

❖ Gas operated refrigerators, although still available, are rarely installed where they would be subject to the permit and inspection process. They usually are used in remote areas such as hunting cabins or in areas where electricity is unavailable or unreliable. At one time, they were used on many farms before electricity became readily available. Clearances are specified in the code for use where manufacturer's installation instructions are not available or do not address clearances (see commentary, Section 501.8).

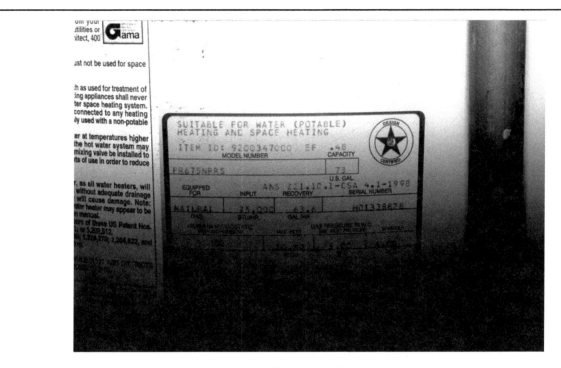

Figure 624.2(2)
WATER HEATER LABEL INDICATING SUITABILITY FOR BOTH POTABLE WATER HEATING AND SPACE HEATING

SECTION 626 (IFGC)
GAS-FIRED TOILETS

626.1 General. Gas-fired toilets shall be tested in accordance with ANSI Z21.61 and installed in accordance with the manufacturer's instructions.

❖ Gas-fired toilets, similar to gas-operated refrigerators, are generally located in remote locations such as hunting or fishing cabins, or in temporary locations such as construction sites. They provide an alternative where water or sewer service is unavailable or where it is not feasible to connect to the sewer for a temporary installation. The firing chamber, upon activation by the user, incinerates the wastes and exhausts the products of combustion through a vent.

626.2 Clearance. A gas-fired toilet shall be installed in accordance with its listing and the manufacturer's instructions, provided that the *clearance* shall in any case be sufficient to afford ready *access* for use, cleanout and necessary servicing.

❖ This section requires installation that complies with the manufacturer's instructions but emphasizes that, regardless of the instructions, room must be provided to allow cleaning, normal maintenance and proper use.

SECTION 627 (IFGC)
AIR-CONDITIONING APPLIANCES

627.1 General. Gas-fired air-conditioning appliances shall be tested in accordance with ANSI Z21.40.1 or ANSI Z21.40.2

and shall be installed in accordance with the manufacturer's instructions.

❖ Gas-fired air conditioning systems include absorption types and internal-combustion-engine-driven machines. The referenced ANSI standards contain requirements for gas-fired air-conditioning equipment. ANSI Z21.40.1 regulates gas-fired absorption systems; ANSI Z21.40.2 regulates work-activated gas-fired air conditioning and heat pump systems. The code relies on the manufacturer's instructions and the referenced standards for installation requirements.

627.2 Independent piping. Gas *piping* serving heating appliances shall be permitted to also serve cooling appliances where such heating and cooling appliances cannot be operated simultaneously (see Section 402).

❖ Where gas supply piping serves both heating and cooling equipment, this section permits the piping to be sized based on the larger gas demand of the two systems rather than sizing based on the total demand of the two systems. This provision is not applicable unless the heating and cooling systems are incapable of operating simultaneously.

627.3 Connection of gas engine-powered air conditioners. To protect against the effects of normal vibration in service, gas engines shall not be rigidly connected to the gas supply *piping*.

❖ Because of the inherent vibration of a gas engine, a connector must be installed between the engine and the gas supply pipe so that the vibration does not place undue stress on the connector or have an adverse effect on the shutoff valve or the gas piping.

Vibration transmitted to the connector and piping could eventually cause joint failure and/or connector failure. The connector must be designed and approved for this application.

627.4 Clearances for indoor installation. Air-conditioning appliances installed in rooms other than alcoves and closets shall be installed with clearances not less than those specified in Section 308.3 except that air-conditioning appliances *listed* for installation at lesser clearances than those specified in Section 308.3 shall be permitted to be installed in accordance with such listing and the manufacturer's instructions and air-conditioning appliances *listed* for installation at greater clearances than those specified in Section 308.3 shall be installed in accordance with such listing and the manufacturer's instructions.

Air-conditioning appliances installed in rooms other than alcoves and closets shall be permitted to be installed with reduced clearances to *combustible material*, provided that the *combustible material* is protected in accordance with Table 308.2.

❖ See the commentary to Section 308.3.

627.5 Alcove and closet installation. Air-conditioning appliances installed in spaces such as alcoves and closets shall be specifically *listed* for such installation and installed in accordance with the terms of such listing. The installation clearances for air-conditioning appliances in alcoves and closets shall not be reduced by the protection methods described in Table 308.2.

❖ Air-conditioning equipment must be specifically listed for installation in rooms not large in comparison with the size of the equipment, such as alcoves and closets.

627.6 Installation. Air-conditioning appliances shall be installed in accordance with the manufacturer's instructions. Unless the *appliance* is *listed* for installation on a combustible surface such as a floor or roof, or unless the surface is protected in an *approved* manner, the *appliance* shall be installed on a surface of noncombustible construction with *noncombustible material* and surface finish and with no *combustible material* against the underside thereof.

❖ The intent of this section is to prohibit the installation of gas-fired air-conditioning equipment on combustible surfaces unless specifically listed for such an installation or the surface is protected in a manner approved by the code official.

627.7 Plenums and air ducts. A plenum supplied as a part of the air-conditioning *appliance* shall be installed in accordance with the *appliance* manufacturer's instructions. Where a plenum is not supplied with the *appliance*, such plenum shall be installed in accordance with the fabrication and installation instructions provided by the plenum and *appliance* manufacturer. The method of connecting supply and return ducts shall facilitate proper circulation of air.

Where the air-conditioning *appliance* is installed within a space separated from the spaces served by the *appliance*, the air circulated by the *appliance* shall be conveyed by ducts

that are sealed to the casing of the *appliance* and that separate the circulating air from the combustion and ventilation air.

❖ This section requires installation of plenums and ducts that are part of an air-conditioning system to comply with the air-conditioning equipment manufacturer's requirements. Further guidance on plenums and ducts can be found in Chapter 6 of the IMC (see commentary, Section 618.4).

627.8 Refrigeration coils. A refrigeration coil shall not be installed in conjunction with a forced-air furnace where circulation of cooled air is provided by the furnace blower, unless the blower has sufficient capacity to overcome the external static resistance imposed by the duct system and cooling coil at the air throughput necessary for heating or cooling, whichever is greater. Furnaces shall not be located upstream from cooling units, unless the cooling unit is designed or equipped so as not to develop excessive temperature or pressure. Refrigeration coils shall be installed in parallel with or on the downstream side of central furnaces to avoid condensation in the heating element, unless the furnace has been specifically *listed* for downstream installation. With a parallel flow arrangement, the dampers or other means used to control flow of air shall be sufficiently tight to prevent any circulation of cooled air through the furnace.

Means shall be provided for disposal of condensate and to prevent dripping of condensate onto the heating element.

❖ A refrigeration coil adds static pressure loss to that of the furnace and duct system. This section requires that the furnace blower be sized to overcome the total resistance.

The location of a cooling coil downstream from a furnace will expose the coil to higher temperatures during the heating cycle. Most systems of this type will be listed, and the manufacturer will design the equipment to prevent the buildup of temperature and pressure in the cooling coils. This may be achieved using a method that allows the coolant to expand into the piping system or an expansion chamber.

Installing cooling coils upstream of the furnace might expose the burners and heat exchanger to condensation formed during a cooling cycle. Only furnaces listed for this application with corrosion-resistant elements are allowed. Refrigeration coils installed in a parallel arrangement with furnaces can also cause condensation problems if the cooled air is allowed to pass over the heat exchanger. Bypass dampers are usually installed in front of the heating unit to prevent this problem. A listed parallel system will have addressed the condensation problem in its design.

A drain pan or other approved method must be incorporated in the installation to capture the cooling coil condensation and prevent it from dripping onto furnace components such as the heat exchanger. Section 307 of the IMC addresses condensate disposal.

627.9 Cooling units used with heating boilers. Boilers, where used in conjunction with refrigeration systems, shall be installed so that the chilled medium is piped in parallel with the heating boiler with appropriate valves to prevent the chilled medium from entering the heating boiler. Where hot water heating boilers are connected to heating coils located in air-handling units where they might be exposed to refrigerated air circulation, such boiler *piping* systems shall be equipped with flow control valves or other automatic means to prevent gravity circulation of the boiler water during the cooling cycle.

❖ Where boilers are used in the same system with cooling units, the code requires that they be in parallel with controls to prevent chilled water from circulating into the boiler. Similarly, if heating coils are located where they could be exposed to cooled air, the code requires valves or other means to prevent chilled water from circulating through the boiler from the heating coils during the cooling cycle. In either case the efficiency of the system will be adversely affected, and boiler damage could result.

627.10 Switches in electrical supply line. Means for interrupting the electrical supply to the air-conditioning *appliance* and to its associated cooling tower (if supplied and installed in a location remote from the air conditioner) shall be provided within sight of and not over 50 feet (15 240 mm) from the air conditioner and cooling tower.

❖ This requirement is a personnel safety issue that is also required by Section 628.

SECTION 628 (IFGC)
ILLUMINATING APPLIANCES

628.1 General. Illuminating appliances shall be tested in accordance with ANSI Z21.42 and shall be installed in accordance with the manufacturer's instructions.

❖ Illuminating appliances must be listed and labeled as required for other appliances regulated by the code. These appliances include gas lamps designed for outdoor use (see commentary, Section 301.4).

628.2 Mounting on buildings. Illuminating appliances designed for wall or ceiling mounting shall be securely attached to substantial structures in such a manner that they are not dependent on the gas *piping* for support.

❖ Gas piping is not designed to act as a support for any appliance. All appliances must be independently supported to prevent stresses and strains on fuel supply connections.

628.3 Mounting on posts. Illuminating appliances designed for post mounting shall be securely and rigidly attached to a post. Posts shall be rigidly mounted. The strength and rigidity of posts greater than 3 feet (914 mm) in height shall be at least equivalent to that of a $2^1/_2$-inch-diameter (64 mm) post constructed of 0.064-inch-thick (1.6-mm) steel or a 1-inch (25.4 mm) Schedule 40 steel pipe. Posts 3 feet (914 mm) or less in height shall not be smaller than a $^3/_4$-inch (19.1 mm) Schedule 40 steel pipe. Drain openings shall be provided near the base of posts where there is a possibility of water collecting inside them.

❖ Regardless of the intended method of mounting and support of an appliance, the installation must be

secure and must not transmit any loading to the fuel supply connection. A requirement is added to provide a drain at the base of the post to prevent collection of water that can deteriorate the post material.

628.4 Appliance pressure regulators. Where an *appliance* pressure regulator is not supplied with an illuminating *appliance* and the service line is not equipped with a service pressure regulator, an *appliance* pressure regulator shall be installed in the line to the illuminating *appliance*. For multiple installations, one regulator of adequate capacity shall be permitted to serve more than one illuminating *appliance*.

❖ Outdoor lighting appliances generally require low-pressure gas flow. Some of these lighting appliances are supplied by the manufacturer with a built-in pressure regulator set for the proper operating pressure. If, however, the regulator is not provided, this section requires the installation of one in the gas line serving the appliance. This regulator must be appropriate for the pressure required by the manufacturer. One pressure regulator with the required capacity is allowed to be installed to serve more than one lighting appliance if the piping pressure losses permit.

SECTION 629 (IFGC)
SMALL CERAMIC KILNS

629.1 General. Kilns shall be installed in accordance with the manufacturer's instructions and the provisions of this code. Kilns shall comply with Section 301.3.

❖ Small ceramic kilns used for hobby and noncommercial purposes operate at high temperatures and therefore need to be regulated by the code as are other heat-producing appliances. The minimum level of safety required for installation of small kilns is compliance with the kiln manufacturer's installation instructions and any applicable requirements of this code such as clearance requirements.

SECTION 630 (IFGC)
INFRARED RADIANT HEATERS

630.1 General. Infrared radiant heaters shall be tested in accordance with ANSI Z83.19 or Z83.20 and shall be installed in accordance with the manufacturer's instructions.

❖ This section addresses radiant heaters, including ceramic element and steel tube-type designs. Infrared heaters are produced in both vented and unvented types and function by creating a very hot surface area from which heat energy is directly radiated. Infrared radiant heaters are usually suspended from ceilings or roofs. These heaters are typically used for "spot" heating in spaces that are otherwise unconditioned or not conditioned to human comfort levels. Radiant heaters have the advantage of being able to heat objects and personnel without heating the surrounding air [see Commentary Figures 630.1(1) through 630.1(3)].

The code relies on the manufacturers' instructions for installation requirements. Maintaining the clear-

ance to combustibles for radiant heaters is of para-
mount importance. Direct radiation is a very effective
method of transferring heat energy, and improperly
located combustible materials can be readily ignited
[see Commentary Figure 630.1(4)].

630.2 Support. Infrared radiant heaters shall be fixed in a
position independent of gas and electric supply lines. Hangers
and brackets shall be of *noncombustible material.*

❖ Supports must prevent radiant heaters from falling. A
fall would mean losing the required clearance to combustibles; putting tension or pressure on electrical,
fuel and vent connections; and dislocating that redi-
rects the radiant output to where a fire hazard would
result. Often, such heaters are improperly hung and
restrained, allowing heater movement (swinging) to
stress flexible (semirigid) gas connectors. Gas connectors must be designed and approved for applica-

Figure 630.1(1)
CERAMIC BURNER INFRARED RADIANT HEATER
(Photo courtesy of Reznor/Thomas & Betts
Corporation)

Figure 630.1(2)
TUBE-TYPE INFRARED RADIANT HEATER
(Photo courtesy of Reznor/Thomas & Betts Corporation)

Figure 630.1(3)
TUBE-TYPE INFRARED RADIANT HEATER
(Photo courtesy of Reznor/Thomas & Betts Corporation)

Figure 630.1(4)
SAMPLE LABEL INFRARED RADIANT TUBE HEATER
(Figure courtesy of Modine Manufacturing Company)

tions in which the appliance moves because of expansion and contraction or lack of restraints. To accommodate expansion/contraction movement, some appliance manufacturers require flexible connectors to be coiled or shaped in a manner that allows limited movement without stressing the metal of the connector, similar to expansion loops and offsets in rigid piping systems.

630.3 (IFGS) Combustion and ventilation air. Where unvented infrared heaters are installed, natural or mechanical means shall provide outdoor ventilation air at a rate of not less than 4 cfm per 1,000 Btu/h (0.38 m³/min/kW) of the aggregate input rating of all such heaters installed in the space. Exhaust openings for removing flue products shall be above the level of the heaters.

❖ Similar to the requirements to Section 612.5, this section requires outdoor ventilation air for any space in which unvented radiant heaters are installed. The introduction of outdoor air below the level of the heaters and the provision of exhaust openings above the level of the heaters would help convey combustion products to the outdoors.

630.4 (IFGS) Installation in commercial garages and aircraft hangars. Overhead infrared heaters installed in garages for more than three motor vehicles or in aircraft hangars shall be installed in accordance with Sections 305.9, 305.10 and 305.11.

❖ See the commentary to Sections 305.9 through 305.11.

SECTION 631 (IFGC)
BOILERS

631.1 Standards. Boilers shall be *listed* in accordance with the requirements of ANSI Z21.13 or UL 795. If applicable, the boiler shall be designed and constructed in accordance with the requirements of ASME CSD-1 and as applicable, the ASME *Boiler and Pressure Vessel Code*, Sections I, II, IV, V and IX and NFPA 85.

❖ The scope of this section includes boilers in all occupancies including power plants; factories; industrial plants; schools; and institutional occupancies such as hospitals, commercial laundries, hotels and residential structures. Boilers are defined in Section 202 of this code [see Commentary Figures 631.1(1) and 631.1(2)].

Boilers are potentially dangerous if not properly designed, constructed and operated, more so than many other appliances because of the potential explosion hazard associated with pressure vessels. In addition to this code, several industry standards are referenced. Manufactured gas-fired boilers must be listed to either ANSI Z21.13 or UL 795. Commentary Figure 202(3) shows a typical label with the required information for a listed boiler. Only the design and construction requirements of boilers are regulated by the referenced standards. The requirements for specific types of boilers can be found in the respective sections of the referenced standards.

631.2 Installation. In addition to the requirements of this code, the installation of boilers shall be in accordance with the manufacturer's instructions and the *International Mechanical Code*. Operating instructions of a permanent type shall be attached to the boiler. Boilers shall have all controls set, adjusted and tested by the installer. A complete control diagram together with complete boiler operating instructions shall be furnished by the installer. The manufacturer's rating data and the nameplate shall be attached to the boiler.

❖ This section governs the installation and commissioning of boilers and their control systems. The mechanical equipment requirements for approval, labeling, installation, maintenance, repair and alteration are regulated by the IMC and the manufacturer's instructions. The IMC contains installation requirements for boilers including provisions for shutoff valves, pressure relief valves, safety valves, electrical control wiring, blowoff valves, expansion tanks and low water cutoff controls [see Commentary Figures 631.1(1) and 631.1(2) and 631.2(1) through 631.2(3)].

Complete operating instructions must be permanently affixed to the boiler upon completion of the installation. Boiler systems can be complex and gen-

Figure 631.1(1)
STEAM HEATING BOILER WITH GAUGE-GLASS,
LOW WATER CUTOFF CONTROL,
PRESSURE GAUGE, PRESSURE LIMIT CONTROL
AND AUTOMATIC VENT DAMPER
(Photo courtesy of Weil-McLain)

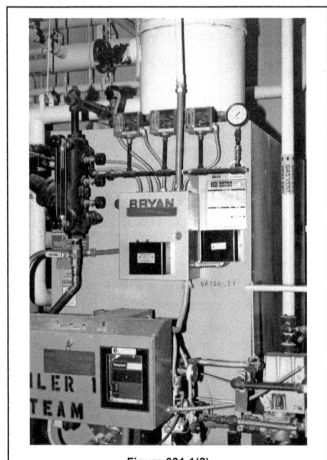

Figure 631.1(2)
STEAM BOILER WITH POWER BURNER, LOW-WATER
CUTOFF CONTROLS AND PRESSURE CONTROLS

erally require coordination of several pieces of equipment. The proper operating procedures, set points, etc., must be specified and included in these instructions. This is usually done in the control system design documentation, including such things as diagrams, system schematics and control sequence descriptions. Typically, an operating and maintenance (O&M) manual is given to the building owner/operator upon completion of the project. Along with operating and control procedures, the O&M manual will clearly identify routine maintenance and calibration information. It is also typical for operation sequences and calibration information to be displayed under glass in a conspicuous place in the boiler room for use by operating and service personnel. The intent is to provide the owner/operator with all operating and control information necessary to properly operate and maintain the boiler and its associated controls and equipment.

Figure 631.2(1)
ELECTRONIC PROBE-TYPE LOW-WATER CUTOFF CONTROL WITH GAUGE GLASS AND TRI-COCKS FOR STEAM BOILER APPLICATION
(Photo courtesy of McDonnell & Miller)

631.3 Clearance to *combustible materials*. Clearances to *combustible materials* shall be in accordance with Section 308.4.

❖ Boilers, like all gas-fired equipment and appliances, operate at high temperatures that require adequate separation from combustibles. This section refers the user to Section 308 of the code for requirements for clearances to combustibles and allowable clearance reduction methods.

SECTION 632 (IFGC)
EQUIPMENT INSTALLED IN EXISTING UNLISTED BOILERS

632.1 General. Gas *equipment* installed in existing unlisted boilers shall comply with Section 631.1 and shall be installed in accordance with the manufacturer's instructions and the *International Mechanical Code*.

❖ This section addresses the replacement of equipment, such as burner assemblies, in existing unlisted boilers. The code relies on the manufacturer's installation instructions, the requirements of the IMC and Section 631 of the code.

SECTION 633 (IFGC)
STATIONARY FUEL-CELL POWER SYSTEMS

[F] 633.1 General. Stationary fuel-cell power systems having a power output not exceeding 10 MW shall be tested in accordance with ANSI CSA America FC 1 and shall be installed in accordance with the manufacturer's instructions, NFPA 853, the *International Building Code* and the *International Fire Code*.

❖ Fuel cell power plants combine hydrogen and oxygen to create electrical and heat energy. Through a process called "reforming," natural gas can be converted to a hydrogen-rich gas that is commonly used as the fuel for stationary fuel cell plants. NFPA 853 is an installation standard that covers site and location requirements and piping, ventilation/exhaust and fire protection requirements (see Chapter 7).

Figure 631.2(2)
ELECTRONIC PROBE-TYPE LOW-WATER CUTOFF CONTROL FOR HOT WATER BOILER APPLICATION
(Photo courtesy of McDonnell & Miller)

Figure 631.2(3)
ELECTRONIC PROBE-TYPE LOW-WATER CUTOFF CONTROL FOR HOT WATER BOILER APPLICATION
(Photo courtesy of McDonnell & Miller)

SECTION 634 (IFGS)
CHIMNEY DAMPER OPENING AREA

634.1 Free opening area of chimney dampers. Where an unlisted decorative *appliance* for installation in a vented *fireplace* is installed, the *fireplace* damper shall have a permanent free opening equal to or greater than specified in Table 634.1.

❖ This section intends to ensure that an adequate fireplace damper opening is installed for all unlisted decorative appliances. The minimum damper opening size is part of the listing for listed appliances, but the installer and the code official might not have access to this information for an unlisted appliance. Section 301.3 requires all appliances to be listed.

The free opening size is a function of the height of the chimney and the appliance input rating. As stated in note "a" of Table 634.1, the opening areas in the table correspond to the cross-sectional area of round chimney flues having nominal sizes from 3 to 8 inches (76 to 203 mm), and the 64 square inches (41 290 mm²) represents the cross-sectional area of a standard 8-inch by 8-inch chimney tile.

SECTION 635 (IFGC)
GASEOUS HYDROGEN SYSTEMS

635.1 Installation. The installation of gaseous hydrogen systems shall be in accordance with the applicable requirements of this code, the *International Fire Code* and the *International Building Code*.

❖ See the commentary to Chapter 7.

SECTION 636 (IFGC)
OUTDOOR DECORATIVE APPLIANCES

636.1 General. Permanently fixed-in-place outdoor decorative appliances shall be tested in accordance with ANSI Z21.97 and shall be installed in accordance with the manufacturer's instructions.

❖ There are different types of gas-fired decorative appliances that are for outdoor use only, such as simulated campfire units and table top decorative display burner units. Such appliances are popular for outdoor living areas, patios, outdoor kitchens, etc. (see Commentary Figure 636.1).

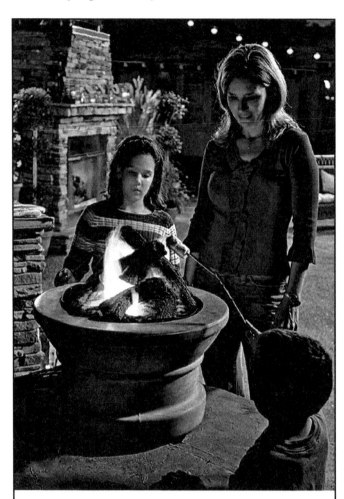

Figure 636.1
OUTDOOR (CAMPFIRE) DECORATIVE APPLIANCE
(Photo courtesy of Hearth and Home Technologies, Inc.)

TABLE 634.1
FREE OPENING AREA OF CHIMNEY DAMPER FOR VENTING FLUE GASES
FROM UNLISTED DECORATIVE APPLIANCES FOR INSTALLATION IN VENTED FIREPLACES

| CHIMNEY HEIGHT (feet) | MINIMUM PERMANENT FREE OPENING (square inches)[a] | | | | | | |
	8	13	20	29	39	51	64
	Appliance input rating (Btu per hour)						
6	7,800	14,000	23,200	34,000	46,400	62,400	80,000
8	8,400	15,200	25,200	37,000	50,400	68,000	86,000
10	9,000	16,800	27,600	40,400	55,800	74,400	96,400
15	9,800	18,200	30,200	44,600	62,400	84,000	108,800
20	10,600	20,200	32,600	50,400	68,400	94,000	122,200
30	11,200	21,600	36,600	55,200	76,800	105,800	138,600

For SI: 1 inch = 25.4 mm, 1 foot = 304.8 mm, 1 square inch = 645.16 m², 1 British thermal unit per hour = 0.2931 W.

a. The first six minimum permanent free openings (8 to 51 square inches) correspond approximately to the cross-sectional areas of chimneys having diameters of 3 through 8 inches, respectively. The 64-square-inch opening corresponds to the cross-sectional area of standard 8-inch by 8-inch chimney tile.

Bibliography

The following resource materials were used in the preparation of the commentary for this chapter of the code.

IBC-2015, *International Building Code.* Washington, DC: International Code Council, 2014.

IMC-2015, *International Mechanical Code.* Washington, DC: International Code Council, 2014.

IPC-2015, *International Plumbing Code.* Washington, DC: International Code Council, 2014.

Chapter 7:
Gaseous Hydrogen Systems

General Comments

Chapter 7 was added to the code in 2003 to cover hydrogen-generating systems, vehicle refueling systems, piping and storage systems and testing of piping systems. Fuel cell power plants are addressed in Section 633. The definition of "Fuel gas" includes hydrogen in addition to the traditional fuel gases such as natural gas and liquefied petroleum gases (LP-gases). Chapter 7 was created to segregate the hydrogen provisions from the provisions applicable to other fuel gases and thus avoid confusion regarding what applies to hydrogen and what does not. The hydrogen coverage is intended to be self-contained in this chapter, except that Chapters 1, 2, 8 and the applicable portions of Chapter 3 also apply in conjunction with Chapter 7.

Hydrogen, symbol H_2, is the lightest of all gases with a vapor density of 0.1, meaning that air is approximately 10 times heavier than hydrogen. At atmospheric pressure, it has a flammability range of 4.0 to 75.0 percent vapor in air by volume. Hydrogen is the most abundant element in the universe and is believed to be the element from which all other elements were created in the cosmos. Because of its small molecular size, hydrogen diffuses rapidly through porous materials. It also dissolves in and diffuses through metals slowly at ambient temperatures and more rapidly at elevated temperatures.

On earth, hydrogen is not found in uncombined form as free hydrogen except in minute traces. The majority of our hydrogen is locked in water molecules, H_2O. Hydrogen is produced by several chemical processes, including the electrolysis of water and the reaction of natural gas or light hydrocarbons with steam in the presence of a catalyst (steam reforming). Most of the hydrogen now produced in the United States is on an industrial scale by the process of steam reforming, or as a byproduct of petroleum refining and chemicals production. However, there is growing interest in two different types of hydrogen-generating appliances to produce hydrogen on-site at the customer's refueling facility or even at a private residence from either electricity or from natural gas, propane or other fuels.

Electrolysis of water breaks it down into its two building blocks, hydrogen and oxygen, by passing an electrical current through electrodes submerged in the water [see Commentary Figures 700(1) and 702.1]. Adding an electrolyte such as a salt improves the conductivity of the water and increases the efficiency of the process. The hydrogen gas is collected and compressed for storage and the oxygen is typically released to the atmosphere.

Although hydrogen holds great promise as a clean, abundant and nonpolluting source of energy, there is no free ride in nature. Before the energy of hydrogen can be used, an equal or greater amount of energy to produce, store and transport the gas must be expended. In other words, energy must be used to tear apart molecules to generate hydrogen before the energy obtained from recombining hydrogen and oxygen atoms in fuel cells can be reaped. It is certainly worth the effort because fuel cells are more efficient and far less polluting than gasoline and diesel engines.

Purpose

Chapter 7 provides for the installation of hydrogen systems to reduce the hazards that are associated with flammable gases and those that are unique to hydrogen. The use of hydrogen as a fuel for power generation and vehicles is expected to grow in the near future as the technology continues to advance and a hydrogen infrastructure is built. In fact, it is the lack of availability of hydrogen, not the lack of equipment that uses hydrogen that is preventing more widespread use of hydrogen fuel [see Commentary Figures 700(2) and 700(3)].

Hydrogen is a colorless, odorless, tasteless, flammable nontoxic gas that burns with an almost invisible bluish flame. A hydrogen flame is very pale and difficult to see in daylight. A major concern with hydrogen is the fact that odorants cannot be added because the chemicals involved would poison the fuel cells, thus hydrogen is not odorized and cannot be detected by any of the human senses.

Water Electrolysis

Figure 700(1)
ELECTROLYSIS OF WATER

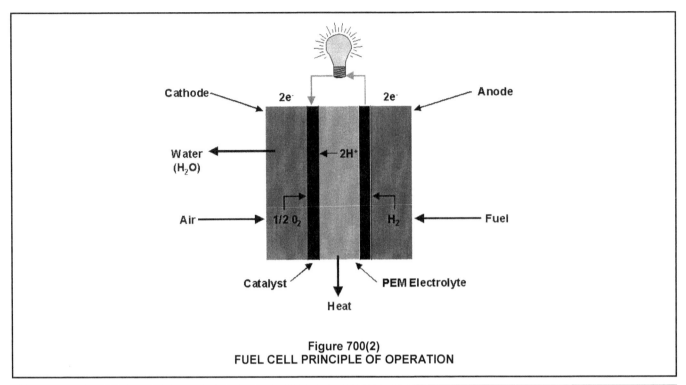

Figure 700(2)
FUEL CELL PRINCIPLE OF OPERATION

Figure 700(3)
FUEL CELL PRINCIPLE OF OPERATION

SECTION 701 (IFGC)
GENERAL

701.1 Scope. The installation of gaseous hydrogen systems shall comply with this chapter and Chapters 53 and 58 of the *International Fire Code*. Compressed gases shall also comply with Chapter 50 of the *International Fire Code* for general requirements.

❖ Chapter 30 of the *International Fire Code®* (IFC®) covers the use, handling and storage of compressed gases and Chapter 35 covers the use and storage of flammable gases. Chapters 30 and 35 of the IFC address the hazards of compressed hydrogen gas systems including stored hydrogen gas in pressure vessels. These referenced code sections are intended to reduce the risk posed by the inadvertent rupture of a pressure vessel and release of its hydrogen gas component or the leakage of the flammable gas associated with a piping rupture.

A reference to the commensurate IFC provisions appropriately serves to limit the quantities of compressed hydrogen gas present on both commercial and residential sites, and considers indoor and outdoor use aspects adequately. Reference to Chapter 41 covering pyrophoric materials is intended to address emerging technology (now being contemplated by the automotive industry) that allows for storage of higher densities of compressed hydrogen gas in containers with relatively lower applied pressures.

701.2 Permits. Permits shall be required as set forth in Section 106 and as required by the *International Fire Code*.

❖ This section refers to the applicable administrative rules in the IFC and code governing the issuance, suspension, renewal or modification of a permit for work affecting gaseous hydrogen service systems.

SECTION 702 (IFGC)
GENERAL DEFINITIONS

702.1 Definitions. The following words and terms shall, for the purposes of this chapter and as used elsewhere in this code, have the meanings shown herein.

❖ Section 202 applies to this chapter. The terms that are unique to this chapter are defined herein for convenience.

[F] GASEOUS HYDROGEN SYSTEM. An assembly of *piping*, devices and apparatus designed to generate, store, contain, distribute or transport a nontoxic, gaseous hydrogen containing mixture having at least 95-percent hydrogen gas by volume and not more than 1-percent oxygen by volume. Gaseous hydrogen systems consist of items such as compressed gas containers, reactors and appurtenances, including pressure regulators, pressure relief devices, manifolds, pumps, compressors and interconnecting *piping* and tubing and controls.

❖ This term includes the source of hydrogen and all piping and devices between the source and the using equipment. The gas in a hydrogen system is above the upper flammability limit and is therefore, "too rich"

to burn. Any leakage, however, will quickly form an explosive mixture in the ambient atmosphere.

[F] HYDROGEN FUEL-GAS ROOM. A room or space that is intended exclusively to house a gaseous hydrogen system.

❖ An enclosed space used exclusively for a gaseous hydrogen system requires construction and protection that are unique to the hazards associated with these areas. The definition itself should not be interpreted to preclude H_2 piping systems from serving distributed hydrogen-using equipment and appliances located elsewhere on site or in the building. However, the appropriate material-specific quantity limitations of the *International Building Code®* (IBC®) and the IFC must be consulted.

HYDROGEN-GENERATING APPLIANCE. A self-contained package or factory-matched packages of integrated systems for generating gaseous hydrogen. Hydrogen-generating appliances utilize electrolysis, reformation, chemical or other processes to generate hydrogen.

❖ Hydrogen generators use water or hydrocarbon fuels as a feedstock for generation of pure hydrogen or a hydrogen-rich gas (see Commentary Figure 702.1). Hydrogen-generating appliances based on chemical reformers separate out the hydrogen from fossil fuels such as natural gas, propane, gasoline, etc. This is the same high-temperature chemical process used at large oil refineries to produce hydrogen. By generating the hydrogen on site at the fueling station or customer's facility, these hydrogen-generating appliances avoid the high cost of either liquefying hydrogen and delivering it by cryogenic tanker truck, or installing a national hydrogen pipeline system that could cost many tens of billions of dollars. In effect, these on-site hydrogen-generating appliances take advantage of one of two existing energy infrastructures: either the natural gas distribution system or the electrical grid.

SECTION 703 (IFGC)
GENERAL REQUIREMENTS

703.1 Hydrogen-generating and refueling operations. Hydrogen-generating and refueling appliances shall be installed and located in accordance with their listing and the manufacturer's instructions. Ventilation shall be required in accordance with Section 703.1.1, 703.1.2 or 703.1.3 in public garages, private garages, repair garages, automotive motor fuel-dispensing facilities and parking garages that contain hydrogen-generating appliances or refueling systems. For the purpose of this section, rooms or spaces that are not part of the living space of a dwelling unit and that communicate directly with a private garage through openings shall be considered to be part of the private garage.

❖ This section intends to minimize the potential for explosions by limiting the source of hydrogen gas and by requiring sufficient ventilation to dissipate any leakage. The last sentence of this section is the same as found in Section 305.3 and is addressed in the commentary for that section.

703.1.1 Natural ventilation. Indoor locations intended for hydrogen-generating or refueling operations shall be limited to a maximum floor area of 850 square feet (79 m²) and shall communicate with the outdoors in accordance with Sections 703.1.1.1 and 703.1.1.2. The maximum rated output capacity of hydrogen generating appliances shall not exceed 4 standard cubic feet per minute (0.00189 m³/s) of hydrogen for each 250 square feet (23.2 m²) of floor area in such spaces. The minimum cross-sectional dimension of air openings shall be 3 inches (76 mm). Where ducts are used, they shall be of the same cross-sectional area as the free area of the openings to which they connect. In such locations, *equipment* and appliances having an *ignition source* shall be located such that the source of ignition is not within 12 inches (305 mm) of the ceiling.

❖ This section parallels the intent of Section 305.3, but, with respect to the ceiling instead of the floor. This section does not apply to spaces that only house vehicles and that do not contain hydrogen-generating or refueling operations. Because it is more buoyant, hydrogen will dissipate more quickly than natural gas, and much more quickly than either propane or gasoline, both of which have vapors that are heavier than air and will linger at an accident site. However, hydrogen and natural gas can both accumulate in unventilated pockets at the top of indoor structures and could represent a risk in such situations.

Thus, ignition sources must be avoided at the top of any unventilated spaces for hydrogen gas. Also, hydrogen is odorless, colorless and burns with a flame that is not generally visible to the human eye. This means that it is unlikely that people will be able to detect unsafe conditions without appropriate instrumentation [similar to carbon monoxide (CO) accumulation in a structure].

Figure 702.1
HYDROGEN-GENERATING APPLIANCE

703.1.1.1 Two openings. Two permanent openings shall be provided within the garage. The upper opening shall be located entirely within 12 inches (305 mm) of the ceiling of the garage. The lower opening shall be located entirely within 12 inches (305 mm) of the floor of the garage. Both openings shall be provided in the same exterior wall. The openings shall communicate directly with the outdoors and shall have a minimum free area of ¹/₂ square foot per 1,000 cubic feet (1 m²/610 m³) of garage volume.

❖ The location requirement will prevent the openings from being more than 12 inches (305 mm) tall because the required openings must be entirely within the 12 inches (305 mm) of wall space measured down from the ceiling and up from the floor. The openings must be in the same wall to help create a gravity flow of gases driven by the natural buoyancy of hydrogen gas. The bottom opening is an air inlet and the top is an air outlet.

703.1.1.2 Louvers and grilles. In calculating the free area required by Section 703.1.1.1, the required size of openings shall be based on the net free area of each opening. If the free area through a design of louver or grille is known, it shall be used in calculating the size opening required to provide the free area specified. If the design and free area are not known, it shall be assumed that wood louvers will have 25-percent free area and metal louvers and grilles will have 75-percent free area. Louvers and grilles shall be fixed in the open position.

❖ This section recognizes that louvers and grilles are usually installed over air inlets and outlets to prevent rain, snow and animals from entering the building. When louvers or grilles are used, the solid portion of the louver or grille must be considered when determining the unobstructed (net clear) area of the opening.

Air openings are sized based on there being a free, unobstructed area for the passage of air into the space. Louvers or grilles placed over these openings reduce the area of the openings because of the area occupied by the solid portions of the grille or louver. The reduction in area must be considered because only the unobstructed area can be credited toward the required opening size.

The reduction in opening area caused by the presence of grilles or louvers will always require openings to be larger than determined from the sizing ratios of this chapter and larger than any duct of the minimum required size that might connect to these openings.

703.1.2 Mechanical ventilation. Indoor locations intended for hydrogen-generating or refueling operations shall be ventilated in accordance with Section 502.16 of the *International Mechanical Code*. In such locations, *equipment* and appliances having an *ignition source* shall be located such that the source of ignition is below the mechanical ventilation *outlet*(s).

❖ Section 502.16 of the *International Mechanical Code®* (IMC®) contains criteria for the design and operation of a mechanical ventilation system in repair garages for natural-gas- and hydrogen-fueled vehicles. This section makes Section 502.16 of the IMC applicable to indoor spaces containing hydrogen-generating and/or refueling operations. Sources of ignition must be located below the exhaust outlets because it is at the exhaust outlets where the highest hydrogen concentration would occur.

703.1.3 Specially engineered installations. As an alternative to the provisions of Sections 703.1.1 and 703.1.2, the necessary supply of air for ventilation and dilution of flammable gases shall be provided by an *approved* engineered system.

❖ The code is not intended to inhibit innovative ideas or technological advances. A comprehensive regulatory document such as a fuel gas code cannot envision and then address all future innovations in the industry. As a result, a performance code must be applicable to and provide a basis for the approval of an increasing number of newly developed, innovative materials, systems and methods for which no code text or referenced standards yet exist. The fact that a material, product or method of construction is not addressed in the code is not an indication that prohibition of the material, product or method is intended. The code official is expected to apply sound technical judgement in accepting materials, systems or methods that, while not anticipated by the drafters of the current code text, can be demonstrated to offer equivalent performance. By virtue of its text, the code regulates new and innovative construction practices while addressing the relative safety of building occupants. The code official is responsible for determining whether a requested alternative provides a level of protection of the public health, safety and welfare as required by the code.

[F] 703.2 Containers, cylinders and tanks. Compressed gas containers, cylinders and tanks shall comply with Chapters 53 and 58 of the *International Fire Code.*

❖ The design and construction of containers must meet the requirements listed in Chapters 30 and 35 of the IFC and the standards referenced there.

[F] 703.2.1 Limitations for indoor storage and use. Flammable gas cylinders in occupancies regulated by the *International Residential Code* shall not exceed 250 cubic feet (7.1 m³) at normal temperature and pressure (NTP).

❖ Flammable gas must not be stored or used where an accident could cause a loss of life. They must be stored and used under preventive safety guidelines. Cylinders with a capacity of 250 cubic feet (7.1 m³) or less are allowed for maintenance of buildings, taking care of patients and equipment operation.

[F] 703.2.2 Design and construction. Compressed gas containers, cylinders and tanks shall be designed, constructed and tested in accordance with Chapter 50 of the *International Fire Code,* ASME *Boiler and Pressure Vessel Code* (Section VIII) or DOTn 49 CFR, Parts 100-180.

❖ Chapter 27 of the IFC and the referenced standards establish rules of safety governing the design, fabrication and testing of storage vessels, including those for gaseous and liquid hydrogen.

[F] 703.3 Pressure relief devices. Pressure relief devices shall be provided in accordance with Sections 703.3.1 through 703.3.8. Pressure relief devices shall be sized and selected in accordance with CGA S-1.1, CGA S-1.2 and CGA S-1.3.

❖ Pressure relief devices are essential for cryogenics because of the high pressures and low temperatures at which cryogenics are maintained. Although storage tanks, other containers and transfer piping are normally well insulated, some heating of the contents will occur with time, causing internal pressures to increase. Pressure relief mechanisms are one method of relieving these overpressures and avoiding a hazardous situation. Three Compressed Gas Association standards that cover the full range of container types from portable to stationary are referenced.

Sections 703.3.1 through 703.3.8 contain requirements related to accessibility for maintenance, general sizing, installation requirements and device integrity.

[F] 703.3.1 Valves between pressure relief devices and containers. Valves including shutoffs, check valves and other mechanical restrictions shall not be installed between the pressure relief device and container being protected by the relief device.

Exception: A locked-open shutoff valve on containers equipped with multiple pressure relief device installations where the arrangement of the valves provides the full required flow through the minimum number of required relief devices at all times.

❖ Any valve or restriction placed between a pressure relief device and the vessel it protects could render the device useless if the valve was totally or partially closed. Restrictions in the inlet to a pressure relief device are dangerous because closing the valve in that location would allow overpressure in the container to build without relief.

[F] 703.3.2 Installation. Valves and other mechanical restrictions shall not be located between the pressure relief device and the point of release to the atmosphere.

❖ Valves and other restrictions in the outlet (discharge) piping from a pressure relief device would have the same effect as valves and restrictions placed upstream of the device (see commentary, Section 703.3.1).

[F] 703.3.3 Containers. Containers shall be provided with pressure relief devices in accordance with the ASME *Boiler and Pressure Vessel Code* (Section VIII), DOTn 49 CFR, Parts 100-180 and Section 703.3.7.

❖ This section states very clearly that pressure relief devices are required for gaseous hydrogen containers. The container standards referred to cover a broad range of container types, portable and stationary. Pressure relief devices are required only as dictated by the specifications to which the container was fabricated (i.e., ASME or DOTn).

[F] 703.3.4 Vessels other than containers. Vessels other than containers shall be protected with pressure relief devices in accordance with the ASME *Boiler and Pressure Vessel Code* (Section VIII), or DOTn 49 CFR, Parts 100-180.

❖ Just as Section 703.3.3 requires pressure relief devices on containers, this section mentions other hydrogen-containing vessels or components in which an overpressure condition could occur.

[F] 703.3.5 Sizing. Pressure relief devices shall be sized in accordance with the specifications to which the container was fabricated. The relief device shall be sized to prevent the maximum design pressure of the container or system from being exceeded.

❖ This section contains only general language that ensures that the pressure relief device is properly designed to fit the needs of the particular container. In most cases the manufacturer will already have the devices installed on the container; however, there are cases when a cryogenic system or container may be constructed to a user specification for a specific purpose. In those cases the relief valves must be sized and installed by the user.

[F] 703.3.6 Protection. Pressure relief devices and any associated vent *piping* shall be designed, installed and located so that their operation will not be affected by water or other debris accumulating inside the vent or obstructing the vent.

❖ As with all such protective devices, the relief devices must be located and protected to prevent submersion and the entrance of water, debris, insects, etc., all of which could impair operation of the device.

[F] 703.3.7 Access. Pressure relief devices shall be located such that they are provided with ready *access* for inspection and repair.

❖ These devices are extremely important safety controls that require periodic inspections and, in some cases, testing.

[F] 703.3.8 Configuration. Pressure relief devices shall be arranged to discharge unobstructed in accordance with Section 2309 of the *International Fire Code*. Discharge shall be directed to the outdoors in such a manner as to prevent impingement of escaping gas on personnel, containers, *equipment* and adjacent structures and to prevent introduction of escaping gas into enclosed spaces. The discharge shall not terminate under eaves or canopies.

> **Exception:** This section shall not apply to DOTn-specified containers with an internal volume of 2 cubic feet (0.057 m³) or less.

❖ This section refers to the IFC regarding requirements for the design and location of hydrogen vent piping. The vent piping must be arranged so that the discharge itself does not cause additional damage simply because of the location chosen for the discharge point.

[F] 703.4 Venting. Relief device vents shall be terminated in an *approved* location in accordance with Section 2309 of the *International Fire Code*.

❖ Because of container locations, vent piping is often used to extend the discharge location of the pressure relief vent. The vent system will prevent excessive buildup of pressure in the system by venting gas to the outdoors. Reference to the IFC addresses such issues as material requirements and joining methods, general support, height and separation distances for the proposed vent outlet location. The location of the vent outlet is critical because the ignition of vented hydrogen has the potential for damage to materials and harm to people from radiant heat exposure.

[F] 703.5 Security. Compressed gas containers, cylinders, tanks and systems shall be secured against accidental dislodgement in accordance with Chapter 53 of the *International Fire Code*.

❖ Pressure vessels and their connection valves and devices can be damaged by falling, rolling and tipping over. Therefore, they must be secured in place. For example, a cylinder containing any gas pressurized to a few thousand pounds per square inch will become a missile if the valve is knocked off during a fall. This is especially dangerous when the gas is flammable.

[F] 703.6 Electrical wiring and equipment. Electrical wiring and *equipment* shall comply with NFPA 70.

❖ Electrical power, control and signal wiring associated with hydrogen systems must comply with NFPA 70. Some areas containing hydrogen system components could be classified as hazardous locations.

SECTION 704 (IFGC)
PIPING, USE AND HANDLING

704.1 Applicability. Use and handling of containers, cylinders, tanks and hydrogen gas systems shall comply with this section. Gaseous hydrogen systems, *equipment* and machinery shall be *listed* or *approved*.

❖ All hydrogen system components must be listed for the application unless specifically approved by the code official. All components must be carefully chosen as suitable for the application, especially because of the extreme flammability, small molecule size, lack of detectable odor and high pressure associated with hydrogen.

704.1.1 Controls. Compressed gas system controls shall be designed to prevent materials from entering or leaving process or reaction systems at other than the intended time, rate or path. Automatic controls shall be designed to be fail safe in accordance with accepted engineering practice.

❖ "Fail-safe" mode means that if and when a device fails, the resulting condition will not create a hazard. For example, if a compressor motor starter fails, the contacts would fail in the open position, thereby stopping the motor or preventing it from starting.

704.1.2 Piping systems. *Piping,* tubing, valves and fittings conveying gaseous hydrogen shall be designed and installed in accordance with Sections 704.1.2.1 through 704.1.2.5.1, Chapter 50 of the *International Fire Code,* and ASME B31.12. Cast-iron pipe, valves and fittings shall not be used.

❖ All apparatus connected and part of the system must comply with this section, Chapter 27 of the IFC and the requirements of the referenced standard applicable to gaseous hydrogen.

704.1.2.1 Sizing. Gaseous hydrogen *piping* shall be sized in accordance with *approved* engineering methods.

❖ Hydrogen piping and manifold systems should be designed and constructed by competent personnel who are thoroughly familiar with the requirements for piping of flammable gases. Sizing information can be obtained from a variety of applicable codes and regulations; however, hydrogen piping systems should be designed to comply with the principles of ASME B31.3, which includes special requirements for hydrogen service.

704.1.2.2 Identification of hydrogen piping systems. Hydrogen *piping* systems shall be marked in accordance with ANSI A13.1. Markings used for *piping* systems shall consist of the name of the contents and shall include a direction-of-flow arrow. Markings shall be provided at all of the following locations:

1. At each valve.

2. At wall, floor and ceiling penetrations.

3. At each change of direction.

4. At intervals not exceeding 20 feet (6096 mm).

❖ This section applies to all piping, regardless of material. ANSI A13.1 details the valve and piping color coding, painting and labeling methods used in the United States.

704.1.2.3 Piping design and construction. *Piping* and tubing materials shall be 300 series stainless steel or materials *listed* or *approved* for hydrogen service and the use intended through the full range of operating conditions to which they will be subjected. *Piping* systems shall be designed and constructed to provide allowance for expansion, contraction, vibration, settlement and fire exposure.

❖ Hydrogen piping systems can consist of structural members, vacuum jackets, valve bodies and valve seats, electrical and thermal insulation, gaskets, seals, lubricants and adhesives, and will involve a multitude of materials. Hydrogen embrittlement involves many variables and can cause significant deterioration in the mechanical properties of certain metals. One must thoroughly review material selection methods and bills-of-lading, quality control procedures and material test reports used during manufacture to verify that the materials are suitable for hydrogen service.

The code specifies stainless steel piping and tubing within the 300 series of alloys or materials listed or approved for gaseous hydrogen service.

Care must be taken to not overstress components in systems because some of these materials may lose ductility if stressed beyond yield.

704.1.2.3.1 Prohibited locations. *Piping* shall not be installed in or through a circulating air duct; clothes chute; chimney or gas vent; ventilating duct; dumbwaiter; or elevator shaft. *Piping* shall not be concealed or covered by the surface of any wall, floor or ceiling.

❖ There are several reasons why hydrogen gas piping is not to be installed in the locations listed in this section as well as in similar areas. Gas piping must not be located where the spread of leaked gas throughout the building would be accelerated. An air supply duct or areas that create a path for the gas to travel are considered to be locations where the problem of a gas leak could be compounded by spreading the gas or any resultant fire throughout the building. Because of the nature of hydrogen, the code prescribes extreme caution in system design and installation and requires hydrogen piping to always be exposed and observable. Also, in some locations, the gas piping could be subject to deterioration because of temperature or corrosive conditions such as in air ducts or chimneys. A circulating air duct is an HVAC supply, return or ventilation air duct that conveys air to and from occupied spaces. An above-ceiling or below-floor plenum is not a circulating air duct; however, hydrogen piping is still prohibited in such plenums by the last sentence of this section.

Locating gas piping in certain areas or atmospheres may cause corrosion of the pipe, which in turn may cause a leak in the piping system. Within a supply air duct, the conditioned air may cause moisture to condense on or within the pipe, thereby causing corrosion of the pipe. Gas piping located within a chimney or a vent will be subjected to high temperatures and the corrosive effects of vent gases.

Additionally, piping located in a dumbwaiter or elevator shaft may be subject to an additional hazard resulting in mechanical damage to the pipe. Damage from a dumbwaiter or elevator impacting the gas piping may not only cause a leak that would allow the gas to escape and travel up the shaft but may also put any occupants within an elevator in immediate danger.

Hydrogen gas piping is prohibited in all concealed spaces including those used as air plenums above a suspended ceiling.

704.1.2.3.2 Interior piping. Except for through penetrations, *piping* located inside of buildings shall be installed in exposed locations and provided with ready *access* for visual inspection.

❖ As a precaution against material failure and leakage, hydrogen piping/tubing installed within a building must not be concealed in any manner and must be in plain view. The only exception to this requirement is where the pipe/tube passes through an assembly such as where a pipe/tube penetrates a wall at an angle perpendicular to the plane of the wall.

704.1.2.3.3 Underground piping. Underground *piping*, including joints and fittings, shall be protected from corrosion and installed in accordance with *approved* engineered methods.

❖ The method of installation and protection from corrosion must be engineered and acceptable to the code official. Buried piping may have to be encased in a conduit to shield it from the backfill.

704.1.2.3.4 Piping through foundation wall. Underground *piping* shall not penetrate the outer foundation or basement wall of a building.

❖ This section does not allow piping to enter or exit a building at any point below ground level. The risk of piping failure from settlement or corrosion is too great to allow underground penetrations; therefore, piping must enter or exit the building above ground.

704.1.2.3.5 Protection against physical damage. Where *piping* other than stainless steel *piping*, stainless steel tubing or black steel is installed through holes or notches in wood studs, joists, rafters or similar members less than $1^1/_2$ inches (38 mm) from the nearest edge of the member, the pipe shall be protected by shield plates. Shield plates shall be a minimum of $^1/_{16}$-inch-thick (1.6 mm) steel, shall cover the area of the pipe where the member is notched or bored and shall extend a minimum of 4 inches (102 mm) above sole plates, below top plates and to each side of a stud, joist or rafter.

❖ See the commentary to Section 404.5.

704.1.2.3.6 Piping outdoors. *Piping* installed above ground, outdoors, shall be securely supported and located where it will be protected from physical damage. *Piping* passing through an exterior wall of a building shall be encased in a protective pipe sleeve. The annular space between the *piping* and the sleeve shall be sealed from the inside such that the sleeve is ventilated to the outdoors. Where passing through an exterior wall of a building, the *piping* shall also be protected against corrosion by coating or wrapping with an inert material. Below-ground *piping* shall be protected against corrosion.

❖ Extraordinary protection of piping is required because of the potential danger of leakage. Wall sleeves are sealed only on the interior end so that any leakage in the sleeve will vent to the outdoors.

704.1.2.3.7 Settlement. *Piping* passing through concrete or masonry walls shall be protected against differential settlement.

❖ Building movement and shifting can put enormous stress on piping; therefore, it must be protected from stress by sleeves, oversized openings, expansion loops and similar measures.

704.1.2.4 Joints. Joints in *piping* and tubing in hydrogen service shall be *listed* as complying with ASME B31.3 to include the use of welded, brazed, flared, socket, slip and compression fittings. Gaskets and sealants used in hydrogen

service shall be *listed* as complying with ASME B31.12. Threaded and flanged connections shall not be used in areas other than hydrogen cutoff rooms and outdoors.

❖ Because hydrogen gas is composed of small molecules (two hydrogen atoms), it is more difficult to contain, especially under the higher pressures anticipated for fuel delivery to vehicles. All joining methods for piping and tubing must be listed as complying with the referenced standards, including welded, brazed, flared, socket, slip or compression fittings. Soft solder joints are not permitted. Threaded or flanged connections must not be used in areas other than hydrogen cutoff rooms or outdoors. Mechanical joints must maintain electrical continuity or otherwise be connected with a bonding jumper to avoid static discharge. Gaskets or sealants must be listed as complying with the referenced standard.

704.1.2.4.1 Brazed joints. Brazing alloys shall have a melting point greater than 1,000°F (538°C).

❖ This is consistent with the definition of "Brazing" and Section 403.10.1.

704.1.2.4.2 Electrical continuity. Mechanical joints shall maintain electrical continuity through the joint or a bonding jumper shall be installed around the joint.

❖ To help control static electricity discharge and provide equi-potential bonding through the metal piping system, electrical conductivity (continuity) must be maintained throughout the system.

704.1.2.5 Valves and piping components. Valves, regulators and *piping* components shall be *listed* or *approved* for hydrogen service, shall be provided with *access* and shall be designed and constructed to withstand the maximum pressure to which such components will be subjected.

❖ Valves, gauges, regulators and other piping components must be listed or approved for hydrogen service for the rated pressure and temperature of the application. The manufacturer or hydrogen supplier should be consulted for valves, regulators and other accessories.

704.1.2.5.1 Shutoff valves on storage containers and tanks. Shutoff valves shall be provided on all storage container and tank connections except for pressure relief devices. Shutoff valves shall be provided with ready *access*.

❖ Containers, tanks and the associated connecting piping must have an accessible manual shutoff valve. Valves must not be installed between a pressure relief device (PRD) and the container protected by the PRD because this would circumvent the purpose of overpressure relief. Shutoff valves on tanks and containers must be conspicuous, and must not be hidden behind panels or access doors because they may have to be accessed quickly in an emergency.

704.2 Upright use. Compressed gas containers, cylinders and tanks, except those with a water volume less than 1.3 gallons (5 L) and those designed for use in a horizontal position, shall be used in an upright position with the valve end up. An upright position shall include conditions where the container, cylinder or tank axis is inclined as much as 45 degrees (0.79 rad) from the vertical.

❖ Containers, cylinders and tanks are to be used in an upright position to protect and provide access to the shutoff valve and to make sure that gaseous hydrogen is taken from the vessel where liquid is also present.

704.3 Material-specific regulations. In addition to the requirements of this section, indoor and outdoor use of hydrogen compressed gas shall comply with the material-specific provisions of Chapters 53 and 58 of the *International Fire Code*.

❖ In addition to meeting the requirements of this section, the indoor and outdoor use and storage of gaseous hydrogen needs to comply with the material-specific provisions of Chapters 30 and 35 of the IFC. For example, the storage and use of a compressed flammable gas system must comply with Chapters 30 and 35 of the IFC in addition to this section.

704.4 Handling. The handling of compressed gas containers, cylinders and tanks shall comply with Chapter 50 of the *International Fire Code*.

❖ The requirements of this section refer to Chapter 27 and the applicable material-specific requirements of the IFC for handling gas containers, cylinders and tanks.

SECTION 705 (IFGC)
TESTING OF HYDROGEN PIPING SYSTEMS

705.1 General. Prior to acceptance and initial operation, all *piping* installations shall be inspected and pressure tested to determine that the materials, design fabrication and installation practices comply with the requirements of this code.

❖ After installation, field-erected piping, tubing, hose and hose assemblies must be tested and proved hydrogen gas tight for the rated pressure and temperature of the gas conveyed in that portion of the system.

Inspection of piping installations is intended to be a visual observation of the system and the testing procedure. The designer may require that welded joints in, for example, very-high-pressure applications be examined by a method that is capable of discovering internal defects that are not detectable by visual observation.

705.2 Inspections. Inspections shall consist of a visual examination of the entire *piping* system installation and a pressure test. Hydrogen *piping* systems shall be inspected in accordance with this code. Inspection methods such as outlined in ASME B31.12 shall be permitted where specified by the design engineer and *approved* by the code official. Inspections shall be conducted or verified by the code official prior to system operation.

❖ In addition to pressure testing, the entire system must be visually inspected to verify that it has been properly installed, supported and protected from damage

and that all required components are present. Recall that the code requires that all piping and tubing be permanently exposed except at through-penetrations.

705.3 Pressure tests. A hydrostatic or pneumatic leak test shall be performed. Testing of hydrogen *piping* systems shall utilize testing procedures identified in ASME B31.12 or other *approved* methods, provided that the testing is performed in accordance with the minimum provisions specified in Sections 705.3.1 through 705.4.1.

❖ Testing and purging procedures must be as described in the referenced standard, ASME B31.3.

Pressure testing in ASME B31.3 may be either by hydrostatic or pneumatic means. Hydrostatic and pneumatic tests are performed at $1^1/_2$ times the maximum allowable working pressure (MAWP), times a temperature correction factor. In some situations, particularly when the hydrogen system contains catalysts or other materials that could be damaged by water, a pneumatic test is preferable to the hydrostatic test. The flexibility is intentional and left to the discretion of the designer.

Considering that some parts of a system might operate at up to 15,000 psi (103.4 MPa), the test pressure could be very high. Extreme care must be used when conducting this testing.

705.3.1 Hydrostatic leak tests. The hydrostatic test pressure shall be not less than one-and-one-half times the maximum working pressure, and not less than 100 psig (689.5 kPa gauge).

❖ See the commentary to Section 705.3.

705.3.2 Pneumatic leak tests. The pneumatic test pressure shall be not less than one-and-one-half times the maximum working pressure for systems less than 125 psig (862 kPa gauge) and not less than 5 psig (34.5 kPa gauge), whichever is greater. For working pressures at or above 125 psig (862 kPa gauge), the pneumatic test pressure shall be not less than 110 percent of the maximum working pressure.

❖ See the commentary to Section 705.3.

705.3.3 Test limits. Where the test pressure exceeds 125 psig (862 kPa gauge), the test pressure shall not exceed a value that produces hoop stress in the *piping* greater than 50 percent of the specified minimum yield strength of the pipe.

❖ This section limits test pressure to not exceed the stress safety factor for the material under test. Hoop stress refers to the pipe/tube's resistance to the force that would cause the pipe/tube to rupture along its circumference. The term is analogous to the hoops of a wooden barrel that hold in the staves.

705.3.4 Test medium. Deionized water shall be utilized to perform hydrostatic pressure testing and shall be obtained from a potable source. The medium utilized to perform pneumatic pressure testing shall be air, nitrogen, carbon dioxide or an inert gas; oxygen shall not be used.

❖ The test medium for any system must not contaminate or corrode the piping/tubing or react with the gas or liquid to be conveyed in the system.

705.3.5 Test duration. The minimum test duration shall be $^1/_2$ hour. The test duration shall be not less than $^1/_2$ hour for each 500 cubic feet (14.2 m^3) of pipe volume or fraction thereof. For *piping* systems having a volume of more than 24,000 cubic feet (680 m^3), the duration of the test shall not be required to exceed 24 hours. The test pressure required in Sections 705.3.1 and 705.3.2 shall be maintained for the entire duration of the test.

❖ As the piping system becomes larger, the test duration must be longer so that any leak can be detected. For example, a leakage rate of 1 cubic foot per hour (28.4 L/m) would produce a more easily detectable pressure drop in a small [10 cubic foot (0.28 m^3)] volume system. It would require a longer test period to detect the same leak in a large volume system because the leakage of 1 cubic foot (0.03 m^3) represents a much smaller fraction of the total gas volume in the system. If a system has a volume of 501 cubic feet (14.2 m^3), the minimum test duration would be 1 hour. To put this in perspective, 500 cubic feet (14.2 m^3) of piping system internal volume would equate to 83,333 feet (25.4 km) of 1-inch (25 mm) Schedule 40 steel pipe or 9,747 feet (2971 m) of 3-inch (76 mm) Schedule 40 steel pipe. At $^1/_2$ hour per each 500 cubic feet (14.2 m^3), a system having 24,000 cubic feet (680 m^3) of volume would require a test duration of 24 hours.

705.3.6 Test gauges. Gauges used for testing shall be as follows:

1. Tests requiring a pressure of 10 psig (68.95 kPa gauge) or less shall utilize a testing gauge having increments of 0.10 psi (0.6895 kPa) or less.

2. Tests requiring a pressure greater than 10 psig (68.98 kPa gauge) but less than or equal to 100 psig (689.5 kPa gauge) shall utilize a testing gauge having increments of 1 psi (6.895 kPa) or less.

3. Tests requiring a pressure greater than 100 psig (689.5 kPa gauge) shall utilize a testing gauge having increments of 2 psi (13.79 kPa) or less.

    Exception: Measuring devices having an equivalent level of accuracy and resolution shall be permitted where specified by the design engineer and *approved* by the code official.

❖ The smaller the increments are (the more detailed), the easier it is to detect a leak. Considering that hydrogen is not odorized, a leak-free system is even more imperative. Leak-indicating manometers and electronic instruments using pressure transducers are available that can read pressure changes with great accuracy, thus allowing detection of even the smallest leaks.

705.3.7 Test preparation. Pipe joints, including welds, shall be left exposed for examination during the test.

❖ Note that Sections 704.1.2.3 and 704.1.2.3.2 do not allow any part of the system to be concealed other than short sections occurring at through-penetrations.

705.3.7.1 Expansion joints. Expansion joints shall be provided with temporary restraints, if required, for the additional thrust load under test.

❖ Expansion joints might have to be clamped, bolted or otherwise restrained to prevent the force of the test pressure from damaging or separating them.

705.3.7.2 Equipment disconnection. Where the *piping* system is connected to appliances, *equipment* or components designed for operating pressures of less than the test pressure, such appliances, *equipment* and components shall be isolated from the *piping* system by disconnecting them and capping the *outlet*(s).

❖ This section has the same intent as Section 406.3.3. Isolation valves must not be depended upon to shield connected equipment and appliances from the test pressure.

705.3.7.3 Equipment isolation. Where the *piping* system is connected to appliances, *equipment* or components designed for operating pressures equal to or greater than the test pressure, such appliances, *equipment* and components shall be isolated from the *piping* system by closing the individual *appliance*, *equipment* or component shutoff valve(s).

❖ This section has the same intent as Section 406.3.4.

705.4 Detection of leaks and defects. The *piping* system shall withstand the test pressure specified for the test duration specified without showing any evidence of leakage or other defects. Any reduction of test pressures as indicated by pressure gauges shall indicate a leak within the system. *Piping* systems shall not be *approved* except where this reduction in pressure is attributed to some other cause.

❖ Any pressure drop, no matter how small, is considered a failure of the test, except where it can be demonstrated that the drop was caused by a change in temperature or other such cause. This would be practically impossible to demonstrate in the field, especially for short test durations. The temperature of the system and test medium should be allowed to stabilize before a test is conducted to lessen the possibility of pressure changes caused by temperature changes.

705.4.1 Corrections. Where leakage or other defects are identified, the affected portions of the *piping* system shall be repaired and retested.

❖ Where a leak is detected by a method addressed in Section 705.3.6 or other accurate testing instrument, the defect must be corrected. Corrections include tightening of fittings or threaded joints and replacing defective pipe, tubing or fittings.

705.5 Purging of gaseous hydrogen piping systems. Purging shall comply with Sections 705.5.1 through 705.5.4.

❖ See the commentary to Sections 705.5.1 through 705.5.4.

705.5.1 Removal from service. Where *piping* is to be opened for servicing, addition or modification, the section to be worked on shall be isolated from the supply at the nearest convenient point and the line pressure vented to the outdoors.

The remaining gas in this section of pipe shall be displaced with an inert gas.

❖ This section is parallel to Section 406.7.1 except that displacement with an inert gas is required regardless of pipe size and length.

705.5.2 Placing in operation. Prior to placing the system into operation, the air in the *piping* system shall be displaced with inert gas. The inert gas flow shall be continued without interruption until the vented gas is free of air. The inert gas shall then be displaced with hydrogen until the vented gas is free of inert gas. The point of discharge shall not be left unattended during purging. After purging, the vent opening shall be closed.

❖ This section is parallel to Section 406.7.2 except that displacement with an inert gas is required regardless of pipe size and length.

705.5.3 Discharge of purged gases. The open end of *piping* systems being purged shall not discharge into confined spaces or areas where there are sources of ignition except where precautions are taken to perform this operation in a safe manner by ventilation of the space, control of purging rate and elimination of all hazardous conditions.

❖ See the commentary to Section 406.7.3.

705.5.3.1 Vent pipe outlets for purging. Vent pipe outlets for purging shall be located such that the inert gas and fuel gas is released outdoors and not less than 8 feet (2438 mm) above the adjacent ground level. Gases shall be discharged upward or horizontally away from adjacent walls to assist in dispersion. Vent outlets shall be located such that the gas will not be trapped by eaves or other obstructions and shall be at least 5 feet (1524 mm) from building openings and lot lines of properties that can be built upon.

❖ The location, height and directional orientation of purge vent pipe outlets are dictated by this section to minimize the hazards of discharging flammable gases.

705.5.4 Placing equipment in operation. After the *piping* has been placed in operation, all *equipment* shall be purged in accordance with Section 707.2 and then placed in operation, as necessary.

❖ This section is parallel to Section 406.7.4 (see also Section 707.2).

SECTION 706 (IFGC)
LOCATION OF GASEOUS HYDROGEN SYSTEMS

[F] 706.1 General. The location and installation of gaseous hydrogen systems shall be in accordance with Sections 706.2 and 706.3.

Exception: Stationary fuel-cell power plants in accordance with Section 633.

❖ See the commentary to Sections 706.2 and 706.3.

[F] 706.2 Indoor gaseous hydrogen systems. Gaseous hydrogen systems shall be located in indoor rooms or areas constructed in accordance with this code, the *International Building Code*, the *International Mechanical Code* or NFPA 2.

❖ The overriding theme of the location requirements is to never permit the maximum concentration of flammable contaminants in air to exceed 25 percent of the lower flammability limit (LFL) for hydrogen during the period that a credible leak exists. This can be accomplished using natural or mechanical means to always maintain adequate ventilation to prevent a hazardous buildup of hydrogen gas in buildings or confined spaces.

[F] 706.3 Outdoor gaseous hydrogen systems. Gaseous hydrogen systems shall be located outdoors in accordance with Section 2309.3.1.1 of the *International Fire Code*.

❖ Section 2209.3.2 of the IFC addresses the location of fuel dispensing, generation, compression and storage relative to buildings, canopies, overhead lines, building openings, air intakes, ignition sources, lot lines, storage of other fuels and many other structures and occupancies.

SECTION 707 (IFGC)
OPERATION AND MAINTENANCE OF GASEOUS HYDROGEN SYSTEMS

[F] 707.1 Maintenance. Gaseous hydrogen systems and detection devices shall be maintained in accordance with the *International Fire Code* and the manufacturer's installation instructions.

❖ This section refers to the IFC for maintenance of hydrogen systems. More specifically, Section 2703.2.6 of the IFC details required maintenance activities for hazardous materials storage and use. This includes maintenance of alarms, cylinders, ventilation systems and other devices needed to ensure safety of the gaseous hydrogen system. Failure to properly maintain any of these systems increases the likelihood of an unanticipated hydrogen release that could compromise the safety of the operation. This section is also consistent with the general maintenance provision of Section 107.1 of the IFC.

Section 107.1 of the IFC does not identify who is responsible for maintenance because that determination should comply with the legal documents created between owners and occupants, such as a lease. However, the owner of a structure or premises is usually the party primarily responsible for its maintenance because the owner stands to gain the most from a well-maintained property. One of the underlying assumptions is that maintaining a commercial property in good condition allows the owner to recoup a substantial portion of his or her investment in maintenance. There are three factors that may influence owners to comply with code requirements:

- Code compliance requires only a small additional investment in the property,
- The owner has a long-term interest in the property and
- The owner expects profitability after incurring the additional expense of complying with the code.

Although all these factors represent economic incentives, code officials should be equally aware of potential disincentives to compliance, such as assessable value; expiring tax credits; or historic, architectural or aesthetic criteria. The code official need not belabor the justifications for compliance but should be prepared to acknowledge the owner's rationalizations for failure to comply.

This section also emphasizes that any "otherwise installed" system that currently exists must be maintained. For example, an existing fire protection system cannot be removed from a building because it is not required in new or existing buildings by current codes.

[F] 707.2 Purging. Purging of gaseous hydrogen systems, other than *piping* systems purged in accordance with Section 705.5, shall be in accordance with Sections 2309.6.3 and 2309.6.3.1 of the *International Fire Code* or in accordance with the system manufacturer's instructions.

❖ This section refers to the applicable requirements of the IFC for the defueling and purging of vehicles powered by gaseous fuels. Section 705.5 covers purging of piping systems whereas this section covers hydrogen storage containers and piping associated with hydrogen-fueled vehicles.

SECTION 708 (IFGC)
DESIGN OF LIQUEFIED HYDROGEN SYSTEMS ASSOCIATED WITH HYDROGEN VAPORIZATION OPERATIONS

[F] 708.1 General. The design of liquefied hydrogen systems shall comply with Chapter 55 of the *International Fire Code*.

❖ The requirements of this section refer to Chapter 32 of the IFC specifically for the design of liquefied hydrogen systems. The standards referenced there dealing with flammable cryogenic fluids must be adhered to in addition to complying with Chapter 32 of the IFC.

Bibliography

The following resource materials were used in the preparation of the commentary for this chapter of the code.

Code Changes Resource Collection. Washington, DC.: International Code Council, 2009.

IBC-2015, *International Building Code.* Washington, DC: International Code Council, 2014.

IFC-2015, *International Fire Code.* Washington, DC: International Code Council, 2014.

IMC-2015, *International Mechanical Code.* Washington, DC: International Code Council, 2014.

Chapter 8:
Referenced Standards

This chapter lists the standards that are referenced in various sections of this document. The standards are listed herein by the promulgating agency of the standard, the standard identification, the effective date and title, and the section or sections of this document that reference the standard. The application of the referenced standards shall be as specified in Section 102.8.

General Comments

Chapter 8 contains a comprehensive list of standards that are referenced in the code. It is organized to make locating specific document references easy.

It is important to understand that not every document related to the design, installation and construction of fuel gas systems, appliances and equipment is qualified to be a referenced standard. The International Code Council® (ICC®) has adopted a criterion that standards referenced in the *International Codes®* (I-Codes®) and standards intended for adoption in the International Codes must meet to qualify as referenced standards. The policy is summarized as follows:

- Code references: The scope and application of the standard must be clearly identified in the code text.

- Standard content: The standard must be written in mandatory language and be appropriate for the subject covered. The standard cannot have the effect of requiring proprietary materials or prescribing a proprietary testing agency.

- Standard promulgation: The standard must be readily available and developed and maintained in a consensus process such as those used by ASTM or ANSI.

The ICC Code Development Procedures, of which the standards policy is a part, are updated periodically. A copy of the latest version can be obtained from the ICC offices.

Once a standard is incorporated into the code through the code development process, it becomes an enforceable part of the code. When the code is adopted by a jurisdiction, the standard also is part of that jurisdiction's adopted code. It is for this reason that the criteria were developed. Compliance with this policy means that documents incorporated into the code are developed through the use of a consensus process, are written in mandatory language, and do not mandate the use of proprietary materials or agencies. The requirement for a standard is representative of the most current body of available knowledge on the subject as determined by a broad range of interested or affected parties without dominance by any single interest group. A true consensus process has many attributes, including but not limited to:

- An open process that has formal (published) procedures that allow for the consideration of all viewpoints,

- A definitive review period that allows for the standard to be updated and/or revised,

- A process of notification to all interested parties and

- An appeals process.

Many available documents related to mechanical system design, installation and construction, though useful, are not standards and are not appropriate for reference in the code. Often, these documents are not developed or written with the intention of being used for regulatory purposes and are unsuitable for use as a standard because of extensive use of recommendations, advisory comments and nonmandatory terms. Typical examples include installation guides and manuals, handbooks and design aids.

The objective of ICC's standards policy is to provide regulations that are clear, concise and enforceable—thus the requirement that standards be written in mandatory language. This requirement is not intended to mean that a standard cannot contain informational or explanatory material that will aid the user of the standard in its application. When the standard's promulgating agency wants such material to be included, however, the information must appear in a nonmandatory location, such as an annex or appendix, and be clearly identified as not being part of the standard.

Overall, standards referenced by the code must be authoritative, relevant, up-to-date and, most important, reasonable and enforceable. Standards that comply with ICC's standards policy fulfill these expectations.

Purpose

This code contains numerous references to documents that are used to regulate materials and methods of construction. The references to these documents within the code text consist of the promulgating agency's initialism and its publication designation (for example ASME B1.20.1) and a further indication that the document being referenced is the one that is listed in Chapter 8. Chapter 8 contains all of the information that is necessary to identify the specific referenced document. Included is the following information on a document's promulgating agency (see Commentary Figure 800):

- The promulgating agency (the agency's title),

- The promulgating agency's initialism, and

- The promulgating agency's address

For example, a reference to an ASME standard within the code indicates that the document is promulgated by the American Society of Mechanical Engineers (ASME), which is located in New York City, New York. Chapter 8 lists the standard agencies alphabetically for ease of identification. Chapter 8 also includes the following information on the referenced document itself (see Commentary Figure 800):

- The document's publication designation,
- The document's edition year,
- The document's title,
- Any addenda or revisions to the document known at the time of the code's publication, and
- Every section of the code in which the document is referenced.

For example, a reference to ASME B1.20.1 indicates this document can be found in Chapter 8 under the heading ASME. The specific standard's designation is B1.20.1. For convenience, these designations are listed in alphanumeric order. Chapter 8 shows that ASME B1.20.1 is titled *Pipe Threads, General Purpose (Inch)*; the applicable edition (that is, its year of publication) is 1992 and it is referenced in one specifically identified section of the code. Chapter 8 also indicates when a document has been discontinued or replaced by its promulgating agency. When a document is replaced by a different one, a note appears to tell the user the designation and title of the new document. The key aspect of the manner in which standards are referenced by the code is that a specific edition of a specific standard is clearly identified. The requirements necessary for compliance are, therefore, established and available to the code official, the mechanical contractor, the designer and the owner.

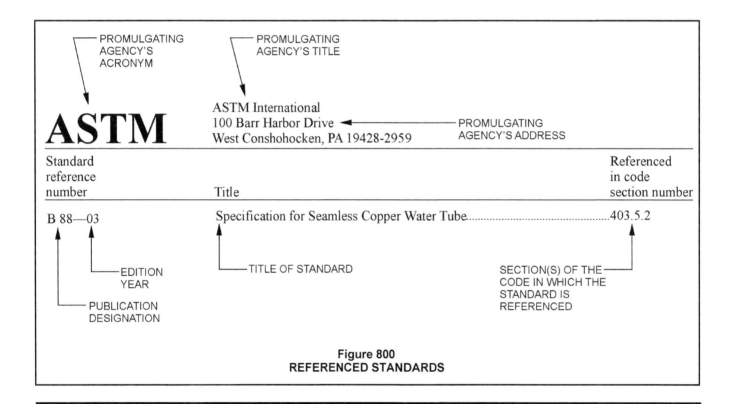

Figure 800
REFERENCED STANDARDS

ANSI

American National Standards Institute
25 West 43rd Street
Fourth Floor
New York, NY 10036

Standard reference number	Title	Referenced in code section number
ANSI A13.1—2007	Scheme for the Identification of Piping Systems	704.1.2.2
ANSI FC 1—2012	Stationery Fuel Cell Power Systems	633.1
LC 1/CSA 6.26—13	Fuel Gas Piping Systems Using Corrugated Stainless Steel Tubing (CSST)	403.5.4
ANSI LC-4/CSA 6.32—12	Press-connect Metallic Fittings for Use In Fuel Gas Distribution Systems	403.10.2
Z21.1—2010	Household Cooking Gas Appliances	623.1
Z21.5.1/CSA 7.1—2014	Gas Clothes Dryers—Volume I—Type 1 Clothes Dryers	613.1
Z21.5.2/CSA 7.2—2014	Gas Clothes Dryers—Volume II—Type 2 Clothes Dryers	613.1, 614.3
Z21.8—94 (R2002)	Installation of Domestic Gas Conversion Burners	619.1
Z21.10.1/CSA 4.1—2012	Gas Water Heaters—Volume I—Storage, Water Heaters with Input Ratings of 75,000 Btu per Hour or Less	624.1
Z21.10.3/CSA 4.3—2011	Gas Water Heaters—Volume III—Storage, Water Heaters with Input Ratings Above 75,000 Btu per hour, Circulating and Instantaneous	624.1
Z21.11.2—2011	Gas-fired Room Heaters-Volume II-Unvented Room Heaters	621.1
Z21.13/CSA 4.9—2011	Gas-fired Low-pressure Steam and Hot Water Boilers	631.1
Z21.15/CGA 9.1—09	Manually Operated Gas Valves for Appliances, Appliance Connector Valves and Hose End Valves	Table 409.1.1
Z21.19/CSA 1.4—02(R2007)	Refrigerators Using Gas (R1999) Fuel	625.1
Z21.24/CGA 6.10—06	Connectors for Gas Appliances	411.1, 411.3
Z21.40.1/CGA 2.91—96 (R2011)	Gas-Fired Heat Activated Air Conditioning and Heat Pump Appliances	627.1
Z21.40.2/CGA 2.92—96 (R2011)	Gas-Fired Work Activated Air Conditioning and Heat Pump Appliances (Internal Combustion)	627.1
Z21.42—2014	Gas-Fired Illuminating Appliances	628.1
Z21.47/CSA 2.3—2012	Gas-Fired Central Furnaces	618.1
Z21.50/CSA 2.22—2012	Vented Gas Fireplaces	604.1
Z21.54—09	Gas Hose Connectors for Portable Outdoor Gas Fired Appliances	411.1
Z21.56/CSA 4.7—2013	Gas-Fired Pool Heaters	617.1
Z21.58/CSA 1.6—2013	Outdoor Cooking Gas Appliances	623.1
Z21.60/CSA 2.26—2012	Decorative Gas Appliances for Installation in Solid-fuel Burning Fireplaces	602.1
Z21.61—83 (R2004)	Gas-fired Toilets	626.1
Z21.69/CSA 6.16—09	Connectors for Movable Gas Appliances	411.1.1, 411.1.4
Z21.75/CSA 6.27—07	Connectors for Outdoor Gas Appliances and Manufactured Homes	411.1, 411.2
Z21.80/CSA 6.22—11	Line Pressure Regulators	410.1
Z21.84—2012	Manually-lighted, Natural Gas Decorative Gas Appliances for Installation in Solid Fuel Burning Fireplaces	602.1, 602.2
Z21.86/CSA 2.32—08	Vented Gas-fired Space Heating Appliances	608.1, 609.1, 622.1
Z21.88/CSA 2.33—2015	Vented Gas Fireplace Heaters	605.1
Z21.91—07	Ventless Firebox Enclosures for Gas-fired Unvented Decorative Room Heaters	621.7.1
Z21.97—2012	Outdoor Decorative Appliances	636.1
Z83.4/CSA 3.7—2012	Nonrecirculating Direct-gas-fired Industrial Air Heaters	611.1
Z83.8/CSA 2.6—09	Gas Unit Heater, Gas Packaged Heater, Gas Utility Heaters and Gas-fired Duct Furnaces	610.1, 620.1
Z83.11/CSA 1.8—2013	Gas Food Service Equipment	623.1
Z83.18—2012	Recirculating Direct Gas-fired Industrial Air Heaters	612.1
Z83.19—01(R2009)	Gas-fired High-intensity Infrared Heaters	630.1
Z83.20—08	Gas-fired Low-intensity Infrared Heaters	630.1

ASME

American Society of Mechanical Engineers
Three Park Avenue
New York, NY 10016-5990

Standard reference number	Title	Referenced in code section number
B1.20.1—83 (Reaffirmed 2006)	Pipe Threads, General Purpose (inch)	403.9
B16.1—2010	Cast-iron Pipe Flanges and Flanged Fittings, Class 25, 125 and 250	403.12.1
B16.5—R2009	Pipe Flanges and Flanged Fittings: NPS $^1/_2$ through NFPS 24 Metric/Inch Standard	403.12.2
B16.24—2011	Cast Copper Alloy Pipe Flanges and Flanged Fittings: Classes 150, 300, 600, 900, 1500 and 2500	403.12.3
B16.42—2011	Ductile Iron Pipe Flanges and Flanged Fittings, Classes 150 and 300	403.12.4
B16.47—2011	Large Diameter Steel Flanges: NPS 26 through NPS 60 Metric/Inch Standard	403.12.2
B16.33—2012	Manually Operated Metallic Gas Valves for Use in Gas Piping Systems up to 125 psig (Sizes ½ through 2)	Table 409.1.1
B16.44—2002 (Reaffirmed 2007)	Manually Operated Metallic Gas Valves for Use in Aboveground Piping Systems Up to 5 psi	Table 409.1.1
B31.3—2012	Process Piping	704.1.2.4
B31.12—2008	Hydrogen Piping and Pipelines	704.1.2, 704.1.2.4, 705.2, 705.3
B36.10M—2004	Welded and Seamless Wrought-steel Pipe	403.4.2
BPVC—2010/2011 addenda	ASME Boiler & Pressure Vessel Code (2007 Edition)	631.1, 703.2.2, 703.3.3, 703.3.4
CSD-1—2011	Controls and Safety Devices for Automatically Fired Boilers	631.1

ASTM

ASTM International
100 Barr Harbor Drive
West Conshohocken, PA 19428-2959

Standard reference number	Title	Referenced in code section number
A53/A53M—12	Specification for Pipe, Steel, Black and Hot Dipped Zinc-coated Welded and Seamless	403.4.2
A106/A106M—11	Specification for Seamless Carbon Steel Pipe for High-temperature Service	403.4.2
A254—97 (2007)	Specification for Copper Brazed Steel Tubing	403.5.1
B88—09	Specification for Seamless Copper Water Tube	403.5.2
B210—12	Specification for Aluminum and Aluminum-alloy Drawn Seamless Tubes	403.5.3
B241/B241M—12E1	Specification for Aluminum and Aluminum-alloy, Seamless Pipe and Seamless Extruded Tube	403.4.4, 403.5.3
B280—08	Standard Specification for Seamless Copper Tube for Air-Conditioning and Refrigeration Field Service	403.5.2
C315—07(2011)	Specification for Clay Flue Liners and Chimney Pots	501.12
D2513—2013E1	Specification for Polyethylene (PE) Gas Pressure Pipe, Tubing and Fittings	403.6, 403.6.1, 403.11, 404.17.2
E136—2012	Standard Test Method for Behavior of Materials in a Vertical Tube Furnace at 750°C	202
F1973—08	Standard Specification for Factory Assembled Anodeless Risers and Transition Fittings in Polyethylene (PE) and Polyamide 11 (PA11) and Polyamide 12 (PA12) Fuel Gas Distribution Systems	404.17.2

CGA

Compressed Gas Association
14501 George Carter Way, Suite 103
Chantilly, VA 20151-2923

Standard reference number	Title	Referenced in code section number
S-1.1—(2011)	Pressure Relief Device Standards—Part 1—Cylinders for Compressed Gases	703.3
S-1.2—(2005)	Pressure Relief Device Standards—Part 2—Cargo and Portable Tanks for Compressed Gases	703.3
S-1.3—(2008)	Pressure Relief Device Standards—Part 3—Stationary Storage Containers for Compressed Gases	703.3

CSA

CSA America Inc.
8501 E. Pleasant Valley Rd.
Cleveland, OH USA 44131-5575

Standard reference number	Title	Referenced in code section number
ANSI CSA America FC1—2012	Stationary Fuel Cell Power Systems	633.1
CSA 8—93	Requirements for Gas-fired Log Lighters for Wood Burning Fireplaces—with Revisions through January 1999	603.1

DOTn

Department of Transportation
400 Seventh St. SW.
Washington, DC 20590

Standard reference number	Title	Referenced in code section number
49 CFR, Parts 192.281(e) & 192.283 (b)—(2009)	Transportation of Natural and Other Gas by Pipeline: Minimum Federal Safety Standards	403.6.1
49 CFR Parts 100—180	Hazardous Materials Regulations	703.2.2, 703.3.3, 703.3.4

ICC

International Code Council, Inc.
500 New Jersey Ave, NW
6th Floor
Washington, DC 20001

Standard reference number	Title	Referenced in code section number
IBC—15	International Building Code®	102.2.1, 201.3, 301.10, 301.11, 301.12, 301.14, 302.1, 302.2, 305.6, 306.5.1, 06.6, 401.1.1, 412.6, 413.3, 413.3.1, 501.1, 501.3, 501.12, 501.15.4, 501.15.4.1, 609.3, 614.10, 633.1, 635.1, 706.2
IECC—15	International Energy Conservation Code®	301.2
IFC—15	International Fire Code®	201.3, 401.2, 412.1, 412.6, 412.7, 412.7.3, 412.8, 413.1, 413.3, 413.3.1, 413.5, 413.9.2.5, 633.1, 701.1, 701.2, 703.2, 703.2.2, 703.3.8, 703.4, 703.5, 704.1.2, 704.3, 704.4, 706.2, 706.3, 707.1, 707.2, 708.1
IMC—15	International Mechanical Code®	101.2.5, 201.3, 301.1.1, 301.13, 304.11, 307.1, 307.5, 501.1, 614.2, 614.10, 618.4, 621.1, 624.1, 631.2, 632.1, 703.1.2
IPC—15	International Plumbing Code®	201.3, 301.6, 307.3, 624.1.1, 624.2
IRC—15	International Residential Code®	101.2, 703.2.1

MSS

Manufacturers Standardization Society of
the Valve and Fittings Industry
127 Park Street, Northeast
Vienna, VA 22180

Standard reference number	Title	Referenced in code section number
SP 58—2009	Pipe Hangers and Supports-Materials, Design and Manufacture	407.2

NFPA

National Fire Protection Association
1 Batterymarch Park
Quincy, MA 02269-9101

Standard reference number	Title	Referenced in code section number
30A—15	Code for Motor Fuel Dispensing Facilities and Repair Garages	305.4, 305.10
37—14	Standard for the Installation and Use of Stationary Combustion Engines and Gas Turbines	616.1
51—13	Design and Installation of Oxygen-fuel Gas Systems for Welding, Cutting and Allied Processes	414.1
58—14	Liquefied Petroleum Gas Code	401.2, 402.6, 403.6.2, 403.11
70—14	National Electrical Code	306.3.1, 306.4.1, 306.5.2, 309.2, 310.1.1.2, 413.9.2.4, 703.6
82—14	Incinerators, Waste and Linen Handling Systems and Equipment	503.2.5, T503.4, 607.1
85—15	Boiler and Combustion Systems Hazards Code	631.1
88A—15	Parking Structures	305.9
211—13	Standard for the Chimneys, Fireplaces, Vents and Solid Fuel-burning Appliances	503.5.2, 503.5.3, 503.5.6.1, 503.5.6.3
409—11	Standard for the Aircraft Hangars	305.11
853—15	Installation of Stationary Fuel Cell Power Systems	633.1

UL

UL LLC
333 Pfingsten Road
Northbrook, IL 60062

Standard reference number	Title	Referenced in code section number
103—2010	Factory-built Chimneys, Residential Type and Building Heating Appliances—with Revisions through July 2012	506.1
127—2011	Factory-built Fireplaces	621.7
441—2010	Gas Vents	502.1
641—2010	Type L Low-temperature Venting Systems—with Revisions through May 2013	502.1
651—2011	Schedule 40 and 80 Rigid PVC Conduit and Fittings—with Revisions through March 2012	403.6.3
795—2011	Commercial-industrial Gas Heating Equipment—with Revisions through September 2012	610.1, 618.1, 631.1
959—2010	Medium Heat Appliance Factory-built Chimneys	506.3
1618—09	Wall Protectors, Floor Protectors and Hearth Extensions—with Revisions through May 2013	308.2
1738—2010	Venting Systems for Gas Burning Appliances, Categories II, III and IV—with Revisions through May 2011	502.1
1777—2007	Chimney Liners—with Revisions through July 2009	501.12, 501.15.4
2200—2012	Stationary Engine Generator Assemblies—with Revisions through June 2013	616.1

The appendices of the *International Fuel Gas Code®* (IFGC®) are explanatory commentaries of the code itself. Thus, there is no additional commentary for them.

Appendix A (IFGS):
Sizing and Capacities of Gas Piping

(This appendix is informative and is not part of the code.)

A.1 General piping considerations. The first goal of determining the pipe sizing for a fuel gas *piping* system is to make sure that there is sufficient gas pressure at the inlet to each *appliance*. The majority of systems are residential and the appliances will all have the same, or nearly the same, requirement for minimum gas pressure at the *appliance* inlet. This pressure will be about 5-inch water column (w.c.) (1.25 kPa), which is enough for proper operation of the *appliance* regulator to deliver about 3.5-inches water column (w.c.) (875 kPa) to the burner itself. The pressure drop in the *piping* is subtracted from the source delivery pressure to verify that the minimum is available at the *appliance*.

There are other systems, however, where the required inlet pressure to the different appliances may be quite varied. In such cases, the greatest inlet pressure required must be satisfied, as well as the farthest *appliance*, which is almost always the critical *appliance* in small systems.

There is an additional requirement to be observed besides the capacity of the system at 100-percent flow. That requirement is that at minimum flow, the pressure at the inlet to any *appliance* does not exceed the pressure rating of the *appliance* regulator. This would seldom be of concern in small systems if the source pressure is $^1/_2$ psi (14-inch w.c.) (3.5 kPa) or less but it should be verified for systems with greater gas pressure at the point of supply.

To determine the size of *piping* used in a gas *piping* system, the following factors must be considered:

(1) Allowable loss in pressure from *point of delivery* to *appliance*.

(2) Maximum gas demand.

(3) Length of *piping* and number of fittings.

(4) Specific gravity of the gas.

(5) Diversity factor.

For any gas *piping* system, or special *appliance*, or for conditions other than those covered by the tables provided in this code, such as longer runs, greater gas demands or greater pressure drops, the size of each gas *piping* system should be determined by standard engineering practices acceptable to the code official.

A.2 Description of tables.

A.2.1 General. The quantity of gas to be provided at each *outlet* should be determined, whenever possible, directly from the manufacturer's gas input Btu/h rating of the *appliance* that will be installed. In case the ratings of the appliances to be installed are not known, Table 402.2 shows the approximate consumption (in Btu per hour) of certain types of typical household appliances.

To obtain the cubic feet per hour of gas required, divide the total Btu/h input of all appliances by the average Btu heating value per cubic feet of the gas. The average Btu per cubic feet of the gas in the area of the installation can be obtained from the serving gas supplier.

A.2.2 Low pressure natural gas tables. Capacities for gas at low pressure [less than 2.0 psig (13.8 kPa gauge)] in cubic feet per hour of 0.60 specific gravity gas for different sizes and lengths are shown in Tables 402.4(1) through 402.4(4) for iron pipe or equivalent rigid pipe; in Tables 402.4(8) through 402.4(11) for smooth wall semirigid tubing; in Tables 402.4(20) through 402.4(24) for polyethylene pipe and tubing; and in Tables 402.4(15) through 402.4(17) for corrugated stainless steel tubing. Tables 402.4(1), 402.4(6) and 402.4(20) are based upon a pressure drop of 0.3-inch w.c. (75 Pa), whereas Tables 402.4(2), 402.4(9), 402.4(15) and 402.4(21) are based upon a pressure drop of 0.5-inch w.c. (125 Pa). Tables 402.4(3), 402.4(4), 402.4(10), 402.4(11), 402.4(16) and 402.4(17) are special low-pressure applications based upon pressure drops greater than 0.5-inch w.c. (125 Pa). In using these tables, an allowance (in equivalent length of pipe) should be considered for any *piping* run with four or more fittings (see Table A.2.2).

A.2.3 Undiluted liquefied petroleum tables. Capacities in thousands of Btu per hour of undiluted liquefied petroleum gases based on a pressure drop of 0.5-inch w.c. (125 Pa) for different sizes and lengths are shown in Table 402.4(28) for iron pipe or equivalent rigid pipe, in Table 402.4(30) for smooth wall semi-rigid tubing, in Table 402.4(32) for corrugated stainless steel tubing, and in Tables 402.4(35) and 402.4(37) for polyethylene plastic pipe and tubing. Tables 402.4(33) and 402.4(34) for corrugated stainless steel tubing and Table 402.4(36) for polyethylene plastic pipe are based on operating pressures greater than $1^1/_2$ pounds per square inch (psi) (3.5 kPa) and pressure drops greater than 0.5-inch w.c. (125 Pa). In using these tables, an allowance (in equivalent length of pipe) should be considered for any *piping* run with four or more fittings (see Table A.2.2).

A.2.4 Natural gas specific gravity. Gas *piping* systems that are to be supplied with gas of a specific gravity of 0.70 or less can be sized directly from the tables provided in this code, unless the code official specifies that a gravity factor be applied. Where the specific gravity of the gas is greater than 0.70, the gravity factor should be applied.

TABLE A.2.2
EQUIVALENT LENGTHS OF PIPE FITTINGS AND VALVES

		SCREWED FITTINGS[1]				90° WELDING ELBOWS AND SMOOTH BENDS[2]					
		45°/Ell	90°/Ell	180°close return bends	Tee	R/d = 1	R/d = 1$\frac{1}{3}$	R/d = 2	R/d = 4	R/d = 6	R/d = 8
k factor =		0.42	0.90	2.00	1.80	0.48	0.36	0.27	0.21	0.27	0.36
L/d' ratio[4] n =		1 4	30	67	60	16	12	9	7	9	12
Nominal pipe size, inches	Inside diameter d, inches, Schedule 40[6]	L = Equivalent Length In Feet of Schedule 40 (Standard-weight) Straight Pipe[6]									
$\frac{1}{2}$	0.622	0.73	1.55	3.47	3.10	0.83	0.62	0.47	0.36	0.47	0.62
$\frac{3}{4}$	0.824	0.96	2.06	4.60	4.12	1.10	0.82	0.62	0.48	0.62	0.82
1	1.049	1.22	2.62	5.82	5.24	1.40	1.05	0.79	0.61	0.79	1.05
1$\frac{1}{4}$	1.380	1.61	3.45	7.66	6.90	1.84	1.38	1.03	0.81	1.03	1.38
1$\frac{1}{2}$	1.610	1.88	4.02	8.95	8.04	2.14	1.61	1.21	0.94	1.21	1.61
2	2.067	2.41	5.17	11.5	10.3	2.76	2.07	1.55	1.21	1.55	2.07
2$\frac{1}{2}$	2.469	2.88	6.16	13.7	12.3	3.29	2.47	1.85	1.44	1.85	2.47
3	3.068	3.58	7.67	17.1	15.3	4.09	3.07	2.30	1.79	2.30	3.07
4	4.026	4.70	10.1	22.4	20.2	5.37	4.03	3.02	2.35	3.02	4.03
5	5.047	5.88	12.6	28.0	25.2	6.72	5.05	3.78	2.94	3.78	5.05
6	6.065	7.07	15.2	33.8	30.4	8.09	6.07	4.55	3.54	4.55	6.07
8	7.981	9.31	20.0	44.6	40.0	10.6	7.98	5.98	4.65	5.98	7.98
10	10.02	11.7	25.0	55.7	50.0	13.3	10.0	7.51	5.85	7.51	10.0
12	11.94	13.9	29.8	66.3	59.6	15.9	11.9	8.95	6.96	8.95	11.9
14	13.13	15.3	32.8	73.0	65.6	17.5	13.1	9.85	7.65	9.85	13.1
16	15.00	17.5	37.5	83.5	75.0	20.0	15.0	11.2	8.75	11.2	15.0
18	16.88	19.7	42.1	93.8	84.2	22.5	16.9	12.7	9.85	12.7	16.9
20	18.81	22.0	47.0	105.0	94.0	25.1	18.8	14.1	11.0	14.1	18.8
24	22.63	26.4	56.6	126.0	113.0	30.2	22.6	17.0	13.2	17.0	22.6

(continued)

TABLE A.2.2—continued
EQUIVALENT LENGTHS OF PIPE FITTINGS AND VALVES

		MITER ELBOWS[3] (No. of miters)					WELDING TEES		VALVES (screwed, flanged, or welded)			
		1-45°	1-60°	1-90°	2-90°[5]	3-90°[5]	Forged	Miter[3]	Gate	Globe	Angle	Swing Check
k factor =		0.45	0.90	1.80	0.60	0.45	1.35	1.80	0.21	10	5.0	2.5
L/d' ratio[4] n =		15	30	60	20	15	45	60	7	333	167	83
Nominal pipe size, inches	Inside diameter d, inches, Schedule 40[6]	L = Equivalent Length In Feet of Schedule 40 (Standard-weight) Straight Pipe[6]										
$1/_2$	0.622	0.78	1.55	3.10	1.04	0.78	2.33	3.10	0.36	17.3	8.65	4.32
$3/_4$	0.824	1.03	2.06	4.12	1.37	1.03	3.09	4.12	0.48	22.9	11.4	5.72
1	1.049	1.31	2.62	5.24	1.75	1.31	3.93	5.24	0.61	29.1	14.6	7.27
$1^1/_4$	1.380	1.72	3.45	6.90	2.30	1.72	5.17	6.90	0.81	38.3	19.1	9.58
$1^1/_2$	1.610	2.01	4.02	8.04	2.68	2.01	6.04	8.04	0.94	44.7	22.4	11.2
2	2.067	2.58	5.17	10.3	3.45	2.58	7.75	10.3	1.21	57.4	28.7	14.4
$2^1/_2$	2.469	3.08	6.16	12.3	4.11	3.08	9.25	12.3	1.44	68.5	34.3	17.1
3	3.068	3.84	7.67	15.3	5.11	3.84	11.5	15.3	1.79	85.2	42.6	21.3
4	4.026	5.04	10.1	20.2	6.71	5.04	15.1	20.2	2.35	112.0	56.0	28.0
5	5.047	6.30	12.6	25.2	8.40	6.30	18.9	25.2	2.94	140.0	70.0	35.0
6	6.065	7.58	15.2	30.4	10.1	7.58	22.8	30.4	3.54	168.0	84.1	42.1
8	7.981	9.97	20.0	40.0	13.3	9.97	29.9	40.0	4.65	22.0	111.0	55.5
10	10.02	12.5	25.0	50.0	16.7	12.5	37.6	50.0	5.85	278.0	139.0	69.5
12	11.94	14.9	29.8	59.6	19.9	14.9	44.8	59.6	6.96	332.0	166.0	83.0
14	13.13	16.4	32.8	65.6	21.9	16.4	49.2	65.6	7.65	364.0	182.0	91.0
16	15.00	18.8	37.5	75.0	25.0	18.8	56.2	75.0	8.75	417.0	208.0	104.0
18	16.88	21.1	42.1	84.2	28.1	21.1	63.2	84.2	9.85	469.0	234.0	117.0
20	18.81	23.5	47.0	94.0	31.4	23.5	70.6	94.0	11.0	522.0	261.0	131.0
24	22.63	28.3	56.6	113.0	37.8	28.3	85.0	113.0	13.2	629.0	314.0	157.0

For SI: 1 foot = 305 mm, 1 degree = 0.01745 rad.

Note: Values for welded fittings are for conditions where bore is not obstructed by weld spatter or backing rings. If appreciably obstructed, use values for "Screwed Fittings."

1. Flanged fittings have three-fourths the resistance of screwed elbows and tees.
2. Tabular figures give the extra resistance due to curvature alone to which should be added the full length of travel.
3. Small size socket-welding fittings are equivalent to miter elbows and tees.
4. Equivalent resistance in number of diameters of straight pipe computed for a value of $(f - 0.0075)$ from the *relation* $(n - k/4f)$.
5. For condition of minimum resistance where the centerline length of each miter is between d and $2^1/_2d$.
6. For pipe having other inside diameters, the equivalent resistance can be computed from the above n values.

Source: Crocker, S. *Piping Handbook*, 4th ed., Table XIV, pp. 100-101. Copyright 1945 by McGraw-Hill, Inc. Used by permission of McGraw-Hill Book Company.

Application of the gravity factor converts the figures given in the tables provided in this code to capacities for another gas of different specific gravity. Such application is accomplished by multiplying the capacities given in the tables by the multipliers shown in Table A.2.4. In case the exact specific gravity does not appear in the table, choose the next higher value specific gravity shown.

TABLE A.2.4
MULTIPLIERS TO BE USED WITH TABLES 402.4(1)
THROUGH 402.4(22) WHERE THE SPECIFIC GRAVITY
OF THE GAS IS OTHER THAN 0.60

SPECIFIC GRAVITY	MULTIPLIER	SPECIFIC GRAVITY	MULTIPLIER
0.35	1.31	1.00	0.78
0.40	1.23	1.10	0.74
0.45	1.16	1.20	0.71
0.50	1.10	1.30	0.68
0.55	1.04	1.40	0.66
0.60	1.00	1.50	0.63
0.65	0.96	1.60	0.61
0.70	0.93	1.70	0.59
0.75	0.90	1.80	0.58
0.80	0.87	1.90	0.56
0.85	0.84	2.00	0.55
0.90	0.82	2.10	0.54

A.2.5 Higher pressure natural gas tables. Capacities for gas at pressures 2.0 psig (13.8 kPa) or greater in cubic feet per hour of 0.60 specific gravity gas for different sizes and lengths are shown in Tables 402.4(5) through 402.4(7) for iron pipe or equivalent rigid pipe; Tables 402.4(12) to 402.4(14) for semirigid tubing; Tables 402.4(18) and 402.4(19) for corrugated stainless steel tubing; and Table 402.4(22) for polyethylene plastic pipe.

A.3 Use of capacity tables.

A.3.1 Longest length method. This sizing method is conservative in its approach by applying the maximum operating conditions in the system as the norm for the system and by setting the length of pipe used to size any given part of the *piping* system to the maximum value.

To determine the size of each section of gas *piping* in a system within the range of the capacity tables, proceed as follows (also see sample calculations included in this Appendix):

(1) Divide the *piping* system into appropriate segments consistent with the presence of tees, branch lines and main runs. For each segment, determine the gas load (assuming all appliances operate simultaneously) and its overall length. An allowance (in equivalent length of pipe) as determined from Table A.2.2 shall be considered for *piping* segments that include four or more fittings.

(2) Determine the gas demand of each *appliance* to be attached to the *piping* system. Where Tables 402.4(1) through 402.4(24) are to be used to select the *piping* size, calculate the gas demand in terms of cubic feet per hour for each *piping* system *outlet*. Where Tables 402.4(25) through 402.4(37) are to be used to select the *piping* size, calculate the gas demand in terms of thousands of Btu per hour for each *piping* system *outlet*.

(3) Where the *piping* system is for use with other than undiluted liquefied petroleum gases, determine the design system pressure, the allowable loss in pressure (pressure drop), and specific gravity of the gas to be used in the *piping* system.

(4) Determine the length of *piping* from the *point of delivery* to the most remote *outlet* in the building/*piping* system.

(5) In the appropriate capacity table, select the row showing the measured length or the next longer length if the table does not give the exact length. This is the only length used in determining the size of any section of gas *piping*. If the gravity factor is to be applied, the values in the selected row of the table are multiplied by the appropriate multiplier from Table A.2.4.

(6) Use this horizontal row to locate ALL gas demand figures for this particular system of *piping*.

(7) Starting at the most remote *outlet*, find the gas demand for that *outlet* in the horizontal row just selected. If the exact figure of demand is not shown, choose the next larger figure left in the row.

(8) Opposite this demand figure, in the first row at the top, the correct size of gas *piping* will be found.

(9) Proceed in a similar manner for each *outlet* and each section of gas *piping*. For each section of *piping*, determine the total gas demand supplied by that section.

Where a large number of *piping* components (such as elbows, tees and valves) are installed in a pipe run, additional pressure loss can be accounted for by the use of equivalent lengths. Pressure loss across any *piping* component can be equated to the pressure drop through a length of pipe. The equivalent length of a combination of only four elbows/tees can result in a jump to the next larger length row, resulting in a significant reduction in capacity. The equivalent lengths in feet shown in Table A.2.2 have been computed on a basis that the inside diameter corresponds to that of Schedule 40 (standard-weight) steel pipe, which is close enough for most purposes involving other schedules of pipe. Where a more specific solution for equivalent length is desired, this can be made by multiplying the actual inside diameter of the pipe in inches by $n/12$, or the actual inside diameter in feet by n (n can be read from the table heading). The equivalent length values can be used with reasonable accuracy for copper or copper alloy fittings and bends although the resistance per foot of copper or copper alloy pipe is less than that of steel. For copper or brass valves, however, the equivalent length of pipe should be taken as 45 percent longer than the values in the table, which are for steel pipe.

A.3.2 Branch length method. This sizing method reduces the amount of conservatism built into the traditional Longest Length Method. The longest length as measured from the meter to the furthest remote *appliance* is only used to size the initial parts of the overall *piping* system. The Branch Length Method is applied in the following manner:

(1) Determine the gas load for each of the connected appliances.

(2) Starting from the meter, divide the *piping* system into a number of connected segments, and determine the length and amount of gas that each segment would carry assuming that all appliances were operated simultaneously. An allowance (in equivalent length of pipe) as determined from Table A.2.2 should be considered for piping segments that include four or more fittings.

(3) Determine the distance from the *outlet* of the gas meter to the *appliance* furthest removed from the meter.

(4) Using the longest distance (found in Step 3), size each *piping* segment from the meter to the most remote *appliance outlet*.

(5) For each of these *piping* segments, use the longest length and the calculated gas load for all of the connected appliances for the segment and begin the sizing process in Steps 6 through 8.

(6) Referring to the appropriate sizing table (based on operating conditions and *piping* material), find the longest length distance in the first column or the next larger distance if the exact distance is not listed. The use of alternative operating pressures or pressure drops will require the use of a different sizing table, but will not alter the sizing methodology. In many cases, the use of alternative operating pressures or pressure drops will require the approval of both the code official and the local gas serving utility.

(7) Trace across this row until the gas load is found or the closest larger capacity if the exact capacity is not listed.

(8) Read up the table column and select the appropriate pipe size in the top row. Repeat Steps 6, 7 and 8 for each pipe segment in the longest run.

(9) Size each remaining section of branch *piping* not previously sized by measuring the distance from the gas meter location to the most remote *outlet* in that branch, using the gas load of attached appliances and following the procedures of Steps 2 through 8.

A.3.3 Hybrid pressure method. The sizing of a 2 psi (13.8 kPa) gas *piping* system is performed using the traditional Longest Length Method but with modifications. The 2 psi (13.8 kPa) system consists of two independent pressure zones, and each zone is sized separately. The Hybrid Pressure Method is applied as follows:

The sizing of the 2 psi (13.8 kPa) section (from the meter to the line regulator) is as follows:

(1) Calculate the gas load (by adding up the name plate ratings) from all connected appliances. (In certain circumstances the installed gas load can be increased up to 50 percent to accommodate future addition of

appliances.) Ensure that the line regulator capacity is adequate for the calculated gas load and that the required pressure drop (across the regulator) for that capacity does not exceed $^3/_4$ psi (5.2 kPa) for a 2 psi (13.8 kPa) system. If the pressure drop across the regulator is too high (for the connected gas load), select a larger regulator.

(2) Measure the distance from the meter to the line regulator located inside the building.

(3) If there are multiple line regulators, measure the distance from the meter to the regulator furthest removed from the meter.

(4) The maximum allowable pressure drop for the 2 psi (13.8 kPa) section is 1 psi (6.9 kPa).

(5) Referring to the appropriate sizing table (based on *piping* material) for 2 psi (13.8 kPa) systems with a 1 psi (6.9 kPa) pressure drop, find this distance in the first column, or the closest larger distance if the exact distance is not listed.

(6) Trace across this row until the gas load is found or the closest larger capacity if the exact capacity is not listed.

(7) Read up the table column to the top row and select the appropriate pipe size.

(8) If there are multiple regulators in this portion of the *piping* system, each line segment must be sized for its actual gas load, but using the longest length previously determined above.

The low pressure section (all *piping* downstream of the line regulator) is sized as follows:

(1) Determine the gas load for each of the connected appliances.

(2) Starting from the line regulator, divide the piping system into a number of connected segments or independent parallel piping segments, and determine the amount of gas that each segment would carry assuming that all appliances were operated simultaneously. An allowance (in equivalent length of pipe) as determined from Table A.2.2 should be considered for piping segments that include four or more fittings.

(3) For each piping segment, use the actual length or longest length (if there are sub-branchlines) and the calculated gas load for that segment and begin the sizing process as follows:

(a) Referring to the appropriate sizing table (based on operating pressure and piping material), find the longest length distance in the first column or the closest larger distance if the exact distance is not listed. The use of alternative operating pressures and/or pressure drops will require the use of a different sizing table, but will not alter the sizing methodology. In many cases, the use of alternative operating pressures and/or pressure drops can require the approval of the code official.

(b) Trace across this row until the appliance gas load is found or the closest larger capacity if the exact capacity is not listed.

(c) Read up the table column to the top row and select the appropriate pipe size.

(d) Repeat this process for each segment of the piping system.

A.3.4 Pressure drop per 100 feet method. This sizing method is less conservative than the others, but it allows the designer to immediately see where the largest pressure drop occurs in the system. With this information, modifications can be made to bring the total drop to the critical *appliance* within the limitations that are presented to the designer.

Follow the procedures described in the Longest Length Method for Steps (1) through (4) and (9).

For each *piping* segment, calculate the pressure drop based on pipe size, length as a percentage of 100 feet (30 480 mm) and gas flow. Table A.3.4 shows pressure drop per 100 feet (30 480 mm) for pipe sizes from $\frac{1}{2}$ inch (12.7 mm) through 2 inches (51 mm). The sum of pressure drops to the critical *appliance* is subtracted from the supply pressure to verify that sufficient pressure will be available. If not, the layout can be examined to find the high drop section(s) and sizing selections modified.

Note: Other values can be obtained by using the following equation:

$$\text{Desired Value} = MBH \times \sqrt{\frac{\text{Desired Drop}}{\text{Table Drop}}}$$

For example, if it is desired to get flow through $\frac{3}{4}$-inch (19.1 mm) pipe at 2 inches/100 feet, multiply the capacity of $\frac{3}{4}$-inch pipe at 1 inch/100 feet by the square root of the pressure ratio:

$$147\ MBH \times \sqrt{\frac{2''\ \text{w.c.}}{1''\ \text{w.c.}}} = 147 \times 1.414 = 208\ MBH$$

$$(MBH = 1000\ \text{Btu/h})$$

A.4 Use of sizing equations. Capacities of smooth wall pipe or tubing can also be determined by using the following formulae:

(1) High Pressure [1.5 psi (10.3 kPa) and above]:

$$Q = 181.6 \sqrt{\frac{D^5 \cdot (P_1^2 - P_2^2) \cdot Y}{C_r \cdot fba \cdot L}}$$

$$= 2237\ D^{2.623} \left[\frac{(P_1^2 - P_2^2) \cdot Y}{C_r \cdot L}\right]^{0.541}$$

(2) Low Pressure [Less than 1.5 psi (10.3 kPa)]:

$$Q = 187.3 \sqrt{\frac{D^5 \cdot \Delta H}{C_r \cdot fba \cdot L}}$$

$$= 2313 D^{2.623} \left(\frac{\Delta H}{C_r \cdot L}\right)^{0.541}$$

where:

Q = Rate, cubic feet per hour at 60°F and 30-inch mercury column

D = Inside diameter of pipe, in.

P_1 = Upstream pressure, psia

P_2 = Downstream pressure, psia

Y = Superexpansibility factor = 1/supercompressibility factor

C_r = Factor for viscosity, density and temperature*

= $0.00354\ ST\left(\frac{Z}{S}\right)^{0.152}$

*Note: See Table 402.4 for Y and C_r for natural gas and propane.

S = Specific gravity of gas at 60°F and 30-inch mercury column (0.60 for natural gas, 1.50 for propane), or = 1488μ

T = Absolute temperature, °F or = $t + 460$

t = Temperature, °F

Z = Viscosity of gas, centipoise (0.012 for natural gas, 0.008 for propane), or = 1488μ

fba = Base friction factor for air at 60°F (CF = 1)

L = Length of pipe, ft

ΔH = Pressure drop, in. w.c. (27.7 in. H_2O = 1 psi)

(For SI, see Section 402.4)

A.5 Pipe and tube diameters. Where the internal diameter is determined by the formulas in Section 402.4, Tables A.5.1 and A.5.2 can be used to select the nominal or standard pipe size based on the calculated internal diameter.

TABLE A.3.4
THOUSANDS OF BTU/H (MBH) OF NATURAL GAS PER 100 FEET OF PIPE AT VARIOUS PRESSURE DROPS AND PIPE DIAMETERS

PRESSURE DROP PER 100 FEET IN INCHES W.C.	PIPE SIZES (inch)					
	$\frac{1}{2}$	$\frac{3}{4}$	1	$1\frac{1}{4}$	$1\frac{1}{2}$	2
0.2	31	64	121	248	372	716
0.3	38	79	148	304	455	877
0.5	50	104	195	400	600	1160
1.0	71	147	276	566	848	1640

For SI: 1 inch = 25.4 mm, 1 foot = 304.8 mm.

TABLE A.5.1
SCHEDULE 40 STEEL PIPE STANDARD SIZES

NOMINAL SIZE (inch)	INTERNAL DIAMETER (inch)	NOMINAL SIZE (inch)	INTERNAL DIAMETER (inch)
$1/4$	0.364	$1^1/_2$	1.610
$3/_8$	0.493	2	2.067
$1/_2$	0.622	$2^1/_2$	2.469
$3/_4$	0.824	3	3.068
1	1.049	$3^1/_2$	3.548
$1^1/_4$	1.380	4	4.026

For SI: 1 inch = 25.4 mm.

A.6 Examples of piping system design and sizing.

A.6.1 Example 1: Longest length method. Determine the required pipe size of each section and *outlet* of the *piping* system shown in Figure A.6.1, with a designated pressure drop of 0.5-inch w.c. (125 Pa) using the Longest Length Method. The gas to be used has 0.60 specific gravity and a heating value of 1,000 Btu/ft³ (37.5 MJ/m³).

Solution:

(1) Maximum gas demand for *Outlet* A:

$$\frac{\text{Consumption (rating plate input, or Table 402.2 if necessary}}{\text{Btu of gas}} =$$

$$\frac{35,000 \text{ Btu per hour rating}}{1,000 \text{ Btu per cubic foot}} = 35 \text{ cubic feet per hour} = 35 \text{ cfh}$$

Maximum gas demand for *Outlet* B:

$$\frac{\text{Consumption}}{\text{Btu of gas}} = \frac{75,000}{1,000} = 75 \text{ cfh}$$

Maximum gas demand for *Outlet* C:

$$\frac{\text{Consumption}}{\text{Btu of gas}} = \frac{35,000}{1,000} = 35 \text{ cfh}$$

Maximum gas demand for *Outlet* D:

$$\frac{\text{Consumption}}{\text{Btu of gas}} = \frac{100,000}{1,000} = 100 \text{ cfh}$$

(2) The length of pipe from the *point of delivery* to the most remote *outlet* (A) is 60 feet (18 288 mm). This is the only distance used.

(3) Using the row marked 60 feet (18 288 mm) in Table 402.4(2):

(a) *Outlet* A, supplying 35 cfh (0.99 m³/hr), requires $1/_2$-inch pipe.

(b) *Outlet* B, supplying 75 cfh (2.12 m³/hr), requires $3/_4$-inch pipe.

(c) Section 1, supplying *Outlets* A and B, or 110 cfh (3.11 m³/hr), requires $3/_4$-inch pipe.

(d) Section 2, supplying *Outlets* C and D, or 135 cfh (3.82 m³/hr), requires $3/_4$-inch pipe.

(e) Section 3, supplying *Outlets* A, B, C and D, or 245 cfh (6.94 m³/hr), requires 1-inch pipe.

(4) If a different gravity factor is applied to this example, the values in the row marked 60 feet (18 288 mm) of Table 402.4(2) would be multiplied by the appropriate multiplier from Table A.2.4 and the resulting cubic feet per hour values would be used to size the *piping*.

TABLE A.5.2
COPPER TUBE STANDARD SIZES

TUBE TYPE	NOMINAL OR STANDARD SIZE (inches)	INTERNAL DIAMETER (inches)
K	$1/_4$	0.305
L	$1/_4$	0.315
ACR (D)	$3/_8$	0.315
ACR (A)	$3/_8$	0.311
K	$3/_8$	0.402
L	$3/_8$	0.430
ACR (D)	$1/_2$	0.430
ACR (A)	$1/_2$	0.436
K	$1/_2$	0.527
L	$1/_2$	0.545
ACR (D)	$5/_8$	0.545
ACR (A)	$5/_8$	0.555
K	$5/_8$	0.652
L	$5/_8$	0.666
ACR (D)	$3/_4$	0.666
ACR (A)	$3/_4$	0.680
K	$3/_4$	0.745
L	$3/_4$	0.785
ACR	$7/_8$	0.785
K	1	0.995
L	1	1.025
ACR	$1^1/_8$	1.025
K	$1^1/_4$	1.245
L	$1^1/_4$	1.265
ACR	$1^3/_8$	1.265
K	$1^1/_2$	1.481
L	$1^1/_2$	1.505
ACR	$1^5/_8$	1.505
K	2	1.959
L	2	1.985
ACR	$2^1/_8$	1.985
K	$2^1/_2$	2.435
L	$2^1/_2$	2.465
ACR	$2^5/_8$	2.465
K	3	2.907
L	3	2.945
ACR	$3^1/_8$	2.945

For SI: 1 inch = 25.4 mm.

A.6.2 Example 2: Hybrid or dual pressure systems. Determine the required CSST size of each section of the *piping* system shown in Figure A.6.2, with a designated pressure drop of 1 psi (6.9 kPa) for the 2 psi (13.8 kPa) section and 3-inch w.c. (0.75 kPa) pressure drop for the 13-inch w.c. (2.49 kPa) section. The gas to be used has 0.60 specific gravity and a heating value of 1,000 Btu/ft^3 (37.5 MJ/m^3).

Solution:

(1) Size 2 psi (13.8 kPa) line using Table 402.4(18).

(2) Size 10-inch w.c. (2.5 kPa) lines using Table 402.4(16).

(3) Using the following, determine if sizing tables can be used.

    (a) Total gas load shown in Figure A.6.2 equals 110 cfh (3.11 m^3/hr).

    (b) Determine pressure drop across regulator [see notes in Table 402.4(18)].

    (c) If pressure drop across regulator exceeds $^3/_4$ psig (5.2 kPa), Table 402.4(18) cannot be used. Note: If pressure drop exceeds $^3/_4$ psi (5.2 kPa), then a larger regulator must be selected or an alternative sizing method must be used.

    (d) Pressure drop across the line regulator [for 110 cfh (3.11 m^3/hr)] is 4-inch w.c. (0.99 kPa) based on manufacturer's performance data.

    (e) Assume the CSST manufacturer has tubing sizes or EHDs of 13, 18, 23 and 30.

FIGURE A.6.1
PIPING PLAN SHOWING A STEEL PIPING SYSTEM

(4) Section A [2 psi (13.8 kPa) zone]

    (a) Distance from meter to regulator = 100 feet (30 480 mm).

    (b) Total load supplied by A = 110 cfh (3.11 m^3/hr) (furnace + water heater + dryer).

    (c) Table 402.4(18) shows that EHD size 18 should be used.

    Note: It is not unusual to oversize the supply line by 25 to 50 percent of the as-installed load. EHD size 18 has a capacity of 189 cfh (5.35 m^3/hr).

(5) Section B (low pressure zone)

    (a) Distance from regulator to furnace is 15 feet (4572 mm).

    (b) Load is 60 cfh (1.70 m^3/hr).

    (c) Table 402.4(16) shows that EHD size 13 should be used.

(6) Section C (low pressure zone)

    (a) Distance from regulator to water heater is 10 feet (3048 mm).

    (b) Load is 30 cfh (0.85 m^3/hr).

    (c) Table 402.4(16) shows that EHD size 13 should be used.

(7) Section D (low pressure zone)

    (a) Distance from regulator to dryer is 25 feet (7620 mm).

    (b) Load is 20 cfh (0.57 m^3/hr).

    (c) Table 402.4(16) shows that EHD size 13 should be used.

A.6.3 Example 3: Branch length method. Determine the required semirigid copper tubing size of each section of the *piping* system shown in Figure A.6.3, with a designated pressure drop of 1-inch w.c. (250 Pa) (using the Branch Length Method). The gas to be used has 0.60 specific gravity and a heating value of 1,000 Btu/ft^3 (37.5 MJ/m^3).

FIGURE A.6.2
PIPING PLAN SHOWING A CSST SYSTEM

Solution:

(1) Section A

  (a) The length of tubing from the *point of delivery* to the most remote *appliance* is 50 feet (15 240 mm), A + C.

  (b) Use this longest length to size Sections A and C.

  (c) Using the row marked 50 feet (15 240 mm) in Table 402.4(10), Section A, supplying 220 cfh (6.2 m³/hr) for four appliances requires 1-inch tubing.

(2) Section B

  (a) The length of tubing from the *point of delivery* to the range/oven at the end of Section B is 30 feet (9144 mm), A + B.

  (b) Use this branch length to size Section B only.

  (c) Using the row marked 30 feet (9144 mm) in Table 402.4(10), Section B, supplying 75 cfh (2.12 m³/hr) for the range/oven requires ¹/₂-inch tubing.

(3) Section C

  (a) The length of tubing from the *point of delivery* to the dryer at the end of Section C is 50 feet (15 240 mm), A + C.

  (b) Use this branch length (which is also the longest length) to size Section C.

  (c) Using the row marked 50 feet (15 240 mm) in Table 402.4(10), Section C, supplying 30 cfh (0.85 m³/hr) for the dryer requires ³/₈-inch tubing.

(4) Section D

  (a) The length of tubing from the *point of delivery* to the water heater at the end of Section D is 30 feet (9144 mm), A + D.

(b) Use this branch length to size Section D only.

(c) Using the row marked 30 feet (9144 mm) in Table 402.4(10), Section D, supplying 35 cfh (0.99 m³/hr) for the water heater requires ³/₈-inch tubing.

(5) Section E

  (a) The length of tubing from the *point of delivery* to the furnace at the end of Section E is 30 feet (9144 mm), A + E.

  (b) Use this branch length to size Section E only.

  (c) Using the row marked 30 feet (9144 mm) in Table 402.4(10), Section E, supplying 80 cfh (2.26 m³/hr) for the furnace requires ¹/₂-inch tubing.

A.6.4 Example 4: Modification to existing piping system. Determine the required CSST size for Section G (retrofit application) of the *piping* system shown in Figure A.6.4, with a designated pressure drop of 0.5-inch w.c. (125 Pa) using the branch length method. The gas to be used has 0.60 specific gravity and a heating value of 1,000 Btu/ft³ (37.5 MJ/m³).

Solution:

(1) The length of pipe and CSST from the *point of delivery* to the retrofit *appliance* (barbecue) at the end of Section G is 40 feet (12 192 mm), A + B + G.

(2) Use this branch length to size Section G.

(3) Assume the CSST manufacturer has tubing sizes or EHDs of 13, 18, 23 and 30.

(4) Using the row marked 40 feet (12 192 mm) in Table 402.4(15), Section G, supplying 40 cfh (1.13 m³/hr) for the barbecue requires EHD 18 CSST.

(5) The sizing of Sections A, B, F and E must be checked to ensure adequate gas carrying capacity since an *appliance* has been added to the *piping* system (see A.6.1 for details).

A.6.5 Example 5: Calculating pressure drops due to temperature changes. A test *piping* system is installed on a warm autumn afternoon when the temperature is 70°F (21°C). In accordance with local custom, the new *piping* system is subjected to an air pressure test at 20 psig (138 kPa). Overnight, the temperature drops and when the inspector shows up first thing in the morning the temperature is 40°F (4°C).

If the volume of the *piping* system is unchanged, then the formula based on Boyle's and Charles' law for determining the new pressure at a reduced temperature is as follows:

$$\frac{T_1}{T_2} = \frac{P_1}{P_2}$$

where:

T_1 = Initial temperature, absolute (T_1 + 459)

T_2 = Final temperature, absolute (T_2 + 459)

P_1 = Initial pressure, psia (P_1 + 14.7)

P_2 = Final pressure, psia (P_2 + 14.7)

FIGURE A.6.3
PIPING PLAN SHOWING A COPPER TUBING SYSTEM

$$\frac{(70+459)}{(40+459)} = \frac{(20+14.7)}{(P_2+14.7)}$$

$$\frac{529}{499} = \frac{34.7}{(P_2+14.7)}$$

$$(P_2+14.7) \times \frac{529}{499} = 34.7$$

$$(P_2+14.7) \times \frac{34.7}{1.060}$$

$$P_2 = 32.7 - 14.7$$

$$P_2 = 18\ psig$$

Therefore, the gauge could be expected to register 18 psig (124 kPa) when the ambient temperature is 40°F (4°C).

FIGURE A.6.4
PIPING PLAN SHOWING A MODIFICATION
TO EXISTING PIPING SYSTEM

A.6.6 Example 6: Pressure drop per 100 feet of pipe method. Using the layout shown in Figure A.6.1 and ΔH = pressure drop, in w.c. (27.7 in. H$_2$O = 1 psi), proceed as follows:

(1) Length to A = 20 feet, with 35,000 Btu/hr.

For $^1/_2$-inch pipe, $\Delta H = {}^{20\ feet}/_{100\ feet} \times 0.3$ inch w.c. = 0.06 in w.c.

(2) Length to B = 15 feet, with 75,000 Btu/hr.

For $^3/_4$-inch pipe, $\Delta H = {}^{15\ feet}/_{100\ feet} \times 0.3$ inch w.c. = 0.045 in w.c.

(3) Section 1 = 10 feet, with 110,000 Btu/hr. Here there is a choice:

For 1 inch pipe: $\Delta H = {}^{10\ feet}/_{100\ feet} \times 0.2$ inch w.c. = 0.02 in w.c.

For $^3/_4$-inch pipe: $\Delta H = {}^{10\ feet}/_{100\ feet} \times [0.5$ inch w.c. + ${}^{(110,000\ Btu/hr-104,000\ Btu/hr)}/_{(147,000\ Btu/hr-104,000\ Btu/hr)} \times (1.0$ inches w.c. - 0.5 inch w.c.)] = 0.1 × 0.57 inch w.c.≈ 0.06 inch w.c.

Note that the pressure drop between 104,000 Btu/hr and 147,000 Btu/hr has been interpolated as 110,000 Btu/hr.

(4) Section 2 = 20 feet, with 135,000 Btu/hr. Here there is a choice:

For 1-inch pipe: $\Delta H = {}^{20\ feet}/_{100\ feet} \times [0.2$ inch w.c. + ${}^{(14,000\ Btu/hr)}/_{(27,000\ Btu/hr)} \times 0.1$ inch w.c.] = 0.05 inch w.c.

For $^3/_4$-inch pipe: $\Delta H = {}^{20\ feet}/_{100\ feet} \times 1.0$ inch w.c. = 0.2 inch w.c.

Note that the pressure drop between 121,000 Btu/hr and 148,000 Btu/hr has been interpolated as 135,000 Btu/hr, but interpolation for the ¾-inch pipe (trivial for 104,000 Btu/hr to 147,000 Btu/hr) was not used.

(5) Section 3 = 30 feet, with 245,000 Btu/hr. Here there is a choice:

For 1-inch pipe: $\Delta H = {}^{30\ feet}/_{100\ feet} \times 1.0$ inches w.c. = 0.3 inch w.c.

For $1^1/_4$-inch pipe: $\Delta H = {}^{30\ feet}/_{100\ feet} \times 0.2$ inch w.c. = 0.06 inch w.c.

Note that interpolation for these options is ignored since the table values are close to the 245,000 Btu/hr carried by that section.

(6) The total pressure drop is the sum of the section approaching A, Sections 1 and 3, or either of the following, depending on whether an absolute minimum is needed or the larger drop can be accommodated.

Minimum pressure drop to farthest *appliance*:

ΔH = 0.06 inch w.c. + 0.02 inch w.c. + 0.06 inch w.c. = 0.14 inch w.c.

Larger pressure drop to the farthest *appliance*:

ΔH = 0.06 inch w.c. + 0.06 inch w.c. + 0.3 inch w.c. = 0.42 inch w.c.

Notice that Section 2 and the run to B do not enter into this calculation, provided that the appliances have similar input pressure requirements.

For SI units: 1 Btu/hr = 0.293 W, 1 cubic foot = 0.028 m^3, 1 foot = 0.305 m, 1 inch w.c. = 249 Pa.

The appendices of the *International Fuel Gas Code®* (IFGC®) are explanatory commentaries of the code itself. Thus, there is no additional commentary for them.

Appendix B (IFGS):
Sizing of Venting Systems Serving Appliances Equipped with Draft Hoods, Category I Appliances and Appliances Listed for Use with Type B Vents

(This appendix is informative and is not part of the code.)

EXAMPLES USING SINGLE APPLIANCE VENTING TABLES

Example 1: Single draft-hood-equipped appliance.

An installer has a 120,000 British thermal unit (Btu) per hour input *appliance* with a 5-inch-diameter draft hood outlet that needs to be vented into a 10-foot-high Type B vent system. What size vent should be used assuming (a) a 5-foot lateral single-wall metal vent connector is used with two 90-degree elbows, or (b) a 5-foot lateral single-wall metal vent connector is used with three 90-degree elbows in the vent system?

Solution:

Table 504.2(2) should be used to solve this problem, because single-wall metal vent connectors are being used with a Type B vent.

(a) Read down the first column in Table 504.2(2) until the row associated with a 10-foot height and 5-foot lateral is found. Read across this row until a vent capacity greater than 120,000 Btu per hour is located in the shaded columns *labeled* "NAT Max" for draft-hood-equipped appliances. In this case, a 5-inch-diameter vent has a capacity of 122,000 Btu per hour and can be used for this application.

(b) If three 90-degree elbows are used in the vent system, then the maximum vent capacity listed in the tables must be reduced by 10 percent (see Section 504.2.3 for single *appliance* vents). This implies that the 5-inch-diameter vent has an adjusted capacity of only 110,000 Btu per hour. In this case, the vent system must be increased to 6 inches in diameter (see calculations below).

For SI: 1 foot = 304.8 mm, 1 British thermal unit per hour = 0.2931 W.

Table 504.2(1) is used when sizing Type B double-wall gas vent connected directly to the appliance.

Note: The appliance can be either Category I draft hood equipped or fan-assisted type.

FIGURE B-1
TYPE B DOUBLE-WALL VENT SYSTEM SERVING A SINGLE APPLIANCE WITH A TYPE B DOUBLE-WALL VENT

For SI: 1 foot = 304.8 mm, 1 British thermal unit per hour = 0.2931 W.

Table 504.2(2) is used when sizing a single-wall metal vent connector attached to a Type B double-wall gas vent.

Note: The appliance can be either Category I draft hood equipped or fan-assisted type.

FIGURE B-2
TYPE B DOUBLE-WALL VENT SYSTEM SERVING A SINGLE APPLIANCE WITH A SINGLE-WALL METAL VENT CONNECTOR

Table 504.2(3) is used when sizing a Type B double-wall gas vent connector attached to a tile-lined masonry chimney.

Note: "A" is the equivalent cross-sectional area of the tile liner.

Note: The appliance can be either Category I draft hood equipped or fan-assisted type.

FIGURE B-3
VENT SYSTEM SERVING A SINGLE APPLIANCE
WITH A MASONRY CHIMNEY OF TYPE B
DOUBLE-WALL VENT CONNECTOR

Table 504.2(4) is used when sizing a single-wall vent connector attached to a tile-lined masonry chimney.

Note: "A" is the equivalent cross-sectional area of the tile liner.

Note: The appliance can be either Category I draft hood equipped or fan-assisted type.

FIGURE B-4
VENT SYSTEM SERVING A SINGLE APPLIANCE
USING A MASONRY CHIMNEY AND A
SINGLE-WALL METAL VENT CONNECTOR

Asbestos cement Type B or single-wall metal vent serving a single draft-hood-equipped appliance [see Table 504.2(5)].

FIGURE B-5
ASBESTOS CEMENT TYPE B OR SINGLE-WALL
METAL VENT SYSTEM SERVING A SINGLE
DRAFT-HOOD-EQUIPPED APPLIANCE

Table 504.3(1) is used when sizing Type B double-wall vent connectors attached to a Type B double-wall common vent.

Note: Each appliance can be either Category I draft hood equipped or fan-assisted type.

FIGURE B-6
VENT SYSTEM SERVING TWO OR MORE APPLIANCES
WITH TYPE B DOUBLE-WALL VENT AND TYPE B
DOUBLE-WALL VENT CONNECTOR

Table 504.3(2) is used when sizing single-wall vent connectors attached to a Type B double-wall common vent.

Note: Each appliance can be either Category I draft hood equipped or fan-assisted type.

FIGURE B-7
VENT SYSTEM SERVING TWO OR MORE APPLIANCES WITH TYPE B DOUBLE-WALL VENT AND SINGLE-WALL METAL VENT CONNECTORS

Table 504.3(3) is used when sizing Type B double-wall vent connectors attached to a tile-lined masonry chimney.

Note: "A" is the equivalent cross-sectional area of the tile liner.

Note: Each appliance can be either Category I draft hood equipped or fan-assisted type.

FIGURE B-8
MASONRY CHIMNEY SERVING TWO OR MORE APPLIANCES WITH TYPE B DOUBLE-WALL VENT CONNECTOR

Table 504.3(4) is used when sizing single-wall metal vent connectors attached to a tile-lined masonry chimney.

Note: "A" is the equivalent cross-sectional area of the tile liner.

Note: Each appliance can be either Category I draft hood equipped or fan-assisted type.

FIGURE B-9
MASONRY CHIMNEY SERVING TWO OR MORE APPLIANCES WITH SINGLE-WALL METAL VENT CONNECTORS

Asbestos cement Type B or single-wall metal pipe vent serving two or more draft-hood-equipped appliances [see Table 504.3(5)].

FIGURE B-10
ASBESTOS CEMENT TYPE B OR SINGLE-WALL METAL VENT SYSTEM SERVING TWO OR MORE DRAFT-HOOD-EQUIPPED APPLIANCES

Example: Manifolded Common Vent Connector LM shall be not greater than 18 times the common vent connector manifold inside diameter; i.e., a 4-inch (102 mm) inside diameter common vent connector manifold shall not exceed 72 inches (1829 mm) in length (see Section 504.3.4).

Note: This is an illustration of a typical manifolded vent connector. Different appliance, vent connector, or common vent types are possible. Consult Section 502.3.

FIGURE B-11
USE OF MANIFOLD COMMON VENT CONNECTOR

Vent connector size depends on:
- Input
- Rise
- Available total height "H"
- Table 504.3(1) connectors

Common vent size depends on:
- Combined inputs
- Available total height "H"
- Table 504.3(1) common vent

FIGURE B-13
MULTISTORY GAS VENT DESIGN PROCEDURE
FOR EACH SEGMENT OF SYSTEM

Example: Offset Common Vent

Note: This is an illustration of a typical offset vent. Different appliance, vent connector, or vent types are possible. Consult Sections 504.2 and 504.3.

FIGURE B-12
USE OF OFFSET COMMON VENT

Principles of design of multistory vents using vent connector and common vent design tables (see Sections 504.3.11 through 504.3.17).

FIGURE B-14
MULTISTORY VENT SYSTEMS

For SI: 1 foot = 304.8 mm, 1 British thermal unit per hour = 0.2931 W.

FIGURE B-15 (EXAMPLE 1)
SINGLE DRAFT-HOOD-EQUIPPED APPLIANCE

122,000 (.90) = 110,000 for 5-inch vent
From Table 504.2(2), Select 6-inch vent
186,000 (.90) = 167,000; This is greater than the required 120,000. Therefore, use a 6-inch vent and connector where three elbows are used.

Example 2: Single fan-assisted appliance.

An installer has an 80,000 Btu per hour input fan-assisted *appliance* that must be installed using 10 feet of lateral connector attached to a 30-foot-high Type B vent. Two 90-degree elbows are needed for the installation. Can a single-wall metal vent connector be used for this application?

Solution:

Table 504.2(2) refers to the use of single-wall metal vent connectors with Type B vent. In the first column find the row associated with a 30-foot height and a 10-foot lateral. Read across this row, looking at the FAN Min and FAN Max columns, to find that a 3-inch-diameter single-wall metal vent connector is not recommended. Moving to the next larger size single wall connector (4 inches), note that a 4-inch-diameter single-wall metal connector has a recommended minimum vent capacity of 91,000 Btu per hour and a recommended maximum vent capacity of 144,000 Btu per hour. The 80,000 Btu per hour fan-assisted *appliance* is outside this range, so the conclusion is that a single-wall metal vent connector cannot be used to vent this *appliance* using 10 feet of lateral for the connector.

For SI: 1 foot = 304.8 mm, 1 British thermal unit per hour = 0.2931 W.

FIGURE B-16 (EXAMPLE 2)
SINGLE FAN-ASSISTED APPLIANCE

However, if the 80,000 Btu per hour input *appliance* could be moved to within 5 feet of the vertical vent, then a 4-inch single-wall metal connector could be used to vent the *appliance*. Table 504.2(2) shows the acceptable range of vent capacities for a 4-inch vent with 5 feet of lateral to be between 72,000 Btu per hour and 157,000 Btu per hour.

If the *appliance* cannot be moved closer to the vertical vent, then Type B vent could be used as the connector material. In this case, Table 504.2(1) shows that for a 30-foot-high vent with 10 feet of lateral, the acceptable range of vent capacities for a 4-inch-diameter vent attached to a fan-assisted *appliance* is between 37,000 Btu per hour and 150,000 Btu per hour.

Example 3: Interpolating between table values.

An installer has an 80,000 Btu per hour input *appliance* with a 4-inch-diameter draft hood outlet that needs to be vented into a 12-foot-high Type B vent. The vent connector has a 5-foot lateral length and is also Type B. Can this *appliance* be vented using a 4-inch-diameter vent?

Solution:

Table 504.2(1) is used in the case of an all Type B vent system. However, since there is no entry in Table 504.2(1) for a height of 12 feet, interpolation must be used. Read down the 4-inch diameter NAT Max column to the row associated with 10-foot height and 5-foot lateral to find the capacity value of 77,000 Btu per hour. Read further down to the 15-foot height, 5-foot lateral row to find the capacity value of 87,000 Btu per hour. The difference between the 15-foot height capacity value and the 10-foot height capacity value is 10,000 Btu per hour. The capacity for a vent system with a 12-foot height is equal to the capacity for a 10-foot height plus $\frac{2}{5}$ of the difference between the 10-foot and 15-foot height values, or $77,000 + \frac{2}{5} (10,000) = 81,000$ Btu per hour. Therefore, a 4-inch-diameter vent can be used in the installation.

EXAMPLES USING COMMON VENTING TABLES

Example 4: Common venting two draft-hood-equipped appliances.

A 35,000 Btu per hour water heater is to be common vented with a 150,000 Btu per hour furnace using a common vent with a total height of 30 feet. The connector rise is 2 feet for the water heater with a horizontal length of 4 feet. The connector rise for the furnace is 3 feet with a horizontal length of 8 feet. Assume single-wall metal connectors will be used with Type B vent. What size connectors and combined vent should be used in this installation?

Solution:

Table 504.3(2) should be used to size single-wall metal vent connectors attached to Type B vertical vents. In the vent connector capacity portion of Table 504.3(2), find the row associated with a 30-foot vent height. For a 2-foot rise on the vent connector for the water heater, read the shaded columns for draft-hood-equipped appliances to find that a 3-inch-diameter vent connector has a capacity of 37,000 Btu per hour. Therefore, a 3-inch single-wall metal vent connector can be used with the water heater. For a draft-hood-equipped furnace with a 3-foot rise, read across the appropriate row to find that a 5-inch-diameter vent connector has a maximum capacity of

120,000 Btu per hour (which is too small for the furnace) and a 6-inch-diameter vent connector has a maximum vent capacity of 172,000 Btu per hour. Therefore, a 6-inch-diameter vent connector should be used with the 150,000 Btu per hour furnace. Since both vent connector horizontal lengths are less than the maximum lengths *listed* in Section 504.3.2, the table values can be used without adjustments.

In the common vent capacity portion of Table 504.3(2), find the row associated with a 30-foot vent height and read over to the NAT + NAT portion of the 6-inch-diameter column to find a maximum combined capacity of 257,000 Btu per hour. Since the two appliances total only 185,000 Btu per hour, a 6-inch common vent can be used.

Example 5a: Common venting a draft-hood-equipped water heater with a fan-assisted furnace into a Type B vent.

In this case, a 35,000 Btu per hour input draft-hood-equipped water heater with a 4-inch-diameter draft hood *outlet*, 2 feet of connector rise, and 4 feet of horizontal length is to be common vented with a 100,000 Btu per hour fan-assisted furnace with a 4-inch-diameter flue collar, 3 feet of connector rise, and 6 feet of horizontal length. The common vent consists of a 30-foot height of Type B vent. What are the recommended vent diameters for each connector and the common vent? The installer would like to use a single-wall metal vent connector.

Solution: [Table 504.3(2)].

Water Heater Vent Connector Diameter. Since the water heater vent connector horizontal length of 4 feet is less than the maximum value listed in Section 504.3.2, the venting

FIGURE B-17 (EXAMPLE 4)
COMMON VENTING TWO DRAFT-
HOOD-EQUIPPED APPLIANCES

FIGURE B-18 (EXAMPLE 5A)
COMMON VENTING A DRAFT HOOD WITH A FAN-ASSISTED
FURNACE INTO A TYPE B DOUBLE-WALL COMMON VENT

table values can be used without adjustments. Using the Vent Connector Capacity portion of Table 504.3(2), read down the Total Vent Height (*H*) column to 30 feet and read across the 2-foot Connector Rise (*R*) row to the first Btu per hour rating in the NAT Max column that is equal to or greater than the water heater input rating. The table shows that a 3-inch vent connector has a maximum input rating of 37,000 Btu per hour. Although this is greater than the water heater input rating, a 3-inch vent connector is prohibited by Section 504.3.21. A 4-inch vent connector has a maximum input rating of 67,000 Btu per hour and is equal to the draft hood *outlet* diameter. A 4-inch vent connector is selected. Since the water heater is equipped with a draft hood, there are no minimum input rating restrictions.

Furnace Vent Connector Diameter. Using the Vent Connector Capacity portion of Table 504.3(2), read down the Total Vent Height (*H*) column to 30 feet and across the 3-foot Connector Rise (*R*) row. Since the furnace has a fan-assisted combustion system, find the first FAN Max column with a Btu per hour rating greater than the furnace input rating. The 4-inch vent connector has a maximum input rating of 119,000 Btu per hour and a minimum input rating of 85,000 Btu per hour. The 100,000 Btu per hour furnace in this example falls within this range, so a 4-inch connector is adequate. Since the furnace vent connector horizontal length of 6 feet does not exceed the maximum value listed in Section 504.3.2, the venting table values can be used without adjustment. If the furnace had an input rating of 80,000 Btu per hour, then a Type B vent connector [see Table 504.3(1)] would be needed in order to meet the minimum capacity limit.

Common Vent Diameter. The total input to the common vent is 135,000 Btu per hour. Using the Common Vent Capacity portion of Table 504.3(2), read down the Total Vent Height (*H*) column to 30 feet and across this row to find the smallest vent diameter in the FAN + NAT column that has a Btu per hour rating equal to or greater than 135,000 Btu per hour. The 4-inch common vent has a capacity of 132,000 Btu per hour and the 5-inch common vent has a capacity of 202,000 Btu per hour. Therefore, the 5-inch common vent should be used in this example.

Summary. In this example, the installer can use a 4-inch-diameter, single-wall metal vent connector for the water heater and a 4-inch-diameter, single-wall metal vent connector for the furnace. The common vent should be a 5-inch-diameter Type B vent.

Example 5b: Common venting into a masonry chimney.

In this case, the water heater and fan-assisted furnace of Example 5a are to be common vented into a clay tile-lined masonry chimney with a 30-foot height. The chimney is not exposed to the outdoors below the roof line. The internal dimensions of the clay tile liner are nominally 8 inches by 12 inches. Assuming the same vent connector heights, laterals, and materials found in Example 5a, what are the recommended vent connector diameters, and is this an acceptable installation?

Solution:

Table 504.3(4) is used to size common venting installations involving single-wall connectors into masonry chimneys.

Water Heater Vent Connector Diameter. Using Table 504.3(4), Vent Connector Capacity, read down the Total Vent Height (*H*) column to 30 feet, and read across the 2-foot Connector Rise (*R*) row to the first Btu per hour rating in the NAT Max column that is equal to or greater than the water heater input rating. The table shows that a 3-inch vent connector has a maximum input of only 31,000 Btu per hour while a 4-inch vent connector has a maximum input of 57,000 Btu per hour. A 4-inch vent connector must therefore be used.

Furnace Vent Connector Diameter. Using the Vent Connector Capacity portion of Table 504.3(4), read down the Total Vent Height (*H*) column to 30 feet and across the 3-foot Connector Rise (*R*) row. Since the furnace has a fan-assisted combustion system, find the first FAN Max column with a Btu per hour rating greater than the furnace input rating. The 4-inch vent connector has a maximum input rating of 127,000 Btu per hour and a minimum input rating of 95,000 Btu per hour. The 100,000 Btu per hour furnace in this example falls within this range, so a 4-inch connector is adequate.

Masonry Chimney. From Table B-1, the equivalent area for a nominal liner size of 8 inches by 12 inches is 63.6 square inches. Using Table 504.3(4), Common Vent Capacity, read down the FAN + NAT column under the Minimum Internal Area of Chimney value of 63 to the row for 30-foot height to find a capacity value of 739,000 Btu per hour. The combined input rating of the furnace and water heater, 135,000 Btu per hour, is less than the table value, so this is an acceptable installation.

Section 504.3.17 requires the common vent area to be not greater than seven times the smallest *listed appliance* categorized vent area, flue collar area, or draft hood outlet area. Both appliances in this installation have 4-inch-diameter outlets. From Table B-1, the equivalent area for an inside diameter of 4 inches is 12.2 square inches. Seven times 12.2 equals 85.4, which is greater than 63.6, so this configuration is acceptable.

Example 5c: Common venting into an exterior masonry chimney.

In this case, the water heater and fan-assisted furnace of Examples 5a and 5b are to be common vented into an exterior masonry chimney. The chimney height, clay tile liner dimensions, and vent connector heights and laterals are the same as in Example 5b. This system is being installed in Charlotte, North Carolina. Does this exterior masonry chimney need to be relined? If so, what corrugated metallic liner size is recommended? What vent connector diameters are recommended?

Solution:

In accordance with Section 504.3.20, Type B vent connectors are required to be used with exterior masonry chimneys. Use Tables 504.3(7a), (7b) to size FAN+NAT common venting installations involving Type-B double wall connectors into exterior masonry chimneys.

The local 99-percent winter design temperature needed to use Table 504.3(7b) can be found in the ASHRAE *Handbook of Fundamentals*. For Charlotte, North Carolina, this design temperature is 19°F.

Chimney Liner Requirement. As in Example 5b, use the 63 square inch Internal Area columns for this size clay tile liner. Read down the 63 square inch column of Table 504.3(7a) to the 30-foot height row to find that the combined *appliance* maximum input is 747,000 Btu per hour. The combined input rating of the appliances in this installation, 135,000 Btu per hour, is less than the maximum value, so this criterion is satisfied. Table 504.3(7b), at a 19°F design temperature, and at the same vent height and internal area used above, shows that the minimum allowable input rating of a space-heating appliance is 470,000 Btu per hour. The furnace input rating of 100,000 Btu per hour is less than this minimum value. So this criterion is not satisfied, and an alternative venting design needs to be used, such as a Type B vent shown in Example 5a or a *listed* chimney liner system shown in the remainder of the example.

In accordance with Section 504.3.19, Table 504.3(1) or 504.3(2) is used for sizing corrugated metallic liners in masonry chimneys, with the maximum common vent capacities reduced by 20 percent. This example will be continued assuming Type B vent connectors.

Water Heater Vent Connector Diameter. Using Table 504.3(1), Vent Connector Capacity, read down the Total Vent Height (H) column to 30 feet, and read across the 2-foot Connector Rise (R) row to the first Btu/h rating in the NAT Max column that is equal to or greater than the water heater input rating. The table shows that a 3-inch vent connector has a maximum capacity of 39,000 Btu/h. Although this rating is greater than the water heater input rating, a 3-inch vent connector is prohibited by Section 504.3.21. A 4-inch vent connector has a maximum input rating of 70,000 Btu/h and is equal to the draft hood outlet diameter. A 4-inch vent connector is selected.

Furnace Vent Connector Diameter. Using Table 504.3(1), Vent Connector Capacity, read down the Vent Height (H) column to 30 feet, and read across the 3-foot Connector Rise (R) row to the first Btu per hour rating in the FAN Max column that is equal to or greater than the furnace input rating. The 100,000 Btu per hour furnace in this example falls within this range, so a 4-inch connector is adequate.

Chimney Liner Diameter. The total input to the common vent is 135,000 Btu per hour. Using the Common Vent Capacity Portion of Table 504.3(1), read down the Vent Height (H) column to 30 feet and across this row to find the smallest vent diameter in the FAN+NAT column that has a Btu per hour rating greater than 135,000 Btu per hour. The 4-inch common vent has a capacity of 138,000 Btu per hour. Reducing the maximum capacity by 20 percent (Section 504.3.19) results in a maximum capacity for a 4-inch corrugated liner of 110,000 Btu per hour, less than the total input of 135,000 Btu per hour. So a larger liner is needed. The 5-inch common vent capacity *listed* in Table 504.3(1) is 210,000 Btu per hour, and after reducing by 20 percent is 168,000 Btu per hour. Therefore, a 5-inch corrugated metal liner should be used in this example.

Single-Wall Connectors. Once it has been established that relining the chimney is necessary, Type B double-wall vent connectors are not specifically required. This example could be redone using Table 504.3(2) for single-wall vent connectors. For this case, the vent connector and liner diameters would be the same as found above with Type B double-wall connectors.

TABLE B-1
MASONRY CHIMNEY LINER DIMENSIONS WITH CIRCULAR EQUIVALENTS[a]

NOMINAL LINER SIZE (inches)	INSIDE DIMENSIONS OF LINER (inches)	INSIDE DIAMETER OR EQUIVALENT DIAMETER (inches)	EQUIVALENT AREA (square inches)
4 × 8	$2^1/_2 \times 6^1/_2$	4	12.2
		5	19.6
		6	28.3
		7	38.3
8 × 8	$6^3/_4 \times 6^3/_4$	7.4	42.7
		8	50.3
8 × 12	$6^1/_2 \times 10^1/_2$	9	63.6
		10	78.5
12 × 12	$9^3/_4 \times 9^3/_4$	10.4	83.3
		11	95
12 × 16	$9^1/_2 \times 13^1/_2$	11.8	107.5
		12	113.0
		14	153.9
16 × 16	$13^1/_4 \times 13^1/_4$	14.5	162.9
		15	176.7
16 × 20	13 × 17	16.2	206.1
		18	254.4
20 × 20	$16^3/_4 \times 16^3/_4$	18.2	260.2
		20	314.1
20 × 24	$16^1/_2 \times 20^1/_2$	20.1	314.2
		22 ×	380.1
24 × 24	$20^1/_4 \times 20^1/_4$	22.1	380.1
		24	452.3
24 × 28	$20^1/_4 \times 20^1/_4$	24.1	456.2
28 × 28	$24^1/_4 \times 24^1/_4$	26.4	543.3
		27	572.5
30 × 30	$25^1/_2 \times 25^1/_2$	27.9	607
		30	706.8
30 × 36	$25^1/_2 \times 31^1/_2$	30.9	749.9
		33	855.3
36 × 36	$31^1/_2 \times 31^1/_2$	34.4	929.4
		36	1017.9

For SI: 1 inch = 25.4 mm, 1 square inch = 645.16 m².

a. Where liner sizes differ dimensionally from those shown in Table B-1, equivalent diameters can be determined from published tables for square and rectangular ducts of equivalent carrying capacity or by other engineering methods.

-10°F (-23°C)

5°F (-15°C)

17°F (-8°C)

27°F (-3°C)

37°F (3°C)

-10°F (-23°C)

-10°F (-23°C)

5°F (-15°C)

17°F (-8°C)

27°F (-3°C)

37°F (3°C)

FIGURE B-19

The appendices of the *International Fuel Gas Code*® (IFGC®) are explanatory commentaries of the code itself. Thus, there is no additional commentary for them.

Appendix C (IFGS): Exit Terminals of Mechanical Draft and Direct-vent Venting Systems

(This appendix is informative and is not part of the code.)

For SI: 1 inch = 25.4 mm, 1 foot = 304.8 mm, 1 British thermal unit per hour = 0.2931 W.

APPENDIX C
EXIT TERMINALS OF MECHANICAL DRAFT AND DIRECT-VENT VENTING SYSTEMS

The appendices of the *International Fuel Gas Code®* (IFGC®) are explanatory commentaries of the code itself. Thus, there is no additional commentary for them.

Appendix D (IFGS): Recommended Procedure for Safety Inspection of an Existing Appliance Installation

This annex is not a part of the requirements of this code and is included for informational purposes only.

D.1 General. The following procedure is intended as a guide to aid in determining that an appliance is properly installed and is in a safe condition for continued use. Where a gas supplier performs an inspection, their written procedures should be followed.

D.1.1 Application. This procedure is intended for existing residential installations of a furnace, boiler, room heater, water heater, cooking appliance, fireplace appliance and clothes dryer. This procedure should be performed prior to any attempt to modify the appliance installation or building envelope.

D.1.2 Weatherization Programs. Before a building envelope is to be modified as part of a weatherization program, the existing appliance installation should be inspected in accordance with these procedures. After all unsafe conditions are repaired, and immediately after the weatherization is complete, the appliance inspections in D.5.2 are to be repeated.

D.1.3 Inspection Procedure. The safety of the building occupant and inspector are to be determined as the first step as described in D.2. Only after the ambient environment is found to be safe should inspections of gas piping and appliances be undertaken. It is recommended that all inspections described in D.3, D.4, and D.6, where the appliance is in the off mode, be completed and any unsafe conditions repaired or corrected before continuing with inspections of an operating appliance described in D.5 and D.6.

D.1.4 Manufacturer Instructions. Where available, the manufacturer's installation and operating instructions for the installed appliances should be used as part of these inspection procedures to determine if it is installed correctly and is operating properly.

D.1.5 Instruments. The inspection procedures include measuring for fuel gas and carbon monoxide (CO) and will require the use of a combustible gas detector (CGD) and a CO detector. It is recommended that both types of detectors be listed. Prior to any inspection, the detectors should be calibrated or tested in accordance with the manufacturer's instructions. In addition, it is recommended that the detectors have the following minimum specifications.

(1) Gas Detector: The CGD should be capable of indicating the presence of the type of fuel gas for which it is to be used (e.g. natural gas or propane). The combustible gas detector should be capable of the following:

   a. *PPM:* Numeric display with a parts per million (ppm) scale from 1ppm to 900 ppm in 1 ppm increments.

   b. *LEL:* Numeric display with a percent lower explosive limit (% LEL) scale from 0 percent to 100 percent in 1 percent increments.

   c. *Audio:* An audio sound feature to locate leaks.

(2) CO Detector: The CO detector should be capable of the following functions and have a numeric display scale as follows:

   a. *PPM:* For measuring ambient room and appliance emissions a display scale in parts per million (ppm) from 0 to 1,000 ppm in 1 ppm increments.

   b. *Alarm:* A sound alarm function where hazardous levels of ambient CO is found (see D.2 for alarm levels)

   c. *Air Free:* Capable of converting CO measurements to an air free level in ppm. Where a CO detector is used without an air free conversion function, the CO air free can be calculated in accordance with footnote 3 in Table D.6.

D.2 Occupant and Inspector Safety. Prior to entering a building, the inspector should have both a combustible gas detector (CGD) and CO detector turned on, calibrated, and operating. Immediately upon entering the building, a sample of the ambient atmosphere should be taken. Based on CGD and CO detector readings, the inspector should take the following actions:

(1) The CO detector indicates a carbon monoxide level of 70 ppm or greater [1]. The inspector should immediately notify the occupant of the need for themselves and any building occupant to evacuate; the inspector shall immediately evacuate and call 911.

[1] U.S. Consumer Product Safety Commission, *Responding to Residential Carbon Monoxide Incidents, Guidelines For Fire and Other Emergency Response Personnel,* Approved 7/23/02

(2) Where the CO detector indicates a reading between 30 ppm and 70 ppm[1]. The inspector should advise the occupant that high CO levels have been found and recommend that all possible sources of CO should be turned off immediately and windows and doors opened. Where it appears that the source of CO is a permanently installed appliance, advise the occupant to keep the appliance off and have the appliance serviced by a qualified servicing agent.

(3) Where CO detector indicates CO below 30 ppm[1] the inspection can continue.

(4) The CGD indicates a combustible gas level of 20% LEL or greater. The inspector should immediately notify the occupant of the need for themselves and any building occupant to evacuate; the inspector shall immediately evacuate and call 911.

(5) The CGD indicates a combustible gas level below 20% LEL, the inspection can continue.

If during the inspection process it is determined a condition exists that could result in unsafe appliance operation, shut off the appliance and advise the owner of the unsafe condition. Where a gas leak is found that could result in an unsafe condition, advise the owner of the unsafe condition and call the gas supplier to turn off the gas supply. The inspector should not continue a safety inspection on an operating appliance, venting system, and piping system until repairs have been made.

D.3 Gas Piping and Connection Inspections.

(1) Leak Checks. Conduct a test for gas leakage using either a non-corrosive leak detection solution or a CGD confirmed with a leak detection solution.

The preferred method for leak checking is by use of gas leak detection solution applied to all joints. This method provides a reliable visual indication of significant leaks.

The use of a CGD in its audio sensing mode can quickly locate suspect leaks but can be overly sensitive indicating insignificant and false leaks. All suspect leaks found through the use of a CGD should be confirmed using a leak detection solution.

Where gas leakage is confirmed, the owner should be notified that repairs must be made. The inspection should include the following components:

a. All gas piping fittings located within the appliance space.

b. Appliance connector fittings.

c. Appliance gas valve/regulator housing and connections.

(2) Appliance Connector. Verify that the appliance connection type is compliant with Section 411 of the *International Fuel Gas Code*. Inspect flexible appliance connections to determine if they are free of cracks, corrosion and signs of damage. Verify that there are no uncoated brass connectors. Where connectors are determined to be unsafe or where an uncoated brass connector is found, the appliance shutoff valve should be placed in the off position and the owner notified that the connector must be replaced.

(3) Piping Support. Inspect piping to determine that it is adequately supported, that there is no undue stress on the piping, and if there are any improperly capped pipe openings.

(4) Bonding. Verify that the electrical bonding of gas piping is compliant with Section 310 of the *International Fuel Gas Code*.

D.4 Inspections to be performed with the Appliance Not Operating. The following safety inspection procedures are performed on appliances that are not operating. These inspections are applicable to all appliance installations.

(1) Preparing for Inspection. Shut off all gas and electrical power to the appliances located in the same room being inspected. For gas supply, use the shutoff valve in the supply line or at the manifold serving each appliance. For electrical power, place the circuit breaker in the off position or remove the fuse that serves each appliance. A lock type device or tag should be installed on each gas shutoff valve and at the electrical panel to indicate that the service has been shut off for inspection purposes.

(2) Vent System Size and Installation. Verify that the existing venting system size and installation are compliant with Chapter 5 of the *International Fuel Gas Code*. The size and installation of venting systems for other than natural draft and Category I appliances should be in compliance with the manufacturer's installation instructions. Inspect the venting system to determine that it is free of blockage, restriction, leakage, corrosion, and other deficiencies that could cause an unsafe condition. Inspect masonry chimneys to determine if they are lined. Inspect plastic venting system to determine that it is free of sagging and it is sloped in an upward direction to the outdoor vent termination.

(3) Combustion Air Supply. Inspect provisions for combustion air as follows:

a. Non-Direct Vent Appliances. Determine that non-direct vent appliance installations are compliant with the combustion air requirements in Section 304 of the *International Fuel Gas Code*. Inspect

[1] U.S. Consumer Product Safety Commission, *Responding to Residential Carbon Monoxide Incidents, Guidelines For Fire and Other Emergency Response Personnel*, Approved 7/23/02

any interior and exterior combustion air openings and any connected combustion air ducts to determine that there is no blockage, restriction, corrosion or damage. Inspect to determine that the upper horizontal combustion air duct is not sloped in a downward direction toward the air supply source.

b. *Direct Vent Appliances.* Verify that the combustion air supply ducts and pipes are securely fastened to direct vent appliance and determine that there are no separations, blockage, restriction, corrosion or other damage. Determine that the combustion air source is located in the outdoors or to areas that freely communicate to the outdoors.

c. *Unvented Appliances.* Verify that the total input of all unvented room heaters and gas-fired refrigerators installed in the same room or rooms that freely communicate with each other does not exceed 20 Btu/hr/ft³.

(4) *Flooded Appliances.* Inspect the appliance for signs that the appliance may have been damaged by flooding. Signs of flooding include a visible water submerge line on the appliance housing, excessive surface or component rust, deposited debris on internal components, and mildew-like odor. Inform the owner that any part of the appliance control system and any appliance gas control that has been under water must be replaced. All flood-damaged plumbing, heating, cooling and electrical appliances should be replaced.

(5) *Flammable Vapors.* Inspect the room/space where the appliance is installed to determine if the area is free of the storage of gasoline or any flammable products such as oil-based solvents, varnishes or adhesives. Where the appliance is installed where flammable products will be stored or used, such as a garage, verify that the appliance burner(s) is a minimum of 18" above the floor unless the appliance is listed as flammable vapor ignition resistant.

(6) *Clearances to Combustibles.* Inspect the immediate location where the appliance is installed to determine if the area is free of rags, paper or other combustibles. Verify that the appliance and venting system are compliant with clearances to combustible building components in accordance with Sections 305.8, 501.15.4, 502.5, 503.6.1, 503.10.5 and other applicable sections of Section 503.

(7) *Appliance Components.* Inspect internal components by removing access panels or other components for the following:

a. Inspect burners and crossovers for blockage and corrosion. The presence of soot, debris, and signs

of excessive heating may indicate incomplete combustion due to blockage or improper burner adjustments.

c. Metallic and non-metallic hoses for signs of cracks, splitting, corrosion, and lose connections.

d. Signs of improper or incomplete repairs

e. Modifications that override controls and safety systems

f. Electrical wiring for loose connections; cracks, missing or worn electrical insulation; and indications of excessive heat or electrical shorting. Appliances requiring an external electrical supply should be inspected for proper electrical connection in accordance with the National Electric Code.

(8) *Placing Appliances Back in Operation.* Return all inspected appliances and systems to their preexisting state by reinstalling any removed access panels and components. Turn on the gas supply and electricity to each appliance found in safe condition. Proceed to the operating inspections in D.5 through D.6.

D.5 Inspections to be performed with the Appliance Operating. The following safety inspection procedures are to be performed on appliances that are operating where there are no unsafe conditions or where corrective repairs have been completed.

D.5.1 General Appliance Operation.

(1) *Initial Startup.* Adjust the thermostat or other control device to start the appliance. Verify that the appliance starts up normally and is operating properly.

Determine that the pilot(s), where provided, is burning properly and that the main burner ignition is satisfactory, by interrupting and re-establishing the electrical supply to the appliance in any convenient manner. If the appliance is equipped with a continuous pilot(s), test all pilot safety devices to determine whether they are operating properly by extinguishing the pilot(s) when the main burner(s) is off and determining, after 3 minutes, that the main burner gas does not flow upon a call for heat. If the appliance is not provided with a pilot(s), test for proper operation of the ignition system in accordance with the appliance manufacturer's lighting and operating instructions.

(2) *Flame Appearance.* Visually inspect the flame appearance for proper color and appearance. Visually determine that the main burner gas is burning properly (i.e., no floating, lifting, or flashback). Adjust the primary air shutter as required. If the appliance is equipped with high and low flame controlling or flame modulation, check for proper main burner operation at low flame.

(3) *Appliance Shutdown.* Adjust the thermostat or other control device to shut down the appliance. Verify that the appliance shuts off properly.

D.5.2 Test for Combustion Air and Vent Drafting for Natural Draft and Category I Appliances. Combustion air and vent draft procedures are for natural draft and category I appliances equipped with a draft hood and connected to a natural draft venting system.

(1) *Preparing for Inspection.* Close all exterior building doors and windows and all interior doors between the space in which the appliance is located and other spaces of the building that can be closed. Turn on any clothes dryer. Turn on any exhaust fans, such as range hoods and bathroom exhausts, so they will operate at maximum speed. Do not operate a summer exhaust fan. Close fireplace dampers and any fireplace doors.

(2) *Placing the Appliance in Operation.* Place the appliance being inspected in operation. Adjust the thermostat or control so the appliance will operate continuously.

(3) *Spillage Test.* Verify that all appliances located within the same room are in their standby mode and ready for operation. Follow lighting instructions for each appliance as necessary. Test for spillage at the draft hood relief opening as follows:

    a. After 5 minutes of main burner operation, check for spillage using smoke.

    b. Immediately after the first check, turn on all other fuel gas burning appliances within the same room so they will operate at their full inputs and repeat the spillage test.

    c. Shut down all appliances to their standby mode and wait for 15 minutes.

    d. Repeat the spillage test steps a through c on each appliance being inspected.

(4) Additional Spillage Tests: Determine if the appliance venting is impacted by other door and air handler settings by performing the following tests.

    a. Set initial test condition in accordance with D.5.2 (1).

    b. Place the appliance(s) being inspected in operation. Adjust the thermostat or control so the appliance(s) will operate continuously.

    c. Open the door between the space in which the appliance(s) is located and the rest of the building. After 5 minutes of main burner operation, check for spillage at each appliance using smoke.

    d. Turn on any other central heating or cooling air handler fan that is located outside of the area where the appliances are being inspected. After 5 minutes of main burner operation, check for spillage at each appliance using smoke. The test should be conducted with the door between the space in which the appliance(s) is located and the rest of the building in the open and in the closed position.

(5) Return doors, windows, exhaust fans, fireplace dampers, and any other fuel gas burning appliance to their previous conditions of use.

(6) If, after completing the spillage test it is believed sufficient combustion air is not available, the owner should be notified that an alternative combustion air source is needed in accordance with Section 304 of the *International Fuel Gas Code.* Where it is believed that the venting system does not provide adequate natural draft, the owner should be notified that alternative vent sizing, design or configuration is needed in accordance with Chapter 5 of the *International Fuel Gas Code.* If spillage occurs, the owner should be notified as to its cause, be instructed as to which position of the door (open or closed) would lessen its impact, and that corrective action by a HVAC professional should be taken.

D.6 Appliance-Specific Inspections. The following appliance-specific inspections are to be performed as part of a complete inspection. These inspections are performed either with the appliance in the off or standby mode (indicated by "*OFF*") or on an appliance that is operating (indicated by "*ON*"). The CO measurements are to be undertaken only after the appliance is determined to be properly venting. The CO detector should be capable of calculating CO emissions in ppm air free.

(1) Forced Air Furnaces:

    a. OFF. Verify that an air filter is installed and that it is not excessively blocked with dust.

    b. OFF. Inspect visible portions of the furnace combustion chamber for cracks, ruptures, holes, and corrosion. A heat exchanger leakage test should be conducted.

    c. ON. Verify both the limit control and the fan control are operating properly. Limit control operation can be checked by blocking the circulating air inlet or temporarily disconnecting the electrical supply to the blower motor and determining that the limit control acts to shut off the main burner gas.

    d. ON. Verify that the blower compartment door is properly installed and can be properly re-secured if opened. Verify that the blower compartment door safety switch operates properly.

    e. *ON.* Check for flame disturbance before and after blower comes on which can indicate heat exchanger leaks.

f. *ON*. Measure the CO in the vent after 5 minutes of main burner operation. The CO should not exceed threshold in Table D.6.

(2) Boilers:

a. OFF and ON. Inspect for evidence of water leaks around boiler and connected piping.

b. ON. Verify that the water pumps are in operating condition. Test low water cutoffs, automatic feed controls, pressure and temperature limit controls, and relief valves in accordance with the manufacturer's recommendations to determine that they are in operating condition.

c. ON. Measure the CO in the vent after 5 minutes of main burner operation. The CO should not exceed threshold in Table D.6.

(3) Water Heaters:

a. *OFF*. Verify that the pressure-temperature relief valve is in operating condition. Water in the heater should be at operating temperature.

b. OFF. Verify that inspection covers, glass, and gaskets are intact and in place on a flammable vapor ignition resistant (FVIR) type water heater.

c. ON. Verify that the thermostat is set in accordance with the manufacturer's operating instructions and measure the water temperature at the closest tub or sink to verify that it is no greater than 120ºF.

d. OFF. Where required by the local building code in earthquake prone locations, inspect that the water heater is secured to the wall studs in two locations (high and low) using appropriate metal strapping and bolts.

e. ON. Measure the CO in the vent after 5 minutes of main burner operation. The CO should not exceed threshold in Table D.6.

(4) Cooking Appliances

a. *OFF*. Inspect oven cavity and range-top exhaust vent for blockage with aluminum foil or other materials.

b. *OFF*. Inspect cook top to verify that it is free from a build-up of grease.

c. *ON*. Measure the CO above each burner and at the oven exhaust vents after 5 minutes of burner operation. The CO should not exceed threshold in Table D.6.

(5) Vented Room Heaters

a. OFF. For built-in room heaters and wall furnaces, inspect that the burner compartment is free of lint and debris.

b. OFF. Inspect that furnishings and combustible building components are not blocking the heater.

a. ON. Measure the CO in the vent after 5 minutes of main burner operation. The CO should not exceed threshold in Table D.6.

(6) Vent-Free (unvented) Heaters

a. OFF. Verify that the heater input is a maximum of 40,000 Btu input, but not more than 10,000 Btu where installed in a bedroom, and 6,000 Btu where installed in a bathroom.

b. OFF. Inspect the ceramic logs provided with gas log type vent free heaters that they are properly located and aligned.

c. OFF. Inspect the heater that it is free of excess lint build-up and debris.

c. OFF. Verify that the oxygen depletion safety shutoff system has not been altered or bypassed.

d. ON. Verify that the main burner shuts down within 3 minutes by extinguishing the pilot light. The test is meant to simulate the operation of the oxygen depletion system (ODS).

e. ON. Measure the CO after 5 minutes of main burner operation. The CO should not exceed threshold in Table D.6.

(7) Gas Log Sets and Gas Fireplaces

a. OFF. For gas logs installed in wood burning fireplaces equipped with a damper, verify that the fireplace damper is in a fixed open position.

b. ON. Measure the CO in the firebox (log sets installed in wood burning fireplaces or in the vent (gas fireplace) after 5 minutes of main burner operation. The CO should not exceed threshold in Table D.6.

(8) Gas Clothes Dryer

a. *OFF*. Where installed in a closet, verify that a source of make-up air is provided and inspect that any make-up air openings, louvers, and ducts are free of blockage.

b. OFF. Inspect for excess amounts of lint around the dryer and on dryer components. Inspect that there is a lint trap properly installed and it does not have holes or tears. Verify that it is in a clean condition.

c. OFF. Inspect visible portions of the exhaust duct and connections for loose fittings and connections, blockage, and signs of corrosion. Verify that the duct termination is not blocked and that it terminates in an outdoor location. Verify that only approved metal vent ducting material is installed (plastic and vinyl materials are not approved for gas dryers).

d. ON. Verify mechanical components including drum and blower are operating properly.

e. ON. Operate the clothes dryer and verify that exhaust system is intact and exhaust is exiting the termination.

f. ON. Measure the CO at the exhaust duct or termination after 5 minutes of main burner operation. The CO should not exceed threshold in Table D.6.

TABLE D.6
CO THRESHOLDS

Central Furnace (all categories)	400 ppm[1] air free[2,3]
Floor Furnace	400 ppm air free
Gravity Furnace	400 ppm air free
Wall Furnace (BIV)	200 ppm air free
Wall Furnace (Direct Vent)	400 ppm air free
Vented Room Heater	200 ppm air free
Vent-Free Room Heater	200 ppm air free
Water Heater	200 ppm air free
Oven/Boiler	225 ppm as measured
Top Burner	25 ppm as measured (per burner)
Clothes Dryer	400 ppm air free
Refrigerator	25 ppm as measured
Gas Log (gas fireplace)	25 ppm as measured in vent
Gas Log (installed in wood burning fireplace)	400 ppm air free in firebox

[1] Parts per million

[2] Air free emission levels are based on a mathematical equation (involving carbon monoxide and oxygen or carbon dioxide readings) to convert an actual diluted flue gas carbon monoxide testing sample to an undiluted air free flue gas carbon monoxide level utilized in the appliance certification standards. For natural gas or propane, using as-measured CO ppm and O2 percentage:

$$CO_{AFppm} = \left(\frac{20.9}{20.9 - O_2}\right) \times CO_{ppm}$$

Where:

CO_{AFppm} = Carbon monoxide, air-free ppm

CO_{ppm} = As-measured combustion gas carbon monoxide ppm

O2 = Percentage of oxygen in combustion gas, as a percentage

[3] An alternate method of calculating the CO air free when access to an oxygen meter is not available:

$$CO_{AFppm} = \left(\frac{UCO_2}{CO_2}\right) \times CO$$

Where:

UCO_2 = Ultimate concentration of carbon dioxide for the fuel being burned in percent for natural gas (12.2 percent) and propane (14.0 percent)

CO_2 = Measured concentration of carbon dioxide in combustion products in percent

CO = Measured concentration of carbon monoxide in combustion products in percent

INDEX

U

V

W

People Helping People Build a Safer World®

2015 I-Code Essentials

This helpful series uses a straightforward, focused approach to explore code requirements with non-code language, allowing readers to gain confidence in their understanding of the material. Each book is an invaluable companion guide to the 2015 IBC, IRC, IFC or IECC for both new and experienced code users.

- Detailed full-color illustrations enhance comprehension of code provisions

- A focused, concise approach facilitates understanding of the essential code provisions

- References to corresponding code sections provide added value as companion guides to the 2015 I-Codes

- Glossary of code and construction terms clarifies their meaning in the context of the code

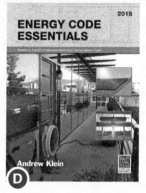

A. BUILDING CODE ESSENTIALS: BASED ON THE 2015 INTERNATIONAL BUILDING CODE
Explores those code provisions essential to understanding the application of the IBC in an easy-to-read format. The user-friendly approach of the text simplifies critical concepts so users can achieve a more complete understanding of the code's intent. Full-color illustrations, examples and simplified tables assist the reader in visualizing the code requirements. This up-to-date, step-by-step guide's topic organization reflects the intent of the IBC and facilitates understanding and application of the code provisions.
SOFT COVER #4031S15
PDF DOWNLOAD #8950P596

B. RESIDENTIAL CODE ESSENTIALS: BASED ON THE 2015 INTERNATIONAL RESIDENTIAL CODE
Explains those code provisions essential to understanding the application of the IRC to the most commonly encountered building practices. The information is presented in a user-friendly manner with an emphasis on technical accuracy and clear non-code language. Full-color illustrations, examples and simplified tables assist the reader in visualizing the code requirements.
SOFT COVER #4131S15
PDF DOWNLOAD #8950P59

C. FIRE CODE ESSENTIALS: BASED ON THE 2015 INTERNATIONAL FIRE CODE
Focuses on the essential provisions of the IFC for controlling ignition sources and fire hazards, fire department access to buildings, building uses and processes, fire protection and life safety systems, special processes and uses and hazardous materials.
SOFT COVER #4431S15
PDF DOWNLOAD #8950P59

D. ENERGY CODE ESSENTIALS: BASED ON THE 2015 INTERNATIONAL ENERGY CONSERVATION CODE®
Designed to provide a straightforward, easy-to-read companion guide to the IECC for both beginning and experienced code users. The text provides accurate information on critical energy code applications in the office and in the field for residential and commercial construction. The focused coverage examines those code provisions essential to understanding the application of the IECC to the most commonly encountered building practices and discusses the reasons behind the requirements.
SOFT COVER #4831S15
PDF DOWNLOAD #8950P651

ORDER YOUR I-CODE ESSENTIALS TODAY! 1-800-786-4452 | www.iccsafe.org/books

14-10387

INTERNATIONAL
CODE
COUNCIL

People Helping People Build a Safer World®

GET IMMEDIATE DOWNLOADS OF THE STANDARDS YOU NEED

Browse hundreds of industry standards adopted by reference. Available to you 24/7!

Count on ICC for standards from a variety of publishers, including:

ACI	CPSC	GYPSUM
AISC	CSA	HUD
ANSI	DOC	ICC
APA	DOJ	ISO
APSP	DOL	NSF
ASHRAE	DOTn	SMACNA
ASTM	FEMA	USC
AWC	GBI	

DOWNLOAD YOUR STANDARDS TODAY!
SHOP.ICCSAFE.ORG